Sensors Applications
Volume 2
Sensors in Intelligent Buildings

Sensors Applications

- Sensors in Manufacturing
- Sensors in Intelligent Buildings
- Sensors in Medicine and Health Care
- Sensors in Automotive Technology
- Sensors in Aerospace Technology
- Sensors in Environmental Technology
- Sensors in Household Appliances

Related Wiley-VCH titles:

W. Göpel, J. Hesse, J. N. Zemel
Sensors Vol. 1–9

ISBN 3-527-26538-4

H. Baltes, W. Göpel, J. Hesse
Sensors Update

ISSN 1432-2404

Sensors Applications
Volume 2

Sensors in Intelligent Buildings

Edited by
O. Gassmann, H. Meixner

Series Editors:
J. Hesse, J. W. Gardner, W. Göpel (†)

WILEY-VCH

Weinheim – New York – Chichester – Brisbane – Singapore – Toronto

Series Editors

Prof. Dr. J. Hesse
Carl Zeiss
Postfach 1380
73447 Oberkochen
Germany

Prof. J. W. Gardner
University of Warwick
Division of Electrical & Electronic Engineering
Coventry CV 7AL
United Kingdom

Prof. Dr. W. Göpel †
Institut für Physikalische
und Theoretische Chemie
Universität Tübingen
Auf der Morgenstelle 8
72076 Tübingen
Germany

Volume Editors

Dr. O. Gassmann
Schindler Elevators & Escalators
R&D Technology Management
Zugerstrasse 13
6030 Ebikon/Luzern
Switzerland

Prof. Dr. H. Meixner
Siemens Corporate Technology
Otto-Hahn-Ring 6
81739 München
Germany

■ This book was carefully produced. Nevertheless, authors, editors and publisher do not warrant the information contained therein to be free of errors. Readers are advised to keep in mind that statements, data, illustrations, procedural details or other items may inadvertently be inaccurate.

Library of Congress Card No.: applied for

British Library Cataloguing-in-Publication Data:
A catalogue record for this book is available from the British Library.

Die Deutsche Bibliothek – CIP-Cataloguing-in-Publication Data
A catalogue record is available from Die Deutsche Bibliothek

© WILEY-VCH Verlag GmbH
D-69469 Weinheim, 2001

All rights reserved (including those of translation in other languages). No part of this book may be reproduced in any form – by photoprinting, microfilm, or any other means – nor transmitted or translated into machine language without written permission from the publishers. Registered names, trademarks, etc. used in this book, even when not specifically marked as such, are not to be considered unprotected by law.

printed in the Federal Republic of Germany
printed on acid-free paper

Composition   K+V Fotosatz GmbH,
D-64743 Beerfelden
Printing   Betz-Druck, D-64291 Darmstadt
Bookbinding   Wilhelm Osswald & Co.,
D-67433 Neustadt

ISBN   3-527-29557-7

Preface to the Series

As the use of microelectronics became increasingly indispensable in measurement and control technology, so there was an increasing need for suitable sensors. From the mid-Seventies onwards sensors technology developed by leaps and bounds and within ten years had reached the point where it seemed desirable to publish a survey of what had been achieved so far. At the request of publishers WILEY-VCH, the task of editing was taken on by Wolfgang Göpel of the University of Tübingen (Germany), Joachim Hesse of Carl Zeiss (Germany) and Jay Zemel of the University of Philadelphia (USA), and between 1989 and 1995 a series called *Sensors* was published in 8 volumes covering the field to date. The material was grouped and presented according to the underlying physical principles and reflected the degree of maturity of the respective methods and products. It was written primarily with researchers and design engineers in mind, and new developments have been published each year in one or two supplementary volumes called *Sensors Update*.

Both the publishers and the series editors, however, were agreed from the start that eventually sensor users would want to see publications only dealing with their own specific technical or scientific fields. Sure enough, during the Nineties we saw significant developments in applications for sensor technology, and it is now an indispensable part of many industrial processes and systems. It is timely, therefore, to launch a new series, *Sensors Applications*. WILEY-VCH again commissioned Wolfgang Göpel and Joachim Hesse to plan the series, but sadly Wolfgang Göpel suffered a fatal accident in June 1999 and did not live to see publication. We are fortunate that Julian Gardner of the University of Warwick has been able to take his place, but Wolfgang Göpel remains a co-editor posthumously and will not be forgotten.

The series of *Sensors Applications* will deal with the use of sensors in the key technical and economic sectors and systems: *Sensors in Manufacturing, Intelligent Buildings, Medicine and Health Care, Automotive Technology, Aerospace Technology, Environmental Technology* and *Household Appliances*. Each volume will be edited by specialists in the field. Individual volumes may differ in certain respects as dictated by the topic, but the emphasis in each case will be on the process or system in question: which sensor is used, where, how and why, and exactly what the benefits are to the user. The process or system itself will of course be outlined and the volume will close with a look ahead to likely developments and applications in the future. Actual sensor functions will only be described where it seems necessary for an understanding of how they relate to the process or system. The basic principles can always be found in the earlier series of *Sensors* and *Sensors Update*.

The series editors would like to express their warm appreciation in the colleagues who have contributed their expertise as volume editors or authors. We are deeply indebted to the publisher and would like to thank in particular Dr. Peter Gregory, Dr. Jörn Ritterbusch and Dr. Claudia Barzen for their constructive assistance both with the editorial detail and the publishing venture in general. We trust that our endeavors will meet with the reader's approval.

Oberkochen and Coventry, November 2000

Joachim Hesse
Julian W. Gardner

Preface to Volume 2 of "Sensors Applications"

In the building control industry, a clear trend towards more 'intelligence' can be observed. In the last two decades, intensive research has been done in the area of intelligent buildings. The concept integrates new technologies from areas such as sensor systems, computer automation, space-age materials, and energy management in order to adjust and adapt to its occupants. Integrated sensor systems judge indoor and outdoor conditions of a building and its devices in order to operate as an integrated system for maximum performance and comfort. Modern buildings become a place of multilateral interaction between the inhabitants and the building entities.

With this volume *Sensors in Intelligent Buildings* of the series *Sensors Applications* the Editors aim to create a work that presents the reader with a competent and comprehensive survey of sensors and sensor systems currently applied in the building industry. The book is primarily aimed at scientists and engineers engaged in research on and the development and application of sensors and searching for detailed references on sensors in the building control area. New system solutions and a wide variety of sensors will be available in all building areas such as energy, HVAC, information, transportation, safety, security, maintenance, and facility management.

The chapters have been contributed by leading scientists in international research institutes, universities, and companies such as the City University of Hong Kong, Coactive Networks, University of Applied Sciences Dortmund, Estia Sàrl, Fraunhofer Institute for Computer Graphics Research, Fraunhofer Institute for Information and Data Processing, Fraunhofer Institute of Microelectronic Circuits and Systems, GMD Institute for Secure Telecooperation, Motorola, Philips Research Laboratories, Purdue University, San José State University, Sauter, Schindler Elevators & Escalators, Siemens Building Technologies, Siemens Corporate Technology, Siemens Energy & Automation, Siemens Landis & Staefa Electronic, Swiss Center for Electronics and Microtechnology (CSEM), Swiss Federal Institute of Technology (EPFL), The Media Laboratory, Massachusetts Institute of Technology (MIT), Technical University Munich, Transtechno, University of the Bundeswehr Munich, University of Vermont, Vienna University of Technology, Viterra Energy Services, Weinzierl Engineering.

The main focus of this book relies on the system principle which is gaining more and more importance throughout the building control industry. Instead of just describing briefly the different types of sensors, it is the aim of the book to illustrate which sensors and sensor systems are used in which subsystem, to explain which reasons were decisive for using especially a particular sensor and to give an outline of future developments. In all chapters a description of the system is followed by a discussion of the sensors currently used. This includes a discussion of their strengths and weaknesses and especially an illustration of the reasons why these sensors made it to industrial building applications. In every area we also provide a short outlook on upcoming sensors and sensor systems – also deduced from the question of what the developments to the subsystem will be and which new sensor types will therefore be needed.

This book has been produced with the contributions of many people. We are very grateful to the authors who spent valuable time to share their research results and experiences with the scientific community. Especially we would like to thank Verena Klaassen, who gave us invaluable assistance in preparing this book. Many thanks go also to the publisher and series editors for their fruitful cooperation.

Lucerne and Munich, February 2001   Oliver Gassmann and Hans Meixner

Contents

List of Contributors *XXI*

1 Introduction

1.1 Sensors in Intelligent Buildings: Overview and Trends *3*
Oliver Gassmann, Hans Meixner
1.1.1 Introduction *3*
1.1.2 Towards the Intelligent Building *4*
1.1.2.1 Reduced Resource Consumption *6*
1.1.2.2 Optimized Convenience and More Comfort *6*
1.1.2.3 Increased Impact of Microsystems Technology *8*
1.1.2.4 Increased Impact of New Communication Systems *10*
1.1.2.5 Development of an Intelligent Home Market *12*
1.1.2.6 More Integrated Buildings: the FuturElife Smart Building *12*
1.1.2.7 Pneumatic Building Structures: Airtecture *14*
1.1.3 Trends in Sensor Systems *16*
1.1.4 Sensor Systems in Intelligent Buildings *21*
1.1.4.1 Energy and HVAC *22*
1.1.4.2 Information and Transportation *22*
1.1.4.3 Safety and Security *23*
1.1.4.4 Maintenance and Facility Management *23*
1.1.4.5 System Technologies *24*
1.1.5 References *24*

2 Energy and HVAC

2.1 Intelligent Air-conditioning Control *29*
Albert T. P. So, Brian W. L. Tse
2.1.1 Introduction *29*
2.1.2 General Specifications of a Sensor *29*
2.1.3 A Quick Review on HVAC Sensors *30*
2.1.3.1 Temperature Sensors *30*

2.1.3.2	Pressure Sensors	31
2.1.3.3	Flow Rate Sensors	33
2.1.3.4	Humidity Sensors	35
2.1.3.5	Comfort Sensors	36
2.1.3.6	Indoor Air Quality Sensors	37
2.1.3.7	Occupancy Sensors	37
2.1.3.8	Smoke Sensors	38
2.1.4	Computer Vision-based HVAC Control	40
2.1.4.1	The Computer Vision System	41
2.1.4.2	Calibration of the Stereoscopic Camera System	41
2.1.4.3	Velocity Field Computation by Optical Flow	42
2.1.4.4	Pixel Correspondence	44
2.1.4.5	Scene Spots Fuzzy Clustering	45
2.1.5	Internet-based HVAC System Monitoring and Control	46
2.1.5.1	Philosophy of Internet-based Building Automation with Image Transfer	47
2.1.5.2	The BAS Web Site	48
2.1.6	PMV-based HVAC Control	50
2.1.6.1	Elements of Comfort-based Control	51
2.1.6.2	Control Algorithms	53
2.1.6.3	Computer Simulation	54
2.1.6.4	Simulation Results	57
2.1.7	Conclusion	60
2.1.8	References	60
2.2	**NEUROBAT – a Self-commissioned Heating Control System Using Neural Networks** 63	
	Jens Krauss, Manuel Bauer, Jürg Bichsel, Nicolas Morel	
2.2.1	Introduction	63
2.2.2	Control Concept	64
2.2.2.1	Methodologies	64
2.2.2.2	Controller Block Diagram	66
2.2.2.3	Optimal Control Algorithm	67
2.2.2.4	Applied Sensors and NEUROBAT Controller Versions	69
2.2.3	Controller Performance Assessment	69
2.2.3.1	Simulation Study	69
2.2.3.2	Comparative Tests Within Office Rooms	72
2.2.4	Prototype Realization with Functional Tests on Residential Buildings	75
2.2.4.1	Industrial NEUROBAT Prototype	75
2.2.4.2	Test Results Heating Season 1999/2000	76
2.2.5	Conclusion	81
2.2.6	References	83

2.3	**Air Quality Measurement and Management** *85*	
	Hanns-Erik Endres	
2.3.1	Introduction *85*	
2.3.2	Substances in Indoor Air *85*	
2.3.3	Sensors for Air Quality Measurements *88*	
2.3.4	Sensor Systems and Arrays for Air Quality Measurement *91*	
2.3.5	Examples of Long-term Air Quality Evaluation *95*	
2.3.6	CO_2 Measurements *96*	
2.3.7	VOC Sensor *98*	
2.3.8	Summary and Future Outlook *99*	
2.3.9	Acknowledgments *101*	
2.3.10	References *101*	
2.4	**Sensor-based Management of Energy and Thermal Comfort** *103*	
	Thomas Bernard, Helge-Björn Kuntze	
2.4.1	Motivation *103*	
2.4.2	Control Concept *104*	
2.4.3	Theoretical Approach of Multi-objective Fuzzy Optimization *106*	
2.4.3.1	The Basic Algorithm *106*	
2.4.3.2	Important Features *107*	
2.4.3.3	Weighting of Different Performance Criteria *107*	
2.4.3.4	Model Equations *108*	
2.4.4	Application to the Supervisory Control of HVAC Systems *109*	
2.4.4.1	Comfort Criteria *109*	
2.4.4.2	Economy Criteria *109*	
2.4.4.3	Optimization of Heating Temperature *110*	
2.4.4.4	Optimization of Air Exchange Rate *113*	
2.4.4.5	Optimization of Blind Position *116*	
2.4.5	Simulations and Measured Results *119*	
2.4.5.1	Supervisory Control of Heating and Ventilation Systems *120*	
2.4.5.2	Supervisory Control of Heating and Blind Systems *123*	
2.4.6	Conclusions *125*	
2.4.7	References *125*	
2.5	**Wireless and M-Bus enabled Metering Devices** *127*	
	Dieter Mrozinski	
2.5.1	Introduction *127*	
2.5.2	Benefits of Remote Reading *128*	
2.5.2.1	User *129*	
2.5.2.2	Energy Supplier and/or Billing Service Provider *129*	
2.5.2.3	Owners and/or Property Management *130*	
2.5.3	Data Transfer via Data Bus *130*	
2.5.3.1	Bus Applications of the Meter Sector and the Resulting Demands on the Data Bus *131*	
2.5.3.2	Available Data Buses for Meter Applications *134*	

2.5.3.3	M-Bus *134*	
2.5.4	Data Transmission via Radio *147*	
2.5.4.1	Data Transmission and Selection Process *147*	
2.5.5	Future Prospects *156*	
2.5.6	References *156*	

2.6 **Sensors in HVAC Systems for Metering and Energy Cost Allocation** *159*
Günter Mügge

2.6.1	Introduction *159*
2.6.2	Possible Implementations of the Energy Allocation *160*
2.6.3	Allocation of Costs for Air Conditioning *161*
2.6.4	Heat Meters *162*
2.6.4.1	Principle of Measurement *162*
2.6.4.2	Temperature Sensors [5, 9, 10] *163*
2.6.4.3	Flow Sensors *164*
2.6.4.4	Application *166*
2.6.5	Heat Cost Allocators (HCAs) *167*
2.6.5.1	Principle of Measurement *167*
2.6.5.2	Evaporative Heat Cost Allocators *167*
2.6.5.3	Electronic Heat Cost Allocators *168*
2.6.6	Reading *169*
2.6.6.1	Visual Reading *169*
2.6.6.2	Automatic Meter Reading *169*
2.6.7	Outlook *170*
2.6.8	References *171*

2.7 **Pressure Sensors in the HVAC Industry** *173*
Yves Lüthi, Rolf Meisinger, Marc Wenzler, Kais Mnif

2.7.1	Introduction *173*
2.7.2	Main Applications and Market Requirements *175*
2.7.2.1	Filter, Fan Monitoring, and Pressure Control *175*
2.7.2.2	Variable Air Volume *176*
2.7.2.3	Summary *178*
2.7.3	Silicon Pressure Sensors *179*
2.7.3.1	Pressure Sensors as Microelectromechanical Systems (MEMS) *179*
2.7.3.2	Could Pressure Sensors be Considered as Standard Electronic Components? *183*
2.7.3.3	Marketing and Application Considerations *185*
2.7.4	Solution: a Flexible, Modular Pressure Sensor for HVAC Applications *187*
2.7.4.1	Concept *187*
2.7.4.2	Autozero Facility *188*
2.7.4.3	Factory Calibration Procedure *190*
2.7.4.4	Characterization of the Sensor Elements *191*

2.7.4.5	Application in the New Damper Actuator from Siemens Building Technologies *196*	
2.7.5	Conclusions *196*	
2.7.6	Acknowledgments *198*	
2.7.7	Glossary *198*	
2.7.8	References *199*	
3	**Information and Transportation**	
3.1	**Fieldbus Systems** *203*	
	Dietmar Dietrich, Thilo Sauter, Peter Fischer, Dietmar Loy	
3.1.1	Introduction *203*	
3.1.2	Abstract View and Definition of the Fieldbus *204*	
3.1.3	Communication Basics for Fieldbus Systems *206*	
3.1.3.1	Decentralization and Hierarchies *206*	
3.1.3.2	The ISO/OSI Model *207*	
3.1.3.3	Topologies *212*	
3.1.4	Historical Aspects *213*	
3.1.4.1	The Roots of Industrial Networks *213*	
3.1.4.2	The Evolution of Fieldbusses *215*	
3.1.5	Examples of Fieldbus Systems *218*	
3.1.5.1	EIB *218*	
3.1.5.2	LonWorks and ANSI/EIA 709 *221*	
3.1.5.3	BACnet *225*	
3.1.5.4	EIBnet *227*	
3.1.6	Fieldbus Systems in Connection with the Internet *230*	
3.1.7	Present and Future Challenges *232*	
3.1.7.1	Interoperability and Profiles *232*	
3.1.7.2	System Complexity and Tools *232*	
3.1.7.3	Management – and Plug and Play *233*	
3.1.7.4	Security *234*	
3.1.7.5	Driving Forces *236*	
3.1.8	Outlook and Conclusion *238*	
3.1.9	References *238*	
3.2	**Wireless In-building Networks** *241*	
	Mike Barnard	
3.2.1	Introduction *241*	
3.2.2	Network Characteristics *241*	
3.2.2.1	Wired vs. Wireless? *241*	
3.2.2.2	Sensor Network Requirements *243*	
3.2.3	Existing and Emerging Standards *250*	
3.2.3.1	Network Standards *251*	
3.2.3.2	Wired Links *252*	

3.2.3.3	Wireless Links	255
3.2.4	Existing and Emerging Wireless Products	258
3.2.4.1	Remote Controls	258
3.2.4.2	Security and Telemetry	258
3.2.4.3	Data Networks	258
3.2.5	References	259
3.3	**Sensor Systems in Modern High-rise Elevators**	**261**
	Enrico Marchesi, Ayman Hamdy, René Kunz	
3.3.1	Elevator System – Overvies	261
3.3.1.1	Functional Description	261
3.3.1.2	Sensor Applications in Elevators	262
3.3.2	Shaft Information System	264
3.3.2.1	Control Sensorics	264
3.3.2.2	Safety Sensorics	267
3.3.2.3	Comments on Currently Used Sensors	270
3.3.3	Present Developments for High-rise Elevators: New Shaft Information System	271
3.3.3.1	The Conflict of High-rise Traction Elevators	271
3.3.3.2	New Challenges of Motion Control	273
3.3.3.3	Specifications of the New Shaft Information System	273
3.3.3.4	Candidate Sensors	276
3.3.3.5	Conclusion for Future Shaft Information Systems	283
3.3.4	Active Ride Control for High-rise Elevators	284
3.3.4.1	Motivation	284
3.3.4.2	Requirements on the Active Damping System	285
3.3.4.3	Concept of the Active Damping System	285
3.3.4.4	Controller Scheme of the Active Damping System	286
3.3.4.5	Sensor Specifications for the Active Damping System	287
3.3.5	Conclusions and Outlook	290
3.4	**Sensing Chair and Floor Using Distributed Contact Sensors**	**293**
	Hong Z. Tan, Alex Pentland, Lynne A. Slivovsky	
3.4.1	Introduction	293
3.4.2	Related Work	294
3.4.3	The Sensing Chair System	295
3.4.3.1	Overview	295
3.4.3.2	The Sensor	297
3.4.3.3	Preprocessing of Pressure Data	299
3.4.3.4	Static Sitting Posture Classification	299
3.4.3.5	Performance Evaluation	300
3.4.4	The Sensing Floor System	301
3.4.4.1	Overview	301
3.4.4.2	The Sensor	302
3.4.4.3	Data Processing	302

3.4.5	The Future	303
3.4.6	Acknowledgments	303
3.4.7	References	304

4 Safety and Security

4.1 Life Safety and Security Systems 307
Marc Thuillard, Peter Ryser, Gustav Pfister

4.1.1	Introduction	307
4.1.2	Fire Sensing	309
4.1.2.1	Fire Physics, Smoke Aerosols, Gases, and Flames	309
4.1.2.2	Smoke Sensing Principles	316
4.1.2.3	Heat/Temperature-sensing Principles	323
4.1.2.4	Flame-sensing Principles	327
4.1.2.5	Multicriteria/Multisensor Detectors	329
4.1.2.6	System Concepts	332
4.1.2.7	Application Concepts and Criteria	334
4.1.2.8	Trends	335
4.1.2.9	Standards	336
4.1.3	Gas Sensing	338
4.1.3.1	Toxic and Combustible and Explosive Gases	338
4.1.3.2	Catalytic Devices (Pellistors)	340
4.1.3.3	Photoacoustic Cells	341
4.1.3.4	Electrochemical Cells	343
4.1.3.5	Metal Oxides	344
4.1.3.6	Application Concepts and Criteria	346
4.1.3.7	Standards	347
4.1.4	Intrusion Sensing	347
4.1.4.1	Passive Sensing Principles	347
4.1.4.2	Active Sensing Principles	356
4.1.4.3	Multisensor Sensing	358
4.1.4.4	System Concepts	359
4.1.4.5	Trends	361
4.1.4.6	Standards	361
4.1.5	Identification Sensing	363
4.1.5.1	PIN Code	363
4.1.5.2	Reading Methods for Identification Cards	364
4.1.5.3	Biometric Reading Principles	368
4.1.5.4	Concepts for Automatic Processing of Card Data	371
4.1.5.5	Trends	373
4.1.5.6	Standards	374
4.1.6	Emergency Handling	374
4.1.6.1	Voice Evacuation Systems	374
4.1.6.2	Fire Extinguishing Systems	376
4.1.6.3	Alarm Receiving Centers	379

4.1.7	Signal Processing 383
4.1.7.1	Intelligent Development Methods 385
4.1.7.2	Application of Multi-resolution and Fuzzy Logic to Fire Detection 387
4.1.8	References 394

4.2 Biometric Authentication for Access Control 399
Christoph Busch

4.2.1	Introduction 399
4.2.2	Access Control 400
4.2.3	Biometric Systems for Access Control 401
4.2.4	Security of Biometric Systems 405
4.2.5	Prospects 408
4.2.6	References 408

4.3 Smart Cameras for Intelligent Buildings 409
Bedrich J. Hosticka

4.3.1	Introduction 409
4.3.2	Technologies for Solid-state Imaging 411
4.3.3	Principles of CMOS Imaging 412
4.3.4	Examples of CMOS Imagers 413
4.3.5	Simple CMOS Occupant Sensors Based on Motion Detection 419
4.3.6	CMOS Imagers and Motion-based Occupant Sensors Using Active Illumination 421
4.3.7	Advanced CMOS Occupant Sensors Based on Shape Recognition 424
4.3.8	Biometric Sensors 425
4.3.9	Summary 425
4.3.10	Acknowledgements 426
4.3.11	References 426

4.4 Load Sensing for Improved Construction Site Safety 427
Peter L. Fuhr, Dryver R. Huston

4.4.1	Introduction 427
4.4.2	Equipment and Data Processing 428
4.4.2.1	Calibration 429
4.4.2.2	Sensor Head Configuration 429
4.4.2.3	Wireless Communication Components 430
4.4.3	Laboratory Work 430
4.4.4	Uplift Monitoring 435
4.4.5	Field Measurements 436
4.4.5.1	Construction Site Specifics 437
4.4.5.2	Logistics of Field Site Work 438
4.4.5.3	Site Data Acquisition 440
4.4.6	Wireless Data Acquisition for Smart Shoring 441
4.4.7	Field Use and Representative Data 443
4.4.8	Conclusion 445
4.4.9	References 446

5 Maintenance and Facility Management

5.1 Maintenance Management in Industrial Installations 451
Jerry Kahn
- 5.1.1 Introduction 451
- 5.1.2 Predictive Maintenance and Condition Monitoring 451
- 5.1.2.1 Vibration 453
- 5.1.2.2 Acoustic and Ultrasonic Monitoring 455
- 5.1.2.3 Lubricant Analysis (Tribology) 455
- 5.1.2.4 Infrared Thermography 456
- 5.1.2.5 Process Parameter Monitoring 457
- 5.1.2.6 Electrical Testing 457
- 5.1.2.7 Sensory Inspection 458
- 5.1.3 Enhancing Condition Monitoring with Expert Systems 458
- 5.1.4 Integration with Plant Systems 460
- 5.1.5 Maintenance Management Methods 461
- 5.1.5.1 Reliability-centered Maintenance (RCM) 462
- 5.1.5.2 Total Productive Maintenance (TPM) 462
- 5.1.6 Future Directions in Maintenance Technology 464
- 5.1.6.1 Wireless and Smart Sensor Development 464
- 5.1.6.2 Human Sensory Sensors 465
- 5.1.6.3 E-Maintenance via the World Wide Web 467
- 5.1.7 Summary 467
- 5.1.8 References 468

5.2 WWFM – Worldwide Facility Management 469
Rolf Reinema
- 5.2.1 Introduction 469
- 5.2.2 Facility Management 470
- 5.2.3 Worldwide Facility Management (WWFM) 472
- 5.2.4 The RoomServer 472
- 5.2.5 Single-chip PCs 475
- 5.2.6 Advanced Architectures 476
- 5.2.7 Security Aspects 477
- 5.2.8 Conclusion 480
- 5.2.9 References 481

6 System Technologies

6.1 Sensor Systems in Intelligent Buildings 485
Hans-Rolf Tränkler, Olfa Kanoun
- 6.1.1 Introduction 485
- 6.1.2 Sensor Applications in Intelligent Buildings 486
- 6.1.3 Requirements for Sensor Systems in Intelligent Buildings 487

6.1.4	Sensor Systems for Safety and Health	488
6.1.4.1	Fire Detection	488
6.1.4.2	Gas Detection	491
6.1.4.3	Intrusion and Person Detection	495
6.1.4.4	Sensor Sytems for Health Safety	500
6.1.5	Sensor Systems for Heating, Ventilation, and Air Conditioning (HVAC) and Comfort	502
6.1.5.1	Convenience and Easy Usability	502
6.1.5.2	Thermal Comfort	503
6.1.5.3	Indoor Air Quality	504
6.1.6	Future Trends for Sensor Systems in Intelligent Buildings	505
6.1.7	Acknowledgements	508
6.1.8	References	508
6.2	**System Technologies for Private Homes**	**511**

Friedrich Schneider, Lars Binternagel, Yuriy Kyselytsya, Wolfgang Müller, Thomas Schlütsmeier, Bernhard Schreyer, Rostislav Stolyar, Kay Werthschulte, Günter Westermeir, Dirk Wölfle, Thomas Weinzierl

6.2.1	Introduction	511
6.2.2	Requirements in Home Automation Systems	512
6.2.3	Microcontroller Level	515
6.2.3.1	Realization	515
6.2.3.2	Choice of the Microcontroller	515
6.2.3.3	Bus Connection with BCU	515
6.2.3.4	Bus Connection via TP-UART	515
6.2.3.5	Bus Coupling with RF-UART	517
6.2.3.6	Operating System ContROS	520
6.2.3.7	Feature Controller	520
6.2.3.8	Intelligent Outlet	521
6.2.4	Operating System's Level	522
6.2.4.1	Introduction	522
6.2.4.2	Interfaces	523
6.2.4.3	The IrDA-EIB Interface	524
6.2.4.4	USB-EIB Interface	525
6.2.4.5	Bluetooth	526
6.2.4.6	The EIB Modem	527
6.2.4.7	Software Interfaces	528
6.2.4.8	Accessing the EIB with Windows CE and Other Operating Sytems	531
6.2.5	Bus Monitoring and Service Programs	532
6.2.5.3	Future Work: Interpretation and Test Management	534
6.2.6	Configuration of Home Automation Systems	534
6.2.6.1	Introduction	534
6.2.6.2	Easy Configuration	536
6.2.6.3	Configuration via the Internet	538

6.2.6.4	The IMOS Tool	538
6.2.7	Visualization and Tele Services	539
6.2.7.1	Possibilities of Visualization	539
6.2.7.2	Video and EIB	540
6.2.7.3	Visualization Software	541
6.2.7.4	Special Applications and Clients for Visualization	543
6.2.7.5	Access Technologies	547
6.2.7.6	Use of PDAs with HTML and CGI	549
6.2.7.7	Standard Browser and EIB. The EIB Web Server	549
6.2.7.8	Security Aspects of Tele-services Using HTTP	553
6.2.7.9	Using Applets, Java in Tele-services	553
6.2.8	Outlook	554
6.2.9	Internet Addresses	556
6.2.10	References	557

List of Symbols and Abbreviations 559

Index 569

List of Contributors

M. E. Barnard
Philips Research Laboratories
Wireless Sector
Cross Oak Lane
Redhill, Surrey RH1 5HA
England

M. Bauer
Estia Sàrl
Parc Scientifique EPFL
1015 Lausanne
Switzerland

T. Bernard
Fraunhofer Institute for Information
and Data Processing IITB
Fraunhofer Strasse 1
76131 Karlsruhe
Germany

J. Bichsel
Sauter SA
Im Surinam 55
4016 Basel
Switzerland

L. Binternagel
Technische Universität München
Lehrstuhl für Messsystem-
und Sensortechnik
80290 München
Germany

C. Busch
Fraunhofer Institute for Computer
Graphics Research IGD
Rundeturmstrasse 6
64283 Darmstadt
Germany

D. Dietrich
Vienna University of Technology
Institute of Computer Technology
Gusshausstrasse 27–29
1040 Wien
Austria

H.-E. Endres
Fraunhofer Institute of Microelectronic
Circuits and Systems IMS
Leonrodstrasse 54
80636 München
Germany

P. Fischer
University of Applied Sciences
Dortmund
Sonnenstrasse 171
44137 Dortmund
Germany

P. L. Fuhr
San José State University
Department of Electrical Engineering
1, Washington Square
San José, CA 95192
USA

O. GASSMANN
Schindler Elevators & Escalators
R&D-Technology Management
Zugerstrasse 13
6030 Ebikon/Luzern
Switzerland

A. HAMDY
Transtechno GmbH
Schädtrütistrasse 67
6006 Luzern
Switzerland

B. HOSTICKA
Fraunhofer Institute of Microelectronic
Circuits and Systems IMS
Finkenstrasse 61
47057 Duisburg
Germany

D. R. HUSTON
University of Vermont
Department of Mechanical
Engineering
Burlington, VT 05405
USA

J. KAHN
Siemens Energy and Automation Inc.
Industrial Systems & Technical
Services Division
900 Westpark Drive, Suite 100
Peachtree City, GA 30269
USA

O. KANOUN
University of the Bundeswehr, München
Institute for Measurement and Control
Werner-Heisenberg-Weg 39
85577 Neubiberg
Germany

J. KRAUSS
Swiss Center for Electronics
and Microtechnology (CSEM)
Industrial Control
Rue Jaquet-Droz 1
2007 Neuchâtel
Switzerland

H.-B. KUNTZE
Fraunhofer Institute for Information
and Data Processing IITB
Fraunhofer Strasse 1
76131 Karlsruhe
Germany

R. KUNZ
Schindler Elevators & Escalators
R&D-Hoistway and Car
Zugerstrasse 13
6030 Ebikon/Luzern
Switzerland

Y. KYSELYTSYA
Technische Universität München
Lehrstuhl für Messsystem-
und Sensortechnik
80290 München
Germany

D. LOY
Coactive Networks
4000 Bridgeway Suite 303
Sausalito, CA 94965
USA

Y. LÜTHI
Siemens Building Technologies AG
Landis & Staefa Division
Gubelstrasse 22
6301 Zug
Switzerland

E. MARCHESI
Schindler Elevators & Escalators
R&D-Technology Management
Zugerstrasse 13
6030 Ebikon/Luzern
Switzerland

R. MEISINGER
Siemens Building Technologies AG
Landis & Staefa Division
Gubelstrasse 22
6301 Zug
Switzerland

H. Meixner
Siemens Corporate Technology
Otto-Hahn-Ring 6
81739 München
Germany

K. Mnif
Motorola SA
Avenue du Général Eisenhower
Le Mirail
31023 Toulouse
France

N. Morel
Swiss Federal Institute of Technology
(EPFL)
LESO-PB
1015 Lausanne
Switzerland

D. Mrozinski
Siemens Landis & Staefa Electronic
GmbH
Sondershäuser Landstr. 27
99974 Mühlhausen
Germany

G. Mügge
Viterra Energy Services AG
Grugaplatz 4
45131 Essen
Germany

W. Müller
Technische Universität München
Lehrstuhl für Messsystem-
und Sensortechnik
80290 München
Germany

A. Pentland
The Media Laboratory Massachusetts
Institute of Technology (MIT)
20 Ames Street
Cambridge, MA 02139
USA

G. Pfister
Siemens Building Technologies AG
Cerberus Products/Alarmcom
Alte Landstrasse 411
8708 Männedorf
Switzerland

R. Reinema
GMD Institute
for Secure Telecooperation
GMD-SIT
Rheinstrasse 75
64295 Darmstadt
Germany

P. Ryser
Swiss Federal Institute of Technology
(EPFL)
Département de Microtechnique
1015 Lausanne
Switzerland

T. Sauter
Vienna University of Technology
Institute of Computer Technology
Gusshausstrasse 27–29
1040 Wien
Austria

F. Schneider
Technische Universität München
Lehrstuhl für Messsystem-
und Sensortechnik
80290 München
Germany

T. Schlütsmeier
Technische Universität München
Lehrstuhl für Messsystem-
und Sensortechnik
80290 München
Germany

B. Schreyer
Technische Universität München
Lehrstuhl für Messsystem-
und Sensortechnik
80290 München
Germany

L. A. Slivovsky
Purdue University
Haptic Interface Research Laboratory
1285 Electrical Engineering Building
West Lafayette, IN 47907-1285
USA

R. Stolyar
Technische Universität München
Lehrstuhl für Messsystem-
und Sensortechnik
80290 München
Germany

A. T. P. So
City University of Hong Kong,
Department of Building
and Construction
Tat Chee Avenue, Kowloon
Hong Kong
China

H. Z. Tan
Purdue University
Haptic Interface Research Laboratory
1285 Electrical Engineering Building
West Lafayette, IN 47907-1285
USA

M. Thuillard
Siemens Building Technologies AG
Cerberus Products/Alarmcom
Alte Landstrasse 411
8708 Männedorf
Switzerland

H.-R. Tränkler
University of the Bundeswehr, München
Institute for Measurement and Control
Werner-Heisenberg-Weg 39
85577 Neubiberg
Germany

B. W. L. Tse
City University of Hong Kong
Department of Building
and Construction
Tat Chee Avenue, Kowloon
Hong Kong
China

T. Weinzierl
Weinzierl Engineering
Bahnhofstrasse 6
84558 Tyrlaching
Germany

M. Wenzler
Siemens Building Technologies AG
Landis & Staefa Division
Gubelstrasse 22
6301 Zug
Switzerland

K. Werthschulte
Technische Universität München
Lehrstuhl für Messsystem-
und Sensortechnik
80290 München
Germany

G. Westermeir
Technische Universität München
Lehrstuhl für Messsystem-
und Sensortechnik
80290 München
Germany

D. Wölfle
Technische Universität München
Lehrstuhl für Messsystem-
und Sensortechnik
80290 München
Germany

1 Introduction

1.1
Sensors in Intelligent Buildings: Overview and Trends

OLIVER GASSMANN, *Schindler Elevators & Escalators, Ebikon/Luzern, Switzerland*
HANS MEIXNER, *Siemens Corporate Technology, München, Germany*

1.1.1
Introduction

Today's home system is influenced by several trends of modern society: distributed work creates geographic distances between home and working place; increased time pressure and dual-income families create the need for efficiency; an aging population and higher requirements regarding health and well-being require modern health care services. At the same time, various external forces act on the aggregate house system: temperature, humidity, and air quality are varying; economic and legal restrictions regarding energy consumption and environmentalism define borders. New technologies open up new opportunities for meeting these trends: new materials as well as new information and communication technologies are the most important enablers for intelligent buildings.

Driven by these needs, we can observe a clear trend towards more 'intelligence' in the building control industry. We use the term 'intelligence', although it is unsatisfactory in an engineering context and no longer has its human or biological meaning. Here, 'intelligence' simply implies that the benefits of buildings for their inhabitants are increased by units which capture the current state of the building and its devices, process those signals and make appropriate adjustments. Although the intellectual capability of human beings has not been reached by a long way, tremendous progress has been made through the use of smart sensors and control systems.

In the last two decades, intensive research has been done in the area of intelligent buildings. The concept integrates new technologies from areas such as sensor systems, computer automation, space-age materials, and energy management in order to adjust and adapt buildings to their occupants. Integrated sensor systems judge indoor and outdoor conditions of a building and its devices in order to operate as an integrated system and achieve maximum performance and comfort levels. Modern buildings thus become places of multilateral interaction between the inhabitants and the buildings themselves.

On the other hand, deficient building environments ('sick buildings') also give rise to high social costs. Poor health and lost productivity associated with office environments alone cost US businesses more than $438 billion per year, according to US Technology, State and Community programs (2000). These costs will increase over time, unless buildings are designed and operated more efficiently and intelligently. The key lies in their ability to handle information and take appropriate actions. In intelligent building systems, research and development efforts are being made to reduce power consumption and emissions, and to achieve longer operation over a wider range of conditions with self-diagnosis and self-adaptation.

New system solutions and a wide variety of sensors will be available for all aspects of a building such as information and communication systems, traffic management, heating, ventilation, and air conditioning (HVAC), energy efficiency, personal safety and security systems, maintenance management, smart house devices, and new intelligent building structures (Figure 1.1-1).

In this section we give an overview of sensor systems in intelligent buildings, providing a brief outline of trends generally observed in sensor systems. We conclude with an overview on sensor systems in intelligent buildings in this book.

1.1.2
Towards the Intelligent Building

Intelligent buildings increase the benefits to their occupants by means of integrated sensor systems, computer automation, information and communication systems, smart home appliance devices, and new materials. The application areas

Fig. 1.1-1 Shape of a brain symbolizes the intelligent building. Source: Neuroscience Institute in Hannover, Germany

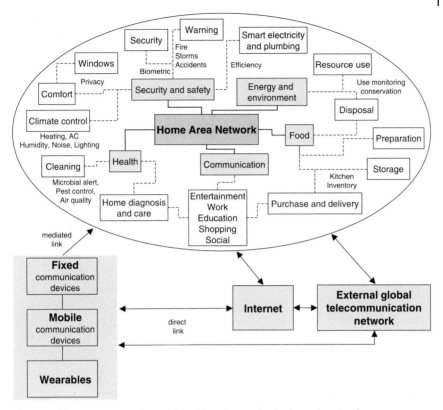

Fig. 1.1-2 Home area network: complex with various technologies and protocols

for new technologies in intelligent buildings are huge. The opportunities have to be evaluated from a functional, technological, and economic perspective. Winston Churchill once said, 'We shape our buildings, and afterwards, our buildings shape us.' Buildings have to be flexible enough to respond to changing human needs.

The complexity of the building environment is characterized by the people in the house, depending on their activities and needs. Different needs can be observed: privacy and comfort, safety and security, quiet, light, atmosphere, sanitation, energy efficiency, ease of disposal, and recycling, as well as ubiquitous communication and entertainment. These different needs lead to a complex structure which can be exemplified by the home area network shown in Figure 1.1-2. This communication system involves different technologies with different communication protocols on different layers. These home appliances can be linked directly to the external networks (eg, the Internet, global telecom network) or mediated by the home area network.

1.1.2.1
Reduced Resource Consumption

Legislation, environmental concerns, and cost awareness suggest that resources should be used efficiently even in private homes. Home and building technology (domotics) offers tremendous potential savings through intelligent controls. Partial measures such as ventilation-on-demand achieve savings in operating costs of up to 70% in this area. Load management, the main function of which is to cap energy consumption peaks, achieves financial savings of up to 50%. Home technology, for example, already meets the requirements for automating buildings. The relevant technology has been used in purpose-built constructions for a long time. To date installation costs in these special constructions have not played a significant role because operating the systems was the real cost factor. This situation is now changing, in part brought about by advanced service concepts (facilities management).

In private homes, the installation costs are the deciding factor. Few owners of a house costing, say, $300000, will be prepared to spend an additional $30000 to equip it with an intelligent home system. The breakthrough is expected to come at a cost of $3000–5000, whereby a modular structure which allows people to select the required functions is an essential prerequisite. Market studies in the USA have come up with evidence to support this view. Expectations are that the market will develop along similar lines to the automobile industry, with fierce competition, large unit production and tight margins, combined with very limited space available (in the domestic appliance or the wall socket). An important requirement is that systems must be flexible, and upgrades or retrofits possible (houses do not rust).

Ultimately, a large and dynamic consumer market will emerge. For example, the market for wireless high-frequency-supported modules for home security systems is expected to grow from 100000 units in 2000 to over 5 million units by 2003. A Europe-wide ESPRIT study has estimated the market volume for new home systems in connection with telecommunications applications at 15 billion euros per year. Affordable package solutions for the home can only be achieved with micro system technologies. Miniaturization is an important step in this direction. It will, for example, be difficult to sell solutions if their components cannot be accommodated in a wall socket or light switch. In some areas, such as individual room temperature control, this has already been achieved.

1.1.2.2
Optimized Convenience and More Comfort

Apart from minimizing consumption costs, private home owners will be interested in higher levels of convenience, safety, and security. The main applications will be as follows:

- Energy management (load management, tariff management, telemetrics, etc.).
- A healthy living environment (heating and ventilation control as well as air conditioning and in the future air quality management).

- Personal safety and security (emergency calls, health monitoring, children's room observation).
- Technical safety (fire, flooding, etc.).
- Increased convenience levels (networking of smart household appliances).
- Entertainment (interactivity from audio and video to hobby tools and toys).

Many innovative products are entering the market for the intelligent home. The consumers are allowed to use a wide variety of online services directly from their homes. There are nearly no limits to the opportunities for optimizing the convenience of the inhabitants: many ideas have been already realized by Bill Gates' famous house: home devices such as the washing machine can be shut down by the mobile phone; an intelligent navigator hooked up to the TV automatically downloads the favorite show; installed wireless technology enables portable devices, such as pocked-sized electronic checkbooks, book-sized reading tablets, and digital kitchen assistants, to interact. Figure 1.1-3 presents examples of today's products in that area.

The increased convenience and technological trends can be exemplified with the *food system*. A popular element of the intelligent building is the smart kitchen. Intelligent handling of food meets the customers' requirements for hygiene, food storage, cooking, and comfort of the user. The basic technology requirements therefore are home area network, flat screens, machine vision and imaging sys-

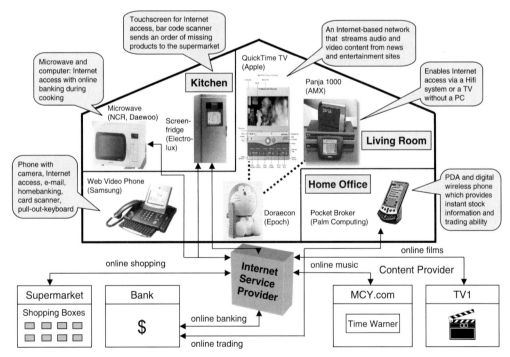

Fig. 1.1-3 Product visions for an intelligent home

tems, advanced sensors with biometric recognition, broadband wired and wireless links to home devices, portables, and wearables as well as voice/gesture recognition and computer voice synthesis.

Today's first passive smart food systems will include functions such as scanning, monitoring, and warning the user and provide database management. Examples are smart storages, eg, pantry or refrigerator and smart microwave ovens. The trash will be able to keep track of what has been thrown out in order to place an online order for delivery of refills. Within the next 5 years, smart stoves and ovens will be available, and microbe alert sensors will be used. In general, the kitchen will change from smart standalone devices to self-identifying, communicative systems with memory. MIT's first smart microwave oven was able to correlate cooking time to weight and content. Current research projects at MIT Media Lab add counter intelligence to the kitchen: this tries to integrate itself into the work habits of the user by serving as an interface between the user, the recipe, and the food being prepared. The cook will be coached in terms of the recipe – sequence, ingredients, possible substitutions – and the food system adapts and learns. Other projects contain the intelligent refrigerator: this keeps track of its contents (location in the refrigerator, date entered, expiration dates) and keeps track of *desired* contents (eg, >2 L of milk, <2 days old). As a result, automatic shopping lists are generated or goods are ordered online to replace staple items [1].

In the long term of 10 years and more, these passive smart food systems will progress towards active ones: The system will make a physical intervention by way of automation and/or robots. Examples are diet management knowbots, automated food preparation, home waste sorting facilities, and robotized cleanup and sanitation. Advanced biometric sensors are a key technology for many such intelligent food systems. Trends are illustrated in Figure 1.1-4.

1.1.2.3
Increased Impact of Microsystems Technology

Microsystems technology will make a significant contribution to the development of such intelligent systems. Current targets from a technical point of view include the following:

- Open, standardized system design with graded options (including software for easy project management, automatic configuration and control).
- Seamless communications capability of all components for the home, integrating the possible physical transmission media, networks, and buses, as well as the necessary gateways and converters.
- Cost-effective, robust and long-lasting sensor systems (eg, for gases, temperature, humidity, pressure, flow) and actuation systems (eg, switches, actuators, valves) for decentralized energy supply of all system components.
- Simple, robust, distributed operating system with the relevant requirements of display and operating units (keyboard entry, touchscreen, voice) using the PC, TV or a handheld device.

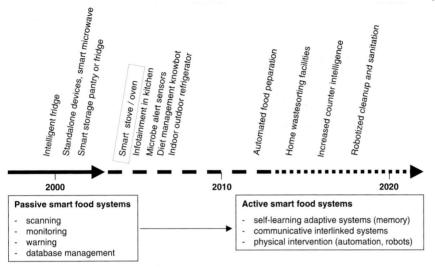

Fig. 1.1-4 Trends in the food system

- Compact intelligent nodes (from wall socket and door to washing machine, television, robot toy or handheld programming device) with the accompanying requirement for build and connection systems (eg, the use of application-specific integrated circuits (ASICs), multi-chip modules), and volume production in the future.
- Software technologies, such as Jini, as enabler for any electronic device to interact and connect to the Internet.

The impact of microsystems technology can be exemplified by monitoring household appliances: intelligent wall sockets (see Figure 1.1-5) can be used in home bus systems to control energy-consuming devices such as washing machines and refrigerators and monitor their functioning. External sensors also connected to the bus system make it possible to detect water damage or changes in temperature. These systems can be programmed from any location using portable wireless operating devices. Other functions to raise comfort levels in the home or to secure private areas against unauthorized access are easy to add, as are features such as control via the home television and data transfer via the Internet.

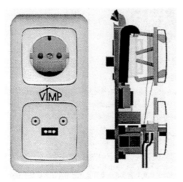

Fig. 1.1-5 Intelligent wall socket for monitoring household appliances. Source: IFAM GmbH

1.1.2.4
Increased Impact of New Communication Systems

Modern home residents are pushed by time stress, multi-income lifestyles and demanding leisure activities, to adopt new information and communication technologies that facilitate complex communications, including transactions, such as shopping and work, from home. Flat-screen monitors will be the key to the wired home, serving as primary conduits of voice and image data.

Today's networking solutions contain phone and coax twisted-pair wiring, which is high speed and business tested, but which is still expensive to install in private homes. New home networking technologies are emerging to allow consumers to link appliances without installing new wires: powerline can outlet everywhere, but is slow and subject to interference. Digital wireless supports portable devices, but at the moment is still connected with high costs (according to Forrester Research, $200–500 per node in 2000). In the future, broadband wireless links to home devices, portables and wearables, and appliances will be extensively spread throughout buildings and there will be a high premium on the rapid evolution of wireless, mobile devices. It is expected that as soon as 2002 most people will no longer access the Web via PCs, but mobile devices. Wireless technologies are not only much more flexible because of the lack of installation, they are also connected with additional advantages: Bluetooth allows the wireless transmission of voice and data in home networks. Advances in wireless technology will allow for the automatic transfer of information between Internet appliances and computers as well as mobile devices.

When considering communication, the future home also includes all mobility aspects around the home area network (HAN). The different activities, such as entertainment, education, contact, shopping, work, and home management are linked via three kind of interfaces: (1) fixed communication devices, eg, touch screens, (2) mobile communication devices, and (3) wearables. All interfaces terminate in the HAN.

For private homes, it is expected that there will be an additional layer between the appliances and the Internet. Reasons are control, security, and costs: fully decentralized solutions are at the moment far more costly than centralizing some of

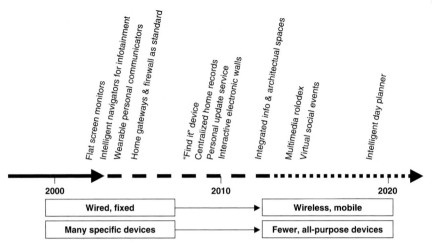

Fig. 1.1-6 Trends in the home communication system

the computing power and most of the security on a hub device. Ericsson developed a device that will act as a firewall and gateway for the networked home (e-box). In a later stage additional services such as health care, remote home monitoring, and remote security control will be conducted through specialized care centers. According to a Siemens forecast, the intelligence of communication devices will grow with 'Moore's law' until the year 2020. Visionary applications of the future are personal update services, centralized home records, multimedia rolodex, virtual social events, and intelligent day planners.

At the moment there are several research projects on improving computer-aided team work, a premise for further extensive teleworking in the private environment. In the future there will be a more integrated design of information and architectual spaces: information will not be limited to displays of desktop computers or portable screens. Buildings themselves with their rooms, walls, floors, ceilings, doors, windows, and furniture constitute rich information spaces, either as concrete, direct information or by providing ambient peripheral information. Electronic walls, developed at the German National Research Center for IT, will act as a touch-sensitive information provider and permit interactivity. Mobile real-time scanning tables and interactive tables will allow augmented reality and ubiquitous computing [2]. New telework practices will come up with stimulating and rich decentralized communication forms. Trends are illustrated in Figure 1.1-6.

1.1.2.5
Development of an Intelligent Home Market

Already today there are many smart house devices, such as the screenfridge, microwaves with Internet access, smart Web video phones, wireless phones and access control systems. In general these are still 'intelligent' standalone devices today. In a subsequent phase, with the increase of home networking components, hubbed homes with external access will become standard and intelligent home services will require an intelligent house with networked components and applications.

The migration of the existing housing situation to intelligent homes will have several phases (see Figure 1.1-7). Today the US market is more advanced than the European one, but new homes can enter directly into developed stages.

1.1.2.6
More Integrated Buildings: the FuturElife Smart Building

Intelligent buildings are mostly the result of common efforts of several companies and industries. Compatible standards, non-proprietary interfaces, and open architectures will be the enabler for those efforts. Whereas the automotive industry is already heavily characterized by modular thinking and open systems, this has only just started in the building control industry. In order to align the research and development efforts of different industries, companies have started to implement show buildings, where the newest technology is implemented. At the same time these buildings serve as pilots for identifying market trends and further explore technological opportunities with their lead user inhabitants.

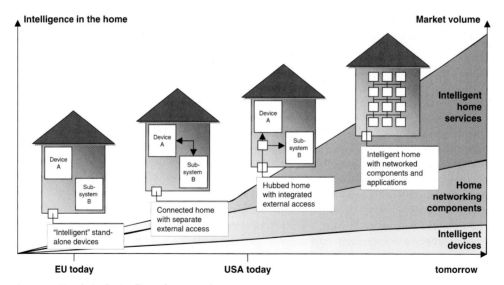

Fig. 1.1-7 Trends in the intelligent home market

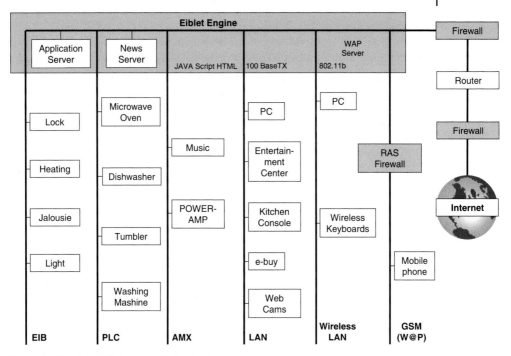

Fig. 1.1-8 The FuturElife house in Switzerland

FuturElife is such an advanced building in Switzerland, which serves as a point of reference for future e-technology-supported living (Figure 1.1-8). The whole building management system of FuturElife is linked to the Internet, including control of HVAC, outdoor curtains, and light. The intelligent security system allows remote monitoring and verification of intrusion via webcams. Internet-enabled cameras are becoming more and more popular. Remote access is possible via the web. Access control can be supported by biometric sensors, such as fingerprint and face recognition. Instead of directly linking each device to the Internet, different communication systems are used to link the house devices as shown in Figure 1.1-1. For example, simple systems such as light or heating are connected via EIB (European Installation Bus) and for the home appliances, eg, the washing machine, connection via a powerline (PLC) is used.

Goods are ordered via the Internet and delivered to the house. If the occupants are not at home during working hours, the goods are delivered to an outside box, secured by an electronic key which only the addressee and the supplier can open. The user will be informed about a delivery by e-mail. Not only the occupants but also the service technicians are able to gain access to these devices. Remote monitoring makes service efficient: remote diagnostics by the service center reduces travelling and, if repairs become necessary, the service technician already has the right spare parts at hand.

1.1.2.7
Pneumatic Building Structures: Airtecture

In the future there will be moves to expand the area of sensor systems in intelligent buildings even as far as active building structures; in the past all buildings were built according to the principles of pressure and tension. According to the OECD's forecast [3], it is expected that future buildings will become ultra-light and will be based on high-performance structures with high-performance and recyclable parts. Adaptive, dynamic structures such as Airtecture by Festo represent the future of modern building structures.

The first building in the world to be constructed with a cubic interior which consists of supporting structures built with air-inflated chambers is an exhibition hall in Germany, which has recently been unveiled by Festo (Figure 1.1-9). The inside of the Airtecture hall is 6 m tall with a floor area of 375 m^2. It has a total volume of 2250 m^3. The exterior of the hall covers a surface area of 800 m^2 with a height of 7.2 m. The structure is considered to be light, having a dead load of 7.5 kg/m^2. The load-bearing structure of the exhibition hall includes 40 Y-shaped columns and 36 wall components along both longitudinal sides; 72 000 distance threads per square meter hold the double-layer walls in place at an operating pressure of 0.5 bar. The slits between the opaque wall components are filled with transparent cushions made of Hostaflon ET; these window sections can be easily replaced by means of slide locks (see also Festo [4–8]).

The load-bearing structure has many new features. These include the use of double-wall fabric as a load-bearing element, flame-inhibiting elastomer coatings and a new translucent ethylene-vinyl acetate coating. Each of the two end walls of the hall consists of two L-shaped elements which surround a space measuring 3×6 m. Two wall openings are used as entrances. Above one of these doors, air conditioning and ventilation ducts enter the hall. They are used to supply and ex-

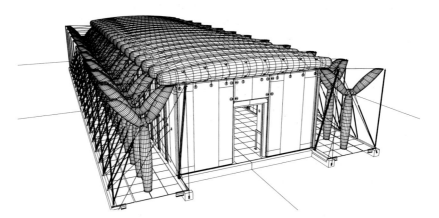

Fig. 1.1-9 The Airtecture building uses an active pneumatic building structure

tract air. A 4 m high PVC membrane, which functions as a roll-up gate, is secured to a steel frame as a separate structure.

The light building provides some special points of interest, such as its unusual structure and its ability to actively respond to wind loads. The columns, arranged in the shape of a saw-tooth pattern in plan view, and the pairs of wall components form a triangular structure that can carry vertical and horizontal loads, such as snow. The horizontal wind load is carried by this space frame together with elastic tension elements (so-called 'fluidic muscles'). When wind hits a side wall, the muscle elements of this wall contract. Both types of computer-controlled muscle elements are made of polyamide fabric with an internal silicone hose. The tensile forces of the outer fabric are in a state of equilibrium with the internal pressure exerted on the membrane. The air pressure in the elastic elements can be varied between 0.3 and 1 bar, thus regulating the axial force continuously and over a wide range.

The horizontal girders, so-called 'airbeams', contribute to the stability of the structure. With a diameter of 0.75 m at the support and 1.25 m in the middle of the span, the airbeams are supported at the connection point of wall components and Y-shaped columns and are fixed in location with textile belts. The airbeams can be pretensioned by varying the internal pressure so that the supports can withstand a snow load of 50 kg/m^2 and, at the same time, a wind speed of 80 km/h.

Translucent membranes connect the upper and lower sides of the airbeams. A vacuum is generated, which stabilizes the membranes. The alternating structural elements of positive (500 mbar) and negative (2.5 mbar) pressure produce a stable roof system. This combination of pressurized and partially evacuated air chambers is being used for the first time in a roof structure. Both end girders on the top of the end walls are braced with a boundary cable made of stainless steel, which carries the wind load in this area and directs it to the foundation via steel struts and guy cables. Furthermore, the cables are used to absorb horizontal forces which act on the slightly evacuated membrane.

The hall consists of approximately 330 individual air elements, most of which come under the six categories wall components, window cushions, Y-shaped columns, roof supports, intermediate membranes, and elastic muscle elements, and have different volumes and internal pressures. For control purposes, the air elements are divided into 10 almost identical sections. Each section is controlled by a valve terminal in order to regulate the operating pressure. The necessary pressure levels are controlled by proportional valves, and the actual pressure in the elements is monitored by sensors. If excess pressure builds up in an air element, this can be balanced out by means of pressure relief valves.

The master control system calculates the required pressure values on the basis of data supplied by a weather station and stabilizes the individual air elements. Integrated into the control concept are the pressure regulation system, the pneumatic control unit, the FPC 405 PLC, the PCS 500 display and control unit and the weather station. The pressure in all air elements is measured by an integrated pressure sensor within each proportional pressure regulator and adjustable within a range of 100–1000 mbar. The external actual-value input provided as standard on the proportional pressure regulators allows optimum regulation of the low-pressure zones in the intermediate roof and wall elements. A special pressure sensor signals the pressure in each air element to the microprocessor and ensures that this is corrected as required. For reasons of cost, a defined number of air elements is assigned to each proportional pressure regulator. A total of 30 proportional pressure regulators are used to maintain pressure within the hall structure.

1.1.3
Trends in Sensor Systems

Sensor systems in advanced intelligent buildings are required to provide comfort and convenience, high performance and automation, energy and resource savings, security, and safety.

According to a Japanese study conducted in 1986 by the Electronic Industries Association of Japan [9], sensors in consumer goods, especially in home appliances, accounted for about 20% of sensor production. For example, a simple audio/video setup contains at least six different sensors: (1) temperature sensors to detect cylinder temperature, (2) magnetic sensors to measure motor revolution and detect tape end and rotor position, (3) ultrasonic sensors for remote control, (4) position sensors to detect loading condition, (4) dew sensors to detect cylinder dew, (5) optical sensors for photo pickup and remote control, and (6) electrical sensors for capacitive pickup [9].

Modern buildings require many kinds of sensors:

- Home control requires sensors for temperature, humidity, wind flow, taste, cooking condition, defrost, frost, water level, dust, rinse, weight, pressure, vision, gas, flames, and gas leaks.
- Home security requires sensors for temperature, infrared radiation, smoke, flammable gas, earthquakes, electric leakage, overheating, vibration, window/door ultrasonics, and face/fingerprint/voice recognition.
- Energy control requires sensing for electric power, voltage/current, flow (gas/water), temperature, water level, water freezing, sun, and illumination.

Remote monitoring and diagnostic systems require the sensing and transmission of several kinds of signals. Washing machines, for example, use at least six different kinds of sensors [9]: (1) optical sensors to detect water dispersion and water levels, (2) pressure sensors to detect water levels, (3) magnetic sensors to detect the rotation speed of agitator and induction motor, (4) rinsing sensors to measure water transparency, (5) load sensors to detect the weight of clothes, and (6) humidity sensors to detect water dispersion in washed clothes.

The development of sensor systems has progressed rapidly in the last two decades. However, the principal classification of sensors by Lion [10] according to their primary and secondary signals is still valid. These principles are mechanical, thermal, electrical, magnetic, radiant and chemical (see Table 1.1-1 for an over-

Tab. 1.1-1 Physical and chemical transduction principles of sensor systems. Source: Ref. [11], p. 6 (modified)

Primary signal	Secondary signal					
	Mechanical	Thermal	Electrical	Magnetic	Radiant	Chemical
Mechanical	(Fluid) mechanical and acoustic effects, eg, diaphragm, gravity balance, echo sounder	Friction effects, eg, friction calorimeter, Cooling effects, eg, thermal flow meters	Piezoelectricity, Piezoresistivity, Resistive, capacitive, and inductive effects	Magnetomechanical effects, eg, piezomagnetic effects	Photoelastic systems, eg, stress-induced birefringence, interferometers, Sagnac effect, Doppler effect	
Thermal	Thermal expansion, eg, bimetallic strip, liquid-in-glass and gas thermometers, resonant frequency Radiometer effect, eg, light mill		Seebeck effect, Thermoresistance, Pyroelectricity, Thermal (Johnsen) noise		Thermo-optical effects, eg, in liquid crystals, radiant emission	Reaction activation, eg, thermal dissociation
Electrical	Electrokinetic and electromechanical effects, eg, piezoelectricity, electrometer, Ampèr's law	Joule (resistive) heating, Peltier effect	Charge collectors, Langmuir probe	Biot-Savart law	Electrooptical effects, eg, Kerr effect, Pockels effect, electroluminescence	Electrolysis, Electromigration
Magnetic	Magnetomechanical effects, eg, magnetostriction, magnetometer	Thermomagnetic effects, eg, Righi-Leduc effect, Galvanomagnetic effects, eg, Ettingshausen effect	Thermomagnetic effects, eg, Ettingshausen-Nernst effect, Galvanomagnetic effects, eg, Hall effect, magnetoresistance		Magnetooptical effects, eg, Faraday effect, Cotton-Mouton effect	

Tab. 1.1-1 (continued)

Primary signal	Secondary signal					
	Mechanical	Thermal	Electrical	Magnetic	Radiant	Chemical
Radiant	Radiation pressure	Bolometer, Thermopile	Photoelectric effects, eg, photovoltaic effect, Photoconductive effect		Photorefractive effects, Optical bistability	Photosynthesis, dissociation
Chemical	Hygrometer, Electrodeposition cell, Photoacoustic effect	Calorimeter thermal conductivity cell	Potentiometry, conductimetry, amperometry, flame ionization, Volta effect, gas-sensitive field effect	Nuclear magnetic resonance	(Emission and absorption) spectroscopy, Chemiluminescence	
Biological	Microgravimetric detection (piezocrystals)	Calorimetry	Amperometry, Conductimetry, Potentiometry		Spectroscopy (UV/VIS, fluorescence), Chemiluminescence, Bioluminescence, Interferometry, Reflectometry	

view). Nevertheless, in terms of industrial applications, other characteristics besides the transduction principle become more important, namely technology, materials, accuracy, and costs. The insides of sensor systems become less important from an application perspective.

All sensor systems are facing a noticeable upward trend in performance requirements for maintenance, down-time, reliability, fault tolerance, fault recovery, and adaptability. Industrial applications in the area of consumer goods are also subject to cost pressure, where the value of electronic products has risen while costs have continuously fallen in past decades.

Current trends in sensor systems can be summarized as follows (see Figure 1.1-10):

1. Higher performance of sensors. Sensor accuracy and linearity, both classical problems of sensor systems, have been improved in all areas. Today's sensor systems also work under better conditions and with higher reliability. Greater demands for environmental protection have led to the development of highly reliable sensors. Development efforts are still focused on maintenance-free sensors with long life expectancy and low electric power dissipation.

New sensors have been developed, such as torque sensors, accelerometers, acoustic sensors, pattern information-sensing devices, and smell sensors. Olfactory sensors and voice recognition combined with synthetic speech generation technology will make remote-controlled home appliances a reality.

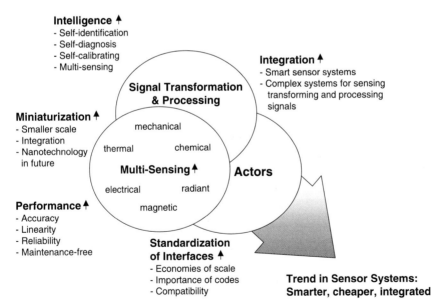

Fig. 1.1-10 Trends in sensor systems

2. *Smart sensor systems.* Smart sensor systems are usually complex systems with integrated intelligence. Sensor technology has developed towards a stronger integration of signal transformation and signal processing. Microprocessors and sensors build an integrated unit for measuring and controlling technology. Sensors and signal processing have to be brought together at the place of measurement. These 'smart sensors' are characterized by an integral electronic design, optimized for defined applications. As a result, higher speeds will permit sensor signal processing. In addition, the integration of sensors and related electronic circuits will lead to lower costs.

Intelligent sensor systems rely on a set of algorithms and rules (knowledge base) which connect the signal enhancement from the sensors to the sensor output. Two kinds of knowledge are important:
- empirical knowledge, which includes model-based calculations, hard rules about the process, and soft (fuzzy) rules based on experience;
- analytical knowledge, which includes extrapolation from the past, involvement of the neighborhood, and learning from success (analytical of neuronal).

3. *Multi-sensor systems.* R&D in sensor technology is tending towards self-identifying, self-diagnosing, and self-calibrating sensors. Today there are already several sensor systems which integrate different transduction principles in a single sensor system, ie, mechanical, thermal, and optical. The output signals of several simple sensors are computed in parallel by a secondary processing unit. These multi-sensor systems create powerful and precise systems with high selectivity. New algorithms for pattern recognition support built-in self-supervision and increase the reliability of the sensor system.

4. *Miniaturization of sensor systems.* Driven by high-tech sectors, such as the aerospace, medical, and entertainment industries, sensor systems have been pushed towards smaller scale devices. Examples are the integration of optical sensors in fiber technology and intelligent noses with biochemical sensors. In the future the importance of nanotechnology for sensor systems will increase further and open up access to new and smaller dimensions.

5. *Integration of sensor-actor systems.* Sensor systems are becoming smaller in size and lighter in weight, which allows further integration of sensor and actor systems. This trend became especially important in the area of mechatronics, a fusion of mechanical and electronic areas at a miniaturized level.

6. *Standardization of sensor interfaces.* Across industries and applications, the importance of standardized interfaces has been on the increase. Economies of scale due to mass production of sensors have led to lower costs. Even complex sensor systems have become commodity items because of standardized interfaces and integrated designs for sensor systems and control systems. Standardization allows easier handling of sensor systems. Some industries support elaborate and advanced standards, such as the CAMAC system in the nuclear industry. The standardiza-

tion constraints, however, mainly apply to the communication aspects of sensor systems (external interfaces); the internal operation of industrial sensor systems remains open to free development [12].

7. Cost reduction in sensor systems. A direct impact of standardization is high-volume production and thus cost reduction. In general, sensor technology has become more smart, integrated, and standardized. Industrialization and the trend towards more functionality at lower cost are the main drivers. The automotive industry has already demonstrated these trends. When the topic of intelligent sensors came up in the field of instrumentation in the late 1970s, it was greeted with some skepticism in industrial communities because of the cost and the current stage of development of computer technology [12]. Today, low-cost systems produced in high volumes provide both intelligence and reliability.

The enablers of these trends are new developments such as material technology, sensor and actuator technology, microelectronics/optoelectronics/information storage, electro-optical and electromechanical transducers, generation/distribution/storage of electrical energy, information engineering, intelligence functions, and software. Major drivers include large efforts in research and development of high-tech industries in the area of microelectronics in the 1970s, in new materials and information technology in the 1980s, and in miniaturization and integration in the 1990s [13]. Huge improvements have been made in the manufacturability of sensor systems. This includes the integration of sensorics, actuators, ASICs, and mechatronics as well as communication elements into one system.

At the same time, the complexity of the sensor-actor systems has increased dramatically. The management of the increased complexity in R&D laboratories will be a critical success factor in the future, according to Bill Raduchel, Chief Strategy Officer of Sun Microsystems: 'The challenge over the next 20 years will not be speed or cost or performance; it will be a question of complexity.' Future sensor systems will be independent, teachable, or adaptable systems, which will extend their geometries into the nano range, include non-electrical signal processing components, and will be based partly on materials from high-temperature electronics [13].

1.1.4
Sensor Systems in Intelligent Buildings

In the following sections of this book we focus on those areas of an intelligent building where we expect the biggest impact of sensor and control technologies. These are

- energy and HVAC;
- information and transportation;
- safety and security;
- maintenance and facility management;
- system technologies.

1.1.4.1
Energy and HVAC

Albert So and *Brian Tse* describe intelligent air-conditioning control systems. After a review of conventional physical sensors, three new concepts and their associated developments in sensing technology are discussed. Computer vision traces people movement and people flow in order to make the most intelligent decision for optimal HVAC control. Internet-based control permits fully remote monitoring and operation of building systems. Comfort-based control is highlighted as the future standard; the authors suggest the use of the predicted mean vote (PMV) as comfort index.

A self-commissioned heating control system, called NEUROBAT, is introduced by *Jens Krauss, Manuel Bauer, Jürg Bichsel,* and *Nicolas Morel.* The heating controller is predictive and adaptive and thus able to achieve energy savings and ensure optimal thermal comfort. Artificial neural networks permit the adaptation of the control model to real conditions, such as climate, building characteristics and user behavior. First empirical results at a real site have shown huge energy savings.

Hanns-Erik Endres focuses on the measurement and management of air quality. Gas sensors play an important role in this.

An optimal sensor-based management of energy and thermal comfort in buildings is discussed by *Thomas Bernard* and *Helge-Björn Kuntze.*

The automatic and remote reading of consumption data recorded by meters installed in residential buildings is an important element of energy management. *Dieter Mrozinski* gives an overview of the opportunities of wireless and M-bus-enabled metering devices.

Since user behavior regarding energy consumption can be positively influenced by metering individual energy consumption, sensors for metering are recommended by the European Union. *Günter Mügge* describes such sensors and the means of energy cost allocation.

Pressure sensors for the HVAC industry are described by *Yves Lüthi, Rolf Meisinger, Marc Wenzler,* and *Kais Mnif.* Special requirements from the HVAC industry are low pressure ranges, simple installation, independence of mounting orientation, low sensitivity to dirt, robustness, a long lifetime, and low costs. Several solutions of pressure sensors which meet these requirements are discussed.

1.1.4.2
Information and Transportation

Fieldbus systems enable decentralized control structures and therefore often lead to lower system costs and higher system reliability. *Dietmar Dietrich, Peter Fischer, Dietmar Loy,* and *Thilo Sauter* give an overview of fieldbusses. This includes the basics of the communication techniques, used in fieldbus systems, as well as a short history and an overview of EIB and LonWorks. Challenges which still remain include interoperability and demands for agent-based systems.

Wireless in-building networks are discussed by *Mike Barnard*, who focuses on low-power/low-bandwidth networks suitable for sensors, telemetry, etc. He also gives an overview of existing and emerging standards and products, related to wired and wireless links from and to sensors.

Sensor systems in modern high-rise elevators are discussed by *Enrico Marchesi, Ayman Hamdy,* and *René Kunz.* In order to travel higher and faster than ever before, elevators need excellent shaft information systems. Modern elevators travel at 10 m/s up to 500 m in height with horizontal vibrations of less than 10 mg. The authors explain an active ride control system, which relies heavily on fast and accurate sensors.

The perceptual user interface of a future intelligent office building includes information on all elements in a room. Opportunities and first empirical results of sensing chair and floor interaction by using distributed sensors are presented by *Hong Z. Tan, Alex Pentland,* and *Lynne A. Slivovsky.* The key problem is the automatic processing and interpretation of touch sensor information and the modeling of user behavior leading to such sensory data.

1.1.4.3
Safety and Security

Marc Thuillard, Peter Ryser, and *Gustav Pfister* give an overview of life safety and security systems in intelligent buildings. These systems provide early warning of dangerous situations in buildings, eg, fire incidents, toxic or explosive gas concentrations, and unwanted intrusion into premises. They discuss technologies and physical principles of those sensors, including the complete alarm systems and emergency systems involved.

Biometric sensing systems represent innovative technologies for access control. *Christoph Busch* describes opportunities of biometric authentication for access control. Pattern recognition is one of the key areas.

Smart cameras and their applications in intelligent buildings are outlined by *Bedrich J. Hosticka.* Smart cameras possess built-in intelligence and allow enhanced imaging under critical light conditions, as well as low-cost image processing. New camera technology, called CMOS-based imaging, is introduced by the author.

Peter L. Fuhr and *Dryver R. Huston* outline new sensing systems and techniques for improved construction site safety. Sensor systems are designed for monitoring construction site shoring and scaffolding. Such a sensor network can provide information about the load distribution on shoring systems and thus increase personal safety. The solutions presented have already been tested on field sites.

1.1.4.4
Maintenance and Facility Management

New forms of maintenance management in industrial installations are described by *Jerry Kahn.* Predictive maintenance and periodical condition monitoring are the key elements. Sensor systems are required to measure and analyze equipment conditions and predict future equipment performance.

Rolf Reinema from the GMD Institute for Secure Telecooperation discusses the opportunities presented by world-wide facility management. Office environments and their interiors are monitored, controlled, and managed from a globally standardized platform, abstracted from particular local technology. A smart, embedded system, called roomServer, bridges the gap between physical and virtual work environments.

1.1.4.5
System Technologies

General trends in sensor systems in intelligent buildings are summarized by *Hans-Rolf Tränkler* and *Olfa Kanoun*. *Friedrich Schneider* et al. outline system technologies for private homes.

Future buildings will become more intelligent. The main trends will be reduced resource consumption, optimized convenience, and more comfort as well as better safety and security. Following other industries such as the automotive industry, the building industry will experience an increased impact of microsystems technologies and new communication systems. Within these trends, new sensor systems with higher performance characteristics will be key technologies and enablers. The performance of these sensor systems will increase regarding accuracy and linearity, maintenance, down-time, reliability, fault tolerance, fault recovery, and adaptability. Sensor systems themselves will become smarter and more integrated. Multi-sensor systems will be increasingly widespread, used with self-identifying, self-diagnosing, and self-calibrating sensors. Within the next 10 years megatrends will be miniaturization, standardization of interfaces, and the close integration of smart sensors with actuators. Higher volumes will lead via economies of scale to low-cost sensor systems. Therefore, most visionary sensor system applications which are described in this book will become tomorrow's commodity products.

1.1.5
References

1 KAYE, J., Working paper; Boston: MIT Media Laboratories, 1998.
2 STREITZ, N.A., GEISSLER, J., HOLMER, T., KONOMI, SH., MÜLLER-TOMFELDE, CH., REISCHL, W., REXROTH, P., SEITZ, P., STEINMETZ, R., in: *Proceedings of the ACM Conference on Human Factors in Computing Systems*; Pittsburgh, PA: May 15–20, 1999.
3 OECD, *21st Century Technologies: Promises and Perils of a Dynamic Future*; Paris: OECD, 1998.
4 FESTO, presented at IASS International Association for Shell and Spatial Structures Symposium, University of Stuttgart, 7–11 October 1996.
5 FESTO, *Portable Architecture*; London: Royal Institute of British Architects, 1997.
6 FESTO, in: *Metropolis*; New York, December 1998, pp. 45–49.
7 FESTO, *Archit. Rec.* (1999) 218.
8 FESTO, *Design Net* **30** (2000) 145.
9 KOBAYASHI, T., in: *Sensors. A Comprehensive Survey*, GÖPEL, W., HESSE, J., ZEMEL, J.N. (eds.); Weinheim: Wiley VCH, 1989, Vol. 1, pp. 425–443.

10 Lion, K. S., *IEEE Trans.* **IECI-16** (1969) 2–5.
11 Grandke, T., Hesse, J., in: *Sensors. A Comprehensive Survey*, Göpel, W., Hesse, J., Zemel, J. N. (eds.); Weinheim: Wiley-VCH, 1989, Vol. 1, pp. 1–16.
12 Brignell, J. E., Smart Sensors, in: *Sensors. A Comprehensive Survey*, Göpel, W., Hesse, J., Zemel, J. N. (eds.); Weinheim: Wiley-VCH, 1989, Vol. 1, pp. 331–353.
13 Meixner, H., in: *Sensors. A Comprehensive Survey*, Göpel, W., Hesse, J., Zemel, J. N. (eds.); Weinheim: Wiley-VCH, 1995, Vol. 8, pp. 2–22.

2 Energy and HVAC

2.1
Intelligent Air-conditioning Control
ALBERT T. P. So and BRIAN W. L. TSE, *City University of Hong Kong, Hong Kong*

2.1.1
Introduction

A precise definition of a sensor (often called a detector in heating, ventilating, and air-conditioning (HVAC) control) has always been elusive. For practical purposes, a sensor in an HVAC system can be thought of as a device which converts a physical property (eg, temperature or humidity) or quantity (eg, flow rate) into a conveniently measurable effect or signal (eg, current, voltage, or number). For HVAC, sensors can be grouped into the following types:

- temperature and humidity (enthalpy);
- pressure;
- flow rate;
- comfort;
- indoor air quality.

Most sensors consist of two 'components', ie, a transducer which converts the raw, measured signal into a 'convenient' signal (usually electrical), and an associated signal conditioner which ensures that the raw signal is converted to a scalable electrical signal which can be calibrated with the raw measured signal. Typically, a linear relationship between the convenient signal and the quantity of the raw, measured signal is preferred. Recent and future generations of sensors have an additional feature, 'intelligence', where a built-in microprocessor enables data to be reported and analyzed besides pure measurement. Comfort and enthalpy 'sensors' are examples of this category.

2.1.2
General Specifications of a Sensor [1]

A sensor can be fully specified with reference to at least 12 performance, practical, and economic factors which can be divided into two classes.

Performance-based Factors

1. Range. The range of the measured variable for which the characteristics are maintained at stated values.
2. Accuracy. The degree of equivalence to which the measured output matches with some known benchmark.
3. Repeatability. The ability of the sensor to reproduce consistently the same output from the same measured value.
4. Sensitivity. The smallest detectable change in measured value that results in outputs change by the sensor.
5. Drift. The degree to which the sensor fails to give a consistent performance throughout its stated life.
6. Linearity. The closeness to linear proportionality between the output and the measured value across the range.
7. Response time. The rate of response with respect to time of the output following an input change (often expressed as a time constant).

Practical and Economic Factors

8. Cost. The cost should include power supply, transducer, the signal conditioner, and the connecting cables. Very often, the cost of installing the sensor consumes a very significant portion within the overall cost.
9. Maintenance. Any special maintenance and re-calibration requirements involving additional labor and expenses.
10. Compatibility. Interoperability and interchangeability with other components and standards.
11. Environment. The ability to withstand harsh or hazardous environments.
12. Interference. Susceptibility to ambient 'noise' such as electromagnetic waves or quasi-stationary electric or magnetic fields.

2.1.3
A Quick Review on HVAC Sensors

A brief description of various sensors installed in HVAC systems is as follows [1–12].

2.1.3.1
Temperature Sensors

Temperature is an important controlled parameter inside an air-conditioned environment because it is closely related to human comfort level as suggested by Fanger [3]. Obviously, temperature sensors are the most common sensors in an air-conditioning system. Usually, three types of temperature sensors are popular, namely thermocouples, resistance temperature detectors (RTDs), and thermistors.

2.1.3.1.1 Thermocouples

Thermocouples make use of the current flowing in a circuit consisting of two dissimilar metals which are joined at different temperatures, a reference temperature and the measured temperature. They are robust but not sensitive, and are normally used for high-temperature measurements such as in a combustion chamber as suggested by Underwood [1].

2.1.3.1.2 Resistance Temperature Detectors

RTDs make use of the variation in electrical resistance of metal with respect to temperature changes. Platinum is the most commonly used metal in RTDs. According to Hordeski [4], it offers a wide measurement range from the triple point of hydrogen (–259 °C) to the freezing point of antimony (630 °C) as its temperature resistance coefficient is linear over this range. The main drawback is its relatively high cost.

2.1.3.1.3 Thermistors

The working principle of thermistors is similar to that of RTDs by making use of the temperature resistance relationship of a semiconductor, which exhibits a negative temperature coefficient of resistance. Typically, oxides of metals such as nickel, manganese, cobalt, copper, and iron are employed. They are highly sensitive compared with RTDs, which have a change in resistance of only 5 mΩ/°C, as mentioned by Elgar [5]. Also, they are relatively inexpensive. Because of these advantages, they are widely used in closed-loop control of air-conditioning devices and for chilled water plant control.

2.1.3.2
Pressure Sensors

According to Chen and Demster [6], most of the large variable air volume (VAV) systems employ the fixed-plenum-pressure method for their normal operation. Precise pressure sensors have to be installed inside the ductwork to monitor the pressure inside. Figure 2.1-1 shows a VAV control scheme, pressures at positions D and D_2 being kept at their set points to maintain the required flow rates of exhaust air and fresh air.

According to Seippel [7], there are five types of pressure sensors available in the industry: capacitive pressure sensors, inductive pressure sensors, piezoelectric pressure sensors, potentiometric pressure sensors and strain gauge pressure sensors. HVAC control usually employs the first two types of sensors and the other three are for other applications. Therefore, we shall focus our discussion on these two types of pressure sensors.

2 Energy and HVAC

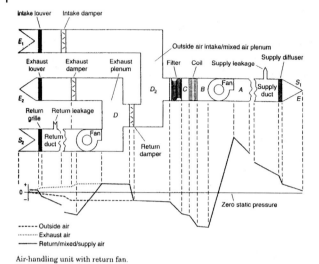

Air-handling unit with return fan.

Fig. 2.1-1 A typical VAV control scheme

2.1.3.2.1 Capacitive Pressure Sensors

A metallic diaphragm for one capacitor plate with the other plate positioned alongside of the diaphragm is shown in Figure 2.1-2. A capacitive pressure sensor can be used as part of a resistive-capacitive (RC) or inductive-capacitive (LC) network in an oscillator or as a reactive element in an ac. bridge. They are small in size with a high frequency response. They can operate under high temperature and allow both static and dynamic measurements. Typically, the range of measurement is from 69 Pa (0.01 psi) to 68 900 kPa (10 000 psi) with an error of 0.25%, as mentioned by Hordeski [4]. Some products with accuracy up to 0.05% are also available. They are usually applied to measure differential pressure difference of ventilation filter and VAV system fan control.

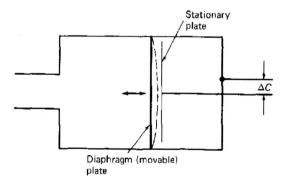

Fig. 2.1-2 Capacitive pressure transducer

2.1.3.2.2 Inductive Pressure Sensors

Inductive pressure sensors sense the pressure by moving a mechanical member which, in turn, changes the inductance. The mechanism is based on the relative motion of a core and the inductive coil. A sensor with two coils is preferred as it can eliminate the temperature sensitivity problem which always occurs with an inductive single-coil sensor. Based on Hordeski [4], a motion of about 0.08 mm produces an output voltage of 100 mV for this kind of sensor. Therefore, these sensors produce a high output, responding to both static and dynamic measurements, providing continuous resolution and having a high signal-to-noise ratio. They are usually employed in relatively low-pressure applications of ventilation systems.

2.1.3.3
Flow Rate Sensors

In HVAC control, measurements of supply air flow rate and chilled water flow rate are very important as their accuracy directly affects the performance of the control actions, as suggested by Sydenham [8]. There are several flow-rate sensors available and each of them serves some specific purposes, namely pitot tubes, orifice plates, venturi meters, hot-wire anemometers, turbine flow meters, vortex-shedding meters, electromagnetic flow meters and ultrasonic flow meters.

2.1.3.3.1 Pitot Tubes

They are essentially for in-duct ventilation applications and are based on two open-ended tubes, one mounted to face the air stream and the other being perpendicular to the air stream. The difference in pressure measured between the two tubes can indicate an air velocity based on the Bernoulli equation. This method is common and highly robust. The accuracy depends very much on the number of sampling measurements taken across the duct and upon the instrument for differential pressure measurement. Griggs et al. [9] discussed in details the placement of these airflow sensors for control.

2.1.3.3.2 Orifice Plates

These are based on a pressure difference across the pipe or duct by the actions of throttling the flow through an orifice and measuring the pressure difference. They are simple but prone to wear when used with liquids, especially those carrying small particles of dirt. They were extensively used in the past for piped liquids. Usually, HVAC reinforcement projects employ most of them for existing installations owing to their simple constructions as suggested by Elgar [5].

2.1.3.3.3 **Venturi Meters**

These have a similar working principle to the orifice plate except that there is a gradual reduction in the pipe or duct bore forming a throat section, instead of an abrupt orifice, and there is gradual enlargement to full bore downstream. Thus, pressure loss at the contraction is almost entirely regained. One distinctive advantage is that a venturi meter is not prone to wear. However, it tends to be large and expensive.

2.1.3.3.4 **Hot-wire Anemometers**

These are essentially for ventilation air flow measurements. The hot-wire anemometer is sensitive enough to detect flow at a very low speed, making it suitable for free air movement measurement and in-duct flow measurement. According to Elgar [5], it can be used over a large range of flows, from very low (eg, 0.03 m/s) to supersonic, and is able to measure unsteady flows. For in-duct flow measurement, it is less robust than the pitot method.

2.1.3.3.5 **Turbine Flow Meters**

These are mainly used in liquid-in-pipe applications and are susceptible to wearing and clogging, and unsuitable for dirty flow streams. Until relatively recently, these meters were the predominant choice for heat meter applications in hot water and chilled water systems, as mentioned by Underwood [1].

2.1.3.3.6 **Vortex-Shedding Meters**

These are suitable for liquids and can be very accurate. The working principle is based on the fact that the frequency of pressure fluctuations caused by the vortices due to fluid impinging on a bluff body is proportional to fluid flow. However, the complicated signal-conditioning requirement makes these types of meters very costly.

2.1.3.3.7 **Electromagnetic Flow Meters**

A magnetic field is established across the flow using a coil wrapped around the pipe or duct through which an alternating current passes. If the fluid is electrically conductive, the magnetic field is cut at a rate which is proportional to the speed of flow. They are suitable for flow measurement of dirty liquids and slurries. Also, they cannot be used with gases, as mentioned by Elgar [5], making them uncommon for HVAC control.

2.1.3.3.8 Ultrasonic Flow Meters
Based on the Doppler effect, they measure the change in frequency of sound waves reflected back from the particles in the stream. The accuracy of this type of meter is questionable. Similarly to electromagnetic flow meters, they cannot be used with gases, as indicated by Elgar [5]. Hence, it is not recommended for HVAC control.

2.1.3.4
Humidity Sensors

These sensors fall into four categories, hygrometers, psychrometers, electronic humidity sensors, and dew-point sensors. Humidity measurement has long been problematic because electromechanical hygrometers have suffered from serious non-linearity and there is a tendency to drift, whilst the more recently developed electronic sensors are prone to contamination in the air stream. Nevertheless, certain electronic sensors are steadily improving in terms of accuracy, long-term stability, and tolerance to contaminants. Hurley and Hasegawa [10] gave a comprehensive overview of these four sensors for HVAC applications as follows.

2.1.3.4.1 Hygrometers
These are based on an earliest developed method by using naturally occurring materials which change dimensionally with moisture absorption and desorption. They are highly non-linear and prone to drift. They are now replaced by electronic devices.

2.1.3.4.2 Psychrometers
Placing a distilled water-wetted wick around a conventional temperature sensor (eg, RTD) results in the reduction of web-bulb temperature, which can be related to the relative humidity. This is robust and potentially accurate. The difficulty is that the speed of air passing the wick must be high enough, makes it not so suitable for HVAC control.

2.1.3.4.3 Resistive and Capacitive Sensors
These are probably the most widespread, commercially available electronic sensors used today and modern devices have improved sensitivity, up to 2% in relative humidity. Another advantage is that this type of sensor tends to have better immunity to contaminated air streams than other electronic types. However, this family of sensors easily suffers from impairment and, in some cases, the damage may be irreparable if operated for a significant period of time at a humidity level higher than 90%, a frequent requirement for in-duct measurement and fresh air measurement.

2.1.3.4.4 Dew-point Sensors

There are generally two types on the market, ie, the chilled mirror sensor and the saturated salt solution device. The chilled mirror sensor has long been popular for dew-point and humidity measurement and it is capable of detecting dew-point with an accuracy up to 1 K. It has a mirror whose temperature is controlled until dew starts to form on its surface. Then, the relativity humidity can be estimated. The saturated salt solution device consists of an RTD surrounded by a wick which is impregnated with lithium chloride solution. A heating element is used to control the temperature of the wick until a stable condition is achieved and the power from the heating element is just enough to supply the latent heat from the wick. At this point, the temperature measured by the RTD is equal to the dew-point temperature.

2.1.3.5
Comfort Sensors

Comfort is not a directly measurable parameter. It is based on Fanger's equation that considers six parameters, ie, metabolic rate of occupants, clothing insulation of occupants, dry bulb temperature, moisture content, wind speed and mean radiant temperature, altogether to estimate a 'predicted mean vote' (PMV) [11]. This PMV is believed to be able to model the sense of comfort of most human beings. A negative value of PMV implies a cool feeling while a positive value implies a warm feeling. The PMV criteria have been included as an international standard, ISO 7730. In the past, instruments that measured the four non-human related parameters were commercially available from Landis and Staefa, but they are usually too bulky to be practical. The first comfort sensor that was practical for HVAC control and integration with the building automation system was manufactured by Yamatake-Honeywell, consisting of the sensor unit and the processor unit (Figure 2.1-3). The sensor detects the perceived temperature, consisting of air temperature, mean radiant temperature, and air speed, while the processor unit con-

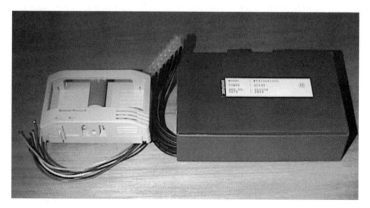

Fig. 2.1-3 A comfort sensor from Yamatake-Honeywell

verts them into an electrical signal. In this sensor, a new PMV equation is developed by simplifying the original PMV equation and this new equation is for the building instead of the human occupants. In this way, the comfort level inside a room can be estimated for real-time control purposes.

2.1.3.6
Indoor Air Quality Sensors

Over the past decade, numerous improvements in commercial building control have been achieved, but today's sophisticated workforce is still far more demanding about its workplace. Such indoor air quality (IAQ) issues as sick building syndrome and building-related illnesses started drawing media attention in the 1980s. A number of IAQ goals were set and energy-wise techniques were deployed to solve existing problems. An indoor air quality sensor, described by Thompson [12], non-selectively measures the concentration of all oxidizable gases collectively. It consists of a tube coated with a thin-film semiconductor, a pair of electrodes and a miniature heating element inside the tube. Maintained at a constant temperature, the semiconductor adsorbs the gases emitted from the occupants, resulting in the release of electrons which in turn alters the resistance across the two electrodes to produce a signal. The process is bi-directional. Although the predominant human emission of carbon dioxide is not measured, the sensor measures a combination of gases emitted from human beings and such combination is proportional to the carbon dioxide content. This instrument is sensitive and has a reasonably fast response time, making it suitable for IAQ applications in comfort ventilation and air-conditioning applications.

2.1.3.7
Occupancy Sensors

As the concept of energy conservation is now the key environmental issue around the world, it is essential to detect the occupancy of a room to make sure electrical appliances such as lighting and air conditioners are only turned on when they are required by the occupants. Therefore, occupancy sensors have two main tasks: keeping the lights or air conditioners on while the room is occupied and, conversely, keeping the lights or air conditioners off when unoccupied. Commercially, there are two kinds of occupancy sensors, namely ultrasonic (US) motion sensors and infrared (IR) motion sensors. Each of them has its own characteristics.

2.1.3.7.1 Ultrasonic Motion Sensor
This sensor makes use of the Doppler effect. It fills the room with continuous high-frequency (ultrasonic) sound waves. Any movement within the sensor's range causes a shift in the originally emitted frequency due to the Doppler effect. The sensor then identifies any change in frequency of the received sound wave by comparing it with that of the emitted wave. High sensitivity to small motion is

the distinctive feature of this sensor. Typical applications include offices, restroom, and small conference rooms where occupants are sitting in place for extended periods of time. However, this kind of sensor is vulnerable to false triggering from air-conditioning currents, corridor activity, and movement of inanimate objects.

2.1.3.7.2 Infrared Motion Sensors

By sensing moving IR heat sources such as people, forklifts, or other heat-emitting objects, the sensor is able to carry out appropriate switching actions on the lights or air conditioners in a room. IR motion detection gives immunity to false triggering due to air currents or fans, ie, a more reliable motion sensor. However, at greater distances, its sensitivity is rather low. Typical applications for this sensor include work areas, storage areas, storerooms, indoor garages, and rooms with pendent fixtures such as overhead fans. Subject to the advantages and drawbacks of the two kinds of sensors, it is advisable to use composite dual tech sensing. Using IR sensing (high error immunity) with US (high sensitivity) sensing provides good performance. The signal strengths are added together to form a composite sum. The advantage of this method is that a weak IR signal plus a strong US signal will turn the appliances on because the sum is sufficient.

2.1.3.7.3 Infrared-based People Counter

Recently, IR systems have become available on the market where an IR beam is projected on to an area where people counting is required. By sensing the reflected beam and using the Doppler effect, the number of people passing through the area will be counted. This type of sensor is useful when it is necessary to know the number of people inside a lobby for elevator control or HVAC control, etc. A more advanced system based on image processing and computer vision will be discussed in a later section.

2.1.3.8
Smoke Sensors

Smoke is one of the most important indicators of hazards inside a building. Fire is developed in four stages, namely incipient, smoldering, flame, and heat, as shown in Figure 2.1-4. Before there is flame in a fire outbreak, smoke is the first visible sign. During every stage of fire development, a particular kind of sensor is required. A sensitive smoke sensor can save a building from complete destruction during a serious fire outbreak. During a fire outbreak, a good smoke extraction system based on the information obtained from the intelligent smoke sensors can save human lives. There are two kinds of smoke sensors commonly used in modern buildings, namely ionization and photoelectronic. Each of them can detect two different stages of fire development.

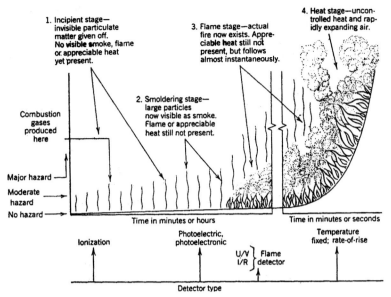

Fig. 2.1-4 Four stages of fire development

2.1.3.8.1 Ionization Smoke Sensors

This kind of sensor can respond first to fast-flaming fires. A flaming fire devours combustibles extremely fast, spreads rapidly, and generates considerable heat with little smoke. Ionization models are best suited for rooms which contain highly combustible materials. These types of materials include cooking fat/grease, flammable liquids, newspapers, paint, and cleaning solutions. Inside the sensor, there is a clean chamber where gas for reference exists and a site chamber where air from the room can enter. A piece of radioactive material is used to emit radiation to the two chambers. If the attenuation of radiation at the site chamber is not excessive, no smoke is identified.

2.1.3.8.2 Photoelectronic Smoke Sensors

A photoelectronic smoke sensor responds first to slow-smoldering fires. A smoldering fire generates large amounts of thick, black smoke with little heat and may smolder for hours before bursting into flames. Photoelectronic models are best suited for living rooms, bedrooms, and kitchens. It is because these rooms often contain large pieces of furniture, such as sofas, chairs, mattresses, counter tops, etc., which will burn slowly and create more smoldering smoke than flames. Photoelectronic smoke sensors are also less prone to nuisance alarms in the kitchen area than ionization smoke alarms. This kind of sensor is based on the principle that light rays will be attenuated if there are smoke particles inside the air.

2.1.4
Computer Vision-based HVAC Control

It is believed that HVAC control will be optimal if the quantity of supply air is adjusted according to what is needed. Who needs HVAC is certainly the human occupants. Therefore, if the presence of occupants can be detected and, furthermore, the number of occupants is known, we have confidence that HVAC control will be excellent. The employment of image processing or computer vision in HVAC control is due to this major reason.

The discipline of image processing and computer vision [13, 14] is an area which, as it continues to expand, becomes more difficult to define or describe in a simple phrase. Often applications are used to define the discipline. One application is to make 'robots' see and a course in this area mainly consists of two essential topics. The first is the development of models underlying the images or imaged scenes that are relevant to the application. The second one is the design and analysis of algorithms and their associated hardware and software requirements which, based on the models, produce useful and usually application-dependent results. There are principally four objectives in computer vision, namely preprocessing, lowest-level feature extraction, intermediate-level feature identification, and high-level scene interpretation via images. Generally, computer vision includes both major features, image processing and image understanding. Application of computer vision in building services engineering is fairly new [15–17].

HVAC can be considered the most important item in a modern building and HVAC control has been a headache for building services engineers for many years. Conventional control for HVAC relies on measuring devices such as thermostats and humidistats to monitor the temperature and humidity of the supply and return air of the air-conditioned space. Various control algorithms, such as PID, adaptive and self-tuning and even fuzzy logic have been incorporated in the control of air-handling units (AHUs) for thermal comfort. However, it is well known that a slow response rate is the major drawback of most commonly used measuring devices. When an ad hoc change of considerable magnitude in the load demand occurs, there is usually a rather long delay before the controller can take any subsequent action, eg, the VAV box only opens after the thermostat confirms an increase in the return air temperature, which is a slow action owing to the high room inertia and the intrinsic delay in the thermostat's response. Although it may not be too serious, bearing in mind the usual non-critical nature required for most commercial and domestic buildings, the long transient period may sometimes imply extra energy consumption in bringing the system back to steady state again.

Here, computer vision has been employed to assist the conventional PID control for HVAC systems. By adopting computer vision that can estimate the number of residents within an air-conditioned space, it is possible to identify any abrupt changes within seconds, which is, in a way, comparable to employing conventional transducers. It has been shown in the computer simulation that there are significant improvements with respect to both response rate and energy saving.

2.1.4.1
The Computer Vision System

People-counting machines based on a sequence of images are available commercially, eg, from ALTAIS and Sentec. The former is a program which traces people movement and people flow. A video camera vertically placed over the spot to be studied continuously records the people movement. The camera is linked to a box that analyses the video data. Every hour, the system builds a map of the flow path with its average density and speed. The latter is a people-counting system which uses standard closed-circuit television (CCTV) security cameras to count accurately pedestrian flows in different types of location, eg, shopping centers, retail stores, railway stations, airports, and convention centers. Both systems face a drawback in that they fail in very crowded environments where there is overlapping of human bodies as seen on the image.

A computer vision system for HVAC control provides two types of services. First, it counts the number of residents within an air-conditioned space. Second, it informs the control system of the distribution of the residents so that real-time zone control becomes possible. By knowing the number of residents, the optimum quantity of supply air can be adjusted to improve the response rate of the air-handling system so that energy saving can be achieved. By knowing the distribution of the residents, the thermal loads of different zones within the conditioned space can be estimated so that the rate of supply air can be adaptively controlled.

A stereoscopic camera system has been used where two charge-coupled device (CCD) cameras are installed at one upper corner of the room so that the whole conditioned space can be monitored. Wide-angle lenses are not recommended since they will make the camera model very non-linear, thus complicating the overall geometry and thus the mathematical equations. It is more preferable to separate camera systems to cover different zones within the conditioned space. Nowadays, cameras for security purpose are standard accessories within a building. Our computer vision-based approach [18] will not greatly increase the capital cost during the installation stage, whereas it can guarantee savings through optimal energy conservation schemes. The two standard CCD optical cameras are connected to a standard high-speed image grabber, through a time-multiplexing interfacing unit so as to save the electronic overheads. The grabber can produce an image file consisting of 512×512 pixels from each camera for onward processing by the computer.

2.1.4.2
Calibration of the Stereoscopic Camera System

Calibration means the accurate transformation of any point in space with dimensioning with respect to a universal world coordinate system to a definite point on the image frame buffer. The pinhole camera model is the most popular representation of a standard optical camera and the calibration procedures have been well developed [19]. Totally, nine parameters need to be calibrated for each camera with respect to the world coordinate system and these parameters cover the geometrical location, the orientation, the focal length, the non-linear distortion of the image grabber, and the lens distortion. The two CCD optical cameras ($i = a$ and b) forming a stereoscopic system are placed side by side. After the calibration, every point in the world coordinate (x_w, y_w, z_w) can be transformed into the two camera-orientated coordinate systems (x_a, y_a, z_a) and (x_b, y_b, z_b) and, further, the frame memory coordinate systems (X_a, Y_a) and (X_b, Y_b), respectively, by the following equations:

$$\begin{bmatrix} x_i \\ y_i \\ z_i \end{bmatrix} = \begin{bmatrix} r_{1i} & r_{2i} & r_{3i} \\ r_{4i} & r_{5i} & r_{6i} \\ r_{7i} & r_{8i} & r_{9i} \end{bmatrix} \begin{bmatrix} x_w \\ y_w \\ z_w \end{bmatrix} + \begin{bmatrix} T_{xi} \\ T_{yi} \\ T_{zi} \end{bmatrix}$$

$$\frac{d_{xi}X_i}{S_{xi}}(1 + k_{1i}\rho_i^2) = f_i \frac{r_{1i}x_w + r_{2i}y_w + r_{3i}z_w + T_{xi}}{r_{7i}x_w + r_{8i}y_w + r_{9i}z_w + T_{zi}}$$

$$d_{yi}Y_i(1 + k_{1i}\rho_i^2) = f_i \frac{r_{4i}x_w + r_{5i}y_w + r_{6i}z_w + T_{yi}}{r_{7i}x_w + r_{8i}y_w + r_{9i}z_w + T_{zi}}$$

where

$$\rho_i = \sqrt{\left(d_{xi}\frac{X_i}{S_{xi}}\right)^2 + (d_{yi}Y_i)^2}$$

To calibrate a camera, a calibration board consisting of 25 square columns is used where each column is of different height but of the same size. The coordinates of the four top corners with respect to the world coordinate system are accurately measured and these form the 100 calibration points on the image frame. By using non-linear optimization on the 100 sets of data, the nine parameters of each camera can be estimated.

2.1.4.3
Velocity Field Computation by Optical Flow

The idea of estimating the number of residents in a conditioned space is to compute the position of each resident with respect to the world coordinates. Pixel correspondence between the two optical cameras is the only way to determine a point in space. One effective way in corresponding pixels is to manipulate the method of 'depth from motion'. Two points from the two cameras are considered correspondence when they have identical velocity vectors. Therefore, before correspondence is executed, the first step is to generate a velocity vector field for each camera and the method of 'optical flow' is employed. Two images of the scene are taken by each camera where the first set of two images from the two cameras is obtained simultaneously. The second set is obtained simultaneously at $\delta t = 0.6$ s later. From the two images taken δt apart by any one camera, a gradient constraint equation can be introduced for each pixel at position (x, y) that relates velocity of a pixel on the image, ie, $(u_{x,y}, v_{x,y})$ to the image brightness function $P(x, y, t)$. The following equation set is based on the assumption that any changes in the scene brightness in the image sequence are due only to motion:

$$P(x, y, t) = P(x + \delta x, y + \delta y, t + \delta t)$$

Hence,

$$0 = P_x u + P_y v + P_t$$

where

$$P_x = \frac{\partial P}{\partial x}, P_y = \frac{\partial P}{\partial y}, P_t = \frac{\partial P}{\partial t}$$

$$u = \frac{\partial x}{\partial t}\bigg|_{x,y}, v = \frac{\partial y}{\partial t}\bigg|_{x,y}$$

The gradient constraint equation has no solution by itself since it only provides one linear constraint for the unknown velocity components. Another smoothness constraint has to be introduced to arrive at a solution, resulting in the following error, E, optimization problem:

$$E^2 = \iint \left\{ [P_x u + P_y v + P_t]^2 + k^2 \left[\left(\frac{\partial u}{\partial x}\right)^2 + \left(\frac{\partial u}{\partial y}\right)^2 + \left(\frac{\partial v}{\partial x}\right)^2 + \left(\frac{\partial v}{\partial y}\right)^2 \right] \right\} dx\, dy$$

where k controls the relative cost of deviations from smoothness and deviations from the motion constraint and it is usually set to 1. The equation set can be solved by discrete approximation and then numerical iteration, resulting in the following expressions for the nth iteration:

$$u^{(n+1)} = \bar{u}^{(n)} - P_x \left[\frac{P_x \bar{u}^{(n)} + P_y \bar{v}^{(n)} + P_t}{3k^2 + P_x^2 + P_y^2} \right]$$

$$v^{(n+1)} = \bar{v}^{(n)} - P_y \left[\frac{P_x \bar{u}^{(n)} + P_y \bar{v}^{(n)} + P_t}{3k^2 + P_x^2 + P_y^2} \right]$$

where

$$\bar{u} = \frac{\left(\sum_{i=-1}^{1} \sum_{j=-1}^{1} u_{x+i,y+j} \right) - u_{x,y}}{8}, \bar{v} = \frac{\left(\sum_{i=-1}^{1} \sum_{j=-1}^{1} v_{x+i,y+j} \right) - v_{x,y}}{8}$$

Iteration is considered a completion when $u^{(n+1)} - u^{(n)} + v^{(n+1)} - v^{(n)}$ is smaller than a threshold value, in our case, 10^{-3}. After this step, each pixel can be assigned a proper velocity vector $(u_{x,y}, v_{xy})$.

2.1.4.4
Pixel Correspondence

It can be seen that the computational loading is very intensive and it is impractical to estimate the velocity vector of each of the 512×512 = 262 144 pixels within an image. In this way, the computation is restricted to a limited number of pixels only. The boundaries between the residents and the background within an image have been chosen for estimating the velocity vectors. A reference image of the vacant conditioned space without any resident is stored inside the computer memory. Image subtraction between a real-time image and the reference image can highlight the patches due to the existence of residents. We are only interested in the edges of the patches and thus the velocity field computation concentrates on the edge pixels only. Edge detection is a well developed technique in image processing and the popular Sobel edge detector is chosen for the application. The job is to find out pixel pairs on images from the two cameras that correspond to the same spot in space. Provided that such pixel pairs are identified, the coordinates of the scene spot in space can be calculated and, hence, its distance from the stereoscopic camera system. Each relevant pixel (X_0, Y_0) on the edge of each patch of an image from camera (a) corresponds to a straight line in space passing through the focal point of the lens and the pixel on the image plane. The coordinates of each point on this line in the world coordinate system can be represented by $[x_{w0}(z_{w0}), y_{w0}(z_{w0}), z_{w0}]$, which is an equation with z_{w0} as the running parameter:

$$\begin{bmatrix} AX_0 T_{za} - T_{xa} \\ BY_0 T_{za} - T_{ya} \end{bmatrix}$$

$$= \begin{bmatrix} f_a r_{1a} - r_{1a} AX_0 & f_a r_{2a} - r_{8a} AX_0 & f_a r_{3a} - r_{9a} AX_0 \\ f_a r_{4a} - r_{7a} BY_0 & f_a r_{5a} - r_{8a} BY_0 & f_a r_{6a} - r_{9a} BY_0 \end{bmatrix} \begin{bmatrix} x_{w0} \\ y_{w0} \\ z_{w0} \end{bmatrix}$$

where

$$A = \frac{d_{xa}}{S_{xa}}(1 + k_{1a}\rho_a^2); \quad B = d_{ya}(1 + k_{1a}\rho_a^2)$$

This line can be mapped on to the image from camera (b) to form an epi-polar line. This epi-polar line will intersect the edges of corresponding patches at various points. The velocity vectors at these points are checked with the velocity vector of (X_0, Y_0). The pixel (X_0', Y_0') with a velocity vector having maximum resemblance with that of (X_0, Y_0) is chosen as the corresponding pixel. Since (X_0, Y_0) corresponds to a straight line in space (X_0', Y_0') corresponds to another straight line in space. The two lines must each have a region where the distance between the two lines is a minimum even if they cannot intersect one another. The midpoint between the two regions is chosen as the scene point and, thus, the scene point is fixed in the world coordinate system.

2.1.4.5
Scene Spots Fuzzy Clustering

After the completion of pixel correspondence, we have a set of points representing the edges of the residents in the world coordinates. These points which are actually on the external surfaces of the residents seen by the cameras are then marked on a horizontal plane by ignoring the z-coordinate. The effect of fuzzy clustering [20] is to identify whether a group of points belongs to one or more residents. Assume that there are n number of points to be clustered into a number of groups. The set of points is defined as

$$U = \{x_1, \ldots, x_n\} \subset \Re^2$$

where $x_i = (x_i, y_i)^T$: coordinates of the ith point.

A relational matrix $S_{n \times n}$ is set up below as:

$$S_{n \times n} = \begin{bmatrix} r_{11} & \cdots & r_{1n} \\ \vdots & & \vdots \\ r_{n1} & \cdots & r_{nn} \end{bmatrix}$$

where $0 \leq r_{ij} \leq 1$: $i, j = 1, \ldots, n$ are the degree of correlation between point i and point j and it can be evaluated from the characteristic values by a number of ways. In our case, since we are using the points to distinguish between residents, the size of a standard adult is a natural choice of the degree of correlation. d_0, the normal diameter of a human being from a top view, which is around 0.4 m, is chosen as a datum. Therefore, r_{ij} can be defined as

$$r_{ij} = 1 - Cd(x_i, x_j)$$

where $d(x_i, x_j) = |x_i - x_j|$

$$= \sqrt{(x_i - x_j)^2 + (y_i - y_j)^2}$$

$$C = \begin{cases} 0 & \text{if } d(x_i, x_j) \geq d_0 \\ \dfrac{1}{d_0} & \text{if } d(x_i, x_j) < d_0 \end{cases}$$

An equivalence relational matrix can be worked out by:

$$(S_{x \times n})^N = S_{n \times n} \circ (S_{n \times n})^{N-1} = (S_{n \times n})^{N+1} = (S_{n \times n})^{N+2} = \cdots$$

such that the operation 'o' is defined as follows:

$$R \circ Q = [r_{ij}] \circ [q_{ij}] = P = [p_{ij}] = \left[\sup_{k=1 \text{ to } n} (\inf(r_{ik}, q_{kj})) \right]$$

By a-cutting the equivalence relational matrix, crisp clustering is achieved. When $r_{ij} > a$, r_{ij} is set to 1 or else it is set to 0. Those elements with $r_{ij} = 1$ imply point i and point j belong to the same group. The clustered points correspond to the surfaces of the residents that are facing the camera system and the other side is invisible. A mirror image is used to supplement the back side before area estimation is carried out. By checking against the size of a group, it is possible to estimate the number of residents seen by the two cameras as well as their distribution in the world coordinate system. Our experimental verification has shown that the average horizontal cross-section area of a human being is around 0.15 m^2.

2.1.5
Internet-based HVAC System Monitoring and Control

The term 'intelligent building' has been dominating the construction industry. One definition of it is a building which provides a productive and cost-effective environment through optimization of its four basic elements, namely structure, systems, services, and management, and the interrelationships between them [21]. Intelligent buildings help building owners, property managers, and occupants realize their goals in the areas of cost, comfort, convenience, safety, long-term flexibility, and marketability. Intelligent buildings rely on integrated building systems and building services. So far, we have been discussing individual physical sensors. Actually, in order to make an intelligent building operate as desirable, we need sensors for the systems. The status and tracked record of each system must be fed back to a control center.

The evolution from the old electrical, mechanical, and manual building services to digitally controlled systems was considered by some people to be significant enough to be called 'smart', although 'automated' might be a more appropriate adjective. The systems being installed make use of modern computer modeling, advanced control techniques, and digital communications technologies. Operator work stations or control consoles are available in different parts of the building, such as operator on-duty room and facility manager's office, etc. Authorized users are able to gain access to all relevant data of each component of the building systems. In addition to data monitoring and alarm generation, there are other functions such as data logging, trend logging, condition-based maintenance and energy auditing and, very often, users need to set instructions, ie, control commands, to certain building systems, such as the temperature set-point, the chilled water flow rate and the illumination level. All these features are available in conventional building automated systems (BASs). However, limitations do exist, as an authorized user finds difficulty in gaining access remotely (ie, beyond the confines of the building) except for some degree of restricted 'telephone dailing-up' services. Even within the same building, additional software and hardware installations may be required to support more than one console. In order to extend the standard monitoring and control functions of a BAS to global coverage, whilst exploiting the potential of modern communication technology, the BAS must be Internet compatible so that every authorized user can keep close contact with the BAS wherever the user is [22].

2.1.5.1
Philosophy of Internet-based Building Automation with Image Transfer

The major objective of the provision of an Internet-based BAS is to turn everything inside the whole building into one sensor. The conventional wiring between the sensor and the control panel is replaced by the Internet. The human operators can stay at home and carry out monitoring and control actions on the building systems. Modern BASs can incorporate thousands of external points or objects for comprehensive building system monitoring and control. However, to provide remote access through the Internet, we are subject to the limitation of channel throughput. Data transfer must be optimized with consideration of usefulness, safety and security. Only critically selected data sets are allowed to be transferred between the BAS and the remote user. Key data that are likely to be accessed remotely can be categorized into five types, namely status, sensors, alarms, trends, and control.

Status
Under the 'status' category, the on/off status of each individual sub-system needs to be indicated, such as chillers, cooling towers, pumps, main valves, fans, air compressors, motor drives, lighting zones, normal and emergency power supplies, including transformers, main switchgears, generators and uninterruptible power supplies, elevators, escalators, air handling units, fan coil units, heat/smoke detectors, control panels, and security systems. Frequently, the on/off status is a binary data (ie, either on or off). However, for certain situations, such as fan operation, the 'on' status can be further divided into 'high speed', 'medium speed', 'low speed' and 'off'.

Sensors
Under the 'sensors' category, sensors, detectors and associated meters are collected together. Readings from temperature sensors are important and they include room air, flue gas, chilled water, condensing water, hot water, and steam. Air flow sensors indicate the amount of air, both upstream and downstream, delivered to the conditioned space. Fuel flow meters are used to indicate the normal operation of diesel engines and oil burners, etc. Water flow meters can be of the pressure difference, turbine, or electromagnetic types and they are used to indicate the amount of water flow through pipes or channels. Electricity meters provide the more significant means of monitoring the power consumption of both individual devices or sub-systems. Pressure sensors can be for water or air, water sensors being used for pump control and air sensors for filter monitoring and variable air volume control, etc. Light sensors can be of the photoconductive or photovoltaic cell type and they provide means for daylight and artificial lighting control.

Alarms
Under the 'alarms' category, special events are logged. Malfunctioning of any sub-system, activation of any fire detectors, sprinklers, hose-reels and hydrants, and detection of unauthorized intruders and improper access fall within this category. As

the user of the Web site cannot do much regarding the alarms because he/she is not recommended to reset the alarm remotely, the information associated with the alarm can be brief. It is just for the remote user's reference so that his/her concern is drawn. The user may contact relevant personnel who are physically inside the captioned building and ask for details via the Internet, fax or telephone lines.

Trends
Under the 'trends' category, the user is able to retrieve the trend logging of some pre-assigned and critical parameters from the 'status' or 'sensors' categories. For example, the user may want to know the trend of electric power consumption of a particular building system, say chillers, for a certain period of time, say 1 month. In this case, the operation manager is able to assess the effectiveness of the work of his/her staff even when he/she is on duty in another continent. The reports can be either in form of text or graphics. As the 'trends' reports are usually fairly lengthy, the number of reports available for retrieval must be kept to a minimum.

Control
Under the 'control' category, the user is able to send control commands from his/her terminal to the BAS for some particular operation of the sub-systems. For example, the user may want to switch on a certain lighting zone before he/she comes back to the office at midnight. This action can be dangerous and, therefore, the scope of control and the grade of user that is allowed to issue commands must be carefully limited. Certain critical equipment, say chillers, air handling units, pumps, and elevators, cannot be controlled from remote sites. Points under this category must be provided with the function of image transfer so that the remote user is able to have a real-time visual sense of the equipment being controlled by him/her.

The prototype was developed in 1996 and we believe that it might perhaps be the first one in the world involving building system control via the Internet with the feature of image transfer. Nowadays, there are various Internet-based BAS available, eg, the iLON manufactured by Echelon Corporation.

2.1.5.2
The BAS Web Site

The hardware requirements for Internet connection include (1) an additional 'ethernet' card on any one operator workstation (OWS) within the BAS, (2) connection to a router, and (3) access via a gateway to the global communication network. The overall structure is shown in Figure 2.1-5.

The operational charges of these hardware components and the connection are comparatively negligible with respect to the management cost of a modern commercial building. Using this approach, information and data from the BAS can be retrieved by any authorized user who has Internet access and vice versa for the transfer of commands from the user back to the BAS. Suitable hypertext markup language (HTML) files are prepared on the Internet server of the BAS, which may

Fig. 2.1-5 Overall structure of an Internet-based BAS

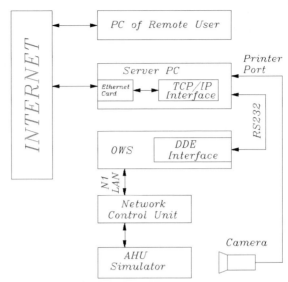

be the OWS itself or an additional personal computer. The publicly accessible address of our trial site is http://www.ibrc.org.hk, which all readers are welcome to access. The username and password are both 'tpso'. What you can play around with is the AHU simulator, as shown in Figure 2.1-6.

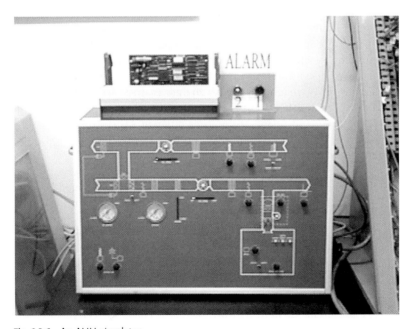

Fig. 2.1-6 An AHU simulator

2.1.6
PMV-based HVAC Control

The availability of the comfort sensor that estimates the PMV of a room has been described in a previous section. With this sensor, PMV-based HVAC control [23] becomes technically possible. From Fanger's equation, the PMV can be derived, which gives an effective measure of human thermal sensation.

According to ISO 7730, it is recommended that the PMV value of occupants in an indoor environment must be within a range –0.5 to +0.5. The process in solving Fanger's equation is extremely complicated so that, in the past, it was impossible, and to obtain this value economically on a real-time basis for real-time control, researchers have long been aiming at the philosophy of comfort-based control of indoor environments. In 1986, MacArthur brought the concept of PMV, the theoretical comfort index, into comfort control [24]. He worked on two strategies, namely PMV-based control and the temperature reset algorithm on the internal humidity level. Once again, an excessively simplified model of the air-conditioned space was used, making it impractical for real implementation. In 1991, a year-round PMV-based control method was proposed [25], where a microprocessor was used to calculate repeatedly the PMV of the conditioned space. The control program also utilized 'expert system' methods to select and control appropriate environmental control devices during the heating, cooling, and intermediate seasons. Scheatzle [25] did not use PMV as the control parameter but used PMV as a criterion to exercise appropriate actions, eg, switching on and off a fan or opening and closing the window, etc. Hence, the PMV of the conditioned space could only be controlled to within a desirable range, say from –0.1 to +0.5. There was no clear relationship between the calculated PMV and the control actions. That paper did not specify the control algorithms that could confine the PMV within the desirable range. In 1992, Henderson et al. [26] assessed the impact of controlling an air-conditioning (AC) system to maintain constant comfort instead of constant temperature. They made use of three different indices, namely effective temperature, PMV, and modified PMV, in their analysis. A building simulation model was used to simulate the impact of comfort control on energy use in a typical residence in Miami and Atlanta using three different AC systems. However, the machine control algorithms were not discussed in detail in the paper. Only the total capacity, the sensible heat ratio, and the energy efficiency ratio were involved in the simulation. We could not be sure that the AC machines worked efficiently inside the simulation process. However, a desirable conclusion could be drawn from that paper that, sometimes, energy consumption could even be saved while a better PMV could be maintained.

In 1993, another paper discussed the issue of energy use with respect to thermal comfort [27]. A rule-based system was devised and different AC configurations were tried. It was also concluded that operating a space under comfort conditions, not temperature, clearly showed that the energy necessary to maintain the required conditions was less than that with conventional systems.

From the above literature survey, we have come to feel that comfort-based control may be the trend of HVAC control in the years to come. However, some researchers had the opinion that under high levels of air movement or high relative humidity, PMV might not be the best index to describe thermal sensation. But at this point in time, there still has not been developed a better index than PMV which is universal enough to describe human thermal sensation satisfactorily under all conditions. Hence PMV is still used as the major control parameter here. Once a newer universal comfort index is available, the control algorithms described here can immediately be adaptable to this new index.

For the comfort sensor, we suggest placing it at the center of the conditioned space, 1.2 m above the floor, ie, the head level of a seated occupant.

2.1.6.1
Elements of Comfort-based Control

Based on Fanger's equation, comfort-based control should involve all six factors. The clothing insulation and metabolic rate are unique to the occupants and so they cannot be controlled. They are merely inputs to the control algorithms by the users, ie, the occupants. Since our target is a commercial office environment, their values can be assumed to be some typical values for personnel working in an office. Even though there may be slight variations in clothing, the occupant can do compensation by fine adjustment of the diffuser as discussed in the section on localized air movement.

Relative Humidity
For relative humidity, its significance on the PMV value is small compared with the other three factors, namely mean radiant temperature, air temperature, and air speed. However, the effort expended on keeping the relative humidity at a precise level is very high. For example, if the relative humidity is very high, the heat absorption capacity of the cooling coil of the AHU must be increased. One conventional way to achieve this goal is to increase the chilled water flow rate significantly to remove excessive moisture content from the return air. However, in order to maintain a reasonable supply air temperature, a heater is needed to heat the overcooled supply air to a desirable set point, resulting in a wastage of energy. Therefore, in the proposed control scheme, relative humidity is not precisely controlled but kept within a desirable range from 30 to 60%.

Mean Radiant Temperature
Under the influence of sunlight, heat gradually conducts through walls and the internal surface temperature of walls will rise. In practice, the mean radiation temperature (sometimes called unirradiated mean radiant temperature, T_{umrt}) is the mean temperature of all surfaces surrounding the air-conditioned space. However, sunlight can easily penetrate through the fenestration area, ie, windows which are common to office environment due to architectural design, striking directly on the occupants inside. This effect causes a much higher mean radiant temperature, in particular when the greenhouse effect is also considered. In order to lower the mean radiant temperature, some researchers proposed the use of a

ventilation fan to cool the surface temperature. However, it is not justified or practical to install a large ventilation fan in an office where aesthetics and limited room space are considered. Normally, occupants will lower the venetian blind to reduce the amount of solar radiation penetrating into the air-conditioned space. However, it is not a necessity while occupants have full freedom in this issue. Hence, our control strategy is to treat it as an uncontrollable factor so that the PMV can be improved by controlling the two major variables, ie, air temperature and air speed, and keeping the relative humidity at an acceptable level. But it should be noted that there are various controllers for standard actuators within the AHU, such as air pressure and air balancing. If an automatic venetian blind system is available, the improvement will be much better. The comfort sensor can immediately confirm the reduction in solar radiation once the venetian blind is lowered. In our simulation, the overcast situation is more or less equivalent to the lowering of the venetian blind.

Air Temperature and Air Speed

The AHU can control the indoor air temperature by adjusting the volume of cool air supplied to the conditioned space under a variable air volume (VAV) control scheme. Room temperature can be adjusted precisely. However, the reliance on pure air temperature control to achieve a good PMV value normally results in a huge amount of energy consumption. Air speed control must be employed to provide a cooling effect in summer. According to the ASHRAE Handbook of Fundamentals (1997), an extension of the upper comfort limit by 1 K is permitted for every 0.275 m s^{-1} change in air speed. The allowable increase in air temperature, ΔT, due to the cooling effect by speed change, v, is

$$\Delta T = 6(v - 0.25) - (v - 0.25)^2$$

According to Auliciems and Szokolay [28], an air speed of 1 m s^{-1} is pleasant to the occupants under hot conditions and even speeds up to 1.5 m s^{-1} are acceptable. However, a draft will exist if a higher speed is encountered. In an existing AHU, it is found that the mean air speed is rather low, a typical value being 0.2 m s^{-1}. One method is to supply an extra volume of air from ceiling diffusers to achieve a higher air speed, but this will cause a higher pressure loss in the ducts, resulting in a higher rating of supply air fan or larger dimensions of ducts and, thus, an increase in the installation cost. So, air speed control at ceiling diffuser level may not be practical. In this way, the supply of cool air is from each ceiling diffuser under a relatively low speed. To boost the air speed for comfort purposes, the job is by a desk fan for each occupant as discussed in the next paragraph.

Localized Air Movement

Subject to the consideration of energy consumption and pressure loss, instead of controlling the air speed generally inside the conditioned space at the ceiling diffusers, localized air movement is provided for each individual. It was reported that people prefer locally controlled air movement under warm conditions, which can be achieved by installing a ventilation fan on the desk of each individual. Since the air outlet is close to the user, around 0.5 m like that of micro-climate control scheme, the size of the fan can be relatively small and its power consumption is negligible compared with the total power of the AHU. The localized fan speed is adjusted by our new control algorithm so as to achieve a satisfactory PMV value, in general. Since we have to admit a variation with predicted percentage dissatisfied (PPD) for a controllable PMV based on the new control scheme, a minor fine-tuning action should be available to the occupants for perfect control. The solution is to install a diffuser for each air outlet so that the exact air speed can be manually adjusted. However, in our simulation, fine adjustment of the diffuser is not allowed.

Indoor Air Quality

In addition to thermal comfort, indoor air quality is also very important and it is incorporated in the PMV-based control scheme. First, a range of desirable relative humidity of 30–60% has been recommended for most air-conditioning applications. Within this range, the hazard of having bacteria inside air ducts can be minimized, thus reducing exposure to occupants. Second, according to ASHARE Standard 62-1989, the fresh air requirement for an office is assumed to be 9.4 L s^{-1} per person who may be smoker. Thus, the total outdoor air required is obtained by multiplying this value by the total number of occupants inside the office. The volume flow rate of outdoor fresh air is controlled by adjusting the fresh air damper in the AHU. In addition, the conditioned space is slightly pressurized to avoid infiltration.

2.1.6.2
Control Algorithms

The VAV control strategy is applied to a standard AHU, ie, the supply air temperature is kept constant and the volume of supply air is adjusted to control the room temperature. There are 10 variables to be controlled, namely PMV value, room temperature (T_{rm}), room relative humidity (RH_{rm}), supply air temperature (T_s), supply air pressure (P_8), exhaust air pressure (P_2), fresh air pressure (P_5), fresh air flow rate (m_f), exhaust air flow rate (m_e), and localized air speed (v_{rm}). Each controlled variable has a corresponding set point value or a desirable range. There are 10 control actuators with one common goal to control the PMV value, namely supply air valve, humidifier, heater, chilled water coil, supply air fan, return air fan, bypass air valve, fresh air valve, exhaust air valve, and the desk fan mounted on each desk.

The PMV value is closely related to the set points of the room temperature and localized air speed. When the PMV exceeds the upper limit, +0.8 for example, the set point of the air speed is raised but that of the room temperature is lowered. The amount of elevation in the localized air speed is comparatively larger than the decrease in the room temperature so as to consume less energy, ie, the gain of the localized air speed controller point is larger with respect to a positive error of the PMV value than that to a negative error. A similar concept is applied to the room temperature set point as its control action is opposite to that of the localized air speed. In this way, the energy consumption of the AHU can be conserved but the same thermal comfort level can be maintained. The designed range of PMV in our case is (–0.8, 0.8). This corresponds to a PPD of 20%. For those 20% people feeling uncomfortable with the controlled environment, they are free to change the localized air speed slightly to attain a more individually favorable comfort level by adjusting diffusers on their own desks if micro-climate control is available.

2.1.6.3
Computer Simulation

A computer simulation was carried out to compare the performances of the comfort-based control algorithms and the conventional constant set-point-based algorithms applying on the same AHU feeding a typical office compartment during a sunny day and a cloudy day. Figure 2.1-7 shows a schematic block diagram of the simulation process. There are four blocks in the simulation, namely Control Scheme, Room Model, AHU Model and Environment Model.

Control Scheme
The main function of this control scheme is to determine a set of control commands for actuators to provide occupants inside the office with a thermally comfortable environment. Besides the two major set points, there are 10 controlled variables associated with 10 controllers whose set points or desirable range of operation are under control. For PMV control, the original designed range is chosen to be (–0.8, 0.8), corresponding to a PPD of 20%. In order to ensure that the PMV falls within this range continuously, the control action is limited to a range of (0, +0.4). The negative portion (–0.8, 0) is considered redundant with a view to energy saving in summer. The high and low limits of localized air speed are chosen to be 1.0 and 0.5 m s^{-1}, respectively. As mentioned before, the highest limit of air speed should be 1.5 m s^{-1} and hence the proposed range can guarantee the localized air speed will not exceed the allowable limit under all circumstances. The high and low limits of room-relative humidity are chosen to be 60% and 30%, re-

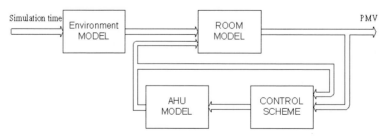

Fig. 2.1-7 Block diagram of the simulation process

spectively. Within this range, the bacterial hazard inside the AHU can be kept to a minimum. The set point of exhaust air flow rate is lower than that of fresh air flow rate to guarantee a positive pressurization of the conditioned space.

Besides the comfort-based control algorithms, conventional control algorithms are required to be implemented for comparison. Under this conventional mode, eight control variables, namely room temperature, room-relative humidity, supply air temperature, supply air pressure, exhaust air pressure, fresh air pressure, fresh air flow rate, and exhaust air flow rate, are controlled by nine control actuators, namely supply air valve, humidifier, heater, chilled water coil, fresh air valve, exhaust air valve, return air fan, bypass air valve, and supply air fan. Again, PID controllers are used to keep all controlled variables at their set-point values or within designed ranges.

Room Model

This model aims at simulating the thermal behavior of an office compartment with reference to the outputs of the AHU model and the environment model. The room modeled is an ordinary office, size being $15 \times 10 \times 3.5$ m, with a window on one wall only. Because of the nature of office work, each person is assumed to have a metabolic rate of 0.0638 kW m^{-2} with mechanical efficiency equal to zero. The clothing is a typical summer business suit of thermal insulation of 1.15 clo. Together with various office equipment, thermal loadings in the room are estimated (ASHRAE 1997). In total, there are 22 occupants, 50 fluorescent lamps (58 W each), 22 microcomputers, four printers and three copiers with a total sensible load of 10.8 kW and a total latent load of 990 W. Solar load is not static and is therefore not included inside the room model.

Solar heat through building walls is significant and generally has a time delay as it needs to propagate through concrete material mainly by conduction. Heat transmitted through windows is divided into parts, namely instantaneous heat and non-instantaneous heat. The former heats up the room air immediately whereas the latter heats up surrounding fixtures such as furniture and floor and this heat is later released back to the room air. This type of heat transfer is dependent on the orientation and configuration of walls.

$$PMV = (0.352\ e^{-0.042met} + 0.032)\{met(1-\eta) - 0.35[43 - 0.061met(1-\eta) - P_a]$$
$$- 0.42[met(1-\eta) - 50] - 0.0023\ met(44 - P_a) - 0.0014met(34 - T_{rm})$$
$$- 3.4 \times 10^{-8}f_{cl}[(T_{cl} + 273)^4 - (T_{mrt} + 273)^4] - f_{cl}h_c[T_{cl} - T_{rm}]\}$$

where *met* is in kcal/h/m^2.

With all thermal loadings applied to the room, the thermal room model is implemented to calculate the room temperature and relatively humidity. Then, the PMV is obtained by solving Fanger's equation as shown below and, once the PMV value has been obtained, the PPD can be found accordingly:

$$PPD = 100 - 95\ e^{-(0.03353\ PMV^4 + 0.2179\ PMV^2)}$$

However, the clothing index, mean radiant temperature, and localized air speed must be calculated first before substituting them into the equations.

AHU Model

Once having got all control commands from the control scheme, this model provides the mass flow rate, temperature, and absolute humidity of the supply air to the room model. The AHU model, shown in Figure 2.1-8, consists of several components, namely supply and return fans, one chilled water coil, one electrical heater, one humidifier, three air dampers, five segments of air duct and two air mixers.

It should be remembered that the speed of supply air from each ceiling diffuser is rather low. The supply air is from the AHU via the air ducts. The exact air speed sensed by each occupant in the air-conditioned space is due to the rotating speed of the desk fan. Although during practical implementation, fine adjustment of the diffuser of each desk fan is allowed for perfect comfort-based control, it is not included in this simulation process. Hence, the air speed from every desk fan is uniformly constant for each occupant to calculate the PMV based on Fanger's equation.

Environment Model

This model is used to estimate the environmental impact on the thermal performance of the room. It provides data inputs to the room model, such as outside air temperature, outside air-relative humidity, direct normal irradiation, vertical diffuse radiation, and vertical reflected radiation. The hottest day in summer in Hong Kong has been used as an illustrative example in the simulation process and the input parameters to this model are given in Table 2.1-1, which shows the outdoor environmental conditions on a sunny day and a cloudy day.

To estimate the outdoor air temperature, an assumption of diurnal temperature variation has been made such that the whole day temperature profile can be viewed as two separate sinusoidal curves. The temperature swing in the table is double the amplitude of the sine curves. Periods of both curves are determined

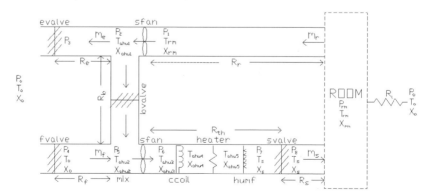

Fig. 2.1-8 The AHU model used in a computer simulation

Tab. 2.1-1 Outdoor environmental conditions

	Sunny day	Cloudy day
Outdoor temperature range (°C)	29–35	27–31
Relative humidity range (%)	70–95	70–88
Solar radiation (%)	100	70

based on the exact time when maximum temperature (around 03.00 p.m.) and minimum temperature (1 h before sunrise) are recorded. The relative humidity can easily be estimated based on the assumption that the moisture content is fairly constant throughout the whole day in peak summer and this situation is very normal in Hong Kong. The simulation consists of four main objects running from the time interval 06.00 a.m. to 07.00 p.m. Two types of control algorithms, the new comfort-based control and the conventional control, have been substituted into the Control Scheme module for comparison.

2.1.6.4
Simulation Results

A sunny day is simulated. Four parameters, ie, PMV, power consumption, room temperature, and mean air speed, are displayed in four separate graphs. The solid line represents comfort-based control and the dotted line represents conventional control. Apart from the total energy consumption, the energy consumption within two time slots, ie, the morning period (07.00 a.m.–12.00 p.m.), the afternoon period (12.01–07.00 p.m.) are calculated and listed. In this way, a detailed energy usage analysis of the two different control schemes using different control algorithms can be investigated precisely.

Figure 2.1-9a shows the variation of PMV versus time, Figure 2.1-9b shows the variation of total power consumption of all hardware components, Figure 2.1-9c shows the variation in the room temperature and Figure 2.1-9d shows the variation of mean air speed.

It can be seen that the comfort-based control can maintain a desirable PMV with the range (0, +0.8) whereas the worst PMV value of the conventional control rises to 2.5, which is totally unacceptable. It should be noted that 76.8% of the occupants will feel uncomfortable when the PMV exceeds 2. In the morning, the new control scheme can even keep a zero value of PMV for 6 h although the conventional control can still give a more or less satisfactory PMV of around –0.3. Since the window is facing west, the mean radiant temperature in the afternoon is extremely high, which is why the conventional control obtains that high PMV. The comfort-based control scheme is able to adjust the room temperature to 20 °C in the afternoon and increase the air speed to the limit to counterbalance the solar radiation. In the morning, the room temperature setting is released to 24 °C, which explains why energy can be saved. In the afternoon, in order to maintain

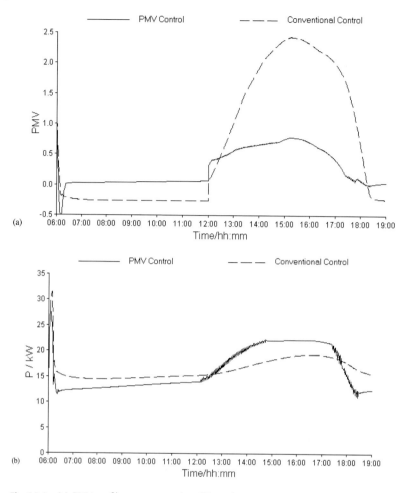

Fig. 2.1-9 (a) PMV profile on a sunny day; (b) total power consumption profile on a sunny day

the desirable PMV, energy needs to be consumed, which explains why the energy curve of comfort-based control is higher than that of the conventional control. Table 2.1-2 shows the total energy consumption of the two control schemes.

In the morning, from 7.00 a.m. to 12.00 p.m., comfort-based control consumes less energy, around 12.4% less compared with the conventional control. In the afternoon, the situation is reversed, as explained above, comfort-based control consuming 7.9% more than the conventional control. However, for the whole day, comfort-based control can still achieve an overall saving of 0.8% and, at the same time, a satisfactory thermal comfort environment can be provided for the occupants.

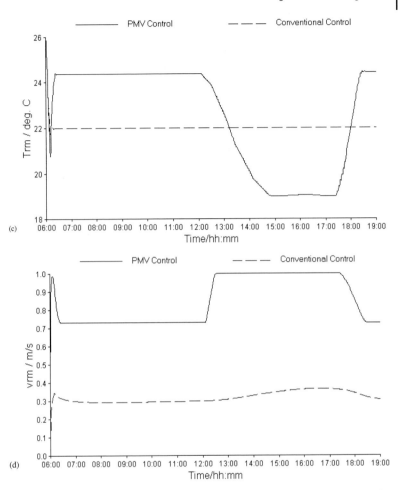

Fig. 2.1-9 (c) room temperature profile on a sunny day and (d) mean air speed profile on a sunny day

Tab. 2.1-2 Total energy consumption of PMV and conventional control schemes

Period	Energy by PMV control (MJ)	Energy by conventional control (MJ)
07.00 a.m.–12.00 p.m.	290	331
12.01 p.m.–07.00 p.m.	478	443
07.00 a.m.–07:00 p.m.	768	774

2.1.7
Conclusion

This section started with an introduction to the specifications of sensors for HVAC control in modern intelligent buildings. Then, a brief review of the characteristics of these conventional physical sensors was presented. Three new concepts and their associated developments in sensing technology were included. Computer vision can give the HVAC control system a pair of eyes to make the most intelligent decision for optimal control. The Internet-based HVAC control system can allow anybody to operate and monitor the building systems within a building anywhere in the world. The building itself becomes a sensor while the cable between the sensor and the human-attended control panel is replaced by the Internet. The last development is the concept of comfort-based control which, we believe, will be the popular approach in the twenty-first century. With the availability of electronic comfort sensors, that are easily installed and integrated with the BAS, comfort control will soon become the trend. However, the index, PMV, that we now rely upon may not be valid in the years to come. The developed system can adapt to any comfort index proposed by researchers in the future.

2.1.8
References

1 UNDERWOOD, C. P., *HVAC Control Systems – Modeling, Analysis and Design*; London: E & FN Spon, 1999.
2 NEWMAN, H. M., *Direct Digital Control of Building Systems*; New York: Wiley, 1994.
3 FANGER, P. O., *Thermal Comfort: Analysis and Applications in Environmental Engineering*; New York: McGraw-Hill, 1972, p. 41.
4 HORDESKI, M. F., *Transducers for Automation*; New York: Van Nostrand Reinhold Company, 1987, Ch. 2–4.
5 ELGAR, P., *Sensors for Measurement and Control*; London: Addison Wesley Longman, 1998, Ch. 5–7.
6 CHEN, S. Y. S., DEMSTER, S. J., *Variable Air Volume Systems for Environmental Quality*; New York: McGraw-Hill, 1995, p. 72.
7 SEIPPEL, R. G., *Transducers, Sensors and Detectors*; Reston, VA: Reston Publishers, 1983, Ch. 1–3.
8 SYDENHAM, P. H., *Transducers in Measurement and Control*; Bristol: Adam Hilger, 1985, Ch. 1–5.
9 GRIGGS, E. I., SWIM, W. B., YOON, H. G., *ASHRAE Trans.* **96** (1990) 523–541.
10 HURLEY, C. W., HASEGAWA, S., in: *Proceedings of the CIB Conference, Trondheim*; 1985, pp. 173–188.
11 FANGER, P. O., *Thermal Comfort*, Malabar: R. E. Krieger, 1982.
12 THOMPSON, D., *Aust. J. Refrig. Air Condit. Heat.* **7** (1992) 529–539.
13 KANADE, T., REDDY, R., *IEEE Spectrum* **20** (1983) 90.
14 LEVINE, M. D., *Vision in Man and Machine*; New York: McGraw-Hill, 1985.
15 SO, A. T. P., CHAN, W. L., in: *Proceedings of International Conference on Advances in Power System Control, Operation and Management*; Hong Kong: IEE, 1991, pp. 335–340.
16 SO, A. T. P., CHAN, W. L., KUOK, H. S., LIU, S. K., in: *Elevator Technology 5*, BARNEY, G. C. (ed.); IAEE, 1993, pp. 203–211.
17 SO, A. T. P., CHAN, W. L., *Architect. Sci. Rev.* **37** (1994) 9–16.
18 SO, A. T. P., CHAN, W. L., CHOW, T. T., *ASHRAE Trans.* **102** (1996) 661–678.
19 TSAI, R. Y., *IEEE J. Robot. Autom.* **RA-3** (1987) 323–344.

20 Zimmermann, H. J., *Fuzzy Set Theory and Its Applications,* 2nd edn.; Boston: Kluwer, 1990, pp. 220–240.
21 So, A. T. P., Wong, A. C. W., Wong, K. C., *Facilities* **17** (1999) 485–491.
22 So, A. T. P., Chan, W. L., Tse, W. L., *ASHRAE Trans.* **104** (1998) 176–191.
23 Tse, W. L., So, A. T. P., *ASHRAE Trans.* **106** (2000) 29–44.
24 MacArthur, J. W., *ASHRAE SF-86-01* **1** (1986) 5–17.
25 Scheatzle, D. G., *ASHRAE Trans.* **97** (1991) 1002–1019.
26 Henderson, H. I., Rengarajan, K., Shirley, D. B., *ASHRAE Trans.* **98** (1992) 104–113.
27 Simmonds, P., *ASHRAE Trans.* **99** (1993) 1037–1048.
28 Auliciems, A., Szokolay, S. V., *Thermal Comfort – Design Tools and Techniques*; PLEA and University of Queensland, 1997.

2.2
NEUROBAT – a Self-commissioned Heating Control System Using Neural Networks

JENS KRAUSS, CSEM, Neuchâtel, Switzerland
MANUEL BAUER, ESTIA Sàrl, Lausanne, Switzerland
JÜRG BICHSEL, Sauter, Basel, Switzerland
NICOLAS MOREL, Swiss Federal Institute of Technology, EPFL, LESO-PB, Lausanne, Switzerland

2.2.1
Introduction

The control strategy of existing water heating systems is usually based on a set of predefined heating curves that determine the nominal flow temperature of the heating fluid as a function of the external temperature. This open-loop control concept leads to poor energy management and reduced thermal comfort and requires a considerable commissioning effort during installation and maintenance [1].

Applying corrective closed-loop control of the indoor air temperature has increased the performances of these so-called 'low-cost' heating controllers. Nevertheless, the extended heating control concept does not meet the requirements of such a global energy management system, since the indoor air temperature control loop acts only in a corrective manner on the momentary thermal state of the building.

Significant advances have been achieved in heating, ventilation, and air conditioning (HVAC) control systems by introducing continuous adaptation of control parameters, optimal start-stop algorithms, and inclusion of passive solar or other free heat gain parameters within the control algorithm. Nevertheless, even the most advanced heating control system available today does not yet operate in an optimal way, since its control law does not take the global energy balance of the building into account and corresponds to a pure temperature control algorithm. The drawbacks of these temperature control systems are not compensated by the installation of thermostatic valves, which reference to the indoor air temperature with only limited accuracy [1, 2].

The choice of a predictive optimal control strategy, combined with the non-linear modeling of the building, the user's behavior, and the weather prediction, has been made to ensure an intelligent management of the free heat gains such as solar and internal gains and therefore to reduce the energy consumption. Moreover, the commissioning time of the new controller is considerably reduced, thanks to the use of self-learning neural algorithms [3].

Like several commercial systems, the new controller is interfaced to four temperature sensors located at the departure of the heating fluid, its return, in a reference room, and outside the building. The new controller has been designed to take advantage of a solar sensor and to anticipate the solar gains. The measurements of the flow and return temperature of the heating fluid allow the estimation of the heat transfer from the heating circuit to the building. The heat transfer from the heating circuit to the building with the prediction of the free gains allows the design of a control algorithm taking the energy balance of the building into consideration.

The controller algorithm has been developed and tested as a collaborative project between the CSEM (Centre Suisse d'Electronique et de Microtechnique, Neuchâtel, Switzerland, project leader), the engineering company Estia Ltd, the industrial partner Sauter and the LESO-PB (Solar Energy and Building Physics Laboratory, EPFL, Lausanne, Switzerland). The project itself has been funded partially by the Swiss Federal Office of Energy (SFOE) and the company Sauter. In this section the results of this collaborative project are presented.

Section 2.2.2 introduces the concept of the NEUROBAT heating controller, detailing its applied methods with reference to the controller block diagram. The core of the NEUROBAT heating controller, the optimal control algorithm, is explained and different controller versions are introduced.

The results of an extensive simulation study are summarized in Section 2.2.3. The comparative test results within two thermal independent office rooms, which have been realized with the help of a PC-based breadboard model of the NEUROBAT heating controller, conclude this section.

The realization of the industrial NEUROBAT prototype and its tests on a residential building are described in Section 2.2.4.

Section 2.2.5 draws conclusions and gives an outlook on future developments based on the NEUROBAT control concept.

2.2.2
Control Concept

2.2.2.1
Methodologies

Nowadays, the most common heating control systems correspond either to an open-loop control, stabilizing the flow temperature of the heating circuit by means of predefined heating curves (eg, central heating controller), or to a closed-loop control, referencing to the indoor air temperature (eg, room controller). Advanced heating control systems reference to both methods by correcting the predefined heating curves according to the measured indoor air temperature and/or the measured solar radiation. Some of the advanced heating controllers integrate functions such as the continuous adaptation of the heating curve parameters on the one hand and the optimal start/stop of the heater on the other. In practice, these functions are applied very little, since the cumbersome parameter setting during installation requires a non-negligible effort [1, 2].

The popularity of these applied control schemes stems mainly from their relative simplicity. Growing environmental concerns regarding the production of energy and the recognition that energy resources are finite are two important motivating factors that provide incentives to develop more extensive and robust optimal control methodologies [4].

At the foundation of an optimal control scheme is the necessity for a representative mathematical description of the physical system, in our case the thermal behavior of a building. For thermal systems, the system equations tend to be highly non-linear and composed of bilinear terms. Bilinear terms consist of products of system and control variables. To meet these non-linear characteristics, artificial neural networks (ANNs) have been applied to describe the behavior of thermal systems. ANNs operate like a 'black box' model, requiring no prior knowledge about the building or subsystem. Instead, ANNs are able to extract key information patterns and non-linear characteristics within multidimensional information domains. The ability to learn building and building system-operating characteristics is one of the ANN features of particular value to a heating control system. Another advantage of using ANNs is their ability to handle overparameterized problems – they seem simply to ignore excess data that are of minimal significance and concentrate instead on the more salient inputs [5].

An equally important feature of the optimal control scheme is a cost function (or performance index) to be minimized or maximized. For optimal control problems involving dynamic effects, an integral cost function is generally utilized because the performance index must be optimized over the duration of the time interval of interest. Typically, cost functions consist of two types of terms, namely terms representing the performance of components within the system and penalty terms used to enforce system constraints or to maintain control system variables near setpoint values. In our case of a thermal system, the cost function would include costs associated with the energy use and a penalty for the room temperature not being at the setpoint temperature. The objective of the optimal control scheme would be to determine the optimal trajectory of energy that yields minimum cost while maintaining thermal comfort conditions [4].

To ensure an optimal management of the free energy gains, the thermal behavior of a building must be regarded as a passive climate system that tries to utilize the outdoor climate as much as possible to reduce the energy consumption of the building. The outdoor climate is used, besides the heating device, for indoor temperature control. It is obvious that the outdoor climate is not always capable of providing the energy to maintain a required level of comfort in a building. However, it might be possible to use the outdoor climate in an advantageous way by storage of energy in the walls. This would require a control system, which is able to predict the future thermal behavior of the building and use this prediction to maximize the outdoor climate contribution to the indoor comfort, simultaneously minimizing the energy consumption. The proposed control system must be able to determine control actions in advance by using prediction of the indoor temperature. This prediction of the indoor temperature will also include prediction of the outdoor climate, especially solar radiation and temperature [4].

As a result, and unlike the traditional commercial systems, the control concept of the NEUROBAT heating controller is based on the optimization of a cost function over a fixed time horizon. The NEUROBAT heating control system optimizes the thermal user comfort and the energy consumption by means of the dynamic programming algorithm on a fixed time horizon of 6 h.

2.2.2.2
Controller Block Diagram

Figure 2.2-1 shows a block diagram of the predictive optimal control of the NEUROBAT controller. As for advanced commercial heating control systems, the applied sensors are outside temperature sensor, indoor air temperature sensor, flow temperature sensor of the heating circuit, return temperature sensor of the heating circuit, and solar radiation sensor. A detailed description of the applied sensors is given in Section 2.2.4.

The NEUROBAT controller comprises a processing-extensive outer control loop, calculating the optimal heating power for the next time step, and a cascaded inner control loop, stabilizing the flow temperature of the heating circuit. The different controller modules of Figure 2.2-1 can be described as follows:

- By means of measurements of the outdoor air temperature and the global solar radiation, the *Climate Module* predicts the outdoor air temperature and the global solar radiation over a fixed time horizon.
- The *Building Module* describes the thermal behavior of the building: with the help of the climate prediction such as outdoor temperature and solar radiation, the current thermal state of the building, and the applied heating power, the thermal evolution of the building is predicted.
- A *User Module* represents the behavior of the users and interfaces the user's inputs to the NEUROBAT control algorithm. The signal processing of the User

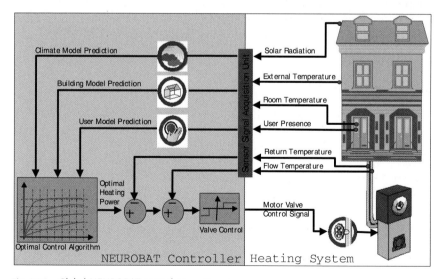

Fig. 2.2-1 Global NEUROBAT control concept

Module comprises a schedule function, determining the user occupancy and a setpoint function, defining the required indoor air temperature.
- The *Optimal Control Module* corresponds to the dynamic programming algorithm, optimizing the cost function with energy consumption and user comfort over a fixed time horizon [6]. The user comfort is quantified by the PMV factor (Predicted Mean Vote: ISO 27730 [7]). A detailed description of the optimal control algorithm and its cost function is given in Section 2.2.3.
- The *Valve Control Module* interfaces the NEUROBAT control algorithm to the commercial HVAC installation with its motor command output to the mixing valve. The optimal heating power, as output of the NEUROBAT control algorithm, determines the flow temperature of the heating fluid by taking into account the return temperature of the heating circuit.

The innovative character of the NEUROBAT control concept not only results from the reduction of the energy consumption, by means of the application of a predictive and model-based controller strategy, but is also based on its cost-efficient commissioning concept. At installation of the NEUROBAT heating controller, only four service parameter initializations are required and the user impact is limited to the definition of the setpoint of the indoor room temperature and the user schedule. The self-learning features of the NEUROBAT control algorithm do not require supplementary parameter definition or parameter adaptation at start-up or as maintenance measures. The NEUROBAT controller adapts and optimizes the building and climate model parameters by means of self-learning procedures with local measurements during operation [3].

The NEUROBAT control parameters will be adjusted during operation, in case of a sub-optimal parameter setting at start-up. Tests on real sites and an extensive simulation study have shown that, after a maximum period of 3 weeks, optimal operation of the NEUROBAT heating controller can be achieved. During the start-up period, the performances of the NEUROBAT heating controller are comparable to those of a conventional commercial heating controller.

2.2.2.3
Optimal Control Algorithm

The optimal control algorithm aims at optimizing thermal comfort and energy consumption over a fixed time horizon (time horizon of 6 h for the NEUROBAT heating controller). The optimization is done through the minimization of a *cost function*, taking into account both aspects, energy use and thermal comfort, and integrated over the time horizon. Several authors (eg, [8–11]) have already applied optimal control theory to building heating systems.

The Optimal Control Module receives the predictions from the building and climate models to elaborate an optimal heating command sequence over a time horizon of 6 h. At each time step k ($k=15$ min), the following input signals are processed to the Optimal Control Module:

- the current and past states of the indoor air temperatures $(T_{in}(k), T_{in}(k-1))$;

- the predicted profile of the solar radiation on the window surface for a time horizon of 6 h ($G_{sol}(n+1), \ldots, G_{sol}(n+6)$);
- the predicted profile of the outdoor air temperature for the time horizon of 6 h, averaged over the last 24 h ($T_{out}(n+1), \ldots, T_{out}(n+6)$).

At each new time step k, the optimal command P_{heat} is the command that minimizes the cost J over the time horizon of 6 h, with

$$H = \sum_{m=k}^{k+n} J(P_{heat(m)}, T_{in(m)}, T_{setpoint(m)})$$

The mathematical expression of the cost function used for the optimal control algorithm is as follows

$$J(P_{heat}, T_{in}, T_{setpoint}) = C_{Pheat} P_{heat} + C_{comf}(\exp(\text{PMV}(T_{in}, T_{setpoint})^2) - 1)$$

where

P_{heat} = heating command (W);
PMV ($T_{in}, T_{setpoint}$) = predicted mean vote (–);
T_{in} = indoor air temperature (°C);
$T_{setpoint}$ = indoor setpoint temperature (°C);
C_{Pheat} = weighting coefficient for the heating energy term;
C_{comf} = weighting coefficient for the thermal discomfort term.

A detailed and complete description of the cost function is given in [11].

The two terms of the right-hand side in the expression for the cost function correspond to the heating energy consumption and to the thermal discomfort felt by an 'average user'. The thermal discomfort is expressed by the deviation from the optimum PMV (Predicted Mean Vote), given by Fanger's formalism [7], on a scale spreading from –3 (very cold) to +3 (very hot), with 0 being the optimal thermal comfort condition.

The correct weighting of the two cost function terms is achieved by using a simple heuristic rule: 'The cost of the energy consumption which is needed to compensate a 0.2 variation on the PMV is equal to the cost of the discomfort resulting from that same PMV variation'. This rule takes into account the effective thermal capacity of the building (or of the considered room).

In order to take into account the user presence, the coefficient C_{Pheat} is fixed to 1 when the user is present and to 0 when the user is not present in the room. In the latter case, there is no need to provide thermal comfort and the only *cost* to optimize is the consumption of energy, ie, an applied heating power of 0.

The calculation of the optimal heating command P_{heat} corresponds to the dynamic programming algorithm, which is described in detail in [6]. The method allows finding a global minimum of the cost function, but requires an extensive processing power. Therefore, a balance must be found between a more detailed discretization of the state variables of the indoor air temperature (T_{in}) and the heating command (P_{heat}) and a too intensive use of the CPU which would not allow to recalculate the optimal command at each time step.

2.2.2.4
Applied Sensors and NEUROBAT Controller Versions

As for advanced commercial heating controllers, the NEUROBAT heating controller applies an outdoor air temperature sensor, flow and return temperature sensors, an indoor air temperature sensor, and a solar radiation sensor.

The applied temperature sensors of the NEUROBAT controller correspond to standard nickel (Ni) elements, changing their resistance with the air temperature [12–14].

The solar radiation sensor device calculates the solar radiation by means of two air temperature measurements, one temperature sensor exposed to the solar radiation and the other shielded from the sun. The difference in the temperature measurements is proportional to the solar radiation [15].

A low-cost version of the NEUROBAT heating controller has been developed, operational without the installation of a solar radiation sensor and without the installation of a room temperature sensor [3]. The low-cost version of the NEUROBAT heating controller aims at the reduction of the installation costs, but results in an increased energy consumption of 11% in comparison with the standard NEUROBAT heating controller (based on a comparable thermal user comfort). With reference to the simulation results in Section 2.2.3, the performance data of the low-cost NEUROBAT heating controller correspond to those of an advanced commercial heating controller.

The low-cost version of the NEUROBAT heating controller estimates the solar radiation with the help of the variation of the outside air temperature. An estimation algorithm, applying the building neural network model, substitutes the indoor air temperature sensor. The estimated indoor air temperature is adjusted with the help of the measurements of the return temperature of the heating fluid.

2.2.3
Controller Performance Assessment

2.2.3.1
Simulation Study

An extensive simulation study on the Matlab platform was completed during the concept phase of the NEUROBAT project. A simulation program was developed simulating the thermal behavior of an office room during a period of a whole year [16].

The program includes a nodal network of 28 nodes equivalent to the simulated office room with an additional four nodes for the heating subsystem and assigned temperature nodes to model the neighboring rooms and the outdoor air temperature. For the simulation, real measured data with a sampling period of 10 min (outdoor air temperature and solar radiation on a horizontal surface) measured in Lausanne, Switzerland, for the years 1981 and 1982 was applied. The simulation program offers the possibility of defining boundary conditions, such as the blind position, the artificial lighting or the internal gains:

- For the blind position (a p) (1 = completely open, 0 = completely closed), a correlation was calculated with the solar radiation incident on the window surface (for $S < 100$ W/m^2, a p = 1; for $S > 450$ W/m^2, a p = 0.2; a p interpolated linearly between these two values). The correlation was established during the test phase of the NEUROBAT project [3], which included measurements on the same office rooms.
- The artificial lighting requirements are evaluated from the illuminance on the user's desk, which is determined by the daylight level and the artificial lighting. The daylight level is evaluated using the 'daylight factor method' [17] and the artificial lighting is used to complement the daylight level, as necessary.
- For the internal gains, a constant value of 100 W during the occupation, corresponding roughly to one person (and no electric appliance, except the artificial lighting) is used. Concerning the occupancy, a fixed schedule 8.00–18.00 during the weekdays, and no occupation during the weekends (Saturdays and Sundays), is considered.

In order to be able to compare and assess the performance of the NEUROBAT heating controller, different commercial heating controllers were implemented and compared with the performance data for that of the NEUROBAT. Table 2.2-1 summarizes the controller types considered for the present simulation study (other controller variants have been taken into account as given in [3] and [18]).

Table 2.2-1 shows the applied sensor signals for the simulation of the different heating controller types such as T_{in} (indoor air temperature), T_{out} (outdoor temperature), T_{flow} (flow temperature of the heating fluid), T_{ret} (return temperature of the heating fluid) and E_{sol} (horizontal global solar radiation). The simulated com-

Tab. 2.2-1 Description of different simulated heating controllers

Controller	Sensor signals					Control algorithm
	T_{in}	T_{out}	T_{flow}	T_{ret}	E_{sol}	
Commercial low-cost controller	No	Yes	Yes	Yes	No	Common PID control algorithm; flow temperature setpoint defined by heating curves
Commercial advanced controller	Yes	Yes	Yes	Yes	Yes	As commercial low-cost controller; adaptation with reference to room temperature; automatic control parameter adaptation; optimal start/stop algorithm
NEUROBAT controller	Yes	Yes	Yes	Yes	Yes	ANN-based predictive optimal controller

mercial controllers are based on the concept of a set of heating curves, with extensions concerning the indoor air temperature adaptation, the start/stop algorithm, the automatic control parameter adaptation, and the inclusion of a solar sensor. These algorithms correspond to a modern and advanced controller design currently on the market and have been provided by a leading Swiss manufacturer of HVAC controllers [19].

The different controllers are assessed with reference to the energy consumption, the user thermal comfort, and the commissioning concept. The evaluation results of the simulation studies are summarized in Figures 2.2-2 and 2.2-3.

The histograms of the indoor air temperatures during user presence (8 a.m. to 6 p.m.) for a simulated heating season (October to March) are shown in Figure 2.2-3. With reference to an indoor air temperature setpoint of 20 °C an unacceptable thermal user comfort can be stated for the commercial 'low-cost' heating controller (top). The thermal user comfort comparison between the NEUROBAT heating controller (bottom) and the advanced commercial controller (center) shows a reduction of the overheating periods and a stabilization around the setpoint temperature for the NEUROBAT heating controller, eg, an optimization of the thermal user comfort.

Figure 2.2-2 shows the comparison of the energy consumption with reference to the different heating controllers for a simulated heating season from October to March. The application of the NEUROBAT heating controller shows a global reduction of the energy consumption of up to 35% in comparison with the commercial 'low-cost' heating controller and of up to 11% in comparison with the advanced commercial heating controller. An efficient application of the heating energy by the NEUROBAT controller can be stated during the in-between seasons

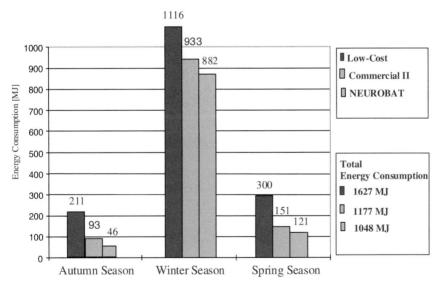

Fig. 2.2-2 Comparative simulation study results with energy consumption (in MJ) per season

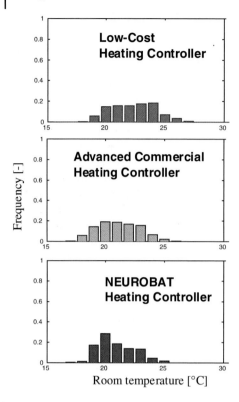

Fig. 2.2-3 Comparative simulation study results with thermal user comfort evaluation during user presence and for one heating season

(autumn (fall) and spring), where energy gains of up to 100% even in comparison with the advanced heating controller result.

Whereas the service parameters of the advanced commercial controller were defined by default, the low-cost commercial controller was tuned to obtain a comparable thermal comfort during the users' presence, thus simulating a rather careful commissioning.

The performance assessment of the NEUROBAT heating controller has shown a reduction in energy consumption in comparison with the commercial heating controllers by optimizing the thermal user comfort with a minimal commissioning effort at start-up.

2.2.3.2
Comparative Tests Within Office Rooms

The performance of the NEUROBAT controller was checked experimentally during the heating seasons 1996/97 and 1997/98 on a real-size-inhabited building, the LESO building of the Federal Institute of Technology in Lausanne (EPFL). The building is made of nine 'thermal units' insulated one from another with each unit having a particular facade, oriented towards plain south [20]. The ther-

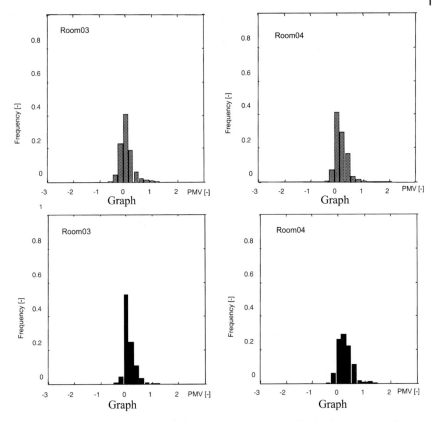

Fig. 2.2-4 Comparative test results with thermal user comfort evaluation during user presence and for one heating season with reference to test room. (a) and (b) summarize the thermal user comfort for the NEUROBAT heating controller and (c) and (d) that of the advanced commercial heating controller

mal unit situated in the center ground floor includes two similar office rooms, separated by an insulated wall, which have been used for the experimental tests. Both rooms were equipped with conventional water radiators with small individual boilers, which simulate the traditional central heating equipment most commonly found in Switzerland. The availability of two independent rooms allows one to make comparisons between the advanced commercial controller Sauter Equitherm QRK 151 (the advanced commercial controller evaluated during the simulation study [19]) and the NEUROBAT heating controller, by the simultaneous use of both controllers. In order to cancel the bias due to different user behaviors, and possibly slightly different thermo-physical room characteristics, both rooms (named Room 03 and Room 04) were regularly interchanged (at about 2–3 week intervals) during the comparative tests. The control parameters of the two heating controllers were adapted for both test rooms and with the interchange of the heating controllers the corresponding parameter set was reloaded. Figures 2.2-4 and 2.2-5 show the comparative test results for a complete heating season (December 1996 to December 1997).

Figure 2.2-4 shows the histograms of the PMV comfort factors [7] of the test-heating season, accumulated during the period of the user occupation (8.00–18.00). Graphs (a) and (b) correspond to the thermal user comfort for the NEUROBAT heating controller and (c) and (d) to the thermal user comfort for the advanced commercial controller. Whereas the parameter set of the advanced commercial heating controller was well adapted to the Room 03 (c), overheating periods can be stated for the user comfort in Room 04 (d). The application of the NEUROBAT heating controller shows for both test rooms a comparable thermal user comfort with an accumulation of the room temperatures around the setpoint temperature of 20 °C, eg, a PMV comfort factor of 0 ((a) and (b)). This optimal user comfort is due to the intelligent energy management of the NEUROBAT heating controller, predicting with the help of its neural network controller models the free heating gains and therefore reducing the overheating periods and at the same time reducing the energy consumption.

Figure 2.2-5a summarizes the test results concerning the user comfort, accumulating the absolute PMV comfort factor during the occupation period for both heating controllers. These comfort costs confirm the optimization of the thermal user comfort by applying the NEUROBAT heating controller.

Figure 2.2-5b shows the total energy consumption in MJ for both heating controllers during the test-heating season. With reference to the simulation study results of Section 2.2.3.1, an energy gain of 13% was measured for the NEUROBAT heating controller in comparison with the advanced commercial heating controller. The test results confirm the promising simulation results with a reduction in

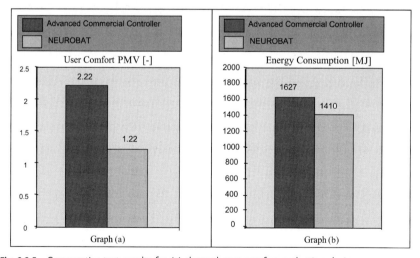

Fig. 2.2-5 Comparative test results for (a) thermal user comfort evaluation during user presence (PMV factor) and (b) energy consumption (in MJ)

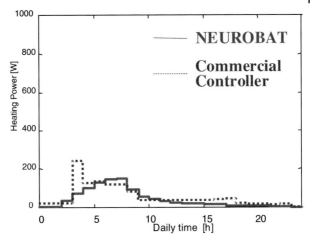

Fig. 2.2-6 Comparison of delivered average heating power per heating season, referenced to daily time

the energy consumption of the NEUROBAT heating controller by optimizing the user comfort.

Figure 2.2-6 emphasizes the predictive character of the NEUROBAT heating controller. The average heating power supply of the two heating controllers during the full test-heating season is compared and referenced to the daily hour. The profiles of the delivered and averag heating power show that the NEUROBAT heating controller is able to anticipate the required comfort. By means of its neural prediction models, which take the free energy gains into account and make reference to the thermal time constant of the building, optimized start–stop heating periods can be obtained. In comparison to with predictive character of the NEUROBAT heating controller, the 'open-loop' concept of the advanced commercial heating controller proves to be inadequate to meet the requirements of optimized start-stop heating periods. The maximum heating power is injected before the required comfort period in the morning, which results in increased energy consumption and possible overheating comfort periods.

2.2.4
Prototype Realization with Functional Tests on Residential Buildings

2.2.4.1
Industrial NEUROBAT Prototype

The industrial feasibility of the NEUROBAT concept was analyzed by realizing an industrial prototype of the NEUROBAT heating controller breadboard. Owing to the processing intensive optimal control algorithm, a powerful processing unit is mandatory. The model-based predictive and neural concept of the NEUROBAT

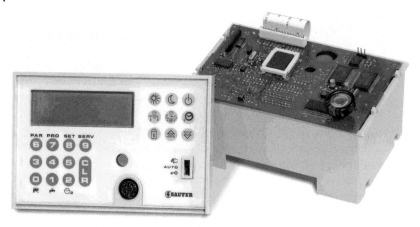

Fig. 2.2-7 Industrial NEUROBAT heating controller prototype

heating controller requires exhaustive volatile and non-volatile memory to save the model parameters in case of a power supply interruption to the heating controller.

The hardware platform of the industrial NEUROBAT prototype is based on a commercial conventional heating controller (Equitherm QRK 151 [14]) with a powerful 16-bit processing unit. The NEUROBAT control algorithm requires a program memory of 128 kbyte ROM and to meet the complexity of the optimal control algorithm a volatile memory of 512 kbyte RAM has been added. The storage of the parameters of the neural network models necessities an EEPROM memory unit of 64 kbyte.

Figure 2.2-7 shows the realized industrial NEUROBAT prototype. The user parameters such as the indoor room temperature setpoint and the comfort periods as well as the service parameters (latitude, longitude, building orientation, and maximum heating power) can be defined via a user-friendly interface. Specific control parameters and measured sensor signals can be downloaded via a serial link from the heating controller to a portable PC.

2.2.4.2
Test Results Heating Season 1999/2000

The industrial NEUROBAT heating controller prototype was installed at the beginning of October 1999 in a three-store residential building in Basel. The east-west oriented test building, shown in Figure 2.2-8, has an energetic reference surface of 100 m^2 and negligible solar gains, as detailed later in this section (see Figure 2.2-12). The building insulation corresponds to a standard for the Swiss property market and the consumed thermal energy per year of the building amounts to 770 MJ/m^2. The distribution of the heating energy, produced by an oil-fired burner of 17–23 kW, is based on a warm-water heating circuit for each floor with a thermostatic valve control on every radiator.

Fig. 2.2-8 East and west facades of residential test building in Basel

In order to obtain a reference for the data evaluation of the industrial NEUROBAT prototype, a conventional commercial heating controller (Equitherm QRK 151 [14]) was installed and evaluated during the heating season 1998/99 on the test site in Basel. A building control expert modified the control parameters of the advanced commercial heating controller continuously on a daily basis and the obtained level of parameter adaptation would never be met in practice. Nevertheless, a non-negligible reduction in the energy consumption and an optimal thermal user comfort can be reached only if the user knows the thermal building characteristics and adapts the control parameters accordingly. The test results for the heating season 1998/99 correspond to an ideal comparison basis for the industrial NEUROBAT prototype, which was tested during the heating season 1999/2000 without realizing any parameter adaptation (plug-and-play principle). To ensure continuous performance monitoring of the heating controllers a monitoring system was established, performing the data acquisition with the help of sensors, independent of the control system.

The comparison of the performance data for the commercial heating controller (heating season 1998/99) and the NEUROBAT heating controller (heating season 1999/2000) was aimed at analyzing the control parameter adaptation and the functioning of the industrial NEUROBAT prototype in general. For the indoor air temperature setpoint definition of the industrial NEUROBAT prototype the physiological thermal sensation was considered. Therefore, with a defined indoor air temperature setpoint of 21.5 °C, an overlapped sine wave variation of the indoor air temperature was introduced with amplitude of ±0.5° and a period of 24 h.

Figures 2.2-9 and 2.2-10 show the test data for 5 days comparing the performance of the commercial heating controller (days 80–82) and the industrial NEUROBAT prototype (days 83–86).

Figure 2.2-9 shows the variable indoor air temperature setpoint of the industrial NEUROBAT prototype (days 83–86) in comparison with the fixed and non-variable indoor air temperature setpoint of the commercial heating controller (days 80–82). In order to be able to anticipate the free gains of the solar radiation and therefore reduce overheating periods and as a result the heating energy consumption, the parameters of the commercial heating controller were optimized by introducing a daily heating reduction from 9 to 11 a.m. Figure 2.2-9a shows the measured outside temperature and the flow and return temperature of the heating fluid. Figure

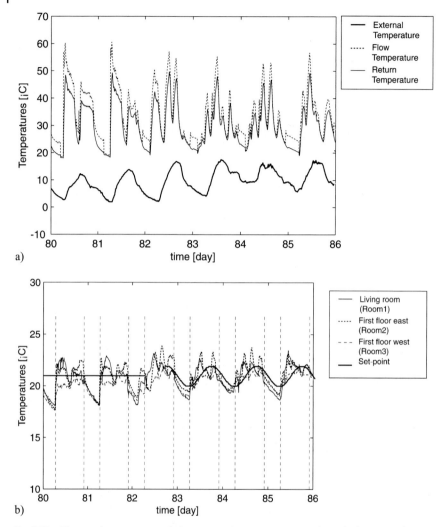

Fig. 2.2-9 Measured temperatures during a test period of 5 days within a residential test building. (a) Outdoor temperature and measured flow and return temperatures of the heating fluid. (b) Indoor air temperatures of the living room, west and east bedrooms, and indoor air temperature setpoint. Days 80–82 correspond to the test period of the advanced heating controller and days 83–86 to the test period of the NEUROBAT heating controller

2.2-9b shows the measured indoor temperatures (within the living room, the east bedroom, and west bedroom on the first floor) with reference to the indoor air temperature setpoint.

The measured room temperatures show a comparable thermal user comfort for both heating controllers. The performance of the NEUROBAT heating controller shows an optimal functioning comparable to the operational data of the advanced commercial heating controller, the control parameters of which were continuously

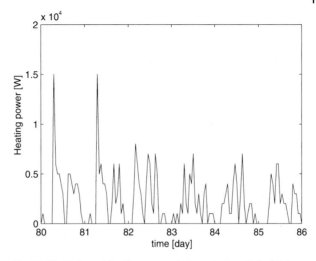

Fig. 2.2-10 Delivered heating power during a test period of 5 days within a residential test building. Days 80–82 correspond to the test period of the advanced heating controller and days 83–86 to the test period of the NEUROBAT heating controller

and manually adapted (ie, programmed heating reduction from 9 to 11 a.m.). Nevertheless, the flow and return temperatures of the heating fluid show a qualitative difference in relation to the delivered heating energy by the two heating controllers. Whereas the commercial heating controller delivers the maximum heating power before the beginning of the comfort period (comfort period from 6.30 am to 10.30 p.m.), the industrial NEUROBAT prototype anticipates the required heating energy in the morning and increases continuously its delivery to the building before the comfort period.

Figure 2.2-10 shows the delivered heating power in watts. The profile of the delivered heating power confirms the optimal start/stop characteristics of the NEUROBAT heating controller, taking the time constant of the building into account. In contrast to the predictive operation of the NEUROBAT heating controller, a maximum heating power before the comfort period at startup is delivered by the advanced commercial heating controller, resulting in a significant decrease in the water temperature within the heater and sub-optimal functioning. The programmed heating reduction of the advanced commercial heating controller from 9 to 11 a.m. can be localized clearly during days 80–82.

Figure 2.2-11 shows the evaluation of the thermal user comfort for the heating season 1998/99 and the heating season 1999/2000 in comparison. Figure 2.2-11 a–c shows the histograms of the room temperatures during the user presence for the commercial heating controller (heating season 1998/99) in comparison with the user comfort for the industrial NEUROBAT prototype in Figure 2.2–11 d–f (heating season 1999/2000). The thermal user comfort for the commercial heating controller (heating season 1998/99) shows an accumulation around the fixed indoor

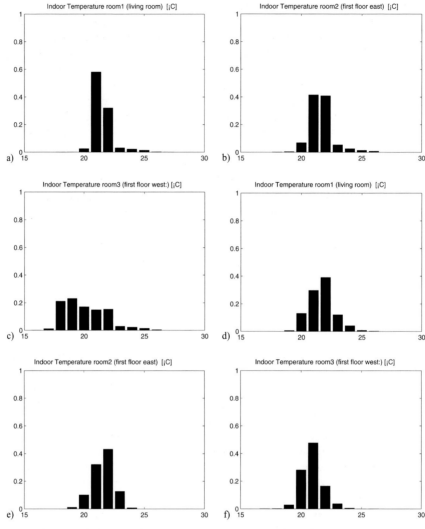

Fig. 2.2-11 Evaluation of the thermal user comfort with the indoor air temperature histograms during user presence. (a)–(c) correspond to the thermal user comfort for the advanced commercial heating controller, and (d)–(f) to the NEUROBAT heating controller

air temperature setpoint of 21 °C except for the west bedroom on the first floor, where the thermostatic valve was desigred to assure a lower room temperature. Owing to the variable indoor air temperature setpoint around 21.5 °C of the industrial NEUROBAT prototype, the room temperatures for the heating season 1999/2000 show a more dispensed distribution around the reference value. Nevertheless, the thermal user comfort for the industrial NEUROBAT prototype corresponds to the required user indoor air temperature setpoint.

The characterization of the energetic efficiency is shown in Figure 2.2-12 with the evaluation of the energetic signature. The energetic signature of the commercial heating controller in Figure 2.2-12a and b shows the energetic signature of the industrial NEUROBAT prototype. The integration of the delivered heating power over a time period of 4 days for the whole heating season results into the characteristic curve of the energetic signature.

The energetic efficiency of the industrial NEUROBAT prototype is comparable to that of the continuously parameterized commercial heating controller. The thermal building coefficient for the commercial heating controller is 238 ± 3 W/K and the non-heated outdoor temperature $19.2 \pm 0.4\,°C$. The corresponding values for the industrial NEUROBAT prototype are 226 ± 10 W/K and $19.5 \pm 1.4\,°C$.

The analysis of the thermal user comfort and the energetic efficiency of the industrial NEUROBAT prototype show optimal functioning without any parameter adaptation at start-up or during operation. The performance of the industrial NEUROBAT prototype is comparable to the operational data for the advanced commercial heating controller with continuous manual parameter adaptation. The NEUROBAT heating controller distinguishes its optimal start/stop characteristics and its installation and maintenance concept (plug-and-play principle).

2.2.5
Conclusion

An optimal control algorithm concept was developed for central heating systems. The so-called NEUROBAT concept was tested with a PC-based breadboard within office rooms and the performance data were compared with those for an advanced commercial heating controller. The tests confirm the results of the simulation study with a reduced energy consumption of 10–15% while simultaneously optimizing user thermal comfort. Moreover, the application of neuro-fuzzy technologies reduces the commissioning efforts both at start-up and during operation.

The development of an industrial heating controller prototype proved the feasibility of the NEUROBAT concept within an industrial context. The industrial prototype was tested during the heating season 1999/2000 on a conventional residential building in Basel. The performance of the industrial NEUROBAT prototype was compared with the data for an advanced commercial heating controller, the control parameters of which were adapted manually on a daily basis by a building control expert. The analysis of the thermal user comfort and the energetic efficiency of the NEUROBAT heating controller confirm its self-commissioning concept (plug-and-play principle) with a reduced parameterization at installation and no further maintenance during its operation. In other words, the best currently available commercial heating controller could not outperform the NEUROBAT heating controller even with additional continuous parameter tuning by a building control expert.

Furthermore, the application of a model-based predictive control concept ensures optimal start/stop heating periods for the NEUROBAT heating controller, taking the free heating gains and the building characteristics into account, and therefore reduces the energy consumption.

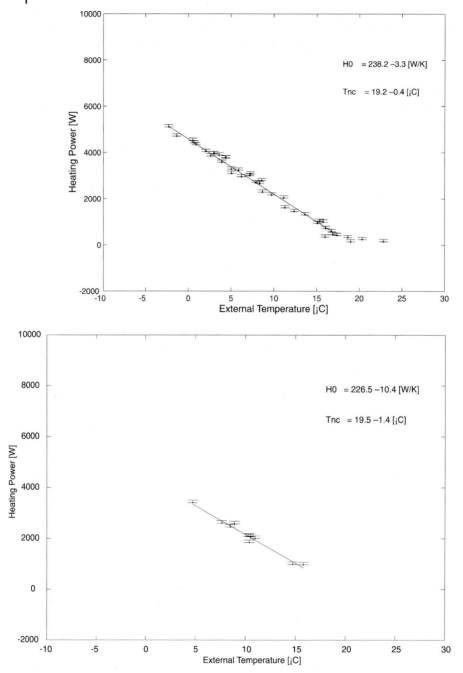

Fig. 2.2-12 Evaluation of (b) the energetic efficiency with the energetic profile of the NEURO-BAT heating controller and (a) the energetic profile of the advanced commercial heating controller

Additional tests are anticipated with the industrial NEUROBAT heating controller (such as on administrative office buildings). Based on this conclusive data evaluation, commercialization of the NEUROBAT heating concept is expected.

2.2.6 References

1. MULL, T. E., *HVAC Principles and Applications Manual*; New York: McGraw-Hill, 1997.
2. NEWMAN, H. M., *Direct Digital Control of Building Systems: Theory and Practice*; New York: Wiley, 1994.
3. KRAUSS, J., BAUER, M., MOREL, N., EL-KHOURY, M., *NEUROBAT – Predictive Neuro-fuzzy Building Control System*; Berne: BfE-Schlussbericht, 1998.
4. MOSCA, E., *Optimal, Predictive and Adaptive Control*; Englewood Cliffs, NJ: Prentice Hall, 1995.
5. HAGAN, DEMUTH, BEALE, *Neural Network Design*; Boston: PWS Publishing, 1996.
6. BERTZEKAS, D., *Dynamic Programming and Stochastic Control*; New York: Academic Press, 1976.
7. FANGER, P. O., *Thermal Comfort*; Malabor: Krieger, 1981.
8. LUTE, P., VAN PAASSEN, D., *IEEE Control Syst. (1995)* 4–9.
9. ROSSET, M. M.; PHD THESIS, UNIVERSITÉ DE PARIS-SUD, CENTRE D'ORSAY, 1986.
10. PARENT, P., *Optimal Control Theory Applied to Dwelling Heating Systems*; Paris: IRCOSE, Agence Francaise pour la Maîtrise de l'Énergie, 1987.
11. NYGARD, A. M., PhD Thesis, EPFL, Lausanne, 1990.
12. *Aussentemperaturfühler EGT301*; Basel: Sauter, 1995.
13. *Raumtemperaturfühler EGT330*; Basel: Sauter, 1995.
14. *Anlegetemperaturfühler EGT310*; Basel: Sauter, 1995.
15. *Sonnenfühler QAS92* Zug: Landis & Staefa, 1995.
16. *LESO-PB/EPFL: DELTA, Final Report*; Lausanne: SFOE, 1996.
17. *Luxmate System Manual*; Dornbirn: Zumtobel Licht, 1995.
18. BAUER, M., PhD Thesis, Lausanne: EPFL, 1998.
19. *Heating Control QRK 201*; Basel: Sauter, 1996.
20. *Dossier Systèmes – Rapport Technique*; Lausanne: LESO-PB/EPFL, 1985.

2.3
Air Quality Measurement and Management
HANNS-ERIK ENDRES, *Fraunhofer Institute of Microelectronic Circuits and Systems IMS, Munich, Germany*

2.3.1
Introduction

A ruling principle for air quality management is, following the German VDMA (Verband Deutscher Maschinen- und Anlagenbau e.V., Frankfurt/Main, Germany) committee 'demand controlled ventilation' [1], the saving of resources at the same time with a better indoor air quality. These recommendations are summarized in the publications VDMA 24772 (Sensoren zur Messung der Raumluftqualität in Innenräumen) [1] and VDMA 24773 (Bedarfsgeregelte Lüftung) [2], which are the basis for this article.

The composition of indoor air is complex and has a high impact on the comfort of inhabitants because the normal citizen spends more than 90% of his/her life in interior rooms. A minimum of air exchange is necessary to achieve comfort of the inhabitants and avoidance of damage to the building material, etc. In older buildings, leakage through wooden windows, etc., assures this minimum air change, whereas in new energy-saving houses controlled ventilation is necessary. HVAC (heating, ventilation, and air conditioning) systems assure a minimum change of air per hour, perhaps with more acceleration steps for conference rooms, etc. This air exchange should be kept to the minimum to preserve energy (heating energy in winter, cooling energy in summer). Time humidity-, and temperature-controlled ventilation is best suited to save energy, because ncomplete data about the other constituents of air force a minimum air exchange. Therefore, an air quality strategy needs air quality sensors to achieve a further reduction of air exchange with the same comfort. Table 2.3-1 gives an overview of the expected values of the energy-saving potential.

2.3.2
Substances in Indoor Air

The definition of a measure of air quality is no easy task, because indoor air contains an enormous amount of substances. In contrast to humidity and temperature, active control of the other substances is neither easy nor cost effective.

Tab. 2.3-1 Expected energy-saving potential with sensor-controlled air conditions [3]

Location	Energy-saving potential (%)
Lecture hall	20–50
Large offices:	
40% inhabitants present	20–30
90% inhabitants present	3–5
Hotel lobby, booking halls, public halls	20–60
Indoor arena, indoor fair	40–70
Theater, cinema, conference room, assembly rooms	20–60

The main constituents of indoor air are oxygen, nitrogen, humidity, carbon dioxide, and an immense number of different hydrocarbons from various sources. Normally, the concentration of nitrogen and oxygen does not vary significantly. Humidity is a special gas, which influences the comfort of inhabitants. Equipment to control the humidity concentration is widespread (humidifier, vaporizer, dryer) and the control of relative humidity is normally no problem for air conditioners. It is well known that carbon dioxide is a significant gas for human indoor air pollution, which was first postulated around 1850 [4] and state of the art until the around 1980. In clean air the carbon dioxide concentration ranges from about 0.035 vol.% in rural regions up to 0.07 vol.% in cities. Carbon dioxide concentrations below 0.1–0.15 vol.% are comfortable and 0.5 vol.% is the threshold limit value (TLV). High concentrations of carbon dioxide are toxic. In addition one may find more than 10000 different gases in a room, most of them different hydrocarbons (volatile organic compounds, VOCs) at very low concentrations. An idea of the minor indoor air components is given in Table 2.3-2.

Some of the constituents may be toxic, carcinogenic, or have a bad smell, which affect the comfort of the inhabitants. Also, one could assume that these emissions may contribute to the 'sick building' syndrome. The concentrations of these VOCs have a range from a few ppb up to ppm. In general, there exist two sources of emission: the inhabitants of the room, which are a changing source of emissions, and fixed building materials such as carpets, wallpaper, electronic devices, etc., which give a constant emission rate. A selective measurement of toxic or carcinogenic substances with sensor methods seems unnecessary for normal indoor air quality, because in the case of an effect on human health the source of the emission should be identified and removed.

The number and type of spurious substances vary from one location to another. A simple measure of the amount of hydrocarbons is the TOC value (total organic carbon). The TOC value is measured with a gas chromatograph. Typical TOC values are given in Table 2.3-3.

The requirements on air quality management are to enhance the comfort of the inhabitants and to remove odors or toxic substances. An empirical measure of indoor air quality is the contentedness of the inhabitants, which is estimated follow-

Tab. 2.3-2 Example of VOCs in indoor air of homes in a city (three samples in one year) [5]

Contaminant	Mean ($\mu g/m^3$)
n-Octane	2.3–5.8
n-Decane	2.0–5.8
n-Undecane	2.7–5.2
n-Dodecane	2.1–2.5
a-Pinene	2.1–6.5
o-Dichlorobenzene	0.3–0.6
1,1,1-Trichloroethane	16–96
m, p-Xylene	11–28
m, p-Dichlorobenzene	5.5–18
Tetrachloroethylene	5.6–16
o-Xylene	4.4–13
Ethylbenzene	3.7–11
Trichloroethylene	3.8–7.8
Styrene	1–3.6
Chloroform	0.6–1.9
Carbon tetrachloride	0.8–1.3
1,2-Dichlorobenzene	0.1–0.5
p-Dioxane	0.2–1.8

Tab. 2.3-3 Mean TOC values during a year (six measurements) [6]

Location	TOC (mg/nm^3)
Office (city)	0.58–0.72
Laboratory (rural region)	0.18–0.38
Laboratory (city)	0.34–0.68

ing Fanger [7]. Fanger defines an emission unit 'olf': 1 olf is the odor and CO_2 emission of a normal man (0.7 showers per day, daily fresh underwear), which is now widely used. The corresponding immission rate is a decipol: in a room where 15% of the test persons are uncontent, there is an immission rate of 1 decipol. A simple calculation gives an immission of 1 decipol for a room occupied with one standard person and a ventilation rate of 10 l/s. Unfortunately, Fanger gives no definition of the size of his standard room.

The Fanger method for estimating air quality is based on a statistical measure.

The conventional experimental setup for measurement following the Fanger method is that a panel of carefully selected persons compares the strength of a probe with a certain standard (mainly diluted acetone). A panel of about 10 persons results in a good quality of decipol measurement. This leads immediately to a classification of indoor air quality (Table 2.3-4). The immission unit decipol leads to an equation to calculate the actual ventilation rate [8]:

Tab. 2.3-4 Classification of air quality after Fanger [7]

High	0.7 decipol	≤10% discontented people
Medium	1.4 decipol	≤20% discontented people
Low	2.5 decipol	≤30% discontented people

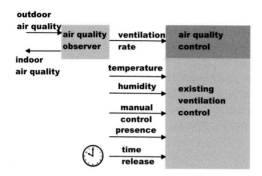

Fig. 2.3-1 Schematic view of the addition of an air quality module to an existing HVAC unit [3]

$$v = 10 \frac{G}{(C_i - C_o)\varepsilon_v} \tag{2.3-1}$$

v = ventilation rate (L/s), G = indoor emission (olf), C_i = expected indoor air quality (decipol), C_o = outdoor air quality (decipol), and ε = ventilation efficiency.

This equation includes the outside pollution, which can be assumed to be 0.1 decipol in cities with high air quality, 0.2 decipol in normal cities and 0.5 decipol in polluted areas.

Equation (2.3-1) leads directly to a scheme for an additional air quality control module on an existing HVAC system, shown in Figure 2.3-1 [3].

Following the VDMA 24772 recommendations it is not necessary to analyze all these constituents to achieve an acceptable indoor air quality. The task is to find a cheap and reliable sensor system which covers most of the demands of air quality measurement.

2.3.3
Sensors for Air Quality Measurements

The goal for gas sensors in air quality management is to give a fast, long-term, stable, reliable signal, where one can estimate a value for CO_2, decipol or other VOC, which have an influence on the air quality. Table 2.3-5 gives an overview of common gas sensors and a brief evaluation of their applicability in air quality metering.

Information about sensors and sensor principles given on a continuing basis in many monographs and journals.

Tab. 2.3-5 Gas sensors (after [9])

Principle	Measurement values	Selectivity	Common name	Application (example)
Liquid-state electrolyte sensors	Voltage U Current I Conductivity σ	0	Electrochemical cells	Toxic gases, safety
Solid-state electrolyte sensors	Voltage U Current I	+	Lambda probe	Automotive
Electronic conductance and capacitance sensors	Resistance R or conductance G Complex impedance X or admittance Y	–	Metal oxide sensor 'Figaro' sensor, conducting polymers	Explosive gases, CO, electronic noses
Field effect sensors	Potential U Work function change $\Delta \varphi$	0		
Calorimetric sensor	Heat of adsorption or reaction Q	–	Pellistor	Explosive gases
Optochemical and photometric sensors	Optical constants ε	+	Infrared cells, fiber optics	CO_2, optically active gases
Mass-sensitive sensors	Mass Δm as frequency shift $\Delta \nu$	0	Quartz microbalance	Electronic noses

In view of the large amount of possible substances in indoor air, the use of selective gas sensors is not meaningful because the composition of air changes from one location to another. The present state of sensor techniques leads to a measurement of carbon dioxide and a summary measurement of hydrocarbons.

Carbon dioxide is well known as a guiding gas for indoor air quality. Available sensors for carbon dioxide include infrared (IR) sensors and electrochemical cells. The IR measurement technique is based on the extinction of an IR beam with small bandwidths which depend on the CO_2 concentration following the Lambert-Beer law. Compared with most other sensors, IR measurement units have excellent stability, depending on the quality of the electronic components. On the other hand, CO_2 sensors have a reasonable energy consumption (IR source) and sensor types for low concentrations and/or high resolution must have a minimum optical length, which limits the miniaturization (Figure 2.3-2).

Electrochemical cells to measure CO_2 are also on the market with a shorter lifetime and lower stability than IR sensors. Electrochemical cells work similarly to a battery and have a low energy consumption and a small size. On the other hand, the baseline drift necessitates regular calibration.

Recently, the first fiber-optic CO_2 sensors with a fluorescent indicator material came under intensive investigation. Their principle is based on the fluorescence of a certain material, which is quenched under the influence of CO_2. Most of these sensors use a transient technology, calculating the CO_2 concentration from

Fig. 2.3-2 Vaisala GM 20W module for measurement of carbon dioxide [10]

the decay time of the fluorescent light of a stimulating pulse. This principle should lead to good stability, but the CO_2 concentrations in indoor air are near the lower detection limit of these devices because the intensity of the fluorescent light is low. Compared with IR cells, these sensors are much more expensive.

IR sensors and electrochemical cells are available for numerous gases, but a problem is their selectivity. The number of different substances present precluders a suitable sensor system for all applications.

One of the best choices for measuring a summary signal for VOCs is metal oxide sensors (Figure 2.3-3). These sensors are based on semiconducting metal ox-

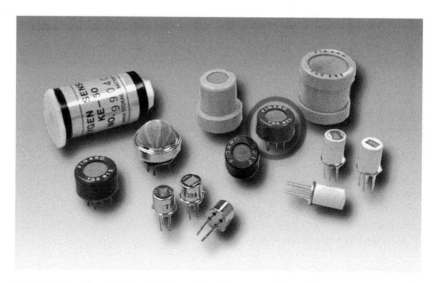

Fig. 2.3-3 Examples of different Figaro metal oxide sensors (air quality sensor TGS 800 similar to the TGS 822 ringed with a circle) [12]

2.3 Air Quality Measurement and Management

Tab. 2.3-6 Technical parameters for indoor air quality sensors (after [1])

Technical specifications	CO_2 sensor	VOC sensor
Measurement range	0–2000 ppm (extension or reduction allowed)	Choice of manufacturer
Measurement accuracy	±10% at 1000 ppm	±20% (related to manufacturer's test gas)
Influence of other gases	Selective (<5%)	Non-selective (100%)
Rise time	<5 min	<5 min
Lower detection limit	<0.5% of range	<0.5% of range
Reproducibility	<1.0% of range	<1.0% of range
Long-term stability	<5% of range/year	<5% of range/year
Influence of relative humidity (RH)	<0.1%/1% RH	<0.3%/1% RH
Humidity (operation)	30–70% RH	30–70% RH

ides. Triggered via a chemical conversion (oxidation or reduction) on the surface of the metal oxide at higher temperatures (ca 200–800 °C), the equilibrium concentration of oxygen at the surface is influenced by the concentration of several gases in the surrounding air. It is well known that this oxygen concentration has a large impact on the conductivity of the semiconductor and can be easily measured. The selectivity is controlled from the temperature and the doping of the metal oxide with (noble) metals and covers many VOCs. Some preparations of the metal oxide are especially recommended for air quality measurements. In general, the sensitivity of metal oxide sensors towards CO_2 is low and they are not suitable for indoor air quality measurements [11].

A large number of VOCs can be measured with metal oxide sensors, but the sensitivity has no decipol scale, because the human recognition of odor is different from the measurement principle of the sensor. Unfortunately, the humidity concentration as OH groups on the sensor's surface also has an important influence on the sensor signal. Therefore, it is not easy to distinguish between changes in the humidity or oxidizing and reducing gases.

The VDMA 24772 gives the recommendations in Table 2.3-6 for indoor air quality sensors.

2.3.4
Sensor Systems and Arrays for Air Quality Measurement

System and array technology is necessary to suppress the baseline drift of sensors and avoid limiting selectivity of the sensors. First, the output of the sensor system can have two different qualities: the detection of events (eg, change of gas concentration indicating leakage of a gas pipe) and the measurement of conditions (eg, a decipol value). In any case, the system is combined with a more or less complicated signal evaluation, chiefly integrated in a microcontroller. The output of an event detection is a binary signal, eg, indication of a leak or a control pulse such as a window open or closed. From a system view, these event detections can be realized via a threshold value of the sensor signal or its first derivative. On the other hand, conditions are a more or less continuous measure of a gas concentration or decipol. Between the two principles is the rough indication of a few steps such as good, slightly polluted, moderately polluted, heavily polluted, polluted.

An important task of the first developments of air quality sensor systems was to compensate for the influence of the floating baseline. For many years it has been well known that a transient measurement mode can reduce drift processes. A transient measurement mode means that a stimulus is given (eg, a temperature ramp or a concentration step) and the relaxation of the sensor signal is recorded and evaluated. Common applications of this technique include flow injection methods for analytical laboratories. Normally this method needs a digital calculation of the signal and is normally not applicable to common analog circuits. The drastic price decrease of microcontrollers has provided a platform to integrate these methods also in inexpensive sensor devices for air quality measurement.

A simple solution is the use of a transient technology with a certain stimulus, which can be a temperature pulse (Figaro). The first commercial realization was the Figaro AM-800 automatic air conditioner, based on a Figaro TGS 800 gas sensor (oxidizing gases, CO, H_2) and a microcontroller-based evaluation. This assesses the air quality in five steps related to a regularly renewed baseline (Figure 2.3-4), has two sensitivity steps and a built-in failure recognition.

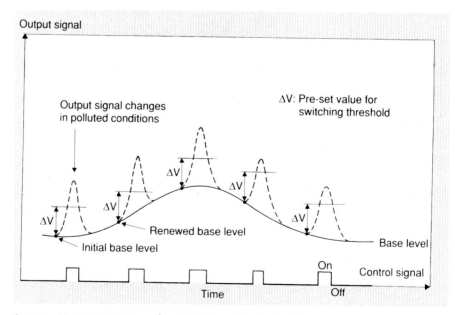

Fig. 2.3-4 Measurement process for AM800 air quality detector [13]

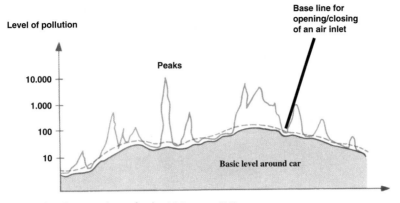

Fig. 2.3-5 Signal evaluation scheme for the AQS sensor [14]

A comparable development was an air quality sensor for the control of automotive ventilation was brought on to the market in 1987 as "AUC" (automatische Umluft Control) made by Kostol and dedicated to upper-range automobiles (first commercial application in a BMW vehicle) [13]. This instrument generates a closing signal for the inlet trap of a car using the deviation of a base pollution line (Figure 2.3-5).

Newer developments of the AQS system (AQS MK III and MK IV) achieved at least the simultaneous measurement of oxidizing and reducing gases with a single sensor (Figure 2.3-6).

Drifting baselines and sensitivities are a problem for all chemical sensors and complicate the measurement of small and slowly changing concentrations of gases, such as emissions from building materials. Transient technologies are base

Fig. 2.3-6 Air quality sensor for automotive use to control the air shutter [14]

Fig. 2.3-7 Landis and Staefa QPA63.2 combined CO_2 and VOC sensor [15] (courtesy: Landis & Staefa)

technologies for the further development to increase the sensitivity and the stability over a large range. The latest developments open up the chance for a lower detection limit in the PPb range (ETR) [13].

Some systems may use a single sensor, but for indoor air quality monitoring applications normally a sensor for CO_2 and another one for VOC are recommended. A specialized development for indoor air quality measurement is the QPA63 of Landis and Staefa, which contains at least a Figaro sensor and a simple signal evaluation to suppress the humidity effect [15] (Figure 2.3-7). This sensor is dedicated to measure especially human emissions, eg, perspiration or smoke, and has a simple output to control two ventilation steps. This sensor has an integrated interface to common building networks.

More sophisticated sensor arrays are the so-called electronic noses, consisting of an array of different, often nonselective sensors. Normally electronic noses are coupled with a probe gathering system and work similar to flow injection analysis (Figure 2.3-8).

This measurement principle is coupled with an advanced signal evaluation method, mainly derived from multivariate statistics (principal component analysis, neural networks, etc.). These algorithms need a calibration to detect certain gases in the environment. On the other hand, these electronic noses can be calibrated to the human odor feeling compared with a human panel of test persons. The selective measurement of certain substances in an unknown environment seems to be impossible, and the recognition of different odors is not simple and needs a high expenditure on calibration [16].

Additionally, many of the VOC have concentrations around 1 ppm, which is nearly at the lower detection limit of chemical sensors and the existence of an-

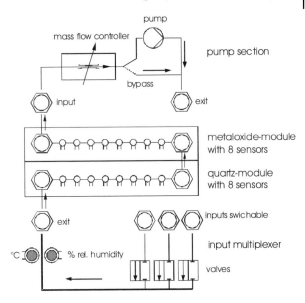

Fig. 2.3-8 Basic scheme of an electronic nose (Moses II) [16]

other gas with a fairly high concentration can mask the gases of interest and prevents their detection. There have been only a few evaluations of indoor air quality with electronic noses with the goal of discriminating between different odors.

One of the first investigations of indoor air quality was performed in the Russian space station MIR. The electronic nose consisted of several differently coated quartz microbalances and different conducting polymers [17]. The evaluation via a partial component analysis gave an activity profile of the astronauts and also the identification of some gases and a first step to fire detection (Figure 2.3-9).

Further investigations with electronic noses for indoor air quality measurements showed, that good comparability of a decipol estimation with a human panel can be achieved [18]. The odor sensation of humans and a sensor array are different and cannot be transferred from one building environment (office) to another (eg, home). In this case the electronic noses were calibrated via a neural net. On the other hand, the electronic noses are too insensitive for a direct measurement of specific unknown contaminants, which may be present at only a few ppm or sub-ppm concentration. This result is a clear hint that at present measurements with a fairly nonselective metal oxide sensor give better results.

2.3.5
Examples of Long-term Air Quality Evaluation

The following measurements were extracted from a database acquired during a period of more than 6 months in two offices and a laboratory of the Fraunhofer Society in Germany. The test platforms were designed for a practical test of newly developed sensor prototypes together with data acquisition of commercially available sensors as

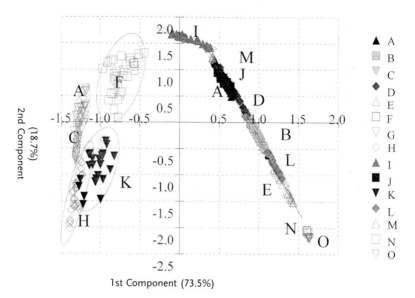

Fig. 2.3-9 Activity profile of the MIR astronauts, derived from electronic nose measurements [17]. A PCA map of some of the patterns associated with representative activities, demonstrating that different atmospheric compositions were recorded by the sensors that could be used as descriptors of those activities. A, Meal; B, Meal; C, Debriefing; D, Meal; E, Sport; F, Sleep; G, Video recording; H, Meal; I, Equipment maintenance check; J, Extravehicular activity; K, Cosmonaut activity; L, Sport; M, Freon test; N, Freon

a comparison and to gather additional information (Figaro TGS 822, Vaisala GMM 11AD CO_2 sensor, Vaisala HMP 233 A80 for relative humidity temperature, HTS-Erni ultrasonic presence sensor, window and door contacts to indicate ventilation). The sensors were regularly calibrated and additionally the TOC value of the indoor air was measured monthly to find a simple contamination index (see Table 2.3-1). The Figaro sensor showed a standard deviation of 10% relative to the test gas CO over 6 months. The standard deviation of the Vaisala CO_2 sensor was less than 1% in the same period. All data channels were sampled in a period of about 1 min, which gave an enormous amount of data, but a fine granulation for further evaluation.

2.3.6
CO_2 Measurements

The main source of CO_2 in a room is human or animal activity. Other CO_2 sources may include gas-fired cooking stoves and, as an accident, a chimney in the case of malfunction, fire, etc. A convenient concentration of CO_2 should be < 1000 ppm, which is

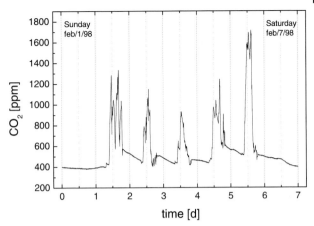

Fig. 2.3-10 CO_2 concentration during a week in an office (maximum three persons present)

one fifth of the threshold limit value of 5000 ppm. Above a concentration of 1000 ppm one can recognize some tiredness and more difficulties with concentration. The investigation was limited to offices and laboratories, where no other CO_2 sources exist. Figure 2.3-10 shows a week in an office with three inhabitants and a mean occupancy of two persons during a week in the autumn period. The window was opened only when necessary. One can easy see that the inhabitants' activity (window opening) limits the CO_2 concentration to a value of <1000 ppm in most cases.

The next example of CO_2 measurement was recorded during a presentation, with an experimental low-cost CO_2 sensor in an older conference room without

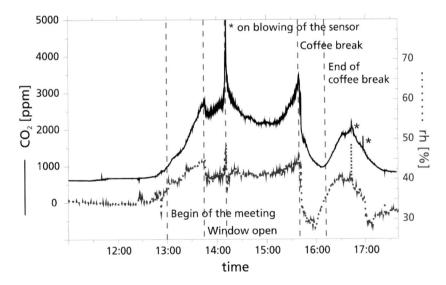

Fig. 2.3-11 CO_2 concentration during a presentation. The participants caused some CO_2 peaks while blowing on the newly presented sensor

an HVAC system [19]. Naturally, during the presentations the opening of a window is not expected. Therefore, one can recognize large increase in CO_2 concentration, until most of the occupants feel the air quality as unacceptable and open the window (Figure 2.3-11).

2.3.7
VOC Sensor

Odor and other hydrocarbons were monitored with some metal oxide sensors. The signal depends on the VOC level and shows a certain higher sensitivity to special gases. The Figaro 822 sensor detects the smoke of a cigarette with a high peak in the signal. It is easily possible to count the number of cigarettes per day (Figure 2.3-12).

Figure 2.3-13 shows a whole month of monitoring air with a metal oxide sensor, from which Figure 2.3-12 is a part. The high peaks are caused by cigarette smoke and other sources during a work day (owing to the small scale of the figure, the measurement curve is smoothed). The VOC background in this laboratory is low. This can be easily understood, because the windows are open during the day and, therefore, the ventilation has a high level. During the night and the vacation week a higher level of contaminants can be detected, because the window is normally closed and also the door is seldom opened.

The measurement curve for an office near a main road may show different behavior.

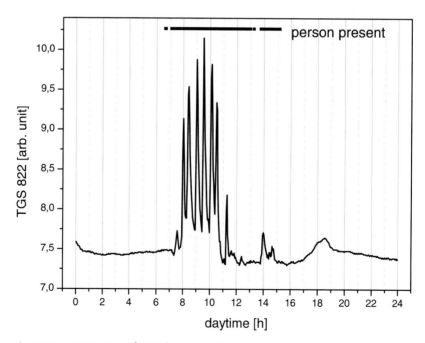

Fig. 2.3-12 Measurement of VOC during a work day in August: The peaks mark cigarette smoke

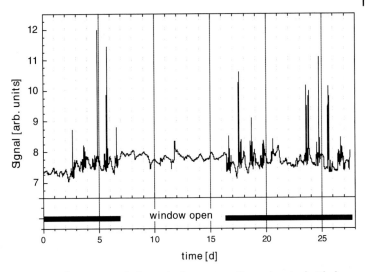

Fig. 2.3-13 VOC monitoring during a month (August), days not exactly synchronized with day time

2.3.8
Summary and Future Outlook

The use of gas sensors, implemented in a home automation network, provides an abundance of new possibilities for use. One of the newer practical applications is the early fire or gas detection. Metal oxide sensors with the possibility of the detection of various gases can detect fire in an early stage [20]. This system, shown in Figure 2.3-14, consists of three gas sensors (main sensitivities: CO, H_2, NO_2, solvents) [21]. A sophisticated signal evaluation system, which uses the sensor signals and also the analysis of the time series, gives the alarm signal.

A similar signal fusion can be carried out with indoor air quality sensors together with other information about the room. An unusual application is presence detection with a gas sensor. Because humans are a major source of emission of various gases, they generate typical patterns on the gas sensor signal, which is different from the continuous emission from other materials. A first survey of the data from the Fraunhofer long-term investigation gave a significant signal for the presence of a person (Figure 2.3-15) [23]. This signal may not be sure enough for burglary detection, but it can be used for an adaptive control of the heating units. This result marks the way to the future, the evaluation of additional information with intelligent systems of a networking cluster of sensors. These results may be not as accurate as the signals of specialized sensors, but the combination with other information delivers additional value for the customer.

Gas sensors, networked in a building automation system, will enhance the comfort, decrease the energy consumption and give the chance of further signal fusion to evaluate more information on the sensor signal. At least the availability of

Fig. 2.3-14 Gas sensor fire detector for industrial applications with high dust concentrations in ambient air: early detection of smoldering fires, eg, in coaling plants, lignite bunkers, waste bunkers [22]

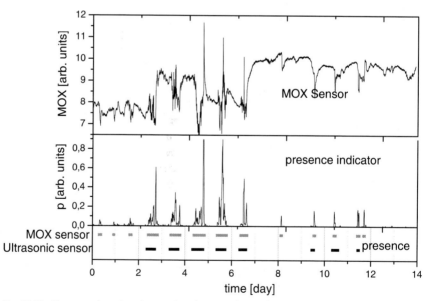

Fig. 2.3-15 Presence detection via signal evaluation of a single sensor signal

economically priced sensors and electronics together with a sophisticated signal evaluation and system technology (networking) is a prerequisite for the introduction of this technique, together with clear regulations for privacy and interoperability of the systems.

2.3.9
Acknowledgments

The author thanks his colleagues R. Hartinger, M. Roth and F. Wendler (IMS, Munich) and V. Grinewitschus and K. Scherer (IMS, Duisburg) and also the partners of other institutes in the Fraunhofer society (IBMT, IBP) in the Project 'intelligent home systems', and the University of Federal Arms in Munich (Prof. Dr. I. Eisele, Prof. Dr. Tränkler, Prof. Dr. T. Doll) in the Project 'Intelligentes Multigassensorsystem', financed by the State of Bavaria. The author also thanks the cited companies for information and figures.

2.3.10
References

1 *VDMA-Einheitsblatt 24772*; Berlin: Beuth Verlag, 1991.
2 *VDMA-Einheitsblatt 24773*; Berlin: Beuth Verlag, 1997.
3 MEIER, S., *Bedarfsgesteuerte Lüftung, TAB – Technik am Bau H.* 6, 1997.
4 PETTENKOFER, M., *Über den Luftwechsel in Wohngebäuden;* Munich: Cotta, 1858.
5 *Introduction to Indoor Air Quality: A Reference Manual*, EPA/400/3-91/003; Washington, DC: Environmental Protection Agency, 1991.
6 Data Source: Fraunhofer IMS and IVV, 1996.
7 FANGER, P.O., *A Comfort Equation for Indoor Air Quality and Ventilation, Healthy Buildings '88*, Stockholm, 1988, Vol. 1, pp. 39–51.
8 MAYER, E., *Deutsche Richtlinie und Europäische Norm zur Bestimmung der Innenraumluftqualität (Sollwerte)*, VDI Berichte 1373, Düsseldorf: VDI, 1998, pp. 33–39.
9 GÖPEL, W. et al., *Definition and Typical Examples in Sensors – A Comprehensive Survey Chemical and Biochemical Sensors*, Vol. 2, Part 1, Weinheim: VCH, 1991.
10 VAISALA, O., P.O. Box 26, FIN-00421 Helsinki; www.vaisala.com.
11 HOEFER, U., et al., *Sens. Actuators B* **22** (1994), 115–119.
12 Figaro Engineering, JP OSAKA 512 and UNITRONIC GmbH, 40472 Düsseldorf.
13 Elektrotechnik Rump GmbH, D-44319 Dortmund-Wickede.
14 Paragon Sensors AG, D-33129 Delbrück, 1999.
15 Siemens Building Technologies, Landis & Staefa Division, D-60388 Frankfurt, 2000.
16 SCHREIBER, F.W., et al., *Electronic Nose: Investigation of the Perceived Air Quality in Indoor Environments, Indoor Air 99*, Edinburgh, 8–13 August 1999; Vol. 2, pp. 624–629.
17 PERSAUD, K.C., et al., *Sens. Actuators B* **55** (1999) 118–126.
18 SCHREIBER, F.W., et al., *Electronic Nose: Investigation of the Perceived Air Quality in Indoor Environments, Indoor Air 99*, Edinburgh, 8–13 August 1999; Vol. 2, pp. 624–629.
19 ENDRES, H.-E., *Sens. Actuators B* **57** (1999) 83–87.

20 Petig, H., et al., *Neuer Gassensor-Brandmelder mit Monitoring System zur Überwachung von Bekohlungsanlagen – Einsatz und Erfahrungen;* Allianz Report 4/99, 1999, 244–250.

21 Kohl, C. D., et al., *Electron Technol. (Poland)* **33** (2000) 13–21.

22 Development and Production: GTE Electronics, Viersen, Germany. Service and sales: Siemens Building Technologies West, Essen, Germany.

23 Endres, H.-E., *Bedarfsgeregelte Lüftung und Intelligentes Multisensorsystem,* Doll, T., et al. (eds.); Rosenheim: Geronimo, 1998.

2.4
Sensor-based Management of Energy and Thermal Comfort
Th. Bernard and H.-B. Kuntze, *Fraunhofer Institute for Information and Data Processing IITB, Karlsruhe, Germany*

2.4.1
Motivation

In recent decades, more and more insulating building materials and construction techniques have been developed and introduced. By these measures remarkably high energy savings have been achieved, but at the cost of a diminished natural air exchange within the buildings. In order to guarantee sufficient air quality and living comfort, it is essential to introduce better controlled ventilation besides controlled heating facilities.

The demand-responsive coordination of control loops for heating, ventilation and blinds is a difficult problem for the average user. On the one hand, he or she is free to choose the nominal commands of heating, ventilation, and blind control to try to ensure that his or her individual cost and comfort criteria are satisfied. On the other hand, the climate state response in the rooms in interaction with the outside climate is very complex and nonlinear. Thus the user can hardly elucidate all the consequences of his or her operations with respect to cost and comfort criteria.

Obviously, there is an increasing demand on the HVAC (heating, ventilating, and air conditioning) market for a user-friendly integrated control and monitoring concept for heating, ventilation, and blind control systems which is optimizable with respect to the individual comfort and economy requirements of the user.

In order to solve this multi-objective online optimization problem, a new fuzzy-logic supervisory control concept has been developed at IITB [1–3], which can also be applied in a modified way to different industrial areas. Especially in the steel and glass industry [4] there is an increasing demand for controlling processes optimally in terms of contradictory performance criteria (eg, productivity versus product quality).

As regards the application of fuzzy methods to multi-criteria optimization problems in the area of operations research, various fuzzy-based optimization concepts have been successfully applied to off-line planning and assistance problems [5, 6]. In the HVAC area, fuzzy-logic approaches are mainly restricted to heating control problems [7].

An important requirement for such a supervisory control concept is flexibility with respect to the number and the properties of the underlying control system.

For example, in many rooms not all control systems (heating, ventilation, blinds) are available. Thus, the supervisory control system also has to cope with such a structural configuration. Another concept requirement is its openness with respect to further performance criteria. For example, in the case of ventilation with tilting windows, the air draught has to be considered as supplementary. It will be discussed in Section 2.4.4 that the proposed concept satisfies both requirements.

2.4.2
Control Concept

The climate dynamics within offices and domestic buildings are more complex than it appears at first glance. Thus, both the comfort perception and the energy consumption depend on the essential climate state variables such as temperature T_i, relative humidity φ_i, the CO_2 concentration $CO2_i$ as reference gas of air quality, and the mean brightness Φ_i in the room. The climate state is disturbed by different measurable or nonmeasurable influences of the outside climate and of the room occupancy. Measurable disturbance inputs are, eg, temperature T_o, relative humidity φ_o, and CO_2 concentration $CO2_o$ outside as well as the presence of persons within the room. Nonmeasurable, mainly stochastic, disturbances are the heating flows, water vapor sources, air draught and CO_2 emissions caused by persons present (Figures 2.4-1 and 2.4-2).

For controlling the room climate in terms of T_i, φ_i, $CO2_i$ and Φ_i, first of all controllable heating, ventilation, and blind facilities have to be installed. As regards the controllability, winter and summer case have to be distinguished. In the winter case, ie, heating is activated, T_i can be selectively controlled, eg, by radiators, but φ_i and $CO2_i$ are strongly coupled with each other. In the summer case, ie, heating deactivated, T_i is controlled by ventilation and blinds. Because of the coupling of φ_i and $CO2_i$ (in the summer case φ_i, $CO2_i$, and T_i), the air exchange rate (AER) which can be controlled by fans or tilting windows has to be introduced as an auxiliary control variable.

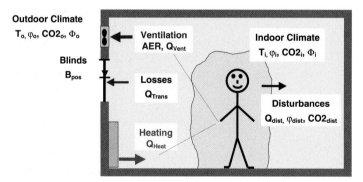

Fig. 2.4-1 Model scheme of a room with the relevant state, disturbances, and manipulated variables

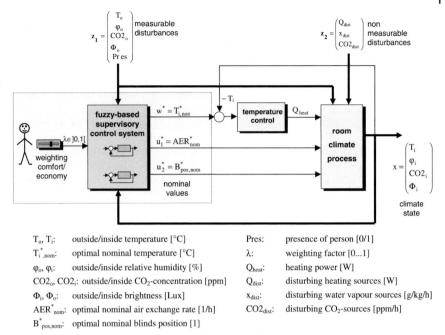

Fig. 2.4-2 Scheme of the fuzzy-based supervisory control and monitoring system

T_o, T_i:	outside/inside temperature [°C]
$T^*_{i,nom}$:	optimal nominal temperature [°C]
φ_o, φ_i:	outside/inside relative humidity [%]
$CO2_o$, $CO2_i$:	outside/inside CO_2-concentration [ppm]
Φ_i, Φ_o:	outside/inside brightness [Lux]
AER^*_{nom}:	optimal nominal air exchange rate [1/h]
$B^*_{pos,nom}$:	optimal nominal blinds position [1]
Pres:	presence of person [0/1]
λ:	weighting factor [0...1]
Q_{heat}:	heating power [W]
Q_{dist}:	disturbing heating sources [W]
x_{dist}:	disturbing water vapour sources [g/kg/h]
$CO2_{dist}$:	disturbing CO_2-sources [ppm/h]

As regards feedback-control loops of T_i in addition to φ_i, $CO2_i$, or Φ_i of rooms, in the past different efficient concepts or products have been proposed [8]. Much less considered has been the supervisory control problem and the integral optimization of the control loops of T_i, φ_i, $CO2_i$, and Φ_i with respect to comfort and economy criteria.

The supervisory control concept introduced here is based on the approach that the user chooses the performance requirements in terms of 'economy' and 'comfort' but not, as usual, the nominal values of heating and ventilation controllers. By means of a simple slide button ('economy-comfort slider'), the user can select the weighting factor λ of his or her individual comfort and economy requirement.

Based on the cost-comfort weighting factor λ, arbitrarily selected by the user, and also on the measured inside climate state (T_i, φ_i, $CO2_i$, Φ_i), outside climate state (T_o, φ_o, $CO2_o$, Φ_o) and the presence of persons in the room (Pres), the optimal nominal values of inside temperature control ($T^*_{i,nom}$), air exchange rate (AER^*_{nom}), and blind position (B^*_{nom}) are computed (Figure 2.4-2). The multi-objective optimization of the three nominal values is based on a fuzzy-algorithm which will be derived below.

In addition to the proposed operation mode above, heuristic control elements can be inserted depending on special day times, seasons, or events. For example, in the absence of persons or during the night time, an economy mode is set automatically and depending on the summer or winter season different comfort criteria are considered (see Section 2.4.4).

An important feature of the proposed supervisory control concept is its modularity. In many practical cases not all control systems (heating, ventilation, and blinds) are available. Within the concept it is possible to obtain sub-optimal results for a reduced number of sub-controllers. For example, if only heating and ventilation are available, the optimal nominal values $T^*_{i,\text{nom}}$ and AER^*_{nom} are computed. This will be demonstrated in the experimental results in Section 2.4.5.1. Another example of sub-optimal results is the sole availability of heating and blinds. This will be illustrated in the simulation results in Section 2.4.5.2.

2.4.3
Theoretical Approach of Multi-objective Fuzzy Optimization

2.4.3.1
The Basic Algorithm

The evaluation of climate performance in a living room by human users incorporates a natural diffusion which is realistically described by fuzzy optimization methods [5, 6]. In general, fuzzy logic provides the opportunity to deal with imprecise information. Whereas in classical logic only the two values 0 or 1 ('yes' or 'no') are defined, in fuzzy logic also smooth transitions between 0 and 1 are allowed [5]. This is done by fuzzy membership functions $\mu_A(x) \in [0,1]$ which indicate the degree of membership of the element x to the set A.

Especially methods of fuzzy decision making, developed by Bellman and Zadeh [9] seem to be well suited to cope with the problem of the optimization of HVAC control systems. In the concept of fuzzy decision making performance criteria and constraints are described as fuzzy membership functions. The fuzzy performance criteria are called *fuzzy goals* $\mu_{G1},\ldots,\mu_{GN}$, the *fuzzy constraints* are denoted by $\mu_{C1},\ldots,\mu_{CM}$. Fuzzy goals and fuzzy constraints are desired to have maximum degree of membership. Consequently, the fuzzy goals and fuzzy constraints are connected by fuzzy-AND-operators (eg, min-operator) to get the *fuzzy decision* μ_D. In the simplest case with only one fuzzy goal μ_G and one fuzzy constraint μ_C and with the min-operator as fuzzy-AND, the *fuzzy decision* μ_D is obtained by

$$\mu_D(x) = \mu_G(x) \wedge \mu_C(x) \wedge = \min . \tag{2.4-1}$$

Equation (2.4-1) can be easily extended for the case of N fuzzy goals $\mu_{G1},\ldots,\mu_{GN}$ and M fuzzy constraints $\mu_{C1},\ldots,\mu_{CM}$:

$$\mu_D(x) = \mu_{G1}(x) \wedge \ldots \wedge \mu_{GN}(x) \wedge \mu_{C1}(x) \wedge \ldots \wedge \mu_{CM}(x) . \tag{2.4-2}$$

The *optimal decision* x^* is calculated by maximizing $\mu_D(x)$, where x is an element in the set of possible solutions X. The algorithm is illustrated in Figure 2.4-3.

$$\mu_D(x^*) = \max_{x \in X} \mu_D(x) . \tag{2.4-3}$$

Fig. 2.4-3 Illustration of the principle of fuzzy decision making with min-operator as fuzzy-AND operator

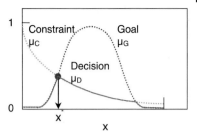

2.4.3.2
Important Features

The method of fuzzy decision making is characterized by three important properties which are of special importance for the given optimization problem of HVAC control systems.

1. Constraints and goals are treated in a absolutely symmetrical way. This is also the case for 'economy' and 'comfort'. From the 'comfort' point of view, 'economy' is only a restriction, and vice versa 'economy' could also be treated as a goal with a minimum of 'comfort' as restriction.
2. As shown by Equation (2.4-2), the algorithm is well suited for multi-criteria optimization problems. Further performance criteria can easily be added. This is useful, eg, for the case when ventilation is applied by tilting windows, where the air draft is another performance criterion.
3. It can easily be shown that values of x which yield zero in *any* performance criteria μ_G or μ_C cannot be the optimal decision x^*. This property can be used to define 'forbidden ranges' of x. In our problem this will be the case for very cold and very hot temperatures inside T_i because it is equivalent to an unacceptable comfort level, as well as for a high CO_2 level (which means very bad air quality) or specific values of humidity φ_i and brightness Φ_i. This will be discussed in detail in Section 2.4.4.1.

2.4.3.3
Weighting of Different Performance Criteria

In many cases the performance criteria are not of equal importance. For example, in the case of optimization with respect to comfort and economy criteria it might be possible to emphasize more the comfort criteria than the economy criteria or vice versa. Thus, a transparent weighting of the individual performance criteria seems to be useful. For the algorithm of fuzzy decision making (Equations (2.4-2) and (2.4-3)) there are many ways to do this [6]. It turned out that the weighting with multipliers λ_i, $0 < \lambda_i \leq 1$, in connection with the min-operator, is an especially transparent way to realize the weighted AND-connection [3]. In Equation (2.4-4) we do not distinguish between fuzzy goals and fuzzy restrictions, the performance criteria are denoted by $\mu_i(x)$. The smaller λ_i is chosen, the *more* the *i*th performance criterion $\mu_i(x)$

will be emphasized. The relative importance of two performance criteria μ_i, μ_k is represented by the ratio λ_i/λ_k. For $\lambda_i/\lambda_k \to 0$, μ_i is much more emphasized than μ_k.

$$\mu_D(x) = \lambda_1\mu_1(x) \wedge \ldots \wedge \lambda_N\mu_N(x), \quad 0 < \lambda_i \leq 1, \quad \wedge = \min. \tag{2.4-4}$$

With Equation (2.4-4) the optimal decision x^* (cf. Equation (2.4-3)) could be ambiguous if there are ranges with constant degree of membership in one or several performance criteria μ_i. A non-ambiguous optimal decision is guaranteed if a correction term is added in Equation (2.4-4):

$$\mu_D(x) = \lambda_1\mu_1(x) \wedge \ldots \wedge \lambda_N\mu_N(x) + \varepsilon\prod_{i=1}^{N}\mu_i(x),$$
$$0 < \lambda_i \leq 1, \quad \wedge = \min, \quad 0 < \varepsilon \ll 1. \tag{2.4-5}$$

In the considered case of two performance criteria μ_1 and μ_2 ('economy' and 'comfort'), the weighting factors λ_1 and λ_2 can be expressed by *one* parameter λ:

$$\lambda_1 = \lambda, \quad \lambda_2 = 1 - \lambda, \quad 0 < \lambda < 1. \tag{2.4-6}$$

$\lambda = 0.5$ corresponds to equal importance of the goals μ_1 and μ_2, and $\lambda \to 0$ corresponds to maximal weighting of μ_1 and $\lambda \to 1$ with maximal weighting of μ_2. Inserting Equation (2.4-6) in Equation (2.4-5) yields

$$\mu_D(x) = \min[\lambda\mu_1(x), (1-\lambda)\mu_2(x)] + \varepsilon\mu_1(x)\mu_2(x), \quad 0 < \lambda < 1, \quad \varepsilon \ll 1. \tag{2.4-7}$$

2.4.3.4
Model Equations

In order to optimize the state variables x with respect to the performance criteria, the controllability of the system dynamics has to be considered. In the problem of optimizing HVAC control systems the nominal values w (eg, $T_{i,\text{nom}}$) or the manipulated variables u (eg, AER_{nom}, B_{nom}) have to be optimized. The performance criteria are defined in dependence on state variables or auxiliary state variables x_i (eg, T_i, φ_i, $CO2_i$, Φ_i). Thus, a dynamic model has to be found which describes the dynamic dependence between u and x, according to

$$\hat{x}_i(k+1) = f_i[x(k), z(k), u(k)]. \tag{2.4-8}$$

For the optimization of nominal values often static models

$$\hat{x}_{j,\text{static}} = f_j[x(k), z(k), w(k)] \tag{2.4-9}$$

are sufficient. Generally, the models depend on some measurable or nonmeasurable disturbance signals, which are summarized by the vector z. In Equations (2.4-8) and (2.4-9) k denotes the time index of the kth sampling interval.

2.4.4
Application to the Supervisory Control of HVAC Systems

2.4.4.1
Comfort Criteria

According to the general concept, in a first step useful performance criteria for comfort and economy depending on climate state variables T_i, φ_i, $CO2_i$, and Φ_i have to be defined, which correspond to the performance criteria μ_i in Equation 2.4-4. Unfortunately, no universal models are available which can realistically describe the human comfort perception. In HVAC technology, however, the limits of comfort in terms of temperature and air quality are well defined [8, 10, 11]. According to these standards, the temperature T_i should be within the range 20–24 °C, the relative humidity φ_i between 30 and 70% and the CO_2 concentration $CO2_i$ up to 1000 ppm. The heat flow of the air which causes a draught q_d should not be much greater than 40 W/m² and the mean brightness, eg, in offices should be in the range 500–2000 lux [8, 10].

Since these parameters are only general recommendations, it is useful to represent them by fuzzy-membership functions, eg, according to Figure 2.4-4. Obviously, the shown fuzzy-membership functions μ_{comf} in terms of T_i, φ_i, $CO2_i$, q_d, and Φ_i represent the human-like comfort evaluation much better than step-like membership functions (dotted lines) of the classical binary logic. Moreover, the characteristic parameters of the membership functions can be easily matched to individual user criteria.

2.4.4.2
Economy Criteria

A membership function μ_{eco} is required which describes the economy rate of the HVAC in terms of heating power. For the considered problem it turns out that a decreasing exponential function which can be easily parameterized provides a sufficient description of the economic rate (cf. Equation (2.4-10) and Figure 2.4-5).

$$\mu_{eco} = \exp(-Q_{heat}/Q_{ref}) \qquad (2.4\text{-}10)$$

Q_{heat} consists of several contributions according to the algebraic equation

$$Q_{heat} = (Q_{vent} + Q_{trans}) - (Q_{sol} + Q_{int}) \qquad (2.4\text{-}11)$$

where Q_{vent} denotes the ventilation losses, Q_{trans} the transmission losses through the walls, Q_{sol} are solar gain and Q_{int} are internal gains caused by machines or humans. The contributions are modeled in the Sections 2.4.4.3.2, 2.4.4.4.2 and 2.4.4.5.2.

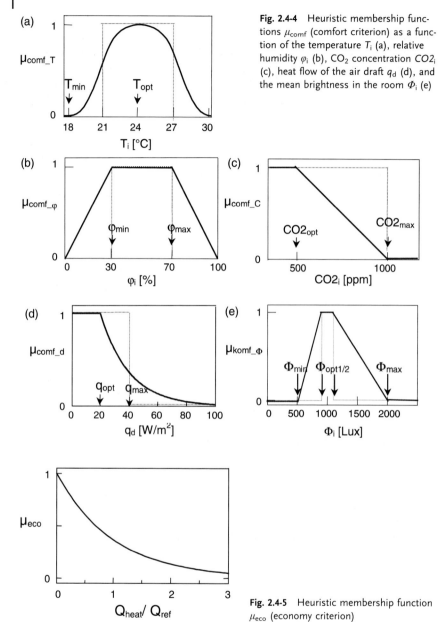

Fig. 2.4-4 Heuristic membership functions μ_{comf} (comfort criterion) as a function of the temperature T_i (a), relative humidity φ_i (b), CO_2 concentration $CO2_i$ (c), heat flow of the air draft q_d (d), and the mean brightness in the room Φ_i (e)

Fig. 2.4-5 Heuristic membership function μ_{eco} (economy criterion)

2.4.4.3
Optimization of Heating Temperature

Using the definition of comfort and economy criteria in Sections 2.4.4.1 and 2.4.4.2, the nominal inside temperature $T_{i,nom}$ can now be optimized.

2.4.4.3.1 Comfort Criterion

As regards the comfort criterion the direct dependence on $T_{i,nom}$ is defined by the membership function μ_{comf_T} (Figure 2.4-4a). It would also be possible to take the *perceived* (operative) temperature T_{op} into consideration, which differs from the air temperature T_i if the temperature of the wall surface is much colder than T_i. In this case T_{op} is calculated as the mean of T_i and the radiation temperature T_r:

$$T_{op} = 0.5(T_i + T_r) \ . \tag{2.4-12}$$

T_r is a weighted mean of the temperatures of all the wall surfaces. Under the assumption that only the outside wall has a contribution to T_r, ie, all the inner walls have the same temperature as the air temperature T_i, T_r can be estimated by the equation

$$T_r = (1 - 2\varepsilon)T_i + 2\varepsilon T_o \tag{2.4-13}$$

where $\varepsilon = kA_S/2aA_{sum}$ is a standardized parameter which indicates the insulation of the outer wall [3]. An ideal insulation corresponds to $\varepsilon = 0$. Typical values are $\varepsilon = 0.01–0.05$ [3], where k is the heat transfer coefficient and A_S the surface area of the outer wall, a is the heat transition coefficient of the air ($a \approx 5$ W/m² K [8]) and A_{sum} is the sum of all surface areas of the wall. Combining Equations (2.4-12) and (2.4-13) yields an equation for T_{op}:

$$T_{op} = (1 - \varepsilon)T_i + \varepsilon T_o \ . \tag{2.4-14}$$

In modern low-energy houses, it can be assumed as a good approximation that $\varepsilon = 0$ [3], which is equivalent to $T_{op} = T_i$ (cf, Equation (2.4-14)). In the following we also consider $\varepsilon = 0$. Results with $\varepsilon > 0$, which corresponds to poorly insulated walls, are presented in [3].

2.4.4.3.2 Economy Criterion

In order to obtain the membership function $\mu_{eco\_T_i}$ the transmission losses Q_{trans} are considered as a representative variable. For quasi-static changes Q_{trans} can be calculated by

$$Q_{trans} = kA(T_i - T_o) \tag{2.4-15}$$

which corresponds to the general static model Equation (2.4-9). Inserting Equation (2.4-15) in the basic definition of μ_{eco} (Equation (2.4-10)) yields the results illustrated in Figure 2.4-6 with T_o as a parameter. The membership degree decreases the higher $T_{i,nom}$ and the lower T_o. This is consistent with the intuitive estimation of the term 'economy'.

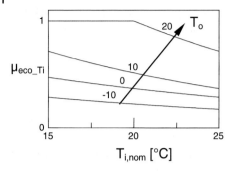

Fig. 2.4-6 Exemplary economy membership function as a function of $T_{i,nom}$ and T_o

2.4.4.3.3 Quasi-static Results

Having defined $\mu_{eco\_T_i}$ and $\mu_{comf\_T_i}$ as a function of $T_{i,nom}$, the fuzzy decision $\mu_{D\_T_i}$ can be calculated (cf, Equation (2.4-7)):

$$\mu_{D\_T_i}(T_{i,nom}) = \min[\lambda \mu_{eco\_T}(T_{i,nom}), (1-\lambda)\mu_{comf\_T}(T_{i,nom})] \\ + \varepsilon \mu_{eco\_T_i}(T_{i,nom}) \mu_{comf\_T}(T_{i,nom}), \\ 0 < \lambda < 1, \quad 0 < \varepsilon \ll 1. \quad (2.4\text{-}16)$$

The optimal temperature $T^*_{i,nom}$ is calculated according to Equation (2.4-3) by

$$\mu_{D\_T_i}(T^*_{i,nom}) = \max \mu_{D\_T_i}(T_{i,nom}). \quad (2.4\text{-}17)$$

The result is a quasi-static set of characteristic curves which are exemplarily shown in Figure 2.4-7. The influence of the characteristic parameters of μ_{comf_T} can be clearly seen. For $\lambda \to 0$, this corresponds to 'max. economy', and it holds that $T^*_{i,nom} = T_{min} = 18\,°C$, and for $\lambda \to 1$, this corresponds to 'max. comfort', with $T^*_{i,nom} = T_{opt} = 24\,°C$ (cf, μ_{comf_T} in Figure 2.4-4a). For $0 < \lambda < 1$, $T^*_{i,nom}$ decreases with

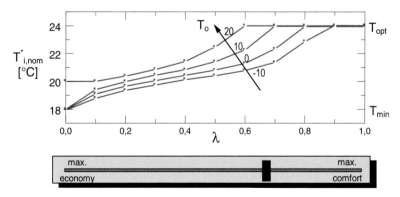

Fig. 2.4-7 The optimal nominal temperature $T^*_{i,nom}$ as a function of the weighting factor λ (slider position) and the outside temperature T_o

lower values of T_o, which corresponds to the higher transmission losses and thus a lower level of the economy membership function $\mu_{eco\_T_i}$ (cf, Figure 2.4-6).

2.4.4.4
Optimization of Air Exchange Rate

2.4.4.4.1 Comfort Criteria
The optimization of the air exchange rate AER_{nom} is more complex than the temperature optimization, for two reasons. First, there are several different comfort criteria depending on $CO2_i$, φ_i, q_d (cf, Figure 2.4-4). Second, the dynamic behavior of $CO2_i$ and φ_i in terms of AER_{nom} which is disturbed by humidity and CO_2 sources, φ_{dist}, and $CO2_{dist}$ (eg, caused by persons), has to be considered in the optimization procedure. The solution to these problems will be discussed later.

In order to obtain only *one* comfort membership function μ_{comf_AER}, the comfort criteria have to be aggregated. This can be achieved in a realistic way by applying a fuzzy AND-operator (eg, min-operator):

$$\mu_{comf\_AER}(AER_{ref}) = \mu_{comf\_C}(CO2_i) \wedge \mu_{comf\_\varphi}(\varphi_i) \wedge \mu_{comf\_d}(q_d) \; . \tag{2.4-18}$$

In the summer AER is the manipulated variable to control T_i. In this case the comfort criterion $\mu_{comf\_T_i}$ has to be considered also:

$$\mu_{comf\_AER}(AER_{ref}) = \mu_{comf\_C}(CO2_i) \wedge \mu_{comf\_\varphi}(\varphi_i) \wedge \mu_{comf\_T}(T_i) \wedge \mu_{comf\_d}(q_d) \; . \tag{2.4-19}$$

Equations (2.4-18) and (2.4-19) can only be computed if there are dynamic and static models:

$$\begin{aligned} \hat{CO2}_i(k+1) &= f[AER_{ref}(k)] \\ \hat{\varphi}_i(k+1) &= g[AER_{ref}(k)] \\ \hat{T}_i(k+1) &= h[AER_{ref}(k)] \\ \hat{q}_d(k) &= i[AER_{ref}(k)] \end{aligned} \tag{2.4-20}$$

which describe $CO2_i$, φ_i, T_i, and q_d in terms of AER_{nom}.

Equations (2.4-20) correspond to the general dynamic and static models of Equations (2.4-8) and (2.4-9). Contrary to the static feedforward optimization of $T_{i,nom}$, in the optimization of AER_{nom} the transition dynamics of $CO2_i$, φ_i and T_i have additionally to be considered, ie, a dynamic feedback optimization is applied to obtain AER_{nom}^*.

By means of an internal predictive model with a reference trajectory, the time response of $CO2_i$, φ_i, and T_i is simulated and optimized at each sampling interval, eg, every 10 s, over a prediction horizon, eg, 10 min. In the case of $CO2_i$ the system behavior is described by the following differential equation:

$$\dot{CO2}_i = -AER(CO2_i - CO2_o) + CO2_{dist} \tag{2.4-21}$$

assuming an ideal mixing of the air. The differential equation is nonlinear due to the multiplication of AER and $CO2_i$. Now a required dynamic response for a transition of the actual value $CO2_i$ to an internal nominal value $CO2_{i,nom}$ is defined by a low-pass first-order function (time constant τ):

$$\dot{CO2}_i \stackrel{!}{=} -\frac{1}{\tau}(CO2_i - CO2_{i,nom}) \,. \tag{2.4-22}$$

By equating Equations (2.4-21) and (2.4-22), the algebraic relation

$$CO2_{i,nom} = CO2_i + \tau(-AER(CO2_i - CO2_o) + CO2_{dist}) \equiv f(AER) \tag{2.4-23}$$

is obtained, which represents the quasi-static model.

In Equation (2.4-23), $CO2_i$, $CO2_o$, and $CO2_{dist}$ are measured or estimated. The applied principle of the prediction with a reference trajectory and an internal nominal value $CO2_{i,nom}$ illustrated in Figure 2.4-8 is similar to the concept of predictive functional control [12, 13].

The functions $\varphi_{i,nom} = g(AER_{nom})$ and $T_i = h(AER_{nom})$ are achieved in the same way as for Equations (2.4-21)–(2.4-23) (for details, see [3]).

For describing the air draft q_d inside a room in terms of AER_{nom} the static model

$$q_d = \gamma(AER_{nom})(T_i - T_o) \quad [W/m^2] \tag{2.4-24}$$

can be assumed, which corresponds to the function $q_d = i(AER_{nom})$ in Equation (2.4-21). The heuristic constant γ can be estimated experimentally. In the IITB test rooms, eg, the value $\gamma \approx 3$ W h/m^2/K has been determined.

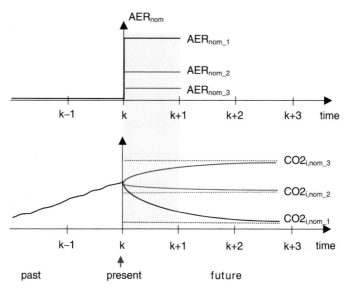

Fig. 2.4-8 Principle of the prediction with a reference trajectory and an internal nominal value $CO2_{i,nom}$

2.4.4.4.2 Economy Criterion

Whereas for the calculation of $\mu_{eco\_T_i}$ the transmission losses Q_{trans} were considered as a characteristic variable, now the ventilation losses Q_{vent} have to be considered. Q_{vent} is estimated by the well known physical relation [8]

$$Q_{vent} = \rho c \, V \, AER_{nom}(T_i - T_o) \qquad (2.4\text{-}25)$$

where c denotes the specific heat capacity, ρ the density of dry air ($c \approx 1000$ J/kg/K, $\rho = 1.2$ kg/m³) and V the volume of the room. The economy membership function μ_{eco_AER} is obtained according to Equation (2.4-10) by

$$\mu_{eco\_AER} = \exp(-Q_{vent}/Q_{ref}) . \qquad (2.4\text{-}26)$$

In Figure 2.4-9, μ_{eco_AER} is illustrated as a function of AER_{nom} with T_o as a parameter of the curve set. The membership degree is equal to 1 when $AER_{nom} = 0$ or $Q_{vent} = 0$ (cf, Equation (2.4-25)). The membership function μ_{eco_AER} decreases with increasing AER_{nom} and decreasing T_o, which is consistent with the intuitive estimation of the term 'economy'.

2.4.4.4.3 Exemplary Quasi-static Results

By the model equations derived in the last section and the aggregation of the different comfort criteria to *one* membership function $\mu_{comf}(AER_{nom})$ with Equations (2.4-18) and (2.4-19) it is possible to calculate the fuzzy decision $\mu_D(AER_{nom})$ and the optimal value AER^*_{nom} corresponding to Equations (2.4-7) and (2.4-3):

$$\mu_{D\_AER}(AER_{nom}) = \min[\lambda \mu_{eco\_AER}(AER_{nom}), (1-\lambda)\mu_{comf\_AER}(AER_{nom})]$$
$$0 < \lambda < 1, \quad \varepsilon \ll 1 \qquad (2.4\text{-}27)$$

$$\mu_{D\_T_i}(AER^*_{nom}) = \max \mu_{D\_T_i}(AER_{nom}) . \qquad (2.4\text{-}28)$$

To simplify matters in Equation (2.4-27), the correction term was neglected (cf, Equation (2.4-5)). Whereas for $T^*_{i,nom}$ the result could be summarized in a set of

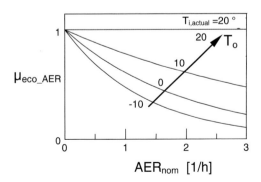

Fig. 2.4-9 Exemplary economy membership function μ_{eco_AER} as a function of AER_{nom} and T_o (assumed actual temperature inside: $T_{i,actual} = 20°C$)

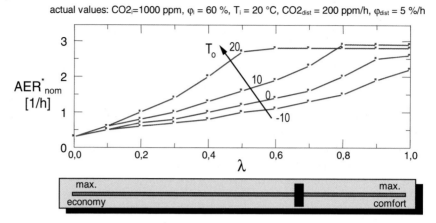

Fig. 2.4-10 The optimal air exchange rate AER^*_{nom} as a function of weighting factor λ (slider position) and outside temperature T_o

characteristic curves (cf, Figure 2.4-7), this is impossible for AER^*_{nom} because it depends not only on λ and T_o but also on the actual inside state of $CO2_i$, φ_i, T_i, $CO2_{dist}$ and φ_{dist}. Thus, in Figure 2.4-10 a 'snapshot' of the state and disturbing variables is shown with actual values $CO2_i = 1000$ ppm, $\varphi_i = 60\%$, $T_i = 20\,°C$, $CO2_{dist} = 200$ ppm/h and $\varphi_{dist} = 5\%/h$. The influence of the weighting factor λ and the outside temperature T_o on AER^*_{nom} can be seen. Since both the relative humidity φ_i and the draft q_d depend strongly on T_o, the saturation limit of AER^*_{nom} for $\lambda \to 1$ also depends on T_o. Just this dependence demonstrates the advantage of the proposed supervisory control concept over the non-coordinated operations of a user which hardly assesses all the consequences of his or her heuristic control actions with respect to economy and comfort. The minimal value $AER^*_{nom} = 0.6/h$ in the case of highest economy ($\lambda \to 0$) results from the limit value $CO2_{max} = 1000$ ppm which was defined in the comfort membership function μ_{comf_C} (cf, Figure 2.4-4c).

2.4.4.5
Optimization of Blind Position

2.4.4.5.1 Comfort and Economy Criteria
Blinds can be used to control both the brightness inside Φ_i and the solar gains in the room Q_{sol}. In the summer season Q_{sol} has a strong impact on the temperature T_i, whereas in the winter season Q_{sol} contributes to reduce the heating power. Hence it seems to be reasonable to distinguish between the winter and the summer cases. In the summer case the performance criteria $\mu_{comf\_\Phi}(\Phi_i)$ and $\mu_{comf\_T}(T_i)$ are the important criteria whereas in the winter case $\mu_{comf\_\Phi}(\Phi_i)$ and $\mu_{eco\_B}(Q_{heat})$ are essential. Thus, according to the generic performance requirement (Equation (2.4-4)) the fuzzy decision μ_D can be defined by

$$\mu_D(B_{pos}) = \begin{cases} \min[\lambda\mu_{comf\_\Phi}(\Phi_i), (1-\lambda)\mu_{eco\_B}(Q_{heat})] & \text{winter case} \\ \min[\lambda\mu_{comf\_\Phi}(\Phi_i), (1-\lambda)\mu_{comf\_T}(T_i)] & \text{summer case} \end{cases} \quad (2.4\text{-}29)$$

To simplify matters, in Equation (2.4-29) the correction term is neglected (cf, Equation (2.4-5)). Note that λ in Equation (2.4-29) has a different meaning in the winter case than in the summer case. Whereas in the winter case the comfort and economy criteria can be weighted with λ, in the summer case the preference in the two comfort criteria is adjusted by λ.

In order to calculate Equation (2.4-29), one has to find models to predict the influence of the blind position B_{pos} on Φ_i, T_i, and Q_{heat}

$$\begin{aligned} T_i &= f(B_{pos}) \\ \Phi_i &= g(B_{pos}) \\ Q_{heat} &= h(B_{pos}), \end{aligned} \quad (2.4\text{-}30)$$

which is the topic of the next section.

2.4.4.5.2 Model Equations

Although the connection between the solar radiation outside $q_{sol,o}$, the blind position B_{pos}, and the solar gain Q_{sol} is described by well-known physical laws the exact modeling of Equation (2.4-30) is very difficult owing to the complex geometrical circumstances [11]. It turnes out that for the design and development of the fuzzy-based supervisory control concept, simple linearized models of temperature and solar gains are sufficient.

Model of Temperature

Although an exact relation between Q_{sol} and T_i can hardly be modeled in practice, it is possible to control T_i by B_{pos} by simple control algorithms (eg, PI controller). However, the limitations of B_{pos} and the dependence of the solar radiation outside $q_{sol,o}$ have to be considered. In the sequel an internal nominal value $T_{i,nom}$ is assumed to be given. With $T_{i,nom}$ the error $e(k)$ can be defined by

$$e(k) = T_{i,nom}(k) - T_i(k) \quad (2.4\text{-}31)$$

where k is the time index. A PI algorithm for B_{pos} is given by

$$B_{pos}(k) = B_{pos}(k-1) + K_P\left[\left(1+\frac{\Delta t}{\tau_I}\right)e(k) - e(k-1)\right] \quad (2.4\text{-}32)$$

where K_P denotes the proportional gain, Δt the sampling time and τ_I the integrator time constant. Resolving Equations (2.4-32) and (2.4-31) to $T_{i,nom}$, the simple prediction model

$$T_{i,\text{nom}}(k) = T_i(k) + \frac{1}{1+\frac{\Delta t}{\tau_I}}\left\{e(k-1) + \frac{1}{K_P}\left[B_{\text{pos}}(k) - B_{\text{pos}}(k-1)\right]\right\} \equiv f(B_{\text{pos}}(k))$$

(2.4-33)

results, which is used for optimization. It corresponds to $T_i = f(B_{\text{pos}})$ (cf. Equation (2.4-30)) and is inserted in Equation (2.4-29) for calculating μ_D.

Model of Brightness
It is assumed that changes of Φ_i are proportional to small changes of B_{pos}:

$$\frac{\Phi_i(k) - \Phi_i(k-1)}{B_{\text{pos}}(k) - B_{\text{pos}}(k-1)} = a(k).$$

(2.4-34)

The proportionality 'constant' $a(k)$ depends on the geometric properties of the room and the blinds and on the solar radiation outside, $q_{\text{sol},o}$. With the plausible assumption that for totally closed blinds ($B_{\text{pos}}=0$) $\Phi_i=0$, $a(k)$ can be estimated by

$$a(k) = \frac{\Phi_i(k)}{B_{\text{pos}}(k)}.$$

(2.4-35)

As this linearized model is valid only for small changes in B_{pos}, the changes in B_{pos} in each sampling period have to be constrained (eg, 10% of the total range).

Model of Solar Gains
In order to estimate the solar gains Q_{sol} it is assumed that the actual value of Q_{heat} is changing by a contribution which is proportional to the change of B_{pos} and the solar radiation $q_{\text{sol},o}$:

$$Q_{\text{heat}}(k+1) = Q_{\text{haet}}(k) - \gamma[B_{\text{pos}}(k) - B_{\text{pos}}(k-1)]q_{\text{sol},o}(k).$$

(2.4-36)

The parameter γ can be estimated depending on the surface area of the window S_w and the transmission coefficient g_w:

$$\gamma = S_w\, g_w.$$

(2.4-37)

2.4.4.5.3 Exemplary Quasi-static Result
In Figure 2.4-11 an exemplary static result for the optimization of the blind position B_{pos} is shown for the winter case. B^*_{pos} depends on the weighting parameter λ as well as on T_o which is considered as a parameter in Figure 2.4-11. For actual values it is assumed that $T_i = 24\,°C$ and $\Phi_i = 2100$ lux. For the slider position 'max. economy' ($\lambda \to 0$) the blinds are almost totally opened in order to maximize the solar gains Q_{sol}. The value $B^*_{\text{pos}} \approx 0.9$ corresponds to $\Phi_i = 2000$ lux, which is defined as the maximum value of Φ_{max} in the comfort membership function μ_{comf_Φ} (cf,

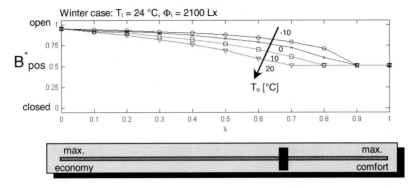

Fig. 2.4-11 The optimal blind position B_{pos} as a function of the slider position λ and outdoor temperature T_o (winter case)

Figure 2.4-4). For the slider position 'max. comfort' ($\lambda \to 1$) the result is $B^*_{pos} \approx 0.5$, which corresponds to the upper limit of the optimal range (Φ_{opt2}, cf. Figure 2.4-4). For λ within the range $0 < \lambda < 1$ the results depend on the outside temperature T_o due to the dependence of the transmission losses on T_o. Summarizing it can be seen that the result shows the transparent behavior of the fuzzy-based control concept according to the defined comfort and economy criteria.

2.4.5
Simulations and Measured Results

In order to investigate the system behavior of the room and performance of the fuzzy-based supervisory control concept under almost realistic conditions, a simulation model was programmed in a MATLAB/Simulink software environment [1–3]. The building physics are characterized by a room volume $V = 50$ m^3, a discretization of walls by five layers, an outside wall consisting of a 20 cm brick layer, an insulation layer of 5 cm ($k = 0.54$ W/m^{2}K), inside walls of 15 cm brick layers ($k = 1.82$ W/m^{2}K) as well as one window ($k = 2.0$ W/m^{2}K).

Furthermore, the fuzzy-based supervisory control concept has been applied to two office rooms at IITB (cf, Figure 2.4-12). In order to demonstrate the robustness of the concept, different sensor-actor configurations and fieldbus systems have been installed in these rooms. For the room ventilation, controllable fans (room 1) and tiltable windows (room 2) have been used. The air quality is measured by a CO_2 sensor (room 1) and by a mixed gas sensor (room 2) able to detect oxidizable components in the air (eg, smoke).

In the following, two characteristic configurations are discussed in order to demonstrate the flexibility of the fuzzy-based supervisory control concept with respect to different HVAC systems. In Section 2.4.5.1 the optimization of heating and ventilation control systems is considered, and Section 2.4.5.2 refers to the op-

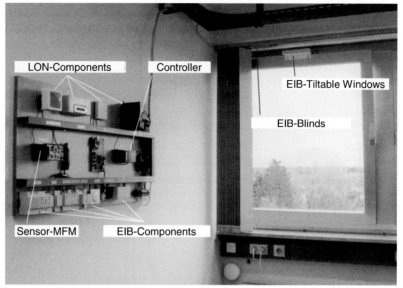

Fig. 2.4-12 Test and demonstration facility for integrated home systems at IITB (room 2)

timization of heating and blind control systems. Both results were obtained for winter conditions (for further details, see [3]).

2.4.5.1
Supervisory Control of Heating and Ventilation Systems

2.4.5.1.1 Simulation Results
From many different simulation scenarios one example represents the course of a typical winter day while assuming three different performance requirements represented by corresponding comfort-cost slider positions. The time responses in Figure 2.4-13 show the inside temperature (actual/nominal) $T_{i,act/nom}$, the number of persons in the room, the air exchange rate (actual/nominal) $AER_{act/nom}$, CO_2 level and relative humidity inside $CO2_i$, φ_i and relative humidity and temperature outside φ_a, T_a. The characteristic parameters of the comfort membership functions (cf, Figure 2.4-4) are plotted as auxiliary lines.

In order to demonstrate the system response with respect to changing room occupancy and to the corresponding disturbances beginning at 8.00 a.m., the room presence is successively increased by one person per 2 h cycle. At 6.00 p.m. all five persons leave the room.

The time response of inside temperature in Figure 2.4-13 underlines the strong influence of different adjustments of the comfort-cost slider on the temperature reference value $T^*_{i,nom}$. It varies within a range which was defined by the chosen comfort membership function (cf, Figure 2.4-4). For the slider positions 'max. comfort', 'medium', and 'max. economy' it means $T^*_{i,nom} = 24$, ~ 21 and $18\,^\circ$C, re-

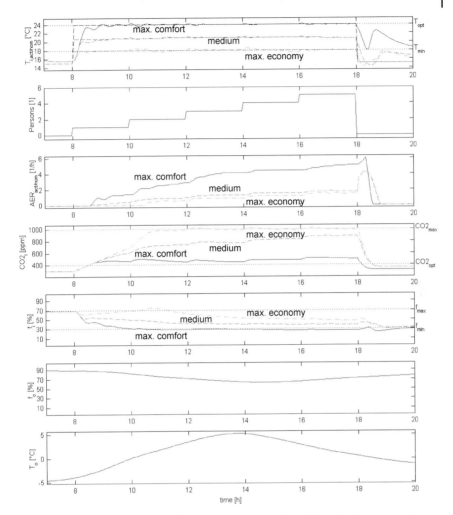

Fig. 2.4-13 Simulation of fuzzy-based supervisory control of heating and ventilation systems of a room in the course of a winter day, assuming three different performance requirements

spectively. Moreover, the influence on the actual room presence can be clearly seen. If the room is empty the fuzzy optimization is deactivated, and a constant set point $T^*_{i,nom} = 15\,°C$ is chosen.

An automatic adaptation with respect to altering room occupancy can be seen from the time response of the air exchange rate AER_{nom}. In the slider position 'max. comfort' the ventilation is activated soon after the presence of the first person in order to maintain the defined CO_2 comfort level $CO2_{opt} = 400$ ppm. The strong relationship between temperature and relative humidity can also be seen. The cold inflowing air from outside becomes considerably less humid when heated. Therefore, the slider position 'max. comfort' represents a tradeoff between the comfort demand with respect to CO_2 concentration and relative humidity while the CO_2 rate increases up to 500 ppm. Correspondingly, in the slider position 'max. economy' the ventilation is not activated before the CO_2 concentration achieves the defined threshold of $CO2_{max} = 1000$ ppm. Then the relative humidity remains in an uncritical range.

Based on numerous simulations of various realistic scenarios of building physics and climate, it was shown that a considerable reduction in energy costs can be achieved by the optimally coordinated fuzzy supervisory control of heating and ventilation systems. In the case considered in Figure 2.4-13 the required heating energy of 13.3 kW h at the slider position 'max. comfort' can be reduced by more than 70% to 3.7 kW h if the slider position 'max. economy' is chosen.

2.4.5.1.2 Measured Results

The measured time responses of weighting factor λ, temperature inside (actual/nominal) $T_{i,act}$, $T_{i,nom}$ and outside T_o, the presence of persons in the room (Pres), the comfort membership μ_{comf_d} in terms of draft q_d, the air exchange rate (actual/nominal) AER_{act}, AER_{nom} and the sensor signal of the mixed gas sensor MG_i in arbitrary units in Figure 2.4-14 were obtained in the winter season and also on 2nd March 1999 in room 2 at IITB.

The corresponding optimal inside temperatures $T^*_{i,nom} = 18$, 21 and 24 °C are controlled by changing the weighting factor λ (slider position) from $\lambda \to 0$ ('max. economy') to $\lambda = 0.5$ ('medium'), and finally to $\lambda \to 1$ ('max. comfort'). In the special case of an empty room (Pres = 0), a constant set point $T^*_{i,nom} = 15$ °C was selected.

The optimized air exchange rate AER^*_{nom} depends not only on the user-controlled λ but also on disturbances such as cigarette smoking. The diagram shows that after cigarette 1 has been smoked, AER^*_{nom} is equal to zero because $\lambda \to 0$ ('max. economy').

When the person leaves the room at 10.00 a.m. the system switches into an absence mode which is characterized by $T^*_{i,nom} = 15$ °C and the internal slider position 'max. comfort' with respect to AER_{nom}. As a consequence, no heating power caused by the enforced ventilation is consumed.

At 12.00 a.m. cigarette 2 is smoked, AER^*_{nom} increases and the window opens. Because of the low outside temperature μ_{comf_d} decreases and a tradeoff between comfort with respect to air quality and comfort with respect to draft is found. During the vacancy starting at 12.30 a.m. the window opens again and closes when the defined level of optimal air quality MG_{opt} is reached.

At 2.30 and 4.00 p.m. when cigarettes 3 and 4 have been smoked, the window is opened for a longer period to the weighting factor $\lambda \to 1$ ('max. comfort'). Again, a tradeoff between the opposing comfort criteria air quality and draft is found resulting in the window closing before optimal air quality has been achieved.

After clearing the room at 5.00 p.m. the window stays open as long as the air quality is equal to MG_{opt}.

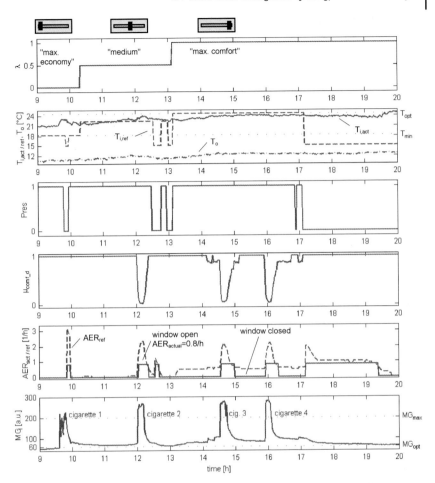

Fig. 2.4-14 Measured results of fuzzy-based supervisory control of heating and ventilation systems

2.4.5.2
Supervisory Control of Heating and Blind Systems

In Figure 2.4-15 the simulation results of a very sunny but also very cold winter day are presented. The diagram shows the time responses of the number of persons occupying the room, the blind position B_{pos} (actual = nominal), the temperature inside (actual/nominal) $T_{i,act/nom}$, the brightness inside Φ_i, the heating power Q_{heat} and the temperature and brightness outside T_o, Φ_o. The characteristic parameters of the comfort membership functions (cf, Figure 2.4-4) are plotted as auxiliary lines. As in Figure 2.4-13 three different performance scenarios are considered which correspond to the slider positions 'max. economy' ($\lambda \rightarrow 0$), 'medium' ($\lambda = 0.5$), and 'max. comfort' ($\lambda \rightarrow 1$).

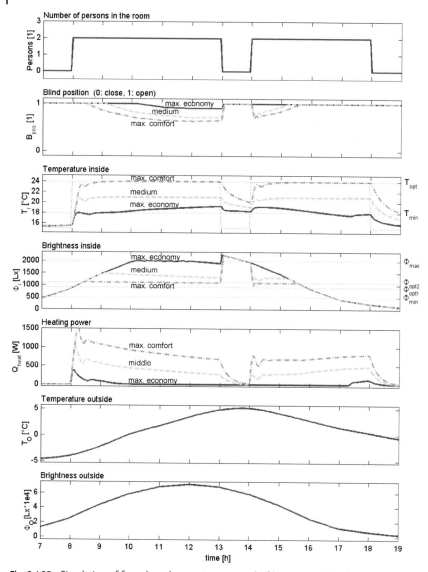

Fig. 2.4-15 Simulation of fuzzy-based supervisory control of heating and blind systems

In the scenario 'max. comfort' B_{pos} the upper value $\Phi_{opt2} = 1100$ lux of the defined optimal range is adjusted exactly, ie, the blinds are set to a lower level with increasing brightness outside Φ_o. In the scenario 'max. economy' the blinds remain opened ($B_{pos} = 1$) until at 10.00 a.m. Φ_i reaches the defined maximal value $\Phi_{opt2} = 2000$ lux. After 10.00 a.m. this value is maintained exactly by B_{pos}. In the scenario 'medium' B_{pos} and Φ_i have values which are between the values of the other scenarios.

When the room is empty (1.00–2.00 a.m. the comfort criterion with respect to brightness ($\mu_{\text{comf\_}\Phi}$) is not considered. As a consequence, the blinds open and more heating energy is saved.

Summarizing, it can be stated that the blinds are precisely controlled according to the defined comfort and economy criteria. Depending on the slider position λ selected by the user a smart tradeoff between energy saving and comfort optimization will be achieved.

2.4.6
Conclusions

A new fuzzy-based supervisory control concept for the optimization of HVAC systems with respect to different partly contradictory performance requirements has been presented. It enables the untrained user to operate easily and optimally his or her home heating, ventilation, and blind control facilities according to his or her individually weighted comfort and economy objectives. The supervisory concept is completely open with respect to different HVAC configurations.

The performance with respect to energy saving and comfort improvement is demonstrated by simulation and experimental results. On-going R&D activities deal with the implementation of the fuzzy concept in a marketable building automation and control system.

The fuzzy-based multi-objective supervisory control concept presented here has a generic character. It can be applied also to completely different complex industrial plants which are characterized by a manifold of different partly contradictory performance and quality requirements.

2.4.7
References

1 BERNARD, TH., KUNTZE, H.-B., *Automatisierungstech. Praxis* **40** (7) (1998) 23.
2 BERNARD, TH., KUNTZE, H.-B., NIRSCHL, G., *Heizung Lüftung/Klima Haustech. (HLH)* **50** (1) (1999) 38.
3 BERNARD, T.H., PhD-Thesis, University of Karlsruhe, Germany, 2000. http://thbernard.leute.server.de/diss.
4 SAJIDMAN, M., KUNTZE, H.-B., in: *Proceedings of the Sixth IEEE International Conference on Fuzzy Systems (FUZZ IEEE'97)*, Barcelona, Spain, July 1997; 1997.
5 ZIMMERMANN, H.-J., *Fuzzy Sets, Decision Making, and Expert Systems*; Boston: Kluwer, 1987.
6 ROMMELFANGER, H., *Fuzzy Decision Support Systeme*; Heidelberg: Springer, 1994.
7 v. ALTROCK, C., AREND, H.-O., KRAUSE, B., STEFFENS, C., BEHRENS-RÖMMLER, E., in: *IEEE Conference on Industrial Fuzzy Control Intelligent Systems*; 1993, pp. 115–119.
8 RECKNAGEL, H., SPRENGER, E., SCHRAMEK, G.R., *Taschenbuch für Heizung und Klimatechnik*; Munich, Germany: Oldenbourg, 1995, pp. 1247–1299.
9 BELLMAN, R.E., ZADEH, L.A., *Manag. Sci.* **17** (1970) 141.
10 *ASHRAE Standard 55-1992*; American Society of Heating, Refrigeration and Air Conditioning Engineers, 1992.

11 *Ventilation for Acceptable Indoor Air Quality, ASHRAE Standard 62-1989;* American Society of Heating, Refrigeration and Air Conditioning Engineers, 1989.

12 Garcia, C.E., Prett, D.M., Morari, M., *Automatica* (1987) 335.

13 Kuntze, H.-B., Richalet, J., et al., *Proceedings of 25th IEEE Conference on Decision and Control CDC'86, Athens, Greece, December 1986;* 1987.

2.5
Wireless and M-Bus enabled Metering Devices
DIETER MROZINSKI, *Siemens Landis & Staefa, Mühlhausen, Germany*

2.5.1
Introduction

The demand for systems which allow the automatic reading of consumption data of meters installed in residential buildings is increasing steadily. At the present time, more than one million household meters including heat cost allocators in Europe and more than eight million in the USA are remotely read. One can assume that this number will double every 2 years.

In some countries, for example, in the USA automatic meter reading is progressing more rapidly than currently in Germany, the reason being that meters are read and billed monthly. A monthly part payment for the consumption of energy, as is generally practiced in Germany, is not permitted in the USA by law. Only the amount that was metered can be billed. In order to limit outstanding accounts, meters are read and billed monthly. The profitability of automatic meter reading for this monthly billing mode is a matter of fact. The deregulation of the energy sector which is also taking place in the USA underlines the trend towards automatic meter reading.

Although automatic meter reading has not yet proved that it is economical as far as the annual reading rota and tariffs involved today are concerned when compared with the manual reading of meters and heat cost allocators, the financial advantages are obvious.

For energy suppliers and/or billing service providers, automatic meter reading can be regarded as a potential for improving services to the customer. Furthermore, journeys to real estates to record all user data can be avoided and intermediate billing when tariffs and tenants change can be carried out at very little additional expense.

In view of the structural changes in the energy sector which was introduced by deregulation, automatic meter reading will gain importance in the future. Free selection and, in all probability, a frequent change of energy suppliers by the user would be virtually impossible without automatic meter reading. Also, some energy suppliers abroad are considering the fact of billing not only according to consumption units alone but also energy and peak loads. This will justify the actual production and distribution costs to a greater extent than simple billing according to energy consumption units. In this case, the large data volume which is required for billing takes automatic data transfer of the meter readings for granted. In the first place, the user will profit from all this and, therefore, it is the user who will encourage it.

A major prerequisite required by automatic remote meter reading to penetrate the market is the rapidly progressing technological development. This applies both to the availability of electronic consumption-measuring devices and the development of transmission methods in local areas via data bus and radio. The availability of new communication methods such as GSM, SMS and Internet services will also play a major role in future for transmitting data at a favorable price from residential buildings to central billing centers and providing extra services for the user.

The transmission of data via radio inside an existing building will gain more importance in the future than the transmission of data via data bus. This is due in the first place to installation costs, which are high particularly in old buildings as a line must be laid for the data bus and this is only acceptable to the parties concerned when the building undergoes a complete overhaul.

Meters which are connected directly to telephone lines via a modem or which transmit their data directly by radio via communication satellites to a monitoring or billing center are not mentioned here; transmission paths of this kind are today only of interest for large-scale meters where consumption and other data are called up at short intervals, and this applies also to the foreseeable future.

Systems with electrical point-to-point connections where pulses of mostly mechanical meters are transmitted to a building central control system and added there are not the object of further observation. These processes do not fulfill today's requirements as regards economical and service-friendly installation, simple retrofitting, and reliable data transmission.

2.5.2
Benefits of Remote Reading

Many decision-makers are of the opinion that investments in a system with remote meter reading, including the required infrastructure and operating costs, greatly exceed the costs of annual manual reading. Therefore, in their opinion, automatic remote reading of meter data is not economical – either now or in the future. This viewpoint is based in the first place on the assumption that reading and billing modalities, which have become established today due to manual reading, will remain so in the future. This is, however, more unlikely which is obvious from subsequent applications (see below and Section 2.5.3.1). Also, this viewpoint does not consider the time that the user must sacrifice to be present during reading or even other requests that the user might have. In other words, the sums are wrong which is typical for some energy suppliers whose present situation in the market is either monopolistic or oligopolistic. Many energy suppliers at home and abroad, however, are starting to take a different approach.

Remote reading already provides today such decisive advantages to both the user and energy suppliers as well as property management that extra expenses for initial equipment can be accepted. The advantages that are given qualitative consideration here can also be regarded as financial advantages.

2.5.2.1
User

- No presence required for reading the various types of meter. The presence of the user – in most cases an employed worker – required during reading of the meter which, depending on the type of energy, takes place at varying times, is becoming a virtually unbearable strain. If users were to present a bill with an appropriate hourly rate for their presence, automatic remote meter reading from this point of view would be economically worthwhile.
- Automatic remote meter reading means that apartments no longer need entering. The protection of the private sphere is rated highly today, particularly in view of criminal offenses. This applies, of course, to criminals who pretend to be meter readers and not to the legitimate meter reader.
- Tenants and tariffs fluctuate frequently in buildings which are supplied by long-distance heating. In these cases, the energy supplier or the service provider makes do with an 'auxiliary bill' (eg, number of degree days) which can be far from the correct billing of actual consumption values. Intermediate billing or cost allocation which is based on actual consumption values is only possible at low cost with the aid of automatic meter reading; the user would profit from this correct form of billing. In the course of deregulation, the necessity for frequent meter reading will increase when the user can profit from frequently changing the energy supplier. Correct billing and the possibility of benefiting from the cost advantages of diverse energy suppliers are the driving forces here behind remote reading.
- In some countries – in the first place Denmark – users receive a statistical evaluation of their consumption behavior from the energy supplier. The users can thus alter their behavior and in turn influence their energy bill – on the condition, of course, that the supplier reads the meter once or several times a day.

2.5.2.2
Energy Supplier and/or Billing Service Provider

- Lower reading and billing costs with automatic remote reading when meters are read more than once a year (changes of tenants and tariffs, increased transparency for the user when meters are read more than once a year; thus improvement of competitiveness).
- Quicker receipt of money as all values are immediately available owing to automatic reading and one does not have to wait for so-called latecomers (user not at home on reading date). This applies, in particular, to a cost allocation where all user data of a real estate are required in order to bill correctly.

- Fewer billing errors – and thus costs – as automatic reading provides all data without multiple manual data reading and realization.
- Automatic meter reading provides energy suppliers with the potential to improve services. In view of deregulation, some energy suppliers are starting to do just that.
- Introduction of a tariff structure which is better aligned to the actual costs: for energy suppliers, fixed costs amount to at least 70% of the overall costs. Energy and/or medium costs (water, long-distance heating, gas, electricity, etc.) constitute only 30% of the overall costs. Almost 90% of the fixed costs are infrastructural costs (power plant/heating plant/reprocessing and distribution network). As peak loads determine the size and configuration of the infrastructure, thought should be given to a tariff structure which corresponds to this situation (with large consumers, namely 'special customers', this is already the case). Payment according to the work price alone (m^3, kW h) does not make allowances for the actual incurred costs. A complex price setting which assists in reducing peak loads is inescapable as the extension of production and distribution capacities are difficult to realize for both political and financial reasons. Appropriate tariffs which are more than just work prices mean more data, yet only automatic transmission can cope with more data.
- Meter monitoring for water and heat meters, which, owing to the medium they measure, are more susceptible to defects than other usage meters.

2.5.2.3
Owners and/or Property Management

- The owner contributes to the satisfaction of the user by investing in the automatic reading of meters. In view of the increasing number of empty apartments, this is a point of major importance.
- Quicker billing. This is an important factor when property management must make advance payments as far as heating is concerned.
- Correct and quick billing when tenants change and thus a better chance that the users who are moving will pay their bill punctually.
- Energy management especially for commercial undertakings and functional buildings.

2.5.3
Data Transfer via Data Bus

Standardized and proprietary bus systems used today must have reached three digital numbers and are increasing continuously. The reason is that there is no bus system at the moment which optimally covers the varying requirements of different applications as far as cost is concerned. Communication requirements are frequently only covered by a hierarchy of buses.

This book does not enter into the complex topic of the 'data bus' in detail. The advantages and disadvantages of the many house buses used today with which meter data can also be transmitted are contained in various publications. Despite the fact that the applications and requirements are still relatively transparent, it is highly improbable that a single standard will prevail on the market.

Many national, European, and international standardization committees (ICE, ISO/OSI, ICI, EIA, CEN, CENELEC) have concerned themselves with the topic of 'house bus' and come to the following results. The standardization suggestions are, to a certain extent, overloaded with requests. This means that efficiency suffers and it is not possible to say whether they will prove their worth in practice. On the other hand, proprietary solutions have prevailed in practice and have been adopted by a large number of industrial companies which are working together in groups.

The transmission of meter data – battery-operated meters and allocators predominate – makes demands of a specific technical and economical nature not only on the data bus but also on the meters themselves. Before the requirements under Section 2.5.3.2 are referred to, the requirements which are in demand today and in the foreseeable future will be considered briefly in Section 2.5.3.1.

2.5.3.1
Bus Applications of the Meter Sector and the Resulting Demands on the Data Bus

In most cases, particularly in old buildings, installation costs/effort involved in laying electrical cables of the meters when data buses are used are considerable. In order to justify these costs/effort, however, it is obvious that a bus system must be designed in such a manner that not only the consumption figures of heat cost allocators but also the data of the remaining meters (gas, water, electricity) can also be read.

- *Central meter reading*: Meter data are transmitted to a central control unit building. In the meantime, a number of European manufacturers are offering usage meters for water, electricity, gas, and heating, whose meter readings (no increments) can be read at the building central control unit via a bus in the form of a protected data protocol. Depending on the make, several meter readings are made available for billing consumption, for example, several consumption values of multi-tariff meters and other measured values such as peak and reactive loads, etc., which are used for more complex billing.
- *Remote meter reading:* This can be regarded as a type of central reading. The meter data for the building which have been collected are transmitted from the building central control unit via the public telephone network or via GSM or SMS to the billing center.
- *Parameterization:* Only meter data and parameters which are irrelevant as far as calibration is concerned can be altered via the data bus, eg, new code for data coding, command for tariff zone switching, new meter number, due date when meter readings can be saved for subsequent reading.
- *Control:* This means the transmission of commands to meters. An example is switching off or disabling of the supply by a stop valve in the meter or a switch.

- *Transmission of auxiliary measured variables:* Many of the meters supplied today have auxiliary measured variables which can be used for the optimization of control processes, in particular for heating systems. Also included is the transmission of data for monitoring and controlling supply networks (load management) as well as peak loads (see Section 2.5.2).

The meters must possess the required 'intelligence' to be able to communicate with the selected data bus. For mechanical meters with pulse output, interface modules are offered which can be used to couple these meters to the data bus.

2.5.3.1.1 Demands on the Data Bus and the Connected Components

The most important requirements for a data bus suitable for meter applications and connected components such as meters and data collectors are as follows.

Topology

The building structures and layout of the apartments vary considerably. As regards the laying of bus cables and the topology, there should be no restrictions, therefore, to avoid higher installation costs. Possible bus topologies are line, tree, and star.

Energy Supply of Electronics

The European approval authorities for meters demand that the energy supply of the calculators of usage meters does not depend on the possibility of supply by the data bus so that, in the case of failure of the bus, an uninterrupted metering of consumption is ensured. To fulfill this requirement at the moment, only a measuring device which is battery operated or supplied by the mains is possible.

Apart from electrical meters, the electronic usage meters and heat cost allocators which are offered today are battery-powered for reasons of cost. The capacity of the batteries integrated in the meters is adapted exclusively to the measuring electronics of the meters; the battery cannot, therefore, provide the electrical energy supply of the bus driver module (interface electronics to the data bus). This must be provided by the bus line itself or by a second additional line. In order for the cross-sections of the supply line to remain acceptable (<1 mm^2), the current consumption of the bus driver should be <5 mA.

Number of Meters Which Can be Connected to the Bus and Network Extension

The number of meters and heat cost allocators that can be connected to a bus must coincide with the requirements in practice. The following meter/allocator configurations are required per apartment: 5 heat cost allocators + 2–3 water meters (if necessary + 1 electrical meter + 1 gas meter) = 7–10 meters. More than 30 apartments in a building complex, possibly consisting of several buildings, means that up to 300 data terminals are connected to the bus network. Furthermore, maximum distances of 1000 m between the central reading/control station and the meters must be given consideration. Depending on the network topology selected in the building, the entire laid cable length may be many times this amount. This means that in view of the addressability and the electrical bus characteristics, it must be possible to connect a large number of meters and data terminals and to bridge distances a minimum of 1000 m between meters and a building central control unit – if necessary, using repeaters (amplifiers).

Addressing Meters
With most utility companies, administration and logistics (meter installation and replacement, reading and billing) are allocated the meter number (eight digits). In order to avoid additional problems for the service provider and/or the energy supplier, it should be possible to address the meters via the meter number. It is also desirable that the bus system can configure itself, ie, the system recognizes all installed devices and automatically creates and manages addressing tables when initially installed.

Supporting Data Types and Structures
The protocol should be open and extendible so that varying data types and structures (fixed and variable) can be supported for future applications.

Transmission Speed
With the above-mentioned applications, the transmission speed is not particularly critical but should nevertheless not be below 2400 bps, as otherwise the times for reading a large number of meters in a building would be too high to realize future applications. As a building can have meters with varying transmission rates, the software and hardware of the central unit must be designed for automatic speed detection.

Transmission Reliability
Taking into consideration that the laying of bus cables in the building can virtually not be affected (possible parallel laying with electrical lines), a high resistance of the data bus to inductive and capacitive field parasitic interference is required. An adequate transmission error recognition by the protocol with a Hamming distance of 4 is also required to ensure an acceptable residual error probability during transmission. This is a prerequisite for correct billing.

Equalization of the Electrical Potential Differences in Buildings
The electrical potential differences in buildings are considerable and vary as follows:

- inside an apartment, virtually no difference;
- between the apartments of a building, several volts (< 10 V);
- between several buildings of an apartment block: several tens of volts.

There is also a risk of a galvanic connection of the measuring electronics with the mains voltage (220 V_{eff}) which must not, however, affect the bus. This should be ensured by constructive measures or by altering the characteristics of the interface modules.

Overvoltage Protection

Protection is required to protect the devices connected to the house bus against possible destruction by overvoltage (eg, lightening). Such overvoltages may result in the destruction of electronic components. When connecting gas meters to the data bus there is a risk of explosion. For example, if a bus cable requires laying outside a building to interconnect several buildings, lightening protection must be provided. Well-proven circuits and processes are available from the telephone and broad-band cable networks.

If necessary, protection against 220 V overvoltage should be provided in the components of the bus system.

Short-circuit Protection

The risk of a line short-circuit in living areas is particularly high. Furthermore, short-circuits can occur in devices which, if they are not intercepted by constructive measures in the components, can affect the bus in such a manner that the localization of this problem in large apartment blocks is extremely difficult. Therefore, a short-circuit in the bus cable must not result in damage to the electronics of the devices connected to the bus. A short-circuit in the electronics of a device should not result in the failure of the entire data bus so that the defective device can be automatically identified and replaced in good time.

2.5.3.2
Available Data Buses for Meter Applications

For meter applications various bus systems such as M-Bus, EIB, Echelon, Batibus, and Euridis were used. The M-Bus was designed specially for the above-mentioned meter applications and has gained acceptance throughout Europe. Euridis has established itself in France for reading E-meters, but, it is hardly suitable for other meters owing to the high current consumption for data transmission and restricted addressing possibilities. The remaining buses mentioned are less suitable for meter applications. For this reason, we will only consider here the M-bus.

2.5.3.3
M-Bus

2.5.3.3.1 Overview
The M-bus (abbreviation for metering bus) was developed initially for central or remote reading of usage meters. Development was focused on:

- support of a large number of meters;
- bridging of long transmission distances at low cost;
- reliable data transmission;
- low hardware costs;
- low costs for planning of systems, installation, maintenance, and supplementation of meter equipment for buildings.

Fig. 2.5-1 Building central control unit OZW 10 (manufacturer: Siemens, Landis & Staefa)
1 Left key block
2 Display
3 Right key block
4 Memory key
5 Operating cards
6 Lock
7 Hole for sealable fastening screw

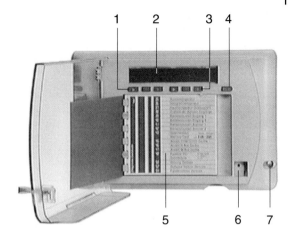

The M-bus is defined in the Standard EN1434-3. Only one communication master is permitted within the system (eg, OZW10 (Figure 2.5-1) with level converter WZC-P250 (Figure 2.5-2)).

All M-bus components comply with the EMC requirements of IEC 801 Part 2–6, severity level 3, and/or En 5008-1 and -2. The bus also complies with the German regulations for high-frequency emission (radiofrequency emission) as per DIN/VDE 0871, Part 20.

Communication is always started by the master. M-bus devices are periodically queried by a so-called master (building central control unit).

Fig. 2.5-2 M-bus level converter (manufacturer: Relay)

Tab. 2.5-1 Application examples

	Maximum distance (m)	Overall length of all lines (m)	Cable diameter (mm)	Number of M bus devices	Maximum transmission rate [1] (baud)
Smaller residential buildings	350	1000	0.8	250	9600
Larger residential buildings	350	4000	0.8	250	2400
				64	9600
Smaller accommodation	1000	4000	0.8	64	2400
Larger accommodation	3000	5000	1.5	64	2400
Town, district	5000	7000	1.5	16	300
Point-to-point connection	10000	10000	1.5	1	300

[1] Maximum cable capacity 150 nF/km

A two-strand twisted-conductor cable is used (eg, J-Y-ST-Y-2*2*0.8). Permissible cable routings are line, tree, and star topology as well as hybrid types. Ring topology is not allowed. A bus terminal is not required.

An expansion of the network and the maximum transmission speed are limited by the number of M-bus devices, suppressor circuits, cable routing, and cable types. The expansion of the transmission network can be subdivided by a so-called 'repeater' (eg, WZC-R250) into segments and extended virtually as required. Table 2.5-1 below contains simple application examples with a level converter.

Terms and Definitions
Access Methods
Bus access is based on the master/slave concept, ie, the meters connected to the M-bus communicate only when requested to do so by the master (polling). The master (eg, building central control unit) queries the slaves (meters); the meter (slave) whose address corresponds with the M-bus device addresses responds.

Transmission Type
The type of transmission is half duplex, ie, data can be transmitted in both directions; however, communication must be made in succession as it is not possible in both directions simultaneously.

Transmission Speed
The bus interface is designed for transmission rates of 300–9600 bit/s. The M-bus standard recommends transmission rates of 300, 2400, and 9600 bit/s.

Level Converter
The level converter forms the interface to the master (building central control unit) and the first bus segment. Depending on the type, 250 meters can be con-

nected to a level converter. When the entire bus network as regards length and meters which require operating is too large, repeaters must be used.

The input of the level converter can be either RS-232 or RS-485. The plug-in unit inserted in the level converter is important.

The output feeds an M-bus segment and thus has similar functions to the repeater.

Repeater

Repeaters are necessary to bridge greater distances and for connecting many meters to the M-bus network. The repeater transmits all data to the meters which are connected to the bus segment which it supplies. Therefore, there is a new M-bus segment at the repeater output.

The repeater amplifies and forms the signals again which could become distorted over greater distances. It also supplies the meters which are connected in the next bus segment with current. The electrical input of a repeater has the same features as a slave. The output of a repeater supplies an M-bus segment.

Expansion

The entire expansion of the bus system is limited by:

- the number of M-bus devices (slaves and/or meters) in the segment (Figure 2.5-3);
- distribution of the device in the bus segment;
- in the segment: resistance values of the used bus line (voltage drop of the bus line);
- required transmission speed; the bandwidth is limited to 9600 bit/s.

As these are the most important points, we will not enter into further details.

All M-bus devices which are connected to the same repeater or level converter belong to the same M-bus segment.

Every segment has its own repeater. Up to 250 data terminals can be connected to one repeater (segment) under worst-case conditions: all data terminals are at the end of a phase and all devices are supplied with current. This high number of terminals means that a bus installation can only consist of one segment which is connected to a master via a repeater.

Topology and Bus Installation Regulations

The M-bus operates in the network topologies star, line, and tree (Figure 2.5-4). Depending on the application, however, there is a 'preferential topology'. A ring topology should not be used in the M-bus networks.

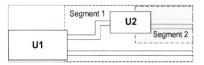

Fig. 2.5-3 Segment

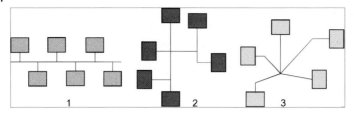

Fig. 2.5-4 Forms of topology. 1 Line; 2 Tree; 3 Star

A four-stranded telephone cable (J-Y-STY-2*2*0.8) is recommended as a transmission medium. It is cheap and easy to acquire. Two of the four strands are used for the bus; the other pair of strands are intended as a stand-by or can be used for another bus. The maximum distance between an end terminal and a repeater depends on several factors. The descriptions below provide something to go by. The entire cable length (all parallel switched lines) in one segment is 1000 m (capacitive load due to the cable: 160 nF).

When using a cable with thicker cross-section, greater distances can be bridged and/or more devices can be connected (compare both previous figures). This association can be seen in Figures 2.5-5 and 2.5-6. The following requirements and/or conditions are parameters for both figures:

- *Row 1:* Theoretical conductor length for equidistant distributed M-bus devices (eg, an M-bus device every 5 m). This is the upper configuration limit and should always have a conductor length of < 4 km.
- *Row 2:* Theoretical conductor length for all M-bus devices at the end of the line (worst case).
- *Row 3:* Same as row 2. It is assumed here that the communication to the bus should still function when a short-circuit has occurred in one of the devices (easy localization of short-circuit).

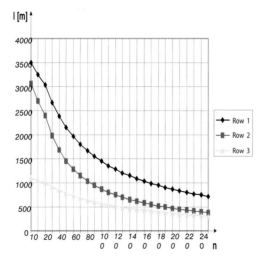

Fig. 2.5-5 Line lengths for conductor diameter 0.8 mm

Fig. 2.5-6 Line lengths for conductor diameter 2.5 mm

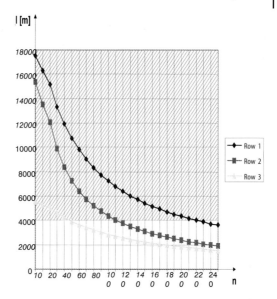

The M-bus therefore does not represent a network. Controllers are not required for 'routing functions' to increase the connection possibilities of data terminals to an acceptable extent as is to be expected in larger residential buildings. Using an 'unintelligent' repeater, widely branched bus structures with an extremely large number of data terminals can be realized (several thousand terminals as long as the required reply times allow it).

Physical Specification

The physical layer is a semi-duplex, asynchronous bit-serial transmission (UART protocol) with baud rates between 300 and 9600 bit/s. Every eleventh bit must be a logic '1'.

The transmission of master to slave is performed with the aid of voltage excursions. A logic '1' (mark state) is represented by a voltage of nominal 36 V and a logic '1' (space state) by a lowering of the voltage by 12 V to a nominal 24 V.

The transmission of data from the slaves to the master is implemented electrically so that no energy is taken from the terminals for this purpose. The terminals model the electrical current which is provided by the master (mains-operated). The bit representation of a message is coded via currents. In the case of a logic '1' (mark) a current of 1.5 mA is removed from the master by the terminal; a logic '0' (space) is displayed to the master by a current consumption increased by 11–20 mA caused by the slave (see Figure 2.5-7). The transmission of a space therefore causes a slight drop in voltage at the repeater.

Explanations of the symbols in Figure 2.5-7 and other electrical parameters of the system components are as follows: $U_{MU,M} = 36$ V, voltage at the master, idle level; $U_{MU,S} = 24$ V, voltage at master, transmission level; $U_{M,M} = 12$ V, voltage at meter, idle level; $U_{M,S} = 11.3$ V, voltage at meter, transmission level; $I_M = 1.5$ mA, supply

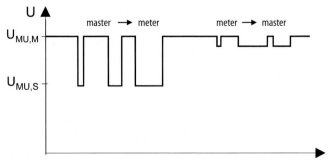

Fig. 2.5-7 Voltage change on M-Bus cable during data transmission

current, idle level; $I_S = 20$ mA, signal current (space level); $n = 1\text{–}250$, number of meters in bus system; $R_S = 440\ \Omega$, maximum safety resistor in each meter; $R_C = \ldots \Omega$, cable resistor; $R_M = 60\ \Omega$, maximum measuring shunt in the master; $R_{CON} = 2\ \Omega$, resistance of all connections.

Owing to the transmission in the response direction (slave to master) with current pulses due to a constant current drain in the bus transceiver (minimum 11 mA) and the voltage difference of 12 V in the opposite direction (master to slave), there is a high degree of insensitivity towards the effect of external interference (capacitive and inductive parasitic interference). Measures preventing interference suppression include, in particular, common-mode rejection, the galvanic decoupling of the repeater, and the terminals (constructive or by the optocoupler).

As a transceiver a customer-specific electric circuit was developed by Texas Instruments (TSS 721). Functions and characteristics of this integrated module are as follows:

- Communication speeds with baud rates of 300–9600 bit/s.
- Remote supply of data terminals. The interface module contains a voltage controller, which generates a controlled voltage for the processor of the data terminal of 3.2 V with maximum 500 µA. As the bus voltage when transmitting data by the master never drops to 0 V (36 V mark; 24 V space), the supply voltage for the electronics of the meters is always available at the output of the TSS 721.
- In case of bus failure and thus failure of the current supply of the terminal by the bus, the TSS 721 transmits a signal to the pin 'PF' which can be used as an interruption for the processor of the terminal to save the data in the EEPROM. Furthermore, with the signal at the output 'VS', a FET (field effect transistor) can be controlled which switches from the current supply by the bus to battery supply. This is important with usage meters which are normally battery-powered. As long as the data bus indicates no fault, they will be supplied by the bus and in case of failure of the bus, supply is continued without interruption by the battery.
- Reverse voltage protection. The connections of the bus lines at the inputs of the transceivers are random. This is achieved by the series-connected bridge rectifier in the TSS 721 at the bus inputs.

- In case of a short-circuit in the terminal, a resistor at the input of the bus (pin: BUSL1 and BUSL2) ensures that the bus is not short-circuited. A bus short-circuit would mean extensive fault locating. This series resistor (430 Ω) ensures that the bus is not short-circuited despite this device fault and still continues to function. The master can detect this short-circuit by querying all devices and demand the replacement of the defect device by giving a warning.
- Protection against overvoltage. The switching circuit is protected against overvoltage to a maximum ±50 V. Furthermore, a circuit as described in Section 2.5.3.1.1 (subsection 'Overvoltage Protection') can protect the transceiver TSS721 also against higher voltages of maximum 220 V. This circuit can ensure protection against explosion, as long as it is installed outside potentially explosive rooms, in front of each gas meter in the bus supply line. The series resistor (430 Ω) is the same as that which provides protection against short-circuits described above.

As a repeater, three marketable versions, mini, midi, and maxi repeaters, for 60, 120, and 250 terminals, respectively, are available. Their major difference is the number of connectable terminals. The repeaters supply the terminals with a maximum current of 300 mA. The repeaters are protected against short-circuit and overvoltages of maximum 220 V. They have a display against overloading (if, for example, too many terminals are connected). A master comprises a repeater and a building control unit which are interconnected via an RS 232 interface. In case meters are only read occasionally, the central control unit must not be installed permanently.

A summary of the most important data is given in Table 2.5-2.

Protocol (Link Layer and Application Layer)

As the bus is not a network and thus requires no connection, layers three to six of the OSI models are empty. In addition to the physical layer, only the link layer and the application layer are equipped with functions.

The protocol is based on the international standard IEC 870-5 which defines transmission protocols for telecontrol systems. This protocol uses asynchronic se-

Tab. 2.5-2 Summaries of the most important data

Transmission speed (bit-rate)	300–9600 bit/s
Master:	
Logic '1' (mark)	36 V nom.
Logic '0' (space)	24 V nom.
Slave:	
Logic '1' (mark)	1.1 mA nom.
Logic '0' (space)	11–20 mA nom.
Supply by the repeater	36 V/24 V; 300 mA
Supply by the bus driver (for processor of device)	500 µA

Tab. 2.5-3 Telegram formats

Individual character	Short frame	Control frame	Long frame
E5h	Start 10 h	Start 68 h	Start 68 h
	C Field	L Field	L Field
	A Field	L Field	L Field
	Test sum	Star 68 h	Start 68 h
	Stop 16 h	C Field	C Field
		A Field	A Field
		CI Field	CI Field
		Test sum	User data (0–252 bytes)
		Stop 16 h	Test sum
			Stop 16 h

rial bit transmission. Synchronization is carried out by start/stop bits at each character. The IEC 870-5 has three different integrity classes I1, I2, and I3. The integrity class is a dimension for the relationship between the rate of non-identified incorrect telegrams and the bit error probability of the transmission. For these integrity classes measures are defined for detecting errors in transmission. For the M-bus protocol, format class FT 1.2 is provided for the transmission of the meter reading (vertical parity bit per character with longitudinal parity total for the message). This selected format has a Hamming distance of four.

The M-bus protocol represents a subset of the IEC-870 protocol. The M-bus protocol can be extended by further functions which this IEC standard offers.

In the format class FT 1.2 there are three different telegram formats which differ from one another by a specific character at the beginning of each data block. The telegram formats are as follows (Table 2.5-3).

Individual Character
Serves to confirm messages (E5h).

Short Frame
The flag (10 h) is followed by the so-called control field and/or function field, the address field, the test sum, and the stop character. The short frame serves to initialize the slaves (normalize) and for requesting the slaves to transmit data which are not time-critical. Initialization serves the purpose of synchronizing the transmitter and receiver of data so that a loss or multiplication of messages is avoided during the subsequent data transfer without having to replace an ACK or NACK after every message (increase in efficiency).

Long Frame
This contains in addition to the fields of the short frame also the identification field CI and the data field with maximum 252 characters. The long frame is vari-

able. A special form of the long frame is the 'control frame', which does not transmit data.

The meaning of the individual fields in the messages is as follows:

- The C field controls the data flow and monitors the correct sequence of messages and avoids the loss or the multiplication of messages. It states whether messages are to be transmitted or received and what priority the messages have.
- The address field permits the addressing of 250 terminals. The addresses 254 (FEh) and 255 (FFh) represent group addresses. The data field is used to extend the addressing space using a meter number of nine bytes (see below).
- The data field is variable and can comprise maximum 252 characters.
- The L field states the length of the data block increased by the number of control characters.
- The application layer was defined in the TC 176 (standard for heat meters). The standardization suggestion is based on the IEC 870-5 standard which defines only the data in the response direction (meter to master) and no data blocks which cause a switching of tariff in the meter.

The data structure in the opposite direction can be seen in Table 2.5-4.

The date field contains further definitions as regards the types of meter (heat meter, gas meter, water meter, electricity meter) which contain (1) the physical units of measurement in question, (2) the type of values, ie, instantaneous values, mean values, minimum values, peak values or meter readings, and (3) to which meter (which index) the data refer if multi-tariff meters are being used. These points will not be referred to in detail; the relevant standards contain all further necessary information.

The application contains the automatic configuration of an extended meter/M-bus installation. This means that the meters with their meter number of 9 bytes identify themselves to the master without a network administrator having to allocate meter numbers and physical bus addresses. This means that an additional meter can be installed in the building at all times or a meter replaced without having to extend or supplement address tables. The energy supplier and/or the service provider can keep to his meter numbers on which his entire logistics and administration are based – for communicating as previously with the meter.

The M-bus standard suggestion for the TC 176 contains at the moment only a part of the IEC 870-5 standard. From the point of view of the physical layer of the

Tab. 2.5-4 M-Bus Data telegramstructure

Identification No.	Manufacturer	Version	Medium	Access No.	Status	Signature	Data
4 bytes	2 bytes	1 byte	1 byte	1 byte	1 byte	max. 243 bytes	max. 243 bytes

M-bus, there are no obstacles or restrictions which prevent adopting further parts of the IEC 870-5 standard, with regard to the link and application layer. The decision for such an extension depends in the end on the product providers who select this bus for their applications.

2.5.3.3.2 Application Examples
With the M-bus, two types of system are applicable. These are: M-bus systems inside the house (in-house systems) and M-bus systems outside the house. Hybrid forms are also possible.

In-house Systems
The in-house systems (Figure 2.5-8) are limited to an apartment house and the bus expansion is usually only slight (eg, consumption billing of real estate).

Systems Outside the House
For systems outside the house (Figure 2.5-9), several independent real estates are usually connected which may be situated far apart (eg, long-distance heating systems).

For greater distances, repeaters are required. The site of the repeater depends on the future extension and local conditions (supply the repeater, accessibility).

Hybrid Systems
In hybrid systems (Figure 2.5-10) several real estates (with in-house bus inside the real estate) are interconnected. Several repeaters may be required.

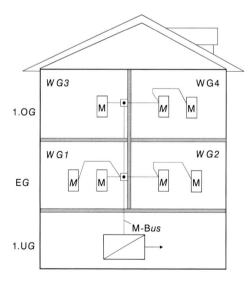

Fig. 2.5-8 In-house system. M-bus device (eg, Megatron 2 (heat meter), Memotron 2 (heat cost allocator), pulse adapter, Volutron 2 (electronic water meter)); repeater or level converter (WZC P250); level converter with connection to central unit (OZW 10) and/or PC; distribution point; apartment No.; basement/ground floor/upper floor. Source: illustrations by Siemens Landis & Staefa

2.5 Wireless and M-Bus enabled Metering Devices | 145

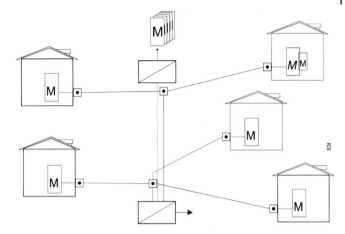

Fig. 2.5-9 Systems outside the house. Details as in Figure 2.5-8

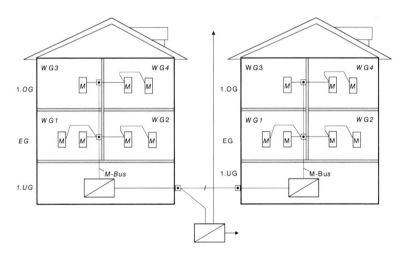

Fig. 2.5-10 Hybrid system. Details as in Figure 2.5-8

Configuration Examples

Below are some configuration examples provided by Landis & Staefa (Figure 2.5-11). The M-bus system is used for consumption cost billing and remote monitoring of district and long-distance systems and apartment houses. The M-bus central unit (OZW10) is the central device at the M-bus. It communicates via the M-bus with the connected usage meters and controllers.

The M-bus central unit can be directly connected with a PC or via a modem. The operating and alarming software and special user programs are installed on the PC.

The following M-bus devices can also be connected:

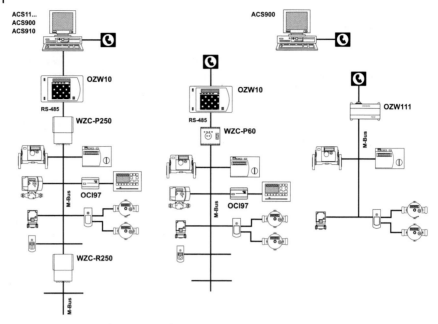

Fig. 2.5-11 M-Bus application examples

Meters:
– Heat meter SONOGYR® energy.
– Heat meter SONOGYR® WSD.
– Heat meter MEGATRON®2 WF.
– Usage meter via pulse adapter AEW21.2.
– Usage meter via pulse adapter RELAY PadpulsM1.
– Electronic heat cost allocator MEMOTRON® WHE21.

Controllers:
– Long-distance heating controller RVD2.
– Long-distance heating controller SIGMAGYR® RVP97.

2.5.3.3.3 Overall Assessment of the M Bus

The M-bus was designed in the first place with meter applications in mind (mains and battery-operated). It represents an extremely efficient and cost-favorable solution for this sector and gives consideration to the special operating conditions and costs of usage meters.

Both technical data and standards exist. A meter manufacturer can choose whether to aim at an integrated solution (measuring functions and protocol in one processor) to reduce manufacturing costs or whether he wishes to complete the protocol procedures in a separate communication processor available on the market.

2.5.4
Data Transmission via Radio

The first larger commercial application of data transmission of household meter readings via radio was introduced in the mid-1980s in the USA. In Germany, legal conditions were not created by the regulating authorities until the beginning of the 1990s when an ISM band of 433 MHz was also released for this application. At the same time, the restriction that at least several users must be mobile for radio applications was also withdrawn.

Since that time, development has accelerated rapidly. Several manufacturers of consumption-measuring devices offer devices and systems for measuring consumption, renowned semiconductor manufacturers offer suitable cost-favorable radio modules, and users, energy suppliers, and property management value the advantages of automatic meter reading via radio. In the meantime, a further frequency band (868 MHz) was released in Europe for the 'home automation' sector and thus also for meter reading. The use of the 868 MHz band contains, however, some worthwhile regulations which did not exist in the 433 MHz band, thus improving the reliability of data transmission.

Unfortunately, a standard does not yet exist for the 433 MHz band for meter data transmission via radio. Since the beginning of 2000, however, intensive efforts are being made to introduce such a standard at the European level in the 868 MHz band.

2.5.4.1
Data Transmission and Selection Process

Since the introduction of data transmission of meter readings via radio, four different data selection and transmission processes have become established, the advantages and disadvantages of which are described briefly below.

The various processes mainly came into being because manufacturers were searching for a compromise for their customers between the comfortable reading of data and subsequent further processing on the one hand and the device and system costs on the other. Each company prepared a solution for this compromise based on its own interests and conditions.

Data Transmission Process
Bi-directional In this mode a dialog takes place between the terminal (meter) and the data collector. The data collector may be a building central control system or a portable data collector. The data collector demands by a command a specific meter (direct addressing) to transmit its data. The receipt of the data in both directions is acknowledged and, in the case of incorrect transmission, data are re-transmitted. Data can also be transmitted to the meter (eg, for parameterization) during bi-directional transmission.

Although this solution is equivalent to the form of communication on the data buses, it has the following disadvantages for the transmission of meter readings via radio when compared with the following solutions:

1. Each device, ie, meter and data collector, must possess a receiver in addition to the transmitter. A receiver is technically more complicated and thus more expensive than a transmitter.
2. The receiver of a meter must 'check' continuously or at short intervals whether it is receiving data. This 'checking' requires a relatively high amount of current (receiving circuit and microprocessor). As most of the meters and heat cost allocators are battery-operated, this mode rapidly reduces the service life. There is also a risk that the battery will rapidly become discharged when many messages at this frequency are received by the meter, even if they are not intended for it. To ensure the measured value, the use of a second battery to supply the radio module with energy with this solution is worthwhile despite additional costs.

With regard to the high costs of a receiver in each meter as well as high current consumption, it should be considered whether this process be used for battery-operated meters.

Uni-directional with wake-up signal This process is distinguished by the fact that the meters are 'woken up' with a radio signal and subsequently transmit data. As this wake-up signal is received by all meters installed in the building, one must, of course, prevent the relevant meters from transmitting their data simultaneously, thus causing a data collision. This can be achieved if every meter transmits its data in its own fixed, allocated time slot or if transmission is stochastically distributed in succession. In the case of a data collision, the repetition algorithm must be chosen in such a manner that the probability that at least one of the repetitions was successful is extremely high. However, with this process also, there is a risk of the battery becoming prematurely discharged.

Uni-directional (stochastical transmission) This process is based on the fact that data only flows from the meter to the data collector and not vice versa and that for billing purposes it suffices when a valid meter reading per day is transmitted. With this process, the meters transmit their data stochastically and several times a day. The reason for transmitting data several times a day is in case a data telegram of a specific meter collides with the data of another meter or an external data terminal. As the telegrams are very short (millionths of a second) and the same data are transmitted several times (five to six times), it can be expected that at least one of the telegrams transmitted by a meter will be received by the data collector.

Such a process assumes that a data collector is already installed and is ready to receive data at all times. Of course, data cannot be transmitted from the collector to the meter.

However, this type of system has the following decisive advantages:

1. The meters do not require a receiver. This means considerable cost reduction.
2. The strain on the battery is reduced considerably as there is no receiver. A transmitter which is only switched on when data is transmitted requires only a

portion of the current of a receiver. This also applies when it is only switched on from time to time to determine whether it is receiving a message (eg, 'Request to transmit').

In view of the fact that most meters, especially heat cost allocators and water meters, are battery-operated and these batteries are supposed to last for 10 years, this process has a considerable advantage compared with those mentioned above.

Selection Process

As far as the form of transmission is concerned, the following possibilities are available in the 433 MHz band:

1. fully installed data collector in a building central control system for uni- or bi-directional operation;
2. fully installed data collector on every floor of a building;
3. hand-held with wake-up pulse;
4. hand-held with wake-up pulse in time window.

These four basic systems, which are described below in detail, allows, the reading of a meter without having to enter the apartment. Only the first-mentioned process provides the possibility of reading the data at all times from a control or EDPC (electronic data processing center) using a wide-area network such as the telephone network.

Despite an expected standard for the transmission of meter readings in the 868 MHz band, one can assume that the process used in the 433 MHz band will in future be applied in the 868 MHz band and that in all probability others will be added.

Fully installed data collectors in a building central control system The main features of this concept compared with the concepts described below are

1. the remote transmission of meter readings via a higher network (PPT or radio network/GSM, etc.) or a data processing central unit,
2. the use of devices with uni- or bi-directional transmission with the above-mentioned advantages.

As there is no need to travel to a real estate to read meters, the advantages are more than obvious:

- The data for all meters can be called up at all times from the billing center without additional costs worth mentioning to complete billing regardless of whether there was a change in tenant or tariff. In case of allocation billing, all user consumption data are required for correct intermediate billing. For both these cases, this can only be regarded as economical when the data are transmitted to the EDPC 'fully automatically'.
- There is a possibility of billing at shorter intervals than yearly without incurring additional costs.
- Meters can be monitored for their functional reliability.

- Meter readings can be called up for statistical purposes.
- Meter readings can be continuously evaluated for facility management.

For this process, however, an infrastructure must be installed in the building. It consists, for example, of several antennas distributed throughout the building and connected to a building central control unit via a cable (signal receiver, data collector and bridge to the public network). The antennas are distributed throughout the building in such a manner that they are within radio range of the devices. The distance between the individual meters and the nearest receiving antenna is often only 20–25 m owing to the low transmission power and the high signal damping in iron-insulated buildings.

Instead of simple antennas, it is also possible to install floor repeaters. These repeaters comprise an antenna, a receiver, a transmitter, a processor, and memory for buffering data. The repeaters should be installed within radio range of the meters as is the case with the antennas. Meter data are transmitted from repeater to repeater by radio to the building central control unit. This means that cables which are required for simple antennas can be omitted.

Collector per floor With this process, meter data which are transmitted via radio from 'intelligent data collectors' which are installed within radio range – usually on every floor – are received and stored. The data must, however, be read from the collector with a terminal, eg, a PC, so that they can then be further processed. On the other hand, mains connection is not required for remote reading and cabling of the selection process described above is not necessary.

As is the case with the above-mentioned selection process, the process described here is also suitable for meters which communicate uni-directionally.

Read-out device with wake-up signal With this portable read-out device, the 'reader' would drive to the real estate in question. The read-out device (frequently referred to as 'hand-held') requests with a command all meters installed in a real estate to transmit their data. The request usually comprises a wake-up pulse which is received by all devices which then transmit their data at different times. The advantages and disadvantages of this process were described in the section 'Data Transmission Process'.

Read-out device with wake-up signal in the time window The disadvantage of the simplified process which has just been described, namely that owing to other devices transmitting in the same frequency band which do not belong to the system the meters can be continuously requested to transmit the meter readings and thus in turn the batteries of the meters are discharged more rapidly, can be reduced by allowing the meters only to be ready for receiving the wake-up signal within a specific time, eg, on a certain day in the year when the 'reader' drives to the real estate with the portable read-out device to take the meter reading. The synchronization of the service with the time window must be ensured, otherwise the meter readings are available for the next programmed reading interval. It

2.5 Wireless and M-Bus enabled Metering Devices | 151

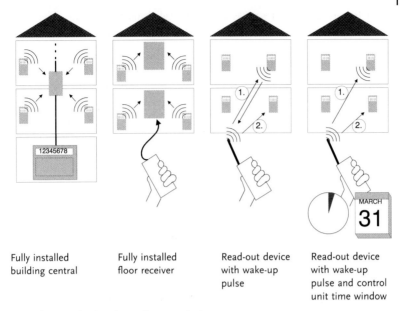

| Fully installed building central | Fully installed floor receiver | Read-out device with wake-up pulse | Read-out device with wake-up pulse and control unit time window |

Fig. 2.5-12 Read-out methods using radio transmission

should also be taken into consideration that, over longer periods of time, not all 'clocks' of the meter still operate synchronously (accuracy and response to temperature changes of the quartz). Corresponding reserves must also be taken into account when defining the time window.

The various devices are illustrated in Figure 2.5-12.

General Requirements with Regard to the Transmission of Meter Readings via Radio

In order to ensure reliable communication via radio, functions must be provided in the hardware and software which take the special conditions in the radio environment into consideration and which go beyond the usual wire-bound transmissions. Restrictions must also be accepted as regards the application which do not apply when using a data bus. Nevertheless, radio transmission is highly suitable for the simple transmission of meter readings as long as the required technical measures given consideration in the conception of the system take the special conditions into account.

Transmission Performance

One of the most important restrictions which should be observed as regards the transmission of meter readings is the fact that most of the consumption-measuring devices are battery-operated. Not only is the energy which is available for the transmission of data within the service life of the device restricted, but also the maximum current which can be drawn over a short period of time from a long-term battery is limited (<30 mA for normal lithium batteries, maximum 1 A h). This restricts the transmission performance to ~0 dB m (1 mW). This performance is also limited by the antenna which can be located in a housing of a measuring device and by the approved maximum output power in a defined band.

Interferences in the Radio Channel

First one must be aware that one is not alone in a radio channel. There are also other 'legal' radio applications in the same frequency band and in the closer environment which can interfere with the transmission of one's own data. Here are some examples:

- Noise is extremely high owing to the varying applications in the same building and applications in adjoining buildings. As the free field damping between buildings is usually low compared with the damping in one's own building, external devices of other buildings may under certain circumstances be received better than the devices belonging to one's own system inside the building. Hence the superimposition and destruction of messages are more probable than when data are transmitted on a data bus. A receiver can usually only determine after receiving a complete telegram whether it has received an external message or not. Whilst receiving a weak telegram from an external system, it is blocked (capture effect) for the receipt of telegrams which originate from telegrams from its own system. Suppression of the so-called 'capture effects' should be aimed at. This means that a receiver should recognize as quickly as possible a telegram belonging to the system, even when it is receiving a telegram with a somewhat weaker level from the adjoining building. The receiver should turn its attention to the newly received telegram and not analyze the weaker one as it was probably already destroyed by the stronger telegram.
- There is a high interference potential in the ISM band of 433 MHz due to an amateur radio in the same band for which extremely high transmission outputs are allowed (maximum 10 W output power).
- In radio communication there is no controlled sequence in which messages are transmitted. Checking whether a transmission has just taken place, such as CSMA/CD processes used with buses, can be ruled out as one will nearly always hear a radio message from the neighborhood and external applications but would never be able to transmit oneself.
- The transmission conditions depend greatly on the physical state of the building and are therefore not easy to forecast. Depending on the manufacturer and application, the transmission processes (physical and link layer) can vary considerably.

Transmission Reliability

Meter readings required for billing purposes must be reliably transmitted. As the interference level with radio transmissions, as mentioned above, is extremely high, errors must be clearly recognized. A Hamming distance of >6 should be aimed at.

2.5 Wireless and M-Bus enabled Metering Devices | 153

Fig. 2.5-13 Application example: apartment house (Source: Siemens Landis & Staefa)
1 Heat cost allocator MEMOTRON® WHE22
2 Pulse adapter AEW22.2
3 Multiple antenna ATW20.2
4 Radio read-out central unit OZW20
5 Coaxial cable (eg, CT100)
6 Supply transformer AC 230/24 V, 10 VA
7 Modem

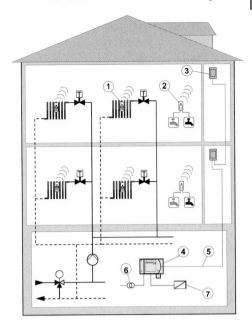

Application Example with a Stationary Receiving System in the 433 MHz Range

After the 433 MHz ISM band in the amateur radio band of 430–440 MHz was released, many buildings were equipped with meters readable by radio. Almost all read-out processes which were described above exist. A great disadvantage, however, is that the various systems differ greatly from one another not only in the system architecture but also as regards the individual layers defined as per the OSI model and especially as regards the 'physical, link, and application layers'. The experience which users have gained with the various system architectures can be regarded as positive with the exception of the above-mentioned restrictions.

A system representing all the others is explained briefly as follows (Figure 2.5-13). It comprises uni-directional communicating devices which transmit data six times a day to an installed receiving system at stochastically distributed time intervals. The receiving system comprises antennas which are installed in the stairway throughout the building and are connected via a coaxial cable with a building central control unit. The consumption figures are collected in the building central control unit where they can be read out via an RS 232 interface per PC, directly by a modem or via the PTT network. An interface is also available to transmit data to a memory card (memory RAM card).

Radio Solutions in the 868 MHz Band

More and more applications are to be found in the ISM 433 MHz band, which causes a high channel load. As the legislators have not stipulated any regulations for this band, one must fear that it will become increasingly more difficult to

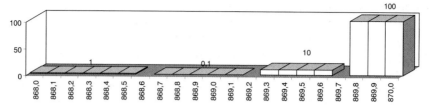

Fig. 2.5-14 Maximum duty cycle in %

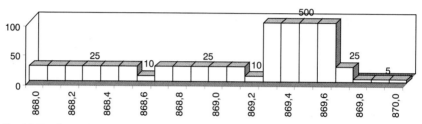

Fig. 2.5-15 Output in mW

guarantee reliable transmission when applications and installations increase. It is therefore worthwhile changing to another band, namely the SRD (short-range devices) in the 868 MHz range. In this range, not only is the output considerably limited but also the duty cycles are stipulated.

Figures 2.5-14 and 2.5-15 show both the transmission rate and the stipulated channel duty cycle.

In the 868 MHz band, the use of 600 kHz in a band is possible. Nevertheless, higher requirements are placed on the tolerances of the frequency-determining structural elements.

The output restriction, however, permits that distances in the building of ~ 25 m distance with normally built walls can be bridged. However, owing to the higher frequency compared with the ISM band of 433 MHz the damping is ~ 10 dB higher, which must be given consideration due to a higher output line of the transmitter in the device. The output is nevertheless very low in this band, so that the emitted energy has no negative effect on the health of humans. The emitted output is equivalent to approximately one thousandth of that of a mobile phone.

Standards Referring to Transmission of Meter Readings

Although the period between releasing this band to the present day is very short, an outline for a standard within the TC 294/WG5 has been agreed. It is, however, too early to go into this matter in detail. Such a step would only impede further activities of this committee. We would, however, like to make the following remarks:

- The consumption-measuring devices should be able to communicate with other components such as mobile data acquisition devices, stationary receivers, data collectors, or system network components.
- For the measuring devices, we have assumed that they can remain in operation without changing the battery for the entire service life (calibration interval) of 3–12 years. For the remaining stationary components such as mobile read-out devices one can reckon with shorter service lives.
- At the moment, five types of telegram are planned which contain on the one hand both uni- and bi-directional communication and on the other hand both portable and stationary systems for receiving data from the meters. The block length is either fixed or variable depending on the type.
- For the above-mentioned various types of telegram, various chip rates are planned. For stationary receiving systems a chip rate of 32 kHz will prevail.
- A 6-bit code (3 of 6 Code, ie, 4 data bits as a 6-bit word) is used. This provides an excellent common-mode rejection and allows suppression of the 'capture effect'.
- The IEC870-5-1/2 will be used as link layer.

Application Example with a Stationary Receiving System in the 868 MHz Band

Whereas in the application example mentioned above in the 433 MHz band it is assumed that a receiving system must be installed in the stairway-comprising antennas which are interconnected by cable, the system described here (Figures 2.5-16) comprises a network of 'intelligent' floor receivers which communicate with one another via radio. The intercommunication of the individual floor receivers is bi-directional. Data which individual measuring devices transmit are stored in the floor receiver which is most favorably located for reception and then forwarded to the next one in the floor receiver. After a certain length of time, the same data are

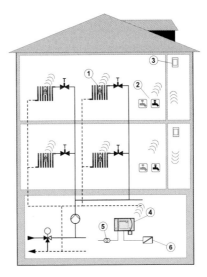

Fig. 2.5-16 Application example: apartment house (Source: Siemens Landis & Staefa)
1 Heat cost allocator MEMOTRON®
2 Electronic water meter VOLUTRON®
3 Floor receiver
4 Central unit
5 Supply transformer
6 Modem

stored in each of the individual floor receivers. All floor receivers form the receiving network. The floor receivers are battery-operated so no cabling and/or wiring is required for either data communication or the current supply.

The communication of the measuring devices is uni-directional, ie, the measuring devices transmit their data unbidden to the receiving network comprising the floor receivers. The transmission of the measuring device is synchronized with the 'temporary receiving window' of one of the floor receivers. Only in this way can the current consumption of the receiver be minimized so that battery operation is possible.

An important feature is that the network 'configures' itself and the devices also make themselves known to the network. This is the only way in which the technical expenditure for the installation of a system can be reduced to a minimum (easy configuration).

One of the installed antennas can be connected to a central control unit. The central control unit can then transmit the data via the PPT or via GSM to an EDPC. For a smaller detached house or apartment house the values stored in a 'floor receiver' can be requested via a hand-held terminal without having to enter the building.

2.5.5
Future Prospects

As already mentioned, it is highly likely that the transmission of meter readings will gain popularity owing to its extensive advantages. Certainly, many other 'home electronic functions' whose data transmission takes place inside the building via radio will without doubt be offered in conjunction with the Internet. It is to be expected that these sectors will finally merge. Attempts at standardization which are to support this merging are already in full swing.

2.5.6
References

1 SCHRÖDER, H., *Elektrische Nachrichtentechnik*, Band 1; Berlin-Borsigwalde: Verlag für Radio-Foto-Kino-Technik.
2 STALLINGS, W., *Data and Computer Communications*, 3rd edn; New York: Macmiillan, 1991.
3 DAVID, K., BENKNER, T., *Digitale Mobilfunksysteme*; Stuttgart: Teubner, 1996.
4 MOULY, M., PAUTET, M.-B., *The GSM System for Mobile Communications*; Lassay-les-Châteaux: Europe Media Duplication, 1993.
5 BARTEE, T.C., *Data Communications, Networks and Systems*; Indianapolis: Howard W. Sams, 1985.
6 BORIES, C., *Beschreibung des M-Bus*; Paderborn: Universität Paderborn, 1998.
7 FÄRBER, G., *Bussysteme*; Munich: Oldenbourg, 1987.
8 GABELE, E., KROLL, M., KREFT, W., *Kommunikation in Rechnernetzen*; Heidelberg: Springer, 1991.
9 *Data Sheet TSS 721*; Texas Instruments Deutschland, 1993.

10 *IEC 870-5-1: Telecontrol Equipment and Systems, Part 5, Transmission Protocols, Section One – Transmission Frame Formats*; 1990.

11 *IEC 870-5-2: Telecontrol Equipment and Systems, Part 5, Transmission Protocols, Section Two – Link Transmission Procedures*; 1992.

2.6
Sensors in HVAC Systems for Metering and Energy Cost Allocation

Günter Mügge, *Viterra Energy Services, Essen, Germany*

2.6.1
Introduction

The registration of the individual energy consumption for building heating is very common. In several countries in Central Europe the registration and account of the heating energy have a long tradition and is partly legally prescribed (eg, in Denmark, Germany, and Austria). In systems for heating, ventilation, and air conditioning (HVAC) of buildings, sofar the energy costs have been registered and accounted for individually in only a small section. A Directive was issued by the European Community [1] which recommends the individual accounting of these energies. This recommendation is based on the realization of the fact that the individual registration of the energy consumption and the consumption-based account has a positive influence on user behavior and so makes a contribution to energy saving [2].

Another aspect of the consumption-based accounting of the energy costs is often to satisfy the feeling of justice of the users. It is a particularity of measuring energy consumption that the interest is not to measure momentary quantities (such as temperature or flow) but to obtain a time-integrated quantity. This time integration will often be done inside the actual measuring instrument. Sometimes it is already implicitly included in the measuring principle.

With the registration of the energy consumption we have to differentiate between a tariff account (metering) and an allocation of the energy costs (submetering). The technical solutions for these two functions are sometimes very similar. With the registration and accounting of heating consumption the principle of allocation is applied in general. It is not necessary to have an absolute accuracy. Instead of measuring accurately the energy consumption (eg, with calibrated heat counters), auxiliary variables are then registered (eg, with heat cost allocators), which are proportional to the energy consumption sufficiently within the respective allocation unit.

2.6.2
Possible Implementations of the Energy Allocation

The registration of the heat consumption can be done with very different methods depending on the considered thermal subsystem and the implemented principle of measurement. Figure 2.6-1 shows schematically a heated building and possible measuring points for the energy registration. Energy is supplied to the heated building, shown here as fuel energy. Usually the fuel consumption is measured during delivery. In the case of gas heating or district heating this measurement will be done continuously. The fuel energy is transformed into thermal energy while burning and will be supplied to the several apartments via radiators over a pipe network. In this example every apartment has its own distribution network. In this case the consumed heat can be measured with a 'classical' heat meter, which means that volume flow and water temperature will be measured. Then the heat consumption is given by

$$Q = \int_t \rho c \dot{V}(\vartheta_S - \vartheta_R) dt \tag{2.6-1}$$

where ρ=density of water, c=specific heat capacity of water, $\dot{V}$=volume flow rate, ϑ_S=supply temperature, and ϑ_R=return temperature.

The same technical measurement principle will be used for the measurement of the building heat consumption with district heating, although in general we have metering with building-wise measurement but submetering or heat cost allocation with measurement of the consumption of an apartment.

Subsequently the heat energy will be transported to the several radiators and transmitted to the rooms. The momentary heat emission of a radiator is determined by the excess temperature of the heating medium (in general water),

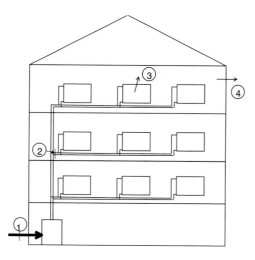

Fig. 2.6-1 Energy flow in a heated building. (1) Fuel supply to the boiler; (2) heat supply to an apartment; (3) heat emission of a radiator; (4) heat loss to the environment

which means the difference between the surface temperature of the radiator and the room temperature. This is given by

$$\dot{Q} = \dot{Q}_0 \left(\frac{\Delta\theta}{\Delta\theta_0}\right)^n \tag{2.6-2}$$

where $\dot{Q}$=heat emission, $\dot{Q}_0$=design heat emission, $\Delta\theta$=mean heating medium excess temperature, $\Delta\theta_0$=design value of the mean heating medium excess temperature, and n=emission exponent.

Usually the emission exponent is about 1.3. The fundamental principle of the common heat cost allocators (HCAs) is based on this law. With registration of the surface temperature of the radiator and time integration, the display of the HCA shows a value proportional to the emitted heat of the corresponding radiator.

The heat emitted by the radiators of an apartment flows due to transmission through the walls and windows and air renewal, at first to the outside air but also to neighboring apartments flats which are heated to lower temperatures. The resulting heat flow is proportional to the respective temperature differences. The heat loss to the outside is given by

$$\dot{Q} = \sum_j [(UA)_j (\theta_i - \theta_o)] + \beta c \rho V(\theta_i - \theta_o) \tag{2.6-3}$$

where U=heat transmission coefficient, A=area of the building component, θ_i=indoor temperature, θ_o=outdoor temperature, β=mean air renewal rate, ρ=density of air, c=specific heat capacity of air, V=heated room volume, and j=number of the building part.

This is also valid for the heat loss to the neighboring apartments with different temperatures (and heat gains from them). Heat cost allocation systems which work on the basis of Equation (2.6-3) are called degree-day systems. Degree-days are the time integrals of the difference between the indoor and outdoor temperatures. If the thermal properties of the building component and the air renewal rate are known then the heat loss can be calculated according to Equation (2.6-3). Usually these values are not known. Therefore, degree-day systems use the area of the apartment as a weighting factor. These systems are more indicative of the thermal comfort than the heat consumption.

2.6.3
Allocation of Costs for Air Conditioning

A big share of the operating costs of modern office buildings is due to the air conditioning, which often exceeds the heating costs. In addition, a more accurate accounting of these costs would lead to a higher transparency, and it could be assumed that a consumption-dependent allocation of these costs will force a more economical user behavior (as it is the case for heat cost allocation).

A fundamental difficulty is the great diversity of the different air conditioning systems. The registration of the energy consumption can be done only partially as with heating systems. In air conditioning systems the energy is not only transported with water as in heating systems but most of the time with air. Whereas the energy transported with water could be determined with heat meters (as was the case for heating), there do not exist any low-price allocators to measure the energy content of the air. Therefore, allocation of air conditioning costs will be realized first in those systems where the energy is transported exclusively or mainly with water.

However, the energy transported with air could be registered principally with physically similar systems.

$$\dot{Q} = \rho c \dot{V}(\theta_o - \theta_s) \tag{2.6-4}$$

where θ_o = outdoor temperature, θ_s = supply temperature, $\dot{V}$ = air flow rate, ρ = density of air, and c = specific heat capacity of air.

An exact registration of the volumetric air flows is comparatively complex. Well-known measurement principles include the measurement of the differential pressure at an orifice plate or at a venturi and the determination of the velocity profile of the air flow with an anemometer [3, 4]. An economic application of these methods for consumption entry has not yet been reported. A large part of air conditioning systems operate with constant volumetric air flows. In these cases the registration of the volumetric air flows would not be necessary, but it could be calculated with design values. In every case the development of an instrumentation system for the registration of the energy consumption remains a complex task.

2.6.4
Heat Meters

2.6.4.1
Principle of Measurement

Heat meters measure simultaneously the volume flow rate through a pipe system and the corresponding supply and return temperatures. Based on these measurements, the heat consumption in this pipe system will be calculated according to Equation (2.6-1) in the calculator and shown by the display. Whereas mechanical measuring instruments were also used in former times, nowadays only electronic heat meters are used in practice [5]. The signals of the temperature and flow sensors are digitally analyzed in a microprocessor of the calculator (Figure 2.6-2). Different measurement principles and sensor types are used for both temperature measurements and flow measurements [6–8].

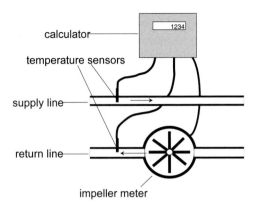

Fig. 2.6-2 Measurement principle for a heat meter with impeller flow meter

2.6.4.2
Temperature Sensors [5, 9, 10]

For temperature measurements, resistance temperature sensors made of platinum are used nearly always today [5]. The precious metal platinum is particularly suitable owing to its high long-term stability. The sensors used in heat meters usually have an electrical resistance of 100, 500 or 1000 Ω at a temperature of 0 °C. The electrical resistance increases almost linearly with rising temperature. At 100 °C the resistance of a Pt100 sensor (ie, $R(0\,°C) = 100\,\Omega$) rises to 138.5 Ω. Whereas in the remaining HVAC technicques, eg, for ambient temperature measurement, Pt100 sensors are usual, the most commonly used sensors for heat meters are Pt500. They have the advantage of a stronger measuring signal but also a higher sensitivity against electromagnetic perturbative fields. Owing to the shorter conduit lengths compared with, eg, ambient temperature sensors, in HVAC systems this aspect is less crucial here.

The actual sensors are implemented mostly in thin-film technology. They consist of layer resistances applied on a ceramic substrate and are arranged in a metallic protective pipe (Figure 2.6-3). The sensor element might have a connection head, at which the cable connections are attached to the calculator. However, the sensor elements are frequently connected inseparably with the cable. An important design feature is that the actual sensor has a heat dissipation over the protective pipe and the cable is as small as possible.

The sensor elements can be built with a sensor pocket or directly into the supply or return pipe. Direct installation has the advantage of better heat transfer to the sensor. This requires special precautions to turn off the piping during the installation. The installation is favorable in a ball valve [5].

Of course, different measurement principles are also possible. The use of thermocouples or NTC sensors (electrical resistances with a negative temperature coefficient) would be possible. However, these measurement principles have no importance for application in heat meters practice.

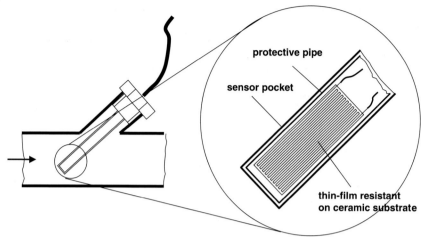

Fig. 2.6-3 Temperature sensor installed in sensor pocket

2.6.4.3
Flow sensors

2.6.4.3.1 **Mechanical Flow Sensors** [5–8]
The measurement of the volume flow can be done using different measurement principles. Most often heat meters for heat cost allocation measure the volume flow mechanically, eg, with single-jet or multi-jet impellers (Figure 2.6-2). Another design of a mechanical flow sensor is the so-called Woltman counter. However, this flow sensor is used only with larger nominal flow rates (> 15 m^3/h). Mechanical flow sensors will give a good proportionality between the flow and the rotation speed of the impeller over a far measurement range. To measure the volume flow, the rotation must be scanned. For this purpose different methods are used. Possibilities are the inductive, conductive, capacitive or optical scanning or ultrasonic scanning. The obtained signal is an impulse rate proportional to the rotation frequency or a modulated high-frequency signal.

2.6.4.3.2 **Static Flow Sensors**

2.6.4.3.2.1 **Ultrasonic Flow Sensor** [5–8, 11]
Today the most important design for static flow meters is the ultrasonic flow meter. The measurement principle is to lead an ultrasonic signal through the flowing heating medium, often in both directions (Figure 2.6-4). The mean velocity of the fluid and thus the flow can be determined from the modification of the signal running time. The signal is led diagonally or axially through the heating medium. The velocity profile in the tube differs depending on the flow form (Reynolds number). This must be considered in the calculation of the mean flow rate. The ultrasonic transmitter and receiver are usually piezoelectric semiconductor components.

Fig. 2.6-4 Ultrasonic flow meter

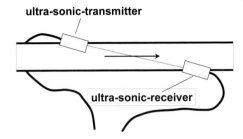

2.6.4.3.2.2 Magnetic-inductive Flow Sensors [5–8, 11]

With the magnetic-inductive measurement principle, a magnetic field is generated perpendicular to the fluid flow (Figure 2.6-5). Because the flowing fluid is a moving electric conductor, an electrical voltage is induced in the fluid and can be measured with two electrodes (Faraday's law). Flow meters based on the magnetic-inductive measurement principle are used today especially for large volume flow rates (eg, district heating stations). For low flow rates mechanical or ultrasonic meters are still much cheaper.

2.6.4.3.2.3 Thermal Flow Sensors [3, 9]

Although other principles are known, they have not been used in practice so far for the registration of heat consumption, eg, thermal flow meters. This principle reduces the flow measurement to a temperature measurement.

A measuring method for the flow in a heating circuit is based on heat exchange between the supply and the return pipe. The medium temperatures before and after the heat transfer are measured in each case. With the known heat transfer conditions the flow and the heating energy used can be determined from the measured temperatures. A fundamental disadvantage of this procedure is the necessity for a direct proximity of the supply and return pipes, which excludes applications in many installation situations (Figure 2.6-6).

A similar caloric method is to measure the rise in temperature in the medium which is produced by defined heating with an electrical heating resistor (Figure 2.6-7). The additional energy consumption for the execution of the measurement is unfavorable, because battery operation is practically excluded.

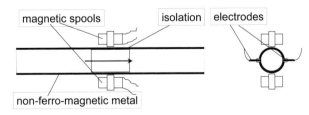

Fig. 2.6-5 Magnetic-inductive flow meter

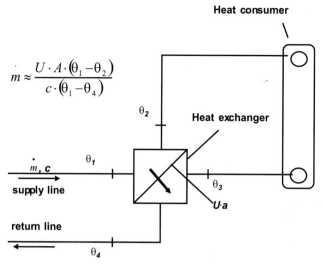

Fig. 2.6-6 Thermal flow meter. Principle of heat exchange. $\dot{m}$, Mass flow; U, heat transmission coefficient; A, area of heat exchange; c, specific heat capacity; θ, temperature

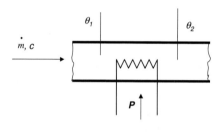

Fig. 2.6-7 Thermal flow meter with electric heating. $\dot{m}$, Mass flow; P, electric power; c, specific heat capacity; θ, temperature

2.6.4.4 Application

Heat meters can be calibrated and have a good measurement accuracy. However, the using of heat meters for heat cost allocation is not appropriate in every case. The radiators of a given apartment receive heating water often from several distribution lines. In those cases it would be necessary to install a heat meter at every radiator, which is not feasible. In the situation shown in Figure 2.6-1, where all radiators of an apartment are supplied through the same distribution line, the use of a heat meter is appropriate. If the apartment is heated with floor heating, a heat meter is the only possible allocation device, because heat cost allocators cannot be used with this heating system.

2.6.5
Heat Cost Allocators (HCAs)

2.6.5.1
Principle of Measurement

The fundamental principle of HCAs is the time integration of the measured radiator temperature. The European standard EN 834 [12] defines: 'The...displayed reading is the approximate value of the time integral of the measured characteristic temperature of the radiator or the time integral of the temperature difference between the radiator surface and the room.'

2.6.5.2
Evaporative Heat Cost Allocators

HCAs based on the evaporation principle were applied already in the 1920s. This measuring instrument integrates the radiator temperature through the evaporation of an organic liquid. The assembly of this instrument type is shown in Figure 2.6-8. The HCA has a heat-conducting metallic lower part mounted on the radiator and a glass tube filled with measurement liquid is fitted on this lower part. The lower part and the tube are covered by the upper part, which is usually made of plastic. The upper part normally has a window or is totally transparent, so that the level of the liquid is visible. Furthermore, the upper part incorporates a scale so that the displayed scale values can be read.

The most common measuring liquid is methyl benzoate, an organic substance which often occurs naturally as a flavor or fragrance and is used in perfumes and soaps. The evaporation of the liquid is very dependent on the temperature (Figure 2.6-9). Residual evaporation with a turned-off radiator and nonlinearity of the evaporation are the most important sources of errors for this type of HCA. However, those errors are leveled by the fact that they occur similarly for all users.

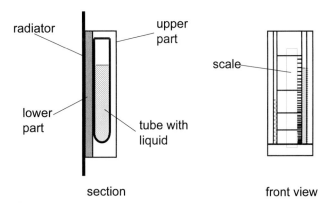

Fig. 2.6-8 Evaporative heat cost allocator

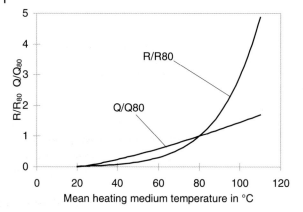

Fig. 2.6-9 Meter characteristic of an evaporative HCA. R, Counting rate; Q, heat emission; *index 80*, heat medium temperature = 80 °C

2.6.5.3
Electronic Heat Cost Allocators

2.6.5.3.1 Construction and Installation

In electronic HCAs (Figure 2.6-10), the tube with measuring liquid is replaced by an electronic sensor. A second sensor for the measurement of the room temperature may also be used. The conversion and processing of the sensor signals are done by a small calculator situated in the housing. The calculated values are normally displayed with a liquid crystal display.

HCAs based on the evaporation principle and also electronic HCAs are mounted on a representative point of the radiator. For most radiator types this mounting position, where the surface temperature is similar to the mean operating heating medium temperature, is located in the middle of the radiator's length and at three-quarters of its height [13]. Other types of electronic HCAs are known which measure the radiator temperature with two sensors. These HCAs measure the supply and the return temperature and calculate the mean heating medium

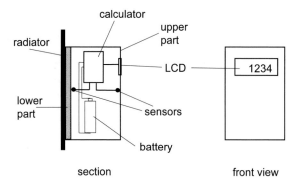

Fig. 2.6-10 Electronic heat cost allocator

temperature. A disadvantage of this system is the high expenditure caused by the installation of the signal lines.

2.6.5.3.2 Temperature Sensors [9, 10]

In HCAs the most commonly used sensors are NTC resistors or semiconductor sensors. Both sensor types are inexpensive. Whereas the NTC resistors use the temperature dependence of the electrical resistance, the semiconductor sensors work with the temperature dependence of the p–n junction.

A few common types of HCA, which measure the heating medium temperature at two positions (near the supply and return connection) and were designed some years ago, use thermocouples. This type of sensor was probably selected because the measured variables (temperature differences) produce proportional voltages, which can be measured easily by a simple analog electronic circuit. Today's HCAs possess a microprocessor and operate digitally, so that nonlinear characteristics can also be analyzed.

The use of semiconductor sensors permitted an interesting development of an electronic HCA, viz, the radiator sensor could be integrated together with the processor and most other electronic components into an ASIC (application-specific integrated circuit), so that the number of necessary electronic components could be substantially reduced.

2.6.6 Reading

2.6.6.1 Visual Reading

Usually the reading of the displayed consumption is done by the service personnel of the billing companies. On the billing date the displayed values of all meters are read and possibly sometimes some service work is done (eg, changing the tubes of evaporative HCAs). If the meters are situated in the private premises of the users (which is the case for HCAs), some problems occur. On the one hand, the inhabitants of an apartment may consider access of service personnel to their premises to be a violation of their private sphere, and, on the other hand, sometimes this access is very difficult owing to the circumstances of modern life (eg, single households, absence of the residents).

2.6.6.2 Automatic Meter Reading

2.6.6.2.1 Cable Network
A high-tech solution is to give the meters the capability to communicate. A very simple possibility is a pulse output for the meter. This is possible already with mechanical water meters. According to the configuration the pulses represent a

determined volume (or quantity of heat) and are transferred to a remote display or a data concentrator. Better communication with less interference can be realized with a serial bus system. The most frequent communication technique for heat meters is the so-called metering bus (m-bus). Modern electronic heat or water meters have interfaces according to this standard.

2.6.6.2.2 Radio Transmission

A great disadvantage of remote reading with a pulse signal line or an m-bus system is the wiring cost. This is a problem especially for communication with HCAs. Central systems for heat cost allocation which were sometimes used previously were not successful for this reason. A consequent further development is the integration of a small radio transmitter into the allocators or meters. This technique is offered by several billing companies on the European market. The radio transmitter sends a very low-power signal especially at 433 MHz frequency a few times a day to a receiver, which is installed in a central location in the building. The received data are stored in a data concentrator which can be read by the service personnel after the billing date. In principle the transmission of the data over a public telecommunication network to the host computer of the billing company is possible.

2.6.7
Outlook

The future development of registration systems for the energy consumption in HVAC systems will be determined by different influences:

- modifications in civil engineering (improved thermal insulation, low-energy building standard);
- new heating systems (eg, wall panel heating);
- increased comfort requests by users;
- progress in electronics and communications technology;
- increasing occurrence of building automation.

The trend towards improved thermal insulation of buildings, which can be seen in the new building sector, leads to reduced heating energy consumption and cost. In these buildings a large user influence on heating energy consumption remains, which makes registration still necessary. However, for reasons of measuring economy, it is desirable that the costs of heating registration are adapted to the decreasing energy costs. This concerns both the costs of the registration system and the related services (reading).

New heating systems (which are possible owing to the reduced heating loads) often cannot be equipped with conventional registration systems. This may be due to the often very low temperature of the heating surfaces such as in wall panel heating. The use of HCAs is not possible here. It depends on the future market share of these new heating systems whether this situation will make new registration systems necessary.

Today the technical development of HCAs and heat counters is following the rapid progress in electronics and in communications technology. A typical example is the increasing miniaturization of electronic components, in particular of integrated circuits. It is possible to integrate most of the electronic components, even the temperature sensor, into an ASIC as described above. Thus both a more compact design of the devices and a lower energy consumption (smaller battery capacity) are obtained.

Today we observe the rapid development of mobile telephones within the area of the communication technology. There is also a trend for the wireless transfer of data, both in larger networks and in the local communication between a PC and its peripheral devices. Wireless transmission paths are also increasingly used within other areas of building engineering, eg, for the regulation of heating systems. It is very probable that there will be a spin-off for registration systems. However, it is still open in which form a mutual influence will take place. The distribution of suitable frequency bands and the development of standards for different applications will play a large role in this field.

The increasing importance of building automation, especially in the home sector, can affect the progress of registration devices. It is very probable that there will be a short-term development of standardized interfaces between the systems for registration of energy consumption and building automation systems. Such an interface could be integrated, eg, into the data concentrator. The direct coupling of the registration devices to the communication bus of the building automation system will also be possible, eg, HCAs could use the same radio protocol as the superordinate building automation system.

2.6.8
References

1 Council Directive 93/76/EEC. *Official Journal 237.* 22/09/1993 p. 0028–0030.
2 HILBERG, M., Einfluß des Verbrauchsverhalten auf den Heizwärmeverbrauch, in: *Jahrbuch 1994*; Düsseldorf: VDI, 1994, p. 90.
3 FIEDLER, O., *Strömungs- und Durchflußmeßtechnik;* Munich: Oldenbourg, 1992.
4 NIEBUHR, J., LINDNER, G., *Physikalische Meßtechnik mit Sensoren*, 4. Auflage; Munich: Oldenbourg, 1996.
5 ECKERT, H. D., *Wärmezähler-Handbuch*, 2nd edn.; Frankfurt/Main: VWEW, 1994.
6 ADUNKA, F., *Handbuch der Wärmeverbrauchsmessung;* Essen: Vulkan, 1991.
7 STUCK, D., in: *Handbuch der Heizkostenabrechnung*, 4th edn., KREUZBERG, J. (ed.); Düsseldorf: Werner, 1997, Chap. 4, p. 179.
8 ADUNKA, F., in: *Heizkosten Richtig Erfassen und Verteilen*, 2nd edn., (KUPPLER, F. (ed.); Ehningen/Böblingen: Expert, 1993, Chap. 2, p. 58.
9 MEJER, G. C. M., VAN HERWAARDEN, A. W. (eds.), *Thermal Sensors*, Bristol: Institute of Physics Publishing, 1994.

10 Ruhm, K., in: *Handbuch der Industriellen Meßtechnik*, 5th edn., Profos, P., Pfeifer, T. (eds.); Munich: Oldenbourg, 1992, Chap. 8.

11 Bonfig, K. W., in: *Handbuch der Industriellen Meßtechnik*, 5th edn., Profos, P., Pfeifer, T. (eds.); Munich: Oldenbourg, 1992, Chap. 6.2.

12 *EN 834: Heat Cost Allocators for the Determination of Room Heating Radiators – Appliances with Electrical Energy Supply*; Brussels: European Committee for Standardization, 1994.

13 Zöllner, G., Bindler, J.-E., *Montageort für Heizkostenverteiler nach dem Verdunstungsprinzip*; in: Heizung Lüftung/Klima Haustechnik 1980, p. 195, Düsseldorf: VDI-Verlag.

2.7
Pressure Sensors in the HVAC Industry
Yves Lüthi, Rolf Meisinger and Marc Wenzler,
Siemens Landis & Staefa, Zug, Switzerland
Kais Mnif, *Motorola, Toulouse, France*

2.7.1
Introduction

The requirements and solutions for pressure sensors in the heating, ventilation, and air conditioning (HVAC) industry are in some cases different to those in other industries, eg, the automotive and medical equipment industries. The reason is that, for example, a specific car model is a clearly defined product which will be manufactured in a series and the responsibility for the design and the manufacturing is undisputed.

The situation for HVAC systems in buildings is different. Any system is more or less a unique solution for a specific building (of course using standard components). Additionally, the responsibility for the design of the system, the design of the control concept, the programming of the software, the installation of the system and mounting of the sensors, the wiring, and finally the commissioning (see Glossary) of the system is shared between many companies. A typical course is shown in Table 2.7-1. It is not surprising that often problems arise between these interfaces.

Often it is not possible to define in the planning phase where exactly the sensors will be located in the plant. This is usually done in an ad hoc manner and depends on the place which is locally available and the accessibility.

We can therefore summarize the recommendations for air pressure sensors in the HVAC industry as follows:

- Very low pressure range. Typically from 0–50 to 0–3000 Pa with an accuracy of 5% full-scale for low-end and 2.5% full-scale for high-end applications. In some cases there are additional recommendations for the zero point stability.
- Very simple installation. The staff who install the equipment are not specially trained on sensor issues. Normally they install the air handling unit (see Glossary) and the air supply tubes in the building and are afterwards also responsible for the mounting of the sensors and actuators on the previously installed equipment.
- Independence of mounting orientation. This point is important for pressure sensors because on most pressure sensors the zero point is influenced by the

Tab. 2.7-1 The responsibility for the installation of an HVAC system is shared between many parties. Often the supplier of the building automation system (BMS) is also the supplier of the controller sensors and actuators. This means it is a bundled business. Finally, the supplier of the BMS is also responsible for bringing the system to full operation

Job	Responsibility
Specification of the HVAC system	HVAC consultants
Specification of the control concept	Measure and control consultants
Installation of the air handling unit and tubes	Supplier of the air handling unit and tubes
Supply of the BMS	BMS supplier
Programming of the control concept	BMS supplier
Supply of the controllers	BMS supplier
Supply of sensors and actuators	BMS supplier
Installation of sensors and actuators	Supplier of the air handling unit and tubes
Wiring	Electro installer
Commissioning (put into operation)	BMS supplier

orientation via gravity. It would be a great drawback if the mounting orientation (eg, horizontally or vertically) had to be dictated. The often heard suggestion of making a special 'zero-button' on the device, which should be pressed after installation, is not practicable; see the previous point and also because there is no power available at this time. The power will not be switched on before the commissioning of the system. A second 'visit' to every sensor affected would add additional cost.
- Rigid stable housing. It cannot be assumed that they are mounted with special care. Also, it can happen that the sensors remain for days on the building site before they are installed.
- Sensitivity to dirt. During the installation process of an air handling facility, much dust and dirt can accumulate in the system. A common procedure is to run the fans at full speed after installation in order to clean the inside of the tubes. If there are sensors in the system, which are sensitive to dirt, then they will be adversely affected by this process.
- Wiring mistakes (exchanged wires) must not damage the sensor. Often wiring errors are not found until the so-called point test is made. A sensor could therefore be wrongly connected for hours.
- Lifetime >7 years. During this time, the specifications of the sensor have to be taken for granted. Any exchange or recalibration in the field would lead to large costs.
- Any information required for configuration should be clearly visible without a de-installation of the sensor or opening of the case (eg, range settings).
- Low cost, due to strong price pressure, particularly in the so-called OEM (original equipment manufacturer) business (compare Section 2.7.2.2).
- In addition to these recommendations, the following typical operating conditions apply:

- Temperature   0–70 °C
- Humidity   0–95% RH
- Voltage   24 V AC
- Output signal   0–10 V DC

Note: For sensors which are operating on a bus, the voltage and signal recommendations may differ.

The application of sensors in systems characterized by small pressure differences has proven both difficult and excessively expensive. Certain types of inexpensive pressure sensors, such as silicon diaphragm sensors, have sufficient sensitivity (see Glossary) to resolve pressure changes of <1 Pa, due to the possibility of fabricating extremely thin membranes. But the main disadvantage is their unreliable performance in the field because of drift problems.

There are, however, several important benefits: the cost factor resulting from the micromachining technique, excellent long-term repeatability due to perfect thermal matching of the piezoresistors with the membrane, fast response time, no flow through, and the continuous progress in micromachining promises improvement potential.

Accordingly, their application to HVAC systems is highly desirable. The main efforts are therefore aimed at the development of an effective compensation of the drift problems.

Here we present a solution based on silicon sensors. After a discussion of the main applications of pressure sensors we will show the extensive advantages offered by silicon sensors. Subsequently we demonstrate the concept together with the development steps needed to obtain an appropriate sensor. Emphasis will be put on a thorough description of the technical hurdles that had to be overcome.

2.7.2
Main Applications and Market Requirements

2.7.2.1
Filter, Fan Monitoring, and Pressure Control

Each of the three applications we describe here needs a pressure sensor with a linear output signal. Figure 2.7-1 shows their usage in a typical HVAC plant.

1. Pressure sensors can be used to monitor air filters. When the filter becomes clogged with dirt, then the pressure drop over the filter will increase. This can be used to monitor the degree of pollution and to trigger an alarm. Clogged filters can lead to contamination of the air and a bad indoor climate.

   The pressure range is typically 500 Pa but can vary with the size of the air handling unit.

2. The correct working of the fan is often monitored by a pressure measurement but the rotary speed of the fan detected by an angular rate sensor is also common. Typical ranges are from 0–100 to 0–2000 Pa.

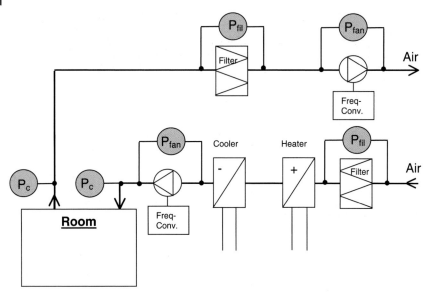

Fig. 2.7-1 Typical applications for pressure sensors with a linear output signal in an air handling facility. P_{fil}, filter monitoring; P_{fan}, fan monitoring; P_c, pressure control of a fan

3. The third application is pressure measurement in the tube. Here the sensor will be used to control the fan speed via a frequency converter. The goal is to reach a specific pressure at the outlet of the air handling unit.

2.7.2.2
Variable Air Volume

The purpose of aeration systems based on the so-called variable air volume (VAV) is to produce and control a comfortable climate by means of a supply of fresh air and the simultaneous removal of stagnant air. The control principle is based on varying the quantity of air supplied and air removal. The critical success factor is the accurate measurement of the flow, and therefore the quality of the sensor.

The main reason for the use of VAV is an energetic one: from this point of view it is better to lower the room temperature by increasing the (cool) air supply instead of cooling the supplied air.

In most of these applications the flow measurement is led back to a determination of a pressure difference. A specially designed pressure probe consisting of an orifice or a Prandtl tube generates a pressure difference bearing in a well-known relation to the volume flow:

$$Q = c\sqrt{\Delta p} \qquad (2.7\text{-}1)$$

where Q is the flow, Δp the pressure difference, and c a parameter that depends on the cross section of the duct and pressure probe design.

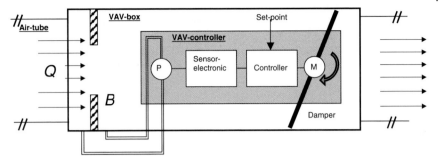

Fig. 2.7-2 VAV box with damper actuator. The volume flow Q creates a pressure drop over the orifice B. From this drop in pressure, the volume flow can be calculated and compared with the given set-point. Depending on the difference between set-point and actual volume flow, the damper position is changed

The application of VAV is different from those described previously. A sensor for VAV applications is in most cases built directly into a damper actuator (Figure 2.7-2). The purpose of such a device is to measure and control the air flow in a tube.

The majority of these VAV damper actuators are sold to OEM customers who mount the devices on their VAV boxes. After that, every single box is checked and tuned in, on a special facility. Any warm-up time needed by the sensor in these boxes will add additional cost for OEM customers. Therefore, it is important to keep the warm-up time short as possible.

The measurement of this pressure difference now has to fulfill extremely stringent requirements. This is perhaps surprising for building comfort applications, where the reference is the well-being of humans, widely known as 'bad sensors'. The main reason that makes the measurement a tricky task lies in the fact that small causes have large effects.

The supply of fresh air should be accompanied by the simultaneous removal of the same amount of stagnant air. Slightly different amounts of inflow and outflow accumulate in time to give a perpetual pressure difference from the neighboring aeration zones and thus circulation. Such a shortcoming would hardly escape human perception in the case of moving paper sheets and slamming doors.

The required pressure measurement accuracy, therefore, follows mainly from the demand that the difference between inflow and outflow does not exceed a certain value. The consequence of this demand will turn out to be salient for our subsequent efforts. The result is shown in Figure 2.7-3. The system is highly sensitive around zero pressure difference. A commonly used pressure range of 100 Pa, for example, yields a required accuracy of 1 Pa. This is really a small value; imagine that already the mere weight of the sensor membrane yields an offset of a multiple of 1 Pa!

Further, the accuracy of 1 Pa has to be maintained over temperatures ranging from 0 to 70 °C. The large temperature range is due, on the one hand, to the refrigerant and, on the other, to the damper actuator being in proximity of the sensor. Hence a suitable temperature compensation has to be provided.

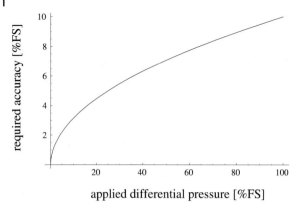

Fig. 2.7-3 Required accuracy of the pressure difference measurement as a function of the applied pressure difference. The units are given as percent of full-scale (FS)

2.7.2.3
Summary

The main recommendations for the different applications are listed in Table 2.7-2. The numbers give only a general impression and they can vary particularly between low-end and high-end applications.

Tab. 2.7-2 Summary of the recommendations for the different applications of air-pressure sensors

Application	Range (Pa)	Output	Accuracy	Mounting independence	Time constant (τ)/ warm-up time (w)
Filter monitoring	0–500	Linear	<±5% FS	Strongly desired	$\tau < 10$ s $w < 15$ m
Fan monitoring	0–250 to 0–3000	Linear	<±5% FS	Strongly desired	$\tau < 10$ s $w < 15$ m
Fan pressure control	0–250 to 0–3000	Linear	<2.5% FS	Strongly desired	$\tau < 0.1$ s $w < 2$ m
Pressure control between rooms	±50 ±100 0–100	Linear	<2.5% FS	Strongly desired	$\tau < 10$ s $w < 15$ m
VAV controller	0–300	Square root	Zero point <0.5% FS or < (0.5% FS + 5% of value)	Essential	$\tau < 0.1$ s $w < 2$ m

2.7.3
Silicon Pressure Sensors

A philosophical question: What are the intellectual abilities of a totally isolated brain? Not much ... Our brain, as an intelligent machine, is unable to learn, and therefore to think and to build a reasoning, without having access to outside information coming from our five senses (or sensors).

Philosophy aside, electronic controls, as 'intelligent' systems, also require outside information to be more responsive and sensitive to their environment. Electronic sensors, and particularly silicon-made sensors (Figure 2.7-4), appear to be the right response to this requirement, since they are constructed in the same material and by using the same techniques as the other surrounding electronic components.

Moreover, the combination of microelectronics and micromachining provides highly integrated products. Control and monitoring systems then become easier to design and more reliable since a decreasing number of components and interconnections are required.

2.7.3.1
Pressure Sensors as Microelectromechanical Systems (MEMS)

MEMS encompasses a broad range of applications. Pressure sensors are one of the most common and best known applications of MEMS, compared with other lesser known devices such as microprobes and micromotors.

In addition to the obvious advantage of down-scaling systems from several centimeters to less than 1 mm, MEMS technology brings to mechanics what microelectronics has brought to electrical engineering – batch processing and very low-cost manufacturing. On the other hand, silicon has excellent mechanical characteristics with a very low hysteresis (see Glossary) and an elastic modulus that is comparable to that of steel [1].

Micromachining could be performed either on the surface of the silicon substrate or deeper in the silicon volume (bulk micromachining). Figure 2.7-5 shows a pressure transducer (see Glossary) cell in both techniques.

In general, the pressure is sensed through a stretchable diaphragm. The resulting deformation is then converted into a measurable signal, the nature of which

Fig. 2.7-4 Silicon-made sensor. Reproduced by courtesy of Motorola

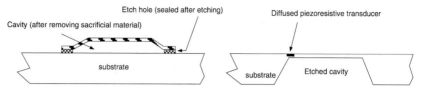

Fig. 2.7-5 Left, surface-machined pressure cell; right, bulk etched pressure cell

depends on the sensing technique utilized MEMS technology offers a set of tools to create cost-effectively silicon diaphragm structures by using semiconductor batch processes based on lithography (see Glossary) and chemical processing.

2.7.3.1.1 Bulk Micromachined Pressure Sensor

Bulk micromachining is usually performed by means of chemical etching. The cavity is generated with an appropriate anisotropic (see Glossary) etchant such as potassium hydroxide (KOH), and the diaphragm thickness obtained is in the range 15–25 µm.

Anisotropic etching consists in corroding one orientation of the 26-face silicon crystal much faster than the others, unlike isotropic (see Glossary) etching where the silicon is etched at the same speed in all directions, providing vaulted structures.

The diaphragm thickness could be controlled simply via the immersion time in the KOH bath, but for better process control, an electrochemical etching (ECE) process is preferable. Figure 2.7-6 shows the principle.

ECE is a batch process where the whole wafer is immersed in the KOH bath, but only the P-type side (see Glossary) is exposed to the etchant solution. When the reverse biased PN junction is reached, the N-type silicon (see Glossary) forms a passivation layer that stops the etching.

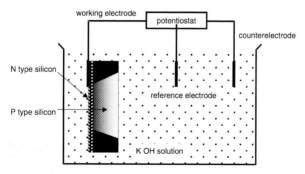

Fig. 2.7-6 IECE process

2.7.3.1.2 Surface Micromachining

Surface micromachining is based on the use of a sacrificial material between two structures. This spacer is then chemically removed. Complicated structures can be obtained simply by building up multiple layers.

Surface micromachining is easy to integrate into a standard complementary metal oxide semiconductor (CMOS) process (see Glossary), and therefore it is the most appropriate technique to design smart single-chip sensors (Figure 2.7-7), including digital functions such as electrical calibration, self-test, and analogue-to-digital conversion.

2.7.3.1.3 Piezoresistive Pressure Transducer

The silicon resistivity changes under mechanical stress. This property is called the piezoresistive effect and it is widely used to make pressure transducers.

The sensing element (see Glossary) is basically a four-resistor bridge (Figure 2.7-8) appropriately positioned relatively to the silicon crystalline orientation. When a stress is applied, the bridge is unbalanced and a differential voltage appears on the output.

In order to optimize the effect, the piezoresistive transducer is diffused on the border of the diaphragm (Figure 2.7-9) where the stress is maximum. Also, the transducer dimension is a critical parameter since the stress optimum is situated in a very limited area. If the transducer is too large, then the sensitivity would be decreased, as it would be averaged with less stressed areas than are around the optimum.

When the transducer is correctly designed, the sensor's response is very linear. Moreover, the linearity (see Glossary) error, for a well-designed pressure sensor, has to be as small as possible because it is the hardest error to compensate. On

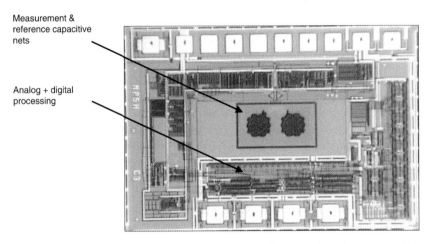

Fig. 2.7-7 CMOS surface-micromachined pressure sensor chip. Reproduced by courtesy of Motorola

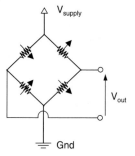

Fig. 2.7-8 The bridge transducer model. The arrows indicate resistance changes when the membrane is bent

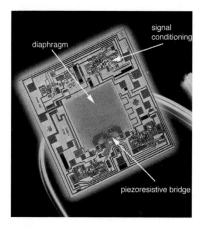

Fig. 2.7-9 Piezoresistive pressure sensor. Reproduced by courtesy of Motorola

the other hand, errors due to thermal effects or offset variability are much easier to compensate through standard calibration techniques.

2.7.3.1.4 Signal Conditioning

To compensate part-to-part variations in offset, sensitivity, and temperature coefficients, the pressure transducer has to be coupled to a signal conditioning stage. Furthermore, the raw output signal is in the range 20–40 mV; thus, an amplification stage is also required to reach 0–5 V levels.

Amplification and calibration stages can be integrated on the same die as the transducer (Figure 2.7-9). Fully conditioned single-chip sensors based on a standard bipolar process (see Glossary) that includes the micromachining have been marketed for a number of years.

Nevertheless, the use of bare transducers allows a better design flexibility as the designer can fit the signal conditioning and the calibration accuracy to the exact application needs.

2.7.3.1.5 Capacitive Pressure Sensors

To reach higher integration levels including digital functions, CMOS technology is better adapted than a bipolar process. Furthermore, CMOS manufacturing processes can integrate surface micromachining and therefore provide single-chip sensing solutions.

The left part of Figure 2.7-5 shows a typical surface-micromachined absolute pressure element; in this case, not the piezoresistive effect, but the capacitive effect is used. The applied pressure changes the space between the two membranes and, thereby, the resulting capacitance. The measurement is usually performed by comparing the output of an active net of capacitive elements with the output of a passivated reference net. This configuration increases the system sensitivity (higher C values) as well as its manufacturability by keeping the element dimensions small enough. The reference net is insensitive to pressure and its role is to simplify the capacitance to voltage (C to V) conversion.

Nevertheless, compared with piezoresistive sensing, this sensing technique requires more complex signal conditioning for linear and accurate C to V conversion. Fortunately, CMOS technology allows easy and inexpensive electronic function integration. Thanks to this combined technology, the pressure sensor becomes really a smart device including features such as autozero (see Glossary), self-test, digital output, and standby mode. Such products already exist in large-scale production, and they are destined to be the norm in the near future for a number of applications.

2.7.3.2
Could Pressure Sensors be Considered as Standard Electronic Components?

Just like any sensor, a pressure sensor is basically a converter. It converts a measurand (see Glossary) into a useful output signal (which is electrical in our case). Consequently, pressure sensors are to be considered as the link between the physical world and the electrical domain (Figure 2.7-10), and therefore they have to comply with the requirements of both domains, which are sometimes contradictory.

2.7.3.2.1 Packaging

One of the most important aspects of pressure sensors is the packaging (Figure 2.7-11). In fact, the package is a protection for the die against mechanical stress or che-

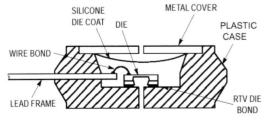

Fig. 2.7-10 Cross-sectional drawing of a pressure sensor

Technically, TPMSs have many specific requirements; the pressure-sensing module is located inside the tire, and powered by a long-life battery, so its electrical consumption therefore has to be as low as possible. Additionally, the device has to withstand the very hard environmental conditions of a running tire.

2.7.3.3.2 HVAC

With the significant cost reductions induced by the improved manufacturing processes and the growing production volumes, silicon pressure sensors became more competitive, either for a first use or as a replacement for a mechanical solution, in many industrial and consumer domains.

Among the industrial applications, HVAC is probably the largest consumer of sensors where silicon is advantageously replacing hot wire-based flowmeters or electromechanical pressure sensors based on metallic or rubber membrane displacement. In fact, silicon sensors offer more stable and repeatable signals.

2.7.3.3.3 Market Trends

Control units used to be mainly mechanical or electromechanical, hence, whenever pressure information was needed, mechanical switches were, and in some cases are still, widely used. The booming development of microcontrollers has changed this situation. Intelligent systems require more than go/no go information, and sensing techniques then started to migrate from mechanics to electronics.

In addition to this 'technical symbiosis' between sensors and microcontrollers, there are principally three market forces that contributed to the expansion of silicon pressure sensors:

1. *Energy saving and environmental concerns* Governments, the general public, and international standards and norms are imposing more and more restrictive criteria on energy consumption and pollutant reject rates. This trend, which started in the automotive industry, is now involving many other sectors such as the appliance and HVAC industries.

2. *Safety and reliability improvement* With the ever-increasing complexity of modern electronic systems, manufacturers are demanding more reliable components to reduce maintenance costs. Also, by extending the mean time to failure, systems are safer, especially in critical applications such as passenger protection and medical equipment.

3. *Smarter user interfaces* The growing complexity of electronic devices should not affect the user interface – on the contrary, it is supposed to simplify it and make it smarter. The washing machine is a typical example of this tendency. Its control module usually has more than one MCU, sometimes using fuzzy logic. Yet the user interface could be reduced to a single start button. The embedded intelligence, thanks to the information provided by the pressure sensor, does the program selection and all the rest automatically [3].

These market trends are mainly driven by end users and consumers – they are, of course, the final decision makers and their wishes are closely observed by the manufacturers.

However, between the suppliers and their final market, designers are an important link in the chain. Silicon pressure sensors represent for them an interesting product to design since their flexibility and high integration level allows a shorter time to market. The magic cocktail made up of half microcontrollers and half silicon sensors makes their designs easier to upgrade and to adapt to diverse platforms.

2.7.4
Solution: a Flexible, Modular Pressure Sensor for HVAC Applications

2.7.4.1
Concept

As could be seen in Section 2.7.2, the recommendations for pressure sensors differ very much and cannot be fulfilled with a single product. There are low-end applications such as filter monitoring and applications that are very demanding such as those described for VAV sensors.

Especially for VAV we did not find a sensor on the market that fulfills all our needs in relation to price, accuracy, and long-term stability. First we started with a project; based on a metallic membrane with an eddy current sensor, a technology we had used with success in other sensor applications, but we failed to reach the recommended zero-point stability.

We then decided to approach the problem in another manner and to use one of the cheapest sensors on the market, even though it did not fulfill the technical recommendations, and to eliminate the drawbacks by a periodic recalibration in the field with a built-in air valve, a method which is not uncommon for pressure sensors. Additional various correction algorithms were used, based on calibration of the sensor and accurate characterization of the sensing element. The concept is shown in Figure 2.7-12.

This concept has proved to be very adaptable and flexible. It is mainly based on software. We are now in a position to tailor sensors for very different applications based on modular building blocks. For example, it is very simple to change the following characteristics:

- range;
- output function (linear or square root);
- accuracy;
- time constant;
- bus interface, proprietary, SPI, RS 485;
- the valve can be omitted for less demanding, low-cost solutions;
- two-point pressure measurement.

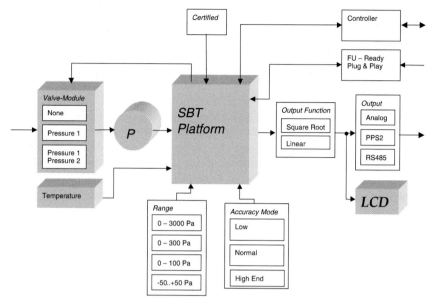

Fig. 2.7-12 Modular concept of the pressure sensor

2.7.4.1.1 **Setup**

The main idea is, roughly speaking, to make such a silicon sensor element as simple as possible, and to put the intelligence in the surrounding. The most important task of the intelligence is to compensate for the essential shortcoming of silicon sensors, namely the drift. The intelligence is based on the three pillars autozero, characterization of the sensor element, and calibration.

The interplay is illustrated in Figure 2.7-13. There are facilities for periodic zero calibration, realized with a valve, and an MCU that triggers the autozero and that contains the individual sensor parameters for different compensation operations. The individual sensor parameters are supplied by a time-consuming calibration.

The main problem to be faced in the design of the surrounding can best be seen when considering the pressure curve as illustrated in Figure 2.7-14. We use only a marginal part of the pressure range. Within the region of interest the pressure-dependent part represents only a small proportion of a few percent of the total output, which makes the measurement system highly susceptible to offset fluctuation. Hence the use of such sensor elements means real twisting!

2.7.4.2
Autozero Facility

The main difficulty that arises with the use of silicon pressure elements is that of offset drift. Although the apparatus can initially be adjusted to provide a particular value of pressure output, this condition is hardly maintained over a longer per-

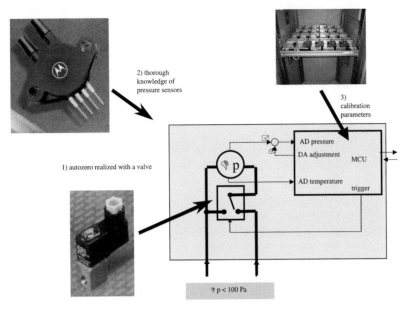

Fig. 2.7-13 Main elements of the sensor system concept at a glance: facility for periodic zero calibration, realized with a valve, an MCU with trigger for the autozero, with the individual sensor parameters and communication interface, and the calibration plant

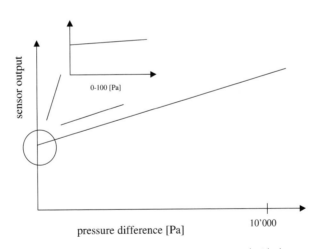

Fig. 2.7-14 Presentation of the effectively used pressure measuring range compared with the whole sensor range

iod. Therefore, we have to provide autozero, allowing the adjustment to be repeated in the field.

The whole autozero procedure consists of the following tasks. The zeroing is periodically triggered by a clock signal, or by a temperature change exceeding a predetermined value. The last pressure value is frozen and output during the whole zeroing procedure. The valve is then switched and produces a shortcut between both pressure areas, thus exposing the strain gage to a zero pressure difference.

The digital-to-analogue converter output of the Δp brings the output of the amplifier within a predetermined interval (as can be seen in Figure 2.7-13) using the successive approximation algorithm. The feedback value is then stored until the next zeroing procedure. The residual deviation is determined and, after the valve is switched back, continually subtracted by digital operation in order to yield an accurate pressure output.

This task splitting between analogue (coarse) and digital (fine) adjustment saves much time: the approximation algorithm is a time-consuming affair when performed at the full measuring resolution.

2.7.4.3
Factory Calibration Procedure

The calibration is critical for the cost and the quality of the device. In addition to the pressure, a temperature calibration of the sensing element is also needed. A setup of the calibration plant is shown in Figure 2.7-15. Because temperature calibration is mostly a time-consuming task, it was decided to calibrate the sensors batchwise (Figure 2.7-16). For the sensor models where a bimetallic valve is needed, the testing of the valve is integrated into the calibration station.

We end by the following recommendations for the calibration station:

1. Easy and fast loading of the sensors to the calibration station.
2. Sufficient capacity to make one day's production quantity with one calibration run.
3. Fully automated procedure, only the start button has to be pressed.
4. Self-test of the calibration facility.
5. Leakage on a specific sensor must not influence others.
6. Function test of valves.
7. Determination of the calibration parameters. Plausibility checking of the calibration parameters (range).
8. Writing of the calibration parameters to the EEPROM (electrically erasable programmable read only memory) of the sensor and to the database for statistical usage.

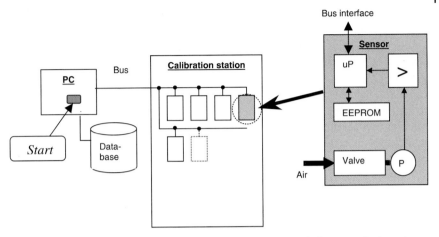

Fig. 2.7-15 Calibration station with sensors. The communication with the sensor is done via a bus interface

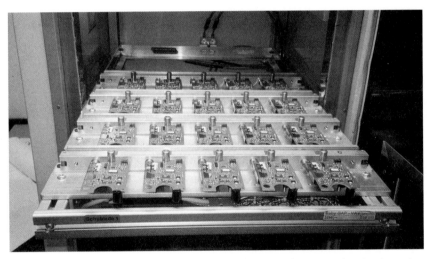

Fig. 2.7-16 The inside of the calibration station, showing one of the racks. One rack can take up to 20 sensors. The electronic prints of the sensors can be connected to the air supply, the bus, and the power with a single 'click'

2.7.4.4
Characterization of the Sensor Elements

The main requirements of the application can roughly be summarized as follows. A high accuracy of 1 Pa is required in the lowest pressure region. The critical success factor, therefore, is the ability to maintain offset stability at a predetermined level, despite temperature changes between 0 and 70 °C.

A thorough analysis of the sensor behavior was performed in order to elaborate the calibration and compensation concept. We restricted ourselves to comments on the critical quantities and therefore only present findings that were of significance for our application: the temperature effect and warm-up effect.

2.7.4.4.1 Temperature Effect

A strong temperature dependence of the offset can already be inferred from the semiconductor nature of the resistive elements. Besides the expected strong temperature dependence, we found in addition a complicated time behavior.

A sufficiently long period after switch-on of the power supply was allowed so that warm-up effects were negligible. A temperature jump from 25 to 50 °C of 1000 s half-life period was performed. Figure 2.7-17 shows the evolution of both offset and sensor temperature (see Glossary). The curves exhibit totally different behavior. Whereas the sensor temperature reaches a plateau after a rapid increase, the offset shows a rapid decrease to a minimum, from where it increases with a diminishing gradient.

Is there really a possibility of implementing powerful temperature compensation in view of this 'disorder'? A signal versus temperature presentation shows us the result. As a striking effect we note in Figure 2.7-18 a short-term region that shows a linear temperature dependence. The linear region extends to over 90% of the temperature range, ie, from the initial temperature to 2.5 °C below the end temperature. The time range of approximately 12 min, however, is rather small.

The idea of the temperature compensation now runs thus: as the temperature coefficient we define and determine the gradient in the short-term, linear region. Such a defined temperature coefficient assures appropriate compensation for rapid temperature change. The badly defined but slowly proceeding long-term drifts are then compensated through the autozero using the valve.

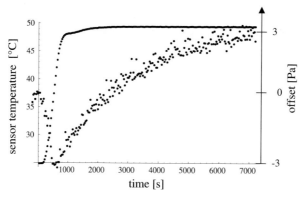

Fig. 2.7-17 Evolution of temperature and offset of a sensor undergoing a temperature jump from 25 to 50 °C

Fig. 2.7-18 The offset as a function of the sensor temperature. The solid line represents the best linear fit in the time window between the start and 12 min

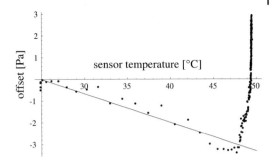

The result of a linear least-squares fitting algorithm, shown as a solid line in Figure 2.7-18, yielded a temperature coefficient of

$$TCO = -0.13 \, \text{Pa}/^\circ\text{C} \pm 0.01 \, \text{Pa}/^\circ\text{C} \,. \tag{2.7-2}$$

The compensation now consists of digitally subtracting the temperature drift of

$$TCO[T_s(t) - T_s(0)] \,, \tag{2.7-3}$$

where $T_s(t)$ and $T_s(0)$ are the sensor temperature at the current instant and at the moment of the last autozero, respectively.

The suggested compensation procedure, of course, applies only when the temperature behavior is symmetrical with respect to the direction of the temperature jump. In order to assure symmetry, a temperature jump back to the initial temperature was performed and compared with the previous curve. Figure 2.7-19 clearly shows the same gradient of the linear regions of both temperature curves.

Figure 2.7-19 points to an interpretation of the offset as a thermodynamic state variable since value is restored after the cycle is concluded. The temperature increase and decrease are simply performed too fast for the sensor offset to follow

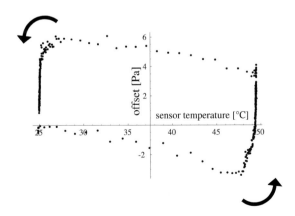

Fig. 2.7-19 Heating and cooling cycle

instantaneously. This non-quasi-static process therefore leads to an important level of temperature hysteresis.

2.7.4.4.2 Warm-up Effect

Warm-up affects the sensor output for a certain period of time after electric power is applied. Figure 2.7-20 shows the evolution of the offset. We can clearly distinguish two regions. There is a rapid increase (initial slope 0.07 Pa/s) in the first 250 s, then the gradient decreases continually until the curve reaches a slightly oblique plateau (with a slope of 10^{-3} Pa/s). As illustrated by the solid line, the curve can be reproduced by a simple function: the sum of an exponential (describing the fast increasing region) and a linear function, yielding

$$p_{\text{off}}(t) = A[1 - \exp(-\gamma t)] + Bt \qquad (2.7\text{-}4)$$

where $A = 6.9$ Pa, $B = 9.2 \times 10^{-4}$ Pa/s, and $\gamma = 9.8 \times 10^{-3}$ s^{-1} are the individual parameters.

The warm-up behavior leads to excessive calibration duration. Since the warm-up behavior is not known *a priori*, one cannot avoid awaiting the whole warm-up period in order to obtain accurate calibration values. Determination of the temperature coefficient takes a longer time of several minutes and is therefore affected even by slow drifts. However, once the calibration is done, the dead time between switch-on and readiness for use can be shortened considerably through increased autozero frequency during the first few hours, or by using a compensation tool working with the previously determined individual calibration parameters A, B, and γ.

One can only speculate at this stage about the causes of the warm-up effect. The role of a manifest candidate, namely sensor temperature change due to electrical power dissipation, is at least controversial. First the total amount of the temperature change is approximately 5 °C. This represents only a fifth of the temperature change performed for the study of the temperature effect as shown in Figure

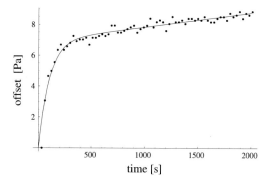

Fig. 2.7-20 Evolution of the offset pressure during the first 2000 s after supply voltage is switched on

Fig. 2.7-21 Steady-state value as a function of the initial drift. The sensor samples were selected so as to complete the whole range of steady-state value

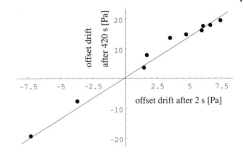

2.7-17. It is therefore hardly imaginable that such a small change can account for such a large effect. In addition, the discussed warm-up behavior is produced by a sample with a negative temperature coefficient. Hence we would expect negative warm-up drift if we assigned a dominant role to the sensor temperature change.

Further thorough analysis could supply a clue to the fundamental principle of this widespread effect, but also help to develop a concept to shorten the calibration process. The latest data for warm-up measurements on a larger number of samples seem to point to a fixed relation between initial and end behavior. This could be used in future to infer the evolution without *a priori* knowledge. Figure 2.7-20 connects the extent of drift of two different time regimes. The drift 420 s after the switch on corresponds to the turning point of the warm-up curve (Figure 2.7-20) and can be regarded as the steady-state value when neglecting the slow drifts responsible for the slight inclination of the plateau. The linear model shown as a solid line in Figure 2.7-21 allows us to anticipate the jump just by monitoring the offset values during the first 2 s.

2.7.4.4.3 Warm-up Versus Temperature Effect

Both warm-up and temperature effect are, in our application, the main causes of offset drift. Once stable operation conditions have been reached, ie, a few hours after the switch-on and at constant operation temperature, the drift per day scarcely exceeds 1 Pa.

It is therefore worthwhile to focus efforts on further studies of these effects. The following cost and quality factors are at stake:

- decrease calibration running time;
- reduce the tuning time at our OEM customers;
- shorten the warm-up time (between switch-on and readiness);
- reduce the autozero frequency (thus enhancing the grade of readiness for use).

There is a basic difference between the causes of warm-up and temperature effect, change of supply voltage and temperature, respectively. Switch-on/off operation is actually performed only three times during the life cycle:

- at factory calibration;
- at the OEM customer, which tunes the VAV damper actuator for the VAV box;

- in the building during the commissioning.

In contrast, temperature changes are perpetual, thus mixing the delayed slow drifts. Although theoretically predictable by convoluting the temperature past life with the known response function, the temperature behavior is almost always flawed with an uncertainty in the range of the hysteresis band.

This difference makes the warm-up the easier handlebar effect (apart from the long calibration duration). We can indeed envisage temperature stabilization thus leaving the switch-on as the unique cause of any drifts. Stable temperature conditions would allow us, for example, to forego the autozero, nevertheless having an effective offset compensation procedure.

2.7.4.5
Application in the New Damper Actuator from Siemens Building Technologies

The new damper actuator shown in Figure 2.7-22 fulfills all the recommendations given in Table 2.7-1. It is very accurate and the zero point is not dependent on the mounting orientation. It has a bus interface, which can be used to change parameters in the field. A special tool based on Labview is available, which greatly facilitate the configuration for the OEM customers. On the cost side it is very helpful that sensors will be used in the flap actuator controller which already uses a microprocessor. Therefore, we obtain the computing capacity and the bus interface more or less for free. Figure 2.7-23 shows the modular concept of the device.

2.7.5
Conclusions

After a general introduction to the needs of the HVAC industries with respect to pressure sensors, we focused on a specific application, the so-called variable air volume (VAV). We noticed the combination of high offset stability (mainly due to

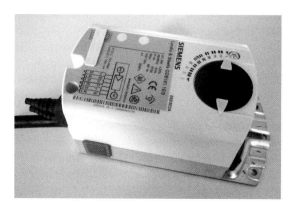

Fig. 2.7-22 VAV damper actuator. The connectors to the air ducts can be seen on the left

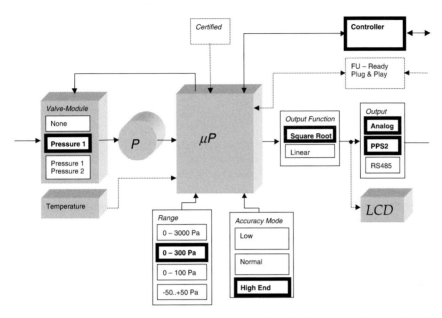

Fig. 2.7-23 Building blocks of the VAV damper actuator. The realized variant is shown with thick lines. The motor is not shown

the square root relation between flow and pressure) at low cost as the featured requirements of VAV.

We presented a solution based on a 'cheap' silicon sensor element where the efforts were focused on compensation and calibration. The result was a modular, flexible, integrated sensor system yielding a versatile technology that enables a wide range of functionality simply selectable as calibration or as equipment variant in the last production step, but also by retrofit.

Temperature and warm-up behavior proved to be handlebar effects but representing an important source of expense. Ongoing efforts to study warm-up behavior and temperature effects and to develop appropriate measures could turn out to be profitable.

Originally specially designed for VAV applications, the technology will soon be used for other applications such as filter monitoring and fan control.

The flexibility can even be interpreted more broadly in the sense that we can adopt other pressure elements, perhaps more suitable for specific applications, without major efforts.

The strategy to put the bulk of the effort into calibration has proven successful, so we can envisage applying the paradigmatic procedure to other measuring quantities also.

2.7.6
Acknowledgments

The authors acknowledge all colleagues who contributed with their work to the success of this project, especially Adrian Andermatt and Jürg Bichel, who gave us the chance and confidence to develop the sensor for their new VAV damper actuator, Josef Jandl and Werner Studer for the design and the development of the bimetallic valve, which allowed a very compact construction at low cost, Bruno Kamm for implementing the various algorithms with very limited processor and memory resources, and Roland Bart and Beat Stocker for the realization and construction of the fully automated calibration facility.

2.7.7
Glossary

Anisotropic	Antonym of isotropic
Autozero	An error-compensation technique that performs calibrations in the field on request, on power up, or periodically, to improve the system accuracy
Bipolar process	A microelectronic manufacturing process based on the transistor effect created by PNP and NPN junctions
CMOS	Complementary metal oxide semiconductor, a microelectronic manufacturing process based on P-channel and N-channel complementary devices
Commissioning	Set of procedures and methods to advance a system from static installation to full working order in accordance with the design intent
HVAC	Heating, ventilation, and air conditioning
Hysteresis	The difference between the output levels when a given measurand value is reached upwardly and then downwardly
Isotropic	Omnidirectional, with no privileged direction
Linearity	The linearity error is the difference at a considered measurand value between the outputs of the actual system and its linear theoretical model
Lithography	An optical process widely used in microelectronics combining photo-sensitive material and chemical etching
MAP	Manifold absolute pressure
MCU	Microcontroller unit, a semiconductor-integrated circuit including a central processing unit, memory, and input/output resources
Measurand	A quantity subjected to measurement
N type silicon	Silicon doped with electron donor impurities
Offset	The output at input (when applicable)
P type silicon	Silicon doped with electron acceptor impurities

Pick & place	An automated assembly technique for electronic components
Sensitivity	The ratio of the output variation and the corresponding input variation
Sensing element	Probing part of a sensor that transduces the measurand into a quantity that can be further processed by the electronics
Sensor	Complete system consisting of sensing element and electronic circuitry
Sensor temperature	Effective temperature of the die, which is determined through the input resistance of the strain gage
Transducer	A device that converts energy from one domain to another

2.7.8
References

1 RISTIC, L., *Sensor Technology and Devices*; Boston: Artech House, 1994.
2 *Automotive Sensors: a Strategic Study of the Global Automotive Sensors Market to 2004,* 2nd edn.; Reed Electronics Research, May 2000.
3 FRANK, R., *Understanding Smart Sensors*; Boston: Artech House, 1996.

3 Information and Transportation

3.1
Fieldbus Systems

DIETMAR DIETRICH, THILO SAUTER, *Vienna University of Technology, Institute of Computer Technology, Vienna, Austria*
PETER FISCHER, *University of Applied Sciences Dortmund, Dortmund, Germany*
DIETMAR LOY, *Coactive Networks, Sausalito, CA, USA*

3.1.1
Introduction

Fieldbus systems play an important role in all automation areas: industrial, process and building automation, ship building, air and space technology, automotive automation, and many others. The expression 'fieldbus', however, originally comes from the process area [1, 2]. Here the term 'field level' designates the lowest of several control levels, where sensors and actuators are connected. Every electronic and even electrical system comprises sensors and actuators. The simplest example in building installation is a switch controlling a lamp (Figure 3.1-1a). Until today lighting in residential properties has mostly been based on very primitive technology, where information and power are not separated. The main disadvantage is the very poor performance (intelligence [3]) of such systems. If one wants to control a more complex environment, one has to distinguish the information system from the power system [4] and has to introduce controllers that process information (the process data) using various algorithms (Figure 3.1-1b).

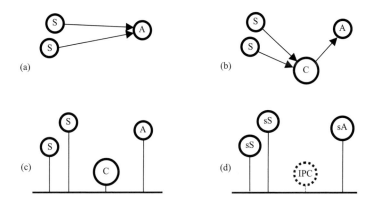

Fig. 3.1-1 Steps from traditional direct electrical process control to modern distributed networks based on embedded systems (S=sensor, A=actuator, C=Computer, sX=smart X, IPC=industrial personal computer)

Another disadvantage of classical cabling soon became apparent, above all in the area of air and space technology. If somebody wants to control a process in a precise and efficient way, a huge amount of sensor data is required, which means a lot of sensors have to be installed and the wiring cost for such a system increases rapidly. In space- and weight-limited applications, for example in an airplane, reducing the amount of cabling is crucial to the success of the project, and the solution leads directly to bus systems (Figure 3.1-1c). Since we are talking about sensors and actuators at the field level, we should be more precise and say that the units (terms such as devices, nodes, and entities are also commonly used) are connected by fieldbus systems. With the rising demand to control more complex processes more accurately and also efficiently, the number of nodes increases. A central controller will no longer be able to serve all sensors and actuators in real time, unless the central unit has an enormous performance, which is much too large and cost prohibitive.

Microelectronic components have evolved from one generation to the next in only a few years. In the most recent change (Figure 3.1-1d) microelectronic technology again enabled designers to integrate a whole computer into one chip: the embedded system was born. Now we can integrate technical intelligence into each node and design smart sensors and actuators. Apparently this will again lead to a dramatic change in control systems. At the moment, however, the change from Figure 3.1-1c to d is taking time because the intelligence in the embedded systems is not sufficient enough, so that most applications need additional computing resources such as IPCs (industrial personal computers) or PLCs (programmable logic controllers). There will be a lot of engineering required to find efficient solutions, structures, and protocols (see Section 3.1-7) before we reach the ultimate goal of fully decentralized systems.

The explanation above is very simplistic and of course there are many other reasons for finding new structures. A very important reason for the success of fieldbus systems is the price, especially for nodes and wiring. As of today, the node cost is one of the biggest problems. The change from one stage to the next is mainly a question of product cost and not of technical nature.

3.1.2
Abstract View and Definition of the Fieldbus

Many standardization organization task groups and user and interest groups, but also scientific institutes all over the world, have worked out definitions for fieldbus systems. Of course, each of them has different interests, goals, and basic knowledge, especially because the fieldbus is used in many different areas such as the process industry, computer industry, air and space technology, etc. The logical consequence is that completely different definitions exist. For instance, some definitions and terms in the European Installation Bus (EIB) protocol specification deviate from those that the International Organization for Standardization (ISO) proposes. For an expert in bus systems this is not acceptable, but building owners or installers traditionally have their own vocabulary, and with the definitions of EIB they work perhaps more easily.

For an analysis of bus systems independent of the applications, it will be best to use a definition based on the functionality of bus systems, because classical distinctions no longer work. Such a classical approach is a classification by the maximum bus length: there are fieldbus systems, implemented in oil fields for fire alarms with an extension of over 60 km, much more than a normal LAN (local area network) usually has. Another example is the data rate: some fieldbus systems have a data rate of over 10 Mbit/s, and this limit will increase with new developments. Finally, a classical differentiation between fieldbus systems and other networks was that fieldbus systems demand real time, whereas other systems do not. Today, a lot of networks are time critical, so this differentiation also no longer works.

A clear definition for bus systems is a hierarchy of their applications, which makes it necessary to send different kinds of data packages and frames and to send them differently. Global area networks (GANs) bridge long distances over the world. They are mostly realized by satellites and submarine cables. The channels of GANs are characterized by high redundancy, a high availability, and complex frame-acknowledge mechanisms. The next lower level is represented by wide area networks (WANs), which deal with the Internet, telecommunication systems, etc. The main characteristics are the large number of terminals and clients and the high costs of cables and transmitter units. They are normally realized by a meshed star topology. The second lower level is the level of the LANs that connect personal computers (PCs) in rooms or buildings. The number of systems is very high, and so there is a lot of competition in this market, which enticed Bill Gates to say in 1999 that each electrical component will have its own Internet address. This statement shows a lack of knowledge about the third and lowest level: the field area networks (FANs) or fieldbus systems. Here all sensors and actuators are situated, which, of course, can be smart, ie, use embedded systems. The demands vary enormously. Some are time critical, eg, FANs replacing car steering. Some of them must have very low maintenance costs, such as FANs that control the components in a household such as the light or the door. A common criterion is the transfer of short information, which must normally be very efficient. This means that the interfaces between the sensor/actuator and the bus must be very inexpensive, much cheaper than in LANs because the price of the whole fieldbus node has to be low. People who think that LANs will supersede FANs do not see two aspects. First, the cost of a bus interface has to be much lower than the cost of the whole node (computer). Second, the number of FAN nodes will dramatically increase in the future, a trend we can already notice in the development of cars.

In some different fields one can find more FAN levels. This means that the FAN level is further divided into the actual FAN and the sensor/actuator levels. These definitions depend on the interests of companies, the applications, etc., and are difficult to generalize. It makes no sense to work out this idea more carefully.

The hierarchy GAN–WAN–LAN–FAN says nothing about the interconnection between the levels. Of course, it makes sense to connect levels that do not lie directly over each other. The networking between the levels is a question of the performance of the nodes and the interfaces, which means that it depends on the cost [5]. However, independently what has been said thus far, the classification between GANs, WANs, LANs, and FANs mirrors the levels of the automation pyramid, which we will talk about in the next section. And there, in general, only connections between consecutive levels exist.

3.1.3
Communication Basics for Fieldbus Systems

Only the most important communication basics for fieldbus systems are explained. For more and detailed information readers should consult [5–7].

3.1.3.1
Decentralization and Hierarchies

Two basics hierarchies exist: the ISO/OSI model and the automation pyramid. The abstract ISO/OSI model, explained in the next section, is a proposal of ISO for designing the communication principles inside a network. It describes mechanisms of parallel processes. The intention of the automation pyramid (Figure 3.1-2), on the other hand, is to structure the information flow required in various automation areas. The original idea was to create a transparent, multi-level network, a basic requirement for CIM (computer-integrated manufacturing). No doubt the goals of CIM were excellent (nowadays it is obvious that without such a networking no modern company, no process, no building would be controllable), but in the beginning

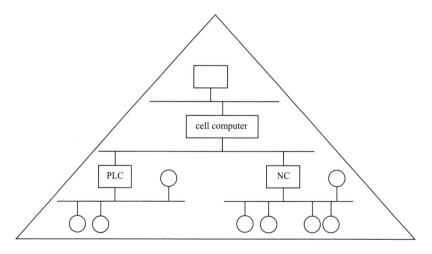

Fig. 3.1-2 Hierarchical network levels in automation. PLC=programmable logic controller, NC=numerical control)

there was not much success. One of the major problems of CIM was that the data acquisition (and this belongs to the networking at the field level) was too expensive. There was no fieldbus available which fulfilled the criteria of low cost and acceptable performance.

The number of levels in a pyramid depends on its application and the designer. With some exceptions there is no rule as to how many levels the user has to define. There are only proposals of organizations and user groups. Sometimes companies define very detailed structures up to six levels, because of the complexity of the process. Smaller systems need only three, sometimes even two, levels. In [8] different pyramids are presented, and here most have only three levels, but five is not unusual. The names of the levels are mostly different. In the production area one will find the definitions factory control level, production level, and field level. If five levels exist, a typical definition is management level, manufacturing management level, process control, monitoring and control and I/O level. For building automation the CEN (Comité Européen de Normalisation) has defined three levels: management level, automation level, and field level. However, in a residential property it will seldom be necessary to take more than two lower levels.

Because of the increasing performance of bus systems, which means the ability of the various bus systems to overlap more and more, the number of levels of the automation pyramid can be reduced.

3.1.3.2
The ISO/OSI Model

The goal of ISO was to find a basic idea for defining communication protocols. This basic idea called OSI (open system interconnection) should be an aid for a developer to find the appropriate communication system for a specific application. The second, very important goal was to define various communication systems that are similar and therefore easy to connect to each other. The ISO/OSI model was defined as a first step to standardize internationally the protocols in the various layers and it deals with connecting open systems (ie, open for communication with other systems), but at that time (1983/84) nobody thought of communication systems in the automation area [6]. And this is the disadvantage we have to deal with today. Nobody gave a thought to real-time ability, or that the protocols have to be implemented into small controllers and embedded systems. The developers of OSI never saw networks with thousands of nodes in one building. They integrated redundancy in various places to guarantee high availability, and they envisioned huge computers that should be networked.

3.1.3.2.1 Layer Structure
It was hard work to agree on a basic model. Communication functions were gathered and assigned to different layers. There was no question that the number of layers had to be limited, otherwise the model would have had too many interfaces between layers, meaning a large overhead in program size and processing time. If

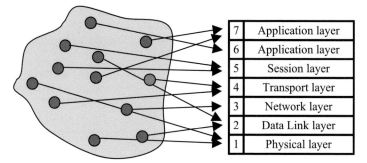

Fig. 3.1-3   ISO/OSI model (seven-layer model)

there were not enough layers, on the other hand, the modularity was not given, and the likeness between different bus systems was too high, resulting in a major effort at the connections between different bus systems.

In the end, the committee found a compromise and chose the lucky number 7 (Figure 3.1-3). They defined a hierarchical system, not a democratic one, because it was easier to handle and promised a high throughput of data. Such a communication column was to be implemented into a unit if it was required to take part in a communication community (Figure 3.1-4).

The designers of the so-called 7-layer model differentiated between two communication principles: horizontal and vertical communication (Figure 3.1-5). Horizontal communication is the protocol, which means that only similar layers are able to communicate with each other, which is logical: only the same layers have the same functions (since they speak the same language). Vertical communication is based on a service principle: The lower layer (SP, service provider) offers services to the higher layer (SU, service user). Different functions transmit their data over SAPs (services access points), as shown in Figure 3.1-6.

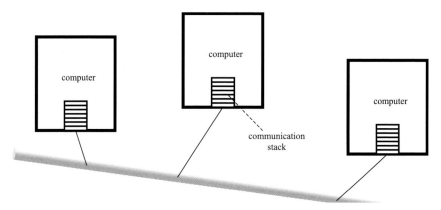

Fig. 3.1-4   Communication principle

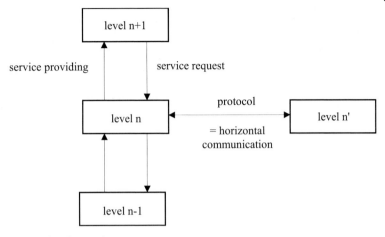

Fig. 3.1-5 Horizontal and vertical communication

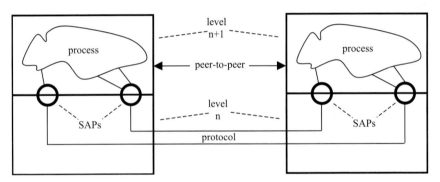

Fig. 3.1-6 Connection opportunities over SAPs (services access points)

The data stream from one unit (communication column) to the other one is shown in Figure 3.1-7. A system application places an order via layer 7 to its communication column. To fulfill this task, layer 7 prepares data for its corresponding layer 7', packages this and requests services of layer 6 to transmit this data. Layer 6 does the same. It prepares data for the corresponding layer 6' to fulfill its task and requests services from the next lower layer 5, and so on all the way down to layer 1, which actually transmits the data to its corresponding layer 1'.

In larger networks there are repeaters, bridges, routers, and gateways, as one can see it in Figure 3.1-7. They are defined by the layers they have implemented, which point out their abilities. A repeater contains only layer 1, a bridge layers 1 and 2, a router the three lower layers, and a gateway all seven layers. This means that if completely different networks have to be connected, a gateway is required.

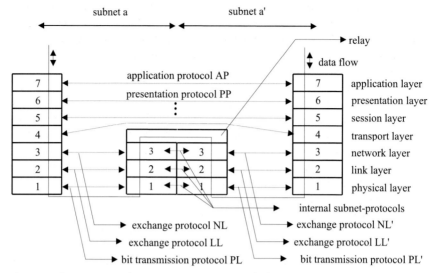

Fig. 3.1-7 Communication between two systems networked over a router

3.1.3.2.2 Functions and Management of the ISO/OSI Layers

Table 3.1-1 shows the assignment of the functions to the different layers. In layer 1 all mechanical and electrical parameters are defined. The whole description of the hardware of a system is part of this layer. In the second layer the medium access control is defined. There are big differences between the various FANs, based on different philosophies. In industrial automation one will often find a strictly organized arbitration method because of the often existing hard real-time requirement. In the automotive area engineers face a completely different situation. Mechanics do not have the necessary training to set up fieldbus systems, so if the car has to be repaired, the exchange of fieldbus nodes must be easy. CAN (controller area network) and the TTP (time triggered protocol) are good examples of such automotive fieldbus systems [9].

Tab. 3.1-1 Layers and functions of the ISO/OSI model

	Name	Functions
7	Application layer	Communication services for the application
6	Presentation layer	Character and language adaptation
5	Session layer	User identification, set-up of sessions
4	Transport layer	Data flow control and other end-to-end functions
3	Network layer	Routing
2	Data link layer	Medium access control, frame configuration, data protection and other point-to-point functions
1	Physical layer	Physical mechanical parameters

The requirements for building automation systems are again different. Such systems have to have high availability and longevity, and support a large number of nodes, but hard real-time requirements are usually not demanded. The designer of the medium access control has more freedom in this application area.

The frame configuration is to be seen in the same way. Of course, the constraints of the different areas play a key role in the decision which principles to implement. Real-time conditions demand short frames, whereas a high node number in a FAN demands a more complex sub-network structure, which results in a longer frame structure. At any rate, frame check sequences are defined in layer 2 to detect transmission errors. Routing is accomplished in layer 3. The nodes in the network have to decide which way and channel the frames have to take. Of course, only huge networks need such functions, and therefore most FANs for industrial automation, where a network typically comprises only a small number of nodes, are realized without this layer. But if routing is necessary, it is not sufficient that the stacks of the different nodes have an efficient routing layer. The network must also have special routing nodes, which are usually complex and expensive. As for the naming conventions, the transmitted pieces of data are called packets from layer 3 up. The term frame is only used for layer 2 information. Finally, the protocols of layers 1–3 are also referred to as point-to-point protocols.

Layer 4 should guarantee the quality of an end-to-end connection for a channel between two nodes. Layer 4 is the first layer which does not see other nodes like routers, bridges, and repeaters, between the actual communication partners. If the quality of transmission is poor, the layer is allowed to change the packet routing. If the packets are too big for the network, this layer has to split the packet into small segments, number the segments, and make sure that the receiver assembles the segments in the right order.

The protocols on layers 1–4 are called transport-oriented protocols, because their functions guarantee the transport of packets with certain quality parameters. The higher layers 5–7 are more application oriented. Especially layers 5 and 6 do not have such an importance in FANs like all the other layers. Layer 5 controls sessions. It allows entities to authenticate each other before establishing dialog sessions and determines the type of the dialog. All these functions will be rarely found in FANs. Layer 6 represents message information, which has nearly no significance in FANs. Layer 7 is the application layer, which means that this layer provides different services that support applications. These services are the available functions for the application like file transfer, remote login, remote job entry, etc. However, note that layer 7, like all other layers, needs the support of the next lower layer to be able to fulfill its tasks (with exception of layer 1).

In the beginning of the ISO/OSI model the management was not defined, but very soon its necessity became obvious, and the designers added further function blocks. One is the management, which principally has to organize the stack, handle all failures, monitor the state process of the communication system and the network, provide network data (address information, timing, data rate, time parameters, etc.), control booting and wakeup processes, maintenance in a general sense, control fault tolerance capabilities, administration tasks, statistics, and

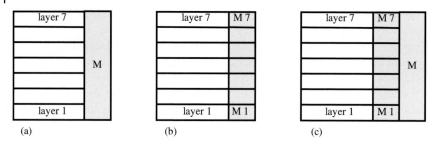

Fig. 3.1-8 Different possibilities of management integration into a communication stack

much more. Of course in home and building automation, this block assumes an enormous significance, because one cannot consider the management of the stack only. If we speak about communication systems, it is important to consider the whole network, and if this network has thousands of nodes, it can only be handled if an efficient management system is implemented. Therefore, many scientists are working in this field, because on the one hand management should be powerful, and on the other FANs are realized by small embedded systems. The space and the channel bandwidth are strictly limited. Management will be and remain a challenge in the years to come.

Depending on the ISO/OSI model, the management blocks are to be seen as Figure 3.1-8 shows. The management unit can be an own block beside the rest of the stack (Figure 3.1-8a), or as parts of each level (Figure 3.1-8b). If the network is rather large and complex and the number of tasks is large, it makes sense to combine both possibilities and to differentiate between several management blocks. A simple example is shown in Figure 3.1-8c.

3.1.3.3
Topologies

In [6, 10, 11] the different topologies and their properties are thoroughly explained. We differentiate six types: ring, star, line, tree, meshed, and heterogeneous, and all of them are used in the area of FANs. An introduction to FANs should explain all the characteristics of a topological type, because in many cases they determine the constraints of a system. This book's emphasis is on building systems, which means for us that only two types are sensible: the line and the tree. All the others have major disadvantages, although some companies and authors have another opinion about it. One important example: LonWorks offers the free topology, which one will never find in other FANs. Some literature defines a ring topology as shown in Figure 3.1-9a. A real ring topology is shown in Figure 3.1-9b. It is characterized by the two interfaces on each node, which in the area of FANs are usually one input and one output. Figure 3.1-9a only uses ring wiring, and has no characteristic ring functionality. Routers can be used to build ring topologies (Figure 3.1-9c), but in this case one has to take care that the packets do not circle the ring forever. The number of messages in the ring would ex-

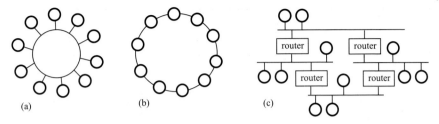

Fig. 3.1-9 Different versions of ring topology: (a) ring wiring, (b) ring bus, and (c) ring between different sub-nets

plode and block the data channel. Such a topology has to be designed carefully and is useful if redundancy is desired.

Why is it that mainly line and tree topologies are used in FANs? First, the distances between nodes are typically long. Second, all nodes should have the same importance. A star topology would be too dangerous because the central node constitutes a single point of failure. Third, the degree of automation in today's tools and management systems is not high enough. Consequently, a complex topology such as a heterogeneous configuration will lead to high integration and maintenance costs. It is very important in home and building automation to reduce the cost to a minimum and this is only possible if simple topologies such as line and tree are used.

3.1.4
Historical Aspects

To understand the problems of FANs and to look forward into the future, it is necessary to take a look back into the history. History will explain many disadvantages that we see today, but also the variety of FANs.

3.1.4.1
The Roots of Industrial Networks

The term 'fieldbus' appeared about 20 years ago, but of course the basic idea of field-level control is much older. The roots of modern fieldbus technology are various [12–15]. One origin is naturally to be seen in classical electrical engineering. Figure 3.1-10 shows this early stage. Telex was the first application to transmit data serially by a worldwide standardized protocol. Other standards followed for data transmission over long-distance telephone lines. Some of them still exist, such as V.21 (data transmission over analog telephone lines) and X.21 (data transmission between digital data network equipment and so-called communication-terminating devices). Large networks made it necessary to divide transmission functions into a telecommunication part and a more abstract one, where the mechanisms of how the entities (terminals, automatic branch exchanges, etc.) exchange their information were defined. These were then defined as protocols, normally

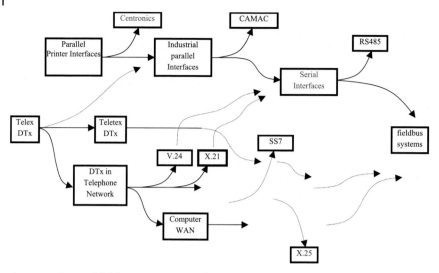

Fig. 3.1-10 Roots of fieldbus systems up to the 1980s

described as a state machine and in principle fairly simple, because of limited electronic 'intelligence' or computing power, a problem that still exists in FANs. The microelectronic developments will change this situation the same way they have dictated the performance of PCs.

Talking about serial communication, we should note that engineers working in different areas and defining the first protocols had a different understanding of the term 'serial' than we have today. For example, in the standard for the serial interface V.24, 25 different signal names are defined but only the receiving and transmitting lines (Rx and Tx) are dedicated to the actual serial data transmission. The remaining signals are control information and the designer can choose which wires he wants to use. So the control information is transmitted in parallel.

The upper part of Figure 3.1-10 shows the early development of industrial bus systems. Many different applications led companies to specify simple electronic I/O interfaces. Most of them had a parallel bus and worked in 'process fields'. This means that early FANs were based on a parallel bus system because of the difficulty of serializing data. The first computer components changed this situation completely. Nowadays the term FAN or fieldbus system stands for serial bus systems, where the application data and the control data are transmitted over the same wires.

Returning to Figure 3.1-10, the areas where hardware engineers defined interfaces for standalone computers ranged from bus systems for printers up to process control and instrumentation equipment. The first applications were the interconnections of measurement devices, such as CAMAC (especially in nuclear physics) and GPIB (IEEE 488). They were still based on parallel bus systems. For long distances, the data was converted into serial data and was transmitted over point-to-point connections (RS 422). The cornerstone in this area was the definition of RS 485, which allowed the realization of multi-drop networks. The first driver chips conquered the market, and RS 485, in an extended definition, is still the base of many FAN standards today.

Then two major evolutionary steps changed the situation. First, computers became more and more significant and the need for internetworking was predominant. The large telecommunication network got competition. The telecommunication principles helped only to transmit data, but not to handle them in a network. That was the beginning of the ISO/OSI model. This model was not defined by CCITT (Comité Consultatif International Télégraphique et Téléphonique) (today ITU-T) but by the competing organization ISO. Later, CCITT adopted this idea, as in X.25 and SS7, which are still used and powerful protocols, where functions such as routing, handling of different channels, fault tolerance, etc., are implemented. This was also the beginning of digital communication systems. Computer systems replaced analog systems and, finally, telecommunication systems were merged with computer systems. In this way, one can see the Internet as the latest step of this merger between two completely different technologies. There is no question that it will still take a long time to integrate many of X.25 or SS7 features into FANs because of the complexity of the functions, but they have had a strong influence on fieldbus systems.

3.1.4.2
The Evolution of Fieldbusses

As in the explanation about the roots of industrial networks, we can also give only a superficial overview of the concrete evolution of fieldbus systems (Figure 3.1-11). Here again, we have to consider the different influences of computer science and electrical engineering. The interesting thing is that the first smart serial bus sys-

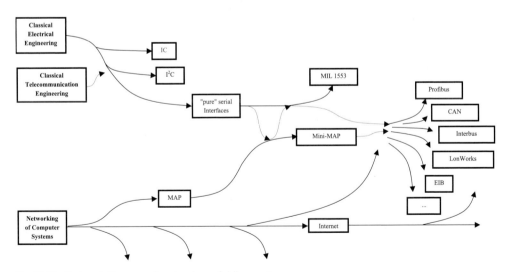

Fig. 3.1-11 Various influences of protocols on fieldbus systems

tems (in those times nobody called them fieldbus systems) such as the IC and the I²C were developed to reduce the pin count of the interfaces between ICs. Consequently, they typically had a central clock to facilitate data exchange, and perhaps dedicated control lines to control the data flow in addition to the data line. There was no abstract communication model to cover several demands independently, and these fieldbus systems were suitable only for very small units. The great leap forward in the complexity of bus systems was a direct consequence of the progress in microelectronics. The development of appropriate single-chip controllers was a prerequisite for the success of bus systems and a first step towards embedded systems. Miniaturization allowed for inexpensive, easy-to-handle nodes and also solved power consumption and EMC (electromagnetic compatibility) problems. Hence these bus systems were widely used. One could find them not only in industrial automation equipment, but also (and mainly) in telephone terminals, radios, TV sets and other electronic circuit boards.

The final breakthrough, however, did not come from electrical engineering or the telecommunications area, but from computer science. The base was the ISO/OSI model, when various companies under the leadership of General Motors tried to define a bus system for industrial automation. They called this standard MAP (manufacturing automation protocol) and it was intended to solve the problems and cover nearly all functions of an industrial network. The complexity and flexibility of this system were enormous – in the final version seven layers were specified – but not feasible. The tightened standard Mini-MAP with only layers 1, 2 and 7 was more successful, but not as anticipated, but it was understood by computer engineers all over the world that a network in the field level cannot consist only of signaling lines, as the electrical engineers wanted. A fieldbus system has to be based on an abstract model with hierarchy levels. Otherwise the complexity is not manageable.

Both independent developments, one coming from the area of electrical engineering and the other from networking of computer systems, led to different fieldbus systems that still exist today. One of the first and very powerful standards was the MIL 1553 for air and space technology, where weight and space were decisive arguments. It is a master/slave system and covers long distances. In rockets and airplanes it was, of course, redundantly integrated, normally as a triple redundant system. Typically the definition was focused on the two lower protocol layers, and the application functions were very simple.

Using the fieldbus systems defined around 1970, the AIRBUS 320 could be controlled by a fly-by-wire system in the early 1990s. Initially, the most appealing advantage was a reduction in wiring and weight. But then, and this is much more important, engineers were able to integrate functions that would have been impossible with mechanical systems – a typical benefit in fieldbus applications. Conventional airplanes fly in an attitude as shown in the upper part of Figure 3.1-12. This means that they work 'against' the air to get enough lift. The fly-by-wire airplanes assume an attitude as shown in the lower part picture of Figure 3.1-12 to save energy. They have only one disadvantage. The nose of the airplane flutters at about 5 Hz, which means that the pilot is no longer able to control such a system

Fig. 3.1-12 Airplane attitudes

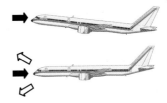

manually. Fieldbus systems are fast enough to enable the whole control system to balance the airplane. The pilot needs only to define the direction, nothing more.

These advantages were also to be seen in industrial automation. European companies, together with scientific institutes, defined bus systems to realize functions that are impossible for mechanical systems, because they do not have an information computing system. But these developers had the disadvantages of MAP and Mini-MAP in the back of their heads, so they saw FANs primarily in connection with PLCs and developed centrally oriented systems, a technology that has had great success in automation up to today. A lot of different FANs were designed.

Another important step was the development of dedicated controllers for fieldbus systems. Early general-purpose controllers such as the INTEL 8051 needed software implementations of the protocols, which in turn meant long delay times and additional costs for memories and maintenance. The design of special controllers for various bus systems, although expensive, increased the performance and reduced the possibility of implementation errors. Paradoxically, the availability of dedicated controllers can also be a hindrance for the advance of fieldbusses. Production volumes, compared with general-purpose ICs such as RAMs or CPUs, are low, and amortization times are long. Therefore, a controller is usually on the market for years, and soon its performance can no longer keep up with rising demands of modern applications. In many cases, the applicability of fieldbus systems is limited not by the fieldbus itself (ie, the functionality and the protocol), but rather by the limited resources supplied by the controllers.

In the beginning the success of fieldbusses was rather poor. The number of different systems was too high, and therefore the profit margin on each of them was too small for the vendors. The customers, on the other hand, wanted to have 'open systems' (which means standardized, as opposed to vendor-specific), so that they would have the freedom to buy their components from different vendors. User groups were organized by companies for various FANs to have platforms for common specification and promotion activities. 'Multi-vendor systems' were installed to prove that the integrated FAN components could play together.

In the course of time, many fieldbus systems disappeared, but several survived the fierce competition and gained substantial market shares. In this complicated situation, the IEC (International Electrical Commission) tried to define a single fieldbus standard for the areas of industrial and process automation. This attempt remained futile for a long time. Those European countries whose industry had done pioneering work in the definition of FANs insisted on their national standards. Therefore, international conflict was unavoidable. Different countries tried alternately to prevent an international IEC standard and were successful up to 1999.

The Europeans, above all the Committee TC 65 of CENELEC, had the goal of being the first with an industrial standard, but countries such as Denmark, France and Germany wanted to have their own fieldbus as European standard. However, the rules of CENELEC were precisely defined: there can be only one system in one standard. To reach the goal and to get the first industrial standard worldwide, they accepted, with justifications difficult to understand, that three different FANs were standardized in one standard [13]. A unique case! But other committees, such as the building automation committee, followed this example, and we now have the situation that a lot of different FANs are standardized for various applications. Still, the market has proven that this decision was right. The success of fieldbusses was great and the users have already invested enormous amounts of money, which is a guarantee that these different systems will also co-exist in the future.

3.1.5
Examples of Fieldbus Systems

In the area of building automation, we currently have EIB and LonWorks in the lowest level of the automation pyramid, and in level 2 again LonWorks, BACnet and EIBnet based on IP networks. The following sections present a brief overview.

3.1.5.1
EIB

The EIB fieldbus [16] is widely used in Europe. It is based on the developments of leading companies in the area of electrical installation. In 1990, these companies decided to join forces and to create a company-independent fieldbus standard. The result of this effort was the EIB specification and the foundation of the user organization EIBA (EIB Association), located in Brussels. The tasks of this organization are to take care of the specification and certification of EIB products (we will come back to this later), the participation in international standardization committees, and the promotion of EIB. Since 1998, EIB has been part of the European pre-standard ENV 13154-2.

3.1.5.1.1 EIB Overview
The main application area of EIB is building automation. Consequently, the topology of EIB resembles the structure of a building. Similarly to a building divided into floors and rooms, the network has zones and lines separated by couplers. A maximum of 15 zones is possible. Depending on the actual implementation, each zone can have 12 or 15 lines, and each line may comprise up to 256 nodes. In total, a network may thus have 57 600 nodes.

The main components of an EIB network are the zone and line couplers. Not only do they separate individual segments, they also act as routers and filter the data traffic. They pass only those data frames that are actually needed on the other side. Provided the network structure was appropriately chosen (ie, nodes that frequently ex-

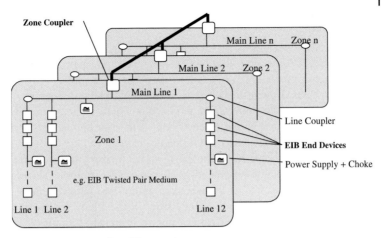

Fig. 3.1-13 Structure of an EIB network

change data are grouped inside a line), this filtering reduces bandwidth consumption and increases the performance. Inside lines, individual segments can be connected via repeaters to obtain the maximum number of (logically addressable) nodes or a cable length of more than 1 km. Note that repeaters do not perform data filtering.

An EIB device usually consists of a BCU (bus coupling unit) and an application module. The BCU forms a universal interface between applications (which can be intelligent field devices, but also simple switches or lamp actuators) and the communication system. It comprises a transmit and receive unit, a communication controller taking care of the fieldbus protocol, and a so-called 'physical external interface' that connects the BCU and the application module. External modules such as switches and actuators are simply plugged on to the BCU, and the BCU automatically identifies them and selects the correct functionality. Apart from these external applications, the controller in the BCU may also host internal applications. However, the computing resources are limited for such applications.

The most commonly used transmission medium for the EIB is a twisted pair line with simultaneous transmission of power and data. For special purposes, other options such as infrared or radiofrequency channels are also available. Transmission over the 230 V powerline is also possible.

3.1.5.1.2 EIB Protocol Characteristics

EIB is a pure peer-to-peer network. The bus access is done by a special mechanism that avoids collisions on the bus. The data rate is 9600 kbps. For the addressing of the nodes, two possibilities exist. During the installation phase of a network, each node is assigned a unique address composed of the zone, line, and device numbers of the node. This address allows a direct communication with individual nodes and is mainly used for network management purposes, ie, initialization, programming, and diagnosis. In normal operation, a group addressing

| Control | Source | Destination | Data Length | User Data | Parity |

Fig. 3.1-14 EIB frame

scheme is used. During installation, the user or system integrator can combine nodes to groups. This permits one to generate logical connections between nodes. An obvious restriction is that a sensor may send data to only one group, whereas an actuator may receive data from different groups.

The EIB frame shown in Figure 3.1-14 shows a structure typical for all fieldbus systems. Apart from the actual data, it consists of control, address, and error control information. Up to 16 user data bytes may be transmitted in a single frame. On the bus line, the individual bytes of the frame are transmitted one by one using the standard UART pattern known from serial interfaces with start bit, stop bit, eight data bits and a parity bit.

To allow the development of interoperable products by different vendors, specifications must exist on how the data are to be interpreted, so that devices can understand each other. In the case of EIB, this has been done by the definition of the so-called EIB Interworking Standards. These standards define data objects tailored to the requirements of building automation. Apart from the objects themselves, they also define the way in which the data types are encoded and transmitted.

3.1.5.1.3 Tools and Certification

Concerning the tools used for the installation of EIB networks, the EIBA pursued from the beginning the goal of having only a single, easy-to-use, and above all company-independent tool. The software package called ETS (EIB Tool Software) was one of the keys to the success of EIB. This software has several modules for the planning, installation, and maintenance of networks. Together with the modular hardware concept (the actual functionality of the hardware is defined by the download of the application software in the course of installation), EIB products have a high degree of flexibility. The existence of only one tool is also a great advantage for the training of the users. It is important to note that the target group for the use of EIB is not highly specialized system integrators, but rather ordinary electricians. This explains the wish to keep things simple.

A second key to the success of EIB is the very strict certification policy of EIBA. Companies who develop EIB components must have their products certified. The certification test ensures the conformity of the product with the EIB specification. Only certified products may be sold under the brand name EIB. This approach is unique in the fieldbus world.

Usually, certification of devices is possible and recommended (also as a marketing argument), but not explicitly required. The good thing about this strict (and expensive) policy is that the EIB is based on a variety of fully tested components. A customer of EIB therefore has the guarantee that all devices will work together, even if they come from different vendors.

3.1.5.2
LonWorks and ANSI/EIA 709

In the early 1990s Echelon, a company based in Palo Alto, CA, introduced LonWorks®, where LON stands for local operating network [10]. Even though the original intention was to have a bus system that can be used for any sort of application, recent developments have focused on home and building automation. In mid-December of 1999, ANSI (American National Standards Institute) adopted LonWorks as the ANSI/EIA 709 standard. In Europe, LonWorks is standardized in ENV 13254-2.

3.1.5.2.1 LonTalk Protocol

One of the characteristics of LonWorks is its unique seven-layer protocol implementation called LonTalk®. Unlike other fieldbus protocols, all seven layers of the ISO/OSI reference model are actually defined and implemented in every single network node. Another characteristic is its elaborated OSI layer 3 that supports a variety of different addressing schemes and advanced routing capabilities as shown in Figure 3.1-15. Every node in the network can be identified with a unique 64-bit physical address (Neuron ID) and with a logical address composed of three addressing elements: the domain ID, the subnet number, and the node

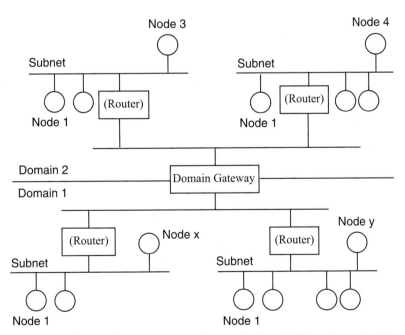

Fig. 3.1-15 Addressing elements in LonWorks netwoks. Up to 32 385 nodes can be addressed within one domain and up to 2^{48} domains are possible

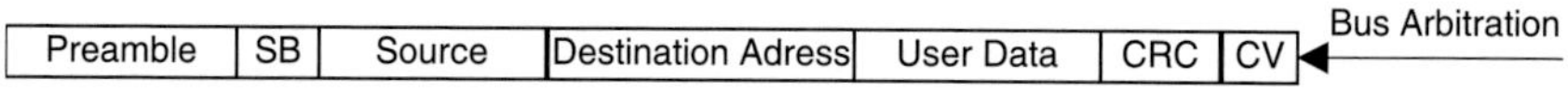

| Preamble | SB | Source | Destination Adress | User Data | CRC | CV | ← Bus Arbitration |

Fig. 3.1-16 LonTalk frame layout

number. Node numbers can be in the range 1–127 and subnet numbers in the range 1–255; hence a maximum of 127×255=32385 nodes can be addressed within a single domain and up to 2^{48} domains are addressable. Domain gateways are used to route packets between multiple domains.

Optionally routers can be used to keep local traffic within the subnet and only forward packets that are addressed to another subnet as shown in Figure 3.1-15. The basic frame layout is sketched in Figure 3.1-16. The preamble allows for bit synchronization at the receiver and the start bit (SB) indicates the beginning of the byte boundary. The source address holds the subnet/node address of the transmitting node and the destination address holds the address of the receiving node(s) followed by the user data that can be between 1 and 228 bytes long. Finally, a 16-bit CRC protects the frame from bit errors (Hamming distance 4) and the code violation (CV) indicates the end of the frame. After the CV a new round of bus arbitration can take place if nodes have data to be sent over the network [10].

3.1.5.2.2 Nodes and Network Topologies

For the last 10 years the Neuron® chip manufactured by Motorola, Toshiba, and Cypress was the only microcontroller that supported the LonTalk protocol. Just recently, with the adoption of the LonTalk protocol as a European and ANSI standard, some platform-independent implementations are available on the market [17]. The traditional LonWorks node is shown in Figure 3.1-17.

The Neuron chip executes the LonTalk protocol and the application program that interfaces to sensors and actuators through the input/output logic. Two types of Neuron chips are available. The 3150 provides an external memory bus interface to expand the internal EEPROM and SRAM with external Flash and SRAM. The 3120 is optimized for low node cost applications and has a built-in 10 kbyte ROM, 2 kbyte EEPROM, and 2 kbyte SRAM but has no external memory interface [18].

An ANSI C derivative language called Neuron C is used to program the Neuron chip. Neuron C uses language extensions to schedule application events and to react to incoming data packets from the network interface. LonWorks supports a variety of different communication media. The most popular media today are 78 kbps or 1.25 Mbps twisted pair communication, 4 kbps power line communication, a 1.25 Mbps fiber optics interface, and RS-485 interfaces at various bit rates between a few hundred bps and 1.25 Mbps.

Depending on the network media and the network transceiver a variety of network topologies are possible with LonWorks nodes. Traditional bus, ring, star, and free topology are supported. Complex networks require networking elements available in LonWorks to separate the local traffic from traffic crossing segment boundaries. Such networking elements include layer 1 repeaters to extend the physical

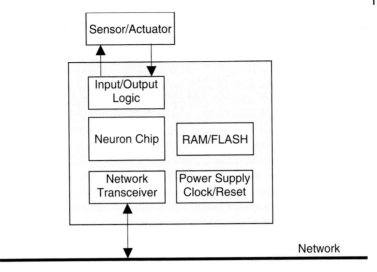

Fig. 3.1-17 Typical LonWorks network node architecture.

length of a network cable, layer 2 bridges and layer 3 routers to decouple individual segments from a networking backbone, and gateways to bridge between the levels of hierarchy in the automation pyramid (see Figure 3.1-15).

3.1.5.2.3 Interoperability and Profiles

Open standards allow individual companies to build single pieces of a bigger puzzle and enable system integrators (SIs) to design complete systems consisting of those building blocks. Node developers must follow certain rules in order to make the bigger picture work. These rules are defined in interoperability guidelines and in profiles. Compatible does not necessarily mean interoperable, let alone plug and play. More elaborate guidelines need to be established for plug and play, interchangeable and interworkable products. Different committees have more or less successfully set up guidelines to define those terms and to establish the rules. The LonMark organization, for example, has published interoperability guidelines for nodes that use the LonTalk protocol [19]. It is important to note that interoperability must be guaranteed on all seven OSI layers.

Still, interoperability on all seven OSI layers is no guarantee of interworkable products. A typical node uses only a subset of all the possible protocol features. Profiles define these subsets based on the intended use of the node. Task groups within LonMark define functional profiles for analog input, analog output, temperature sensor, humidity sensor, CO_2 sensor, VAV controller, fan coil unit, chiller, thermostat, damper actuator, and many more. As of today, LonMark hosts different task groups for fire, home/utility, petrol station, HVAC, lighting, sun blinds, security, and other applications.

3.1.5.2.4 Tools

In order to program the Neuron chip, the user has two tool options. For node manufacturers it might be sufficient to buy the NodeBuilder from Echelon, the more advanced and also more expensive tool is the LonBuilder from Echelon. Both tools allow writing Neuron C programs, to compile and link them and download the final application into node hardware. The LonBuilder supports simultaneous debugging of multiple nodes, whereas the NodeBuilder only supports debugging of one node at the time. The LonBuilder has a built-in protocol analyzer and a network binder to create communication relationships between network nodes.

SIs need tools to design the network and define the communication relationships before the nodes are actually installed in the field. Similarly to creating a schematic of resistors, capacitors, integrated circuits (ICs), or diodes, the SI creates a floorplan of all the nodes in the network.

After physically installing the nodes in the network, a network management tool is required to commission the nodes. During commissioning nodes are assigned a logical address and the binding is created. Binding is the name of the process to establish the communication relationships between sensors, actuators, and controller nodes. Protocol analyzers are used to debug communication problems and to gather traffic and error statistics. During the maintenance phase of the installation, the network management tool also helps to replace faulty nodes and to extend the network with additional nodes and changes in the communication relationships.

3.1.5.2.5 ANSI/EIA 709 Outlook

As mentioned earlier, until 1999 the Neuron chip was the only vehicle to run the LonTalk protocol. Owing to the limited resources available on the Neuron chip (CPU performance, memory space, I/O capabilities), many applications could not be realized with the existing technology. Recent hardware platform-independent implementations of the ANSI/EIA 709 protocol allow for good scalability of the CPU performance and memory requirements tailored for the specific application. Furthermore, simulation of the network behavior before actual hardware exists is now possible as shown in Figure 3.1-18.

Application programs run on top of the ANSI/EIA 709 protocol stack that talks to an interface layer (LDI), which communicates with the virtual network devices. As shown on the right-hand side in Figure 3.1-18, the LDI can also talk to a real physical network, hence existing nodes in a network can be part of the system simulation. The application programs on top will not see a difference if the node already exists as physical hardware or if it is still a virtual device simulated on a PC or workstation [20].

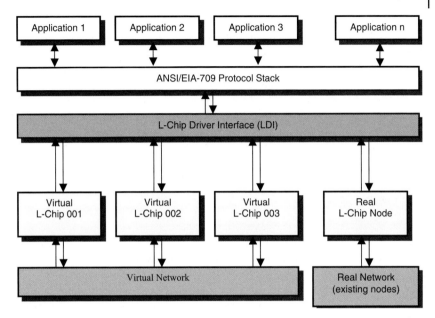

Fig. 3.1-18 Platform-independent ANSI/EIA 709 implementation. Simulation environment with virtual nodes linked to a physical network

3.1.5.3
BACnet

In mid-1987 the ASHRAE (American Society of Heating, Refrigerating and Air Conditioning Engineers) Standard Project Committee (SPC) 135P was established for developing a communication standard for building automation and control networks – BACnet. The first draft was published in 1991. To work on the 507 remarks and comments took a while, so the second public draft was published in March 1994. After a third public draft in spring 1995, the BACnet standard was issued by ANSI as ANSI/ASHRAE on 19 December 1995 [21]. The CEN Technical Committee 247 'Controls for Mechanical Building Services' adopted and published BACnet as a European pre-standard ENV 1805-1 for the management level in spring 1997 and as ENV 13321-1 for automation level in December 1998. As mentioned before, there are different automation pyramids with different numbers of levels. In the building automation and control area the three levels management, automation, and field were defined and BACnet became a European pre-standard for the upper two levels. However, a look at the BACnet protocol stack shows the flexibility of this communication system and in most current projects a two-level architecture is sufficient.

BACnet is a communication protocol for the exchange of data in multi-vendor building automation and control systems and networks. It satisfies the requirements of heating, cooling, ventilation, and control but may also be the basis for integration of applications like lighting, security, and fire control.

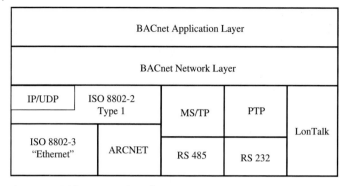

Fig. 3.1-19 BACnet protocol stack

The BACnet protocol stack has a collapsed architecture with the layers 1–3 and 7 of the ISO/OSI model (see Figure 3.1-19). The network layer is the most important one because it connects different lower layers to different communication protocols and because of the routing capability of the network layer. Many fieldbus systems with only two lower layers have difficulties in communicating via a telephone link, eg, for remote control. If the communication system should be flexible and expandable, ie, if routing mechanisms are necessary, then the network layer is needed.

The protocol stack in Figure 3.1-19 shows the broad range of communication media from high-speed LAN communication using the standards ISO 8802-3 (MAC sublayer) and ISO 8802-2 (LLC sublayer) to a slow point-to-point communication link based on the RS 232 standard. For a fieldbus environment there are three possibilities within the BACnet protocol stack: 'Ethernet', MS/TP, and LonTalk. The use of Ethernet in fieldbus systems is growing thanks to the decreasing costs of the hardware interfaces. For the higher layers IP tunneling can be used as well as BACnet/IP, which is recommended. BACnet/IP is specified and published in the normative annex Addendum 135a to the BACnet standard and uses TCP/UDP in a so-called BACnet virtual link layer. The fundamental concept behind a virtual link layer (VLL) is to present a view of some network topology and function to the existing BACnet network layer, taking advantage of whatever functionality is built into the new protocol and adding functions as necessary to maintain the BACnet network viewpoint in the existing standard. The current presentation is limited to IPv4 (the current IP protocol version), but a VLL could easily be developed for any other type of network [22].

MS/TP means master-slave/token passing communication. It is a data-link layer protocol based on the RS 485 standard for the physical layer and was specified as a cheap solution for simple BACnet devices. LonTalk is the communication part of LonWorks. Of the seven-layer architecture of the LonTalk protocol only the lowest two layers are used to obtain a fast and inexpensive communication interface.

Tab. 3.1-2 List of BACnet objects

Generic	Complex
Analog input	Averaging
Analog output	Calendar
Analog value	Command
Binary input	Device
Binary output	Event enrollment
Binary value	Notification class
Multi-state input	File
Multi-state output	Group
Multi-state value	Loop
	Program
	Schedule
	Trend log

The BACnet application layer comprises 21 objects (Table 3.1-2) and 38 application services divided into five groups specified in the ANSI/ASHRAE Standard 135-1995 and the normative Addendum 135b.

The generic objects describe simple sensors and actuators with mandatory and optional properties depending on the hardware and software used. The other objects are closely linked to a specific functionality such as the loop object that describes the properties of a closed-loop controller independently of whether it is a real fieldbus device or a software in an automation station. The five groups for the application services are Alarm and Event Services, Object Access Services, Remote Device Management Services, and Virtual Terminal Services. The most important ones are the Alarm and Event Services and the Object Access Services. They are used for the information exchange between (field) devices and/or operator stations. With a view to the number of object properties and services available, we recognize that BACnet was not specified for the field level. But if the field devices have enough processing power (concerning CPU performance, etc.) and are at least in a mid-price range, then BACnet can be used as a fieldbus system. Otherwise BACnet is able to communicate via gateways with EIB (see Section 3.1.5.1) and LonWorks (see Section 3.1.5.2).

3.1.5.4
EIBnet

The path from a LonWorks application on the field net level to a BACnet application in the management net level via the automation net level could be realized using BACnet with LonTalk (see Figure 3.1-19). For an EIB application a possible solution could be to use the same path but with additional effort regarding protocol conversion. To fill this gap in the information flow from the field level to the management level, the 'EIB on Automation Net' (EIBnet) was specified and pub-

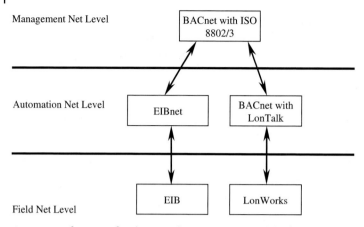

Fig. 3.1-20 Information flow between the management and the field level in BAC (building automation and control) systems

lished by CEN as European pre-standard ENV 13321-2:2000 [23]. EIBnet is able to handle the EIB information coming from the field very easily. The path to the management level is slightly more intricate. BACnet seems to be totally different from EIB. However, in an informative annex of the EIBnet standard an interface to BACnet is described in detail. The relationship between EIB and BACnet object types and properties is shown as well as the conversion from EIB objects to BACnet objects. Figure 3.1-20 shows the possible information flow between the management and the field level via EIBnet.

EIBnet offers the possibility to use the EIB on faster media. EIB devices for the automation net level use an addressing method consistent with the field net level. This allows the use of the EIB with automation level devices such as the Application Specific Controller (ASC) and programmable controllers in home and commercial building environments. EIB offers full compatibility for process communication and for monitoring, engineering, and commanding of applications. Furthermore, EIBnet is well suited for the interconnection of EIB networks (Figure 3.1-21) since it supports the structured wiring approach.

For the physical layer and the MAC sublayer all standards using the ISO/IEC 8802-2 logical link control layer can be used (see Figure 3.1-22). Other types are under investigation. The EIB data link layer for automation net level uses the ISO/IEC 8802-2 with the DL-UNITDATA primitive only. The EIB network layer controls the number of routers and bridges a frame with a group destination address passes on its way from the source to the destination. Layer 3 offers more detailed services to the transport layer by encoding and decoding data services of layer 2 which are mapped to data, broadcast and group data services of the network layer. The transport layer provides a reliable data transmission over communication relationships. Communication relationships are logical channels connecting users of layer 4 with each other. Four different types of communication relationships are provided:

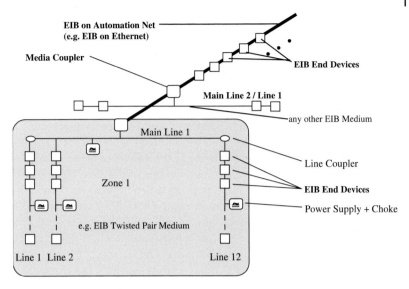

Fig. 3.1-21 Relationship between EIB and EIBnet

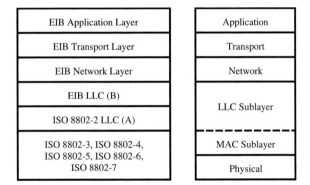

Fig. 3.1-22 EIBnet protocol stack

- one-to-many connection-less (multicast);
- one-to-all connection-less (broadcast);
- one-to-one connection-less;
- one-to-one connection-oriented.

Every communication relationship type provides specific transport layer services. The application layer provides a large variety of application services to the application process. Application processes in different EIB end devices interoperate by using services of layer 7 over communication relationships. According to the transport layer the same four different kinds of communication relationships as in layer 4 exist. Depending on the type of the communication relationship, different layer 7 services are offered.

3.1.6
Fieldbus Systems in Connection with the Internet

FANs as standalone systems are similar to standalone PCs. Both reach top performance when being networked over the Internet. Applications such as remote control, remote monitoring or smart facility management will be possible in a broad spectrum. These fields will become of huge importance in the near future. As we see in today's projects, new ideas are being born permanently.

The idea is tempting but the realization implies problems [24]. The difference between Internet and LANs is not as big as the difference between Internet/LANs and FANs. As explained in the previous sections, FANs have other requirements and constraints and therefore use completely different protocols. One of these differences is the peer-to-peer structure in many FANs. They are organized in a flat hierarchy. Nodes in CAN, in LonWorks, or in the EIB are able to send data directly to other nodes. They are able to perform transactions. On the other hand, the Internet mostly relies on a client-server structure in which the clients initiate a transaction. The server has only to fulfill services the client has requested.

Of course, the Internet also has peer-to-peer protocols such as TCP (Transmission Control Protocol) or UDP (User Datagram Protocol), however, higher protocols such as HTTP (Hypertext Transfer Protocol), SMTP (Simple Mail Transfer Protocol), FTP (File Transfer Protocol), or TELNET have the client-server structure. Likewise, there are FANs that have client-server structures, but those are more commonly used in the industrial and process fields.

The main question is: How can we connect FANs to the Internet or integrate FANs into an Intranet [25]?

Three different approaches are possible, as shown in Figure 3.1-23. We will give a brief introduction to this broad field, owing to its importance in the future and the big difference between principles and solutions.

The first possibility is to tunnel the FAN protocol over the Internet: the communication partners speak the same FAN language, which is absolutely independent of the Internet protocol. In this case two aspects are important: (i) if one wants to integrate a remote service the remote station has to understand the FAN protocol, which means that additional functions in the remote station are required; (ii) the Internet systems have no information about the FAN's data and vice versa. Therefore, this kind of connection is only a solution for rather special applications such as the connection of distant segments of the same FAN.

The second approach is widely discussed today. The idea is to tunnel IP (and consequently also higher layer protocols such as HTTP) over the fieldbus to the nodes. The advantage is that services known from the Internet (such as web access) could be implemented directly on the fieldbus node. However, this needs computing resources that might be available on high-end nodes, but not on smart and low-cost sensors and actuators. In addition, the tunneling of IP over FAN protocols requires appropriate segmentation strategies, because IP packets are too long for conventional FAN data frames.

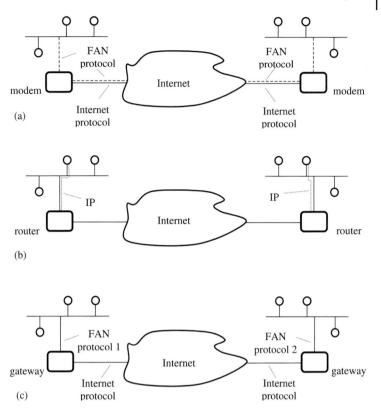

Fig. 3.1-23 FAN over Internet: (a) tunneling FANs protocol over Internet; (b) taking IP as a transport medium; (c) gateway solution

The third option is to connect different networks over gateways: this does not sound tempting, because of the complexity of a gateway, but it is in principle the best solution. The functions of the two protocols (eg, the FAN and an application protocol on the Internet side) have to be converted into one another and a direct mapping is often not possible. However, if only standardized and common protocols and technologies are used, such as Jini, HTTP, SNMP, OLE, CORBA, OPC, or the like [26–29], it is reasonable and doable for various applications.

The Internet is to be seen not only as a network to transmit FAN data and poll functions, but also as a large pool of software designers worldwide. This means that there is a lot of knowledge we can also use for the development of new FAN principles. Of course, the basic properties of the Internet and FANs are different, and the evolution of FANs is much slower than that of the Internet. But still, the approaches pursued in modern Internet programming could be applied to FANs in order to adapt the functionality to the requirements of tomorrow's automation systems.

3.1.7
Present and Future Challenges

Today, FANs are well established, but we must keep in mind that fieldbus technology is very young. We find ourselves at the very beginning of the development. We have borrowed the majority of our ideas from other techniques, especially other networks. For the future, we have to understand the new requirements and constraints that fieldbus technology demands. The main goal is to think about what is possible with this new technology and not about the improvement of 'small networks'.

3.1.7.1
Interoperability and Profiles

As mentioned above, the breakthrough of fieldbus systems came through standardization and improvements that allowed vendors to interconnect their components to multi-vendor systems. However, this standardization suffices in the mechanical world but not in the world of networks, where the complexity of information systems is too high. Standards leave room for interpretation. On the one hand, it is often not possible to describe a system precisely. On the other hand, companies do not want too tight standards in order to retain competitiveness and free space to integrate new own ideas. Certification of fieldbus devices is a suitable way to reduce the problem.

There is still another interoperability problem. The decisive step in Figure 3.1-11 between the early 'fieldbusses', like IC, I^2C, and those of today was the definition of the ISO/OSI model. Today, we are confronted with similar problems and new definitions are needed, as the ISO/OSI model is not enough. In fact, it is less a problem of communication but of application definitions. The first step was the development of profiles for different, particular applications, which limits the freedom of variations of fieldbus devices [30]. Profiles specify communication objects, data types, and encoding. They can be seen as an additional layer of the ISO/OSI model. The only problem is that all the standardization organizations and user groups define these profiles independently of each other. As FANs in different areas tend to be interconnected, however, engineers have the problem of adapting various profiles. It becomes a necessity to define clear object-oriented structures for profiles with inherited objects and also to harmonize the profiles for different fieldbus systems. To solve all these problems several projects have been launched such as NOAH [31], RACKS [32], and SIIA [33].

3.1.7.2
System Complexity and Tools

Figure 3.1-24 reflects the trend of fieldbus technology and thereby also their characteristic problem. Some 15 years ago, when fieldbus technology was discovered as an interesting and efficient technology in industrial automation, systems had

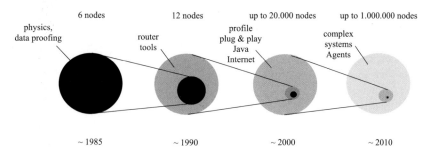

Fig. 3.1-24 Increasing number of nodes in one fieldbus system

average node numbers of 3–6 and later up to 12. Today, in the area of building automation we find numbers up to 20 000 and more. We claim we can increase the number up to one million by 2010 if we extrapolate the experience from the field of computer technology. One reason will be the falling prices of nodes and sensors and the increasing performance of the nodes.

All experts know what complexity means during the integration phase of a network. Today's tools are not appropriate for the setup and management of huge networks. Current principles of network descriptions, the ways of monitoring, controlling, and maintaining networks, and the methods to search and identify failures and bottlenecks are not powerful enough. We need new answers such as the agent concept known from DAI (distributed artificial intelligence), where autonomous units (the agents) move in the network to fulfill specific services. It is necessary to apply these new ideas to FANs [34], although it is clear that the nodes in a FAN are not comparable with Internet nodes in size and performance. It is indeed a challenge we have to face.

3.1.7.3
Management – and Plug and Play

The number of electrical and electronic components in a car has increased to 500 and more over the past 20 years. The only economic way to handle such a high number of units is to network and to control them with computers. The same applies to building automation. But even if we use computers, large networks are difficult to handle, especially if we have node numbers of more than 1000 or even 10 000. The keyword here is 'smart management' and therein the wish for 'plug and play'. What does this mean?

Today, we have problems collecting information because of high equipment costs, development costs, etc. Our electrical components are designed to do only the bare minimum. A light switch only serves to switch on a particular light. It has no additional monitoring and maintenance functionality for the saving of energy or for safety and personal security. People are surprised and amazed if the controller of their car is able to find failures such as defective lights or worn-out brake shoes. We must realize, however, that these additional maintenance functions are just the beginning of this development. In order to get more information about the system, we could and should have thousands of networked sensors in our cars, machines and buildings. This would enable us to design more sophisticated control systems. This is the great challenge of tomorrow. We must keep in mind what kind of information we require and how to collect it in order to construct higher automated management systems. Today, many scientists are working in this field and are likely to continue doing so because of its complexity.

'Plug and play' is to be seen as a part of the management system. If in the future we want to integrate economically thousands of nodes into one system, it will be practical only if we learn to 'plug and play'. Let us consider an easy example. The turning on of a switch of a machine or a stove today works without any redundancy. The user has to check beforehand whether it is safe to turn it on. This check could be automated. The user then no longer actually switches something on, he just initiates a more complex sequence. The control system (technical intelligence) has to test: Is the switch really being used or is the system just reading incorrect information? Is the environment safe and secure? Only then the controller actually turns on the machine and monitors the process as long as it is operating. These principles are becoming more and more common in modern electronic systems.

To be ready for the requirements of complex networks, modern management systems should be equipped with:

- redundant technology;
- built-in test functions;
- reasonable checks;
- learning ability;
- agents ability.

Is that affordable? We think so, if life cycle costs and not only purchase costs are considered.

3.1.7.4
Security

For remote control, smart facility management, and a lot of other things, it is necessary to network FANs over the Internet. Although doubters are few, many hesitate to do so owing to security problems. The Internet is the medium for hackers. How can we prevent them from breaking into our system if the fieldbus is connected to the Internet? Let us again consider an example.

In Europe, most of the utility companies have the need to reduce their costs dramatically because of global competition. In the near future, every European resident will be able to buy energy all over the continent. Therefore, utility companies are seeking solutions to save money. Interesting aspects are demand side management and automatic metering. The daily diagram of power consumption shows high peaks not only at special times but also in the morning, at lunchtime, and in the evening. However, utility companies have to guarantee that customers

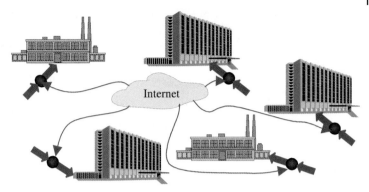

Fig. 3.1-25 Demand side management

have enough power at all times. This means that power plants must produce much more energy, which the consumer does not need but has to pay for. For example, if companies or offices were able to integrate fieldbus nodes into all their machines and to network them through the Internet (Figure 3.1-25), then these nodes could speak and negotiate about energy consumption as well as the individual demands of energy. In this way a lot of energy could be saved. In private homes we can realize the same principle. Afterwards, all utility companies will be able to collect their metering data over FANs and the Internet as well as offering add-on services. These could include the monitoring of burglar or fire alarms, the controlling of refrigerators and the water levels in basements, etc.

How can security be guaranteed by utility companies, especially if we consider the possibility of hackers inside these companies, if all their meters in homes and buildings can easily be accessed via the Internet?

A simple principle for security – other then firewalls and other typical technologies, which are often efficient but not enough for a fieldbus application – is the point-to-point security reached by smart cards. Figure 3.1-26 shows this principle. The hacker can attack different areas: the communication channel over the Internet, the utility company itself (or another company) or the network in the house itself. We have to prevent all of these attacks. This is possible if we take over the electronic money principle based on smart cards. The principle is simple and also easy to handle. In each important electronic component of the house, a smart card for the security part will be integrated. Each of these cards has a matching piece or is a member of a strictly defined smart card group (all of them are smart cards or virtual smart cards). Only these grouped smart cards are able to talk to each other. Highly sophisticated cryptographic algorithms guarantee a high quality of security as in the field of the electronic purse. The utility company can only talk with the nodes of the energy meter. The energy nodes of the different electronic devices which talk about energy consumption are only able to talk to each other and no one else outside the group is able to talk with one of them. This solution for security in FANs is feasible and affordable [35]. The fear of being hacked is baseless.

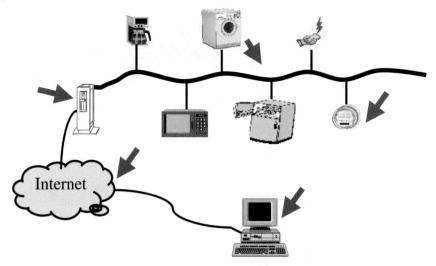

Fig. 3.1-26 Possible attacks on a fieldbus connected to the Internet

3.1.7.5
Driving Forces

Historically, the essential driving forces for integrating FANs were the reduction of wiring, weight, and cost. The main enabling technology of microelectronics could drastically reduce the big EMC (electromagnetic compatibility) and temperature problems of the electronic devices. Meanwhile the focus has completely changed, as new architectures, principles, etc., have been worked out and ideas of new applications have been created, influencing other fields such as sensor technology.

Conventional sensors transform the physical into electrical information, whereas new smart sensors (Figure 3.1-27) have an integrated computing system transforming a simple sensor into a complex unit with the following functions:

- transformation of physical information into electrical data;
- compensation of nonlinearities of components and computing compensated values;
- interface function between sensor kernel functions and fieldbusses;
- parameterization;
- built-in test functions;
- plug and play functions.

Especially the second point shows the direct influence of computer technology: in former times it was the great challenge for sensor producers to develop sensitive and linear sensors. Today, sensors have a much higher sensitivity, as it is possible to integrate electronic amplifiers directly into the sensor itself. This in addition reduces the compensation problems.

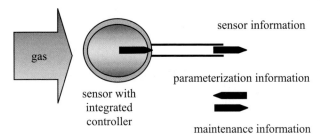

Fig. 3.1-27 Modern interface principle for sensors and actuators

However, the essential change is the direct integration of the controller with the sensors on the same silicon, so engineers now can implement complex mathematical algorithms to compute a value with high precision. An example is simple pressure sensors that use basic formulas, often based on linearized models and ignoring additional influences such as temperature. If an additional temperature sensor and a controller were to be integrated into the pressure sensor on the same silicon, complex mathematical algorithms could be used to compute high-precision pressure values. This can mean cost reduction, as the sensor components can be cheap, because the integrated controller can numerically compensate for all deficiencies.

The last two items above point into the future. As already explained in this section, plug and play is a great challenge and is expected to offer interesting solutions. However, today this area is still under intense scientific investigation [36, 37].

In general, the influence of microelectronics, computer technology, and fieldbus systems on sensors will lead to cheaper and more precise sensors that are easy to install in fieldbus networks. This will increase the number of sensors and also the number of fieldbus installations.

In addition, other important factors will be responsible for an enormous growth in the number of fieldbus nodes. Micromechanics has opened up a new world of sensors, which have reached a highly mature state of development. This gave a great chance to the automotive industry for a new kind of low-priced sensors. As an example consider the fully integrated airbag sensor. Similar sensors will follow for rotation, brake pressure, and all kinds of gas identification. These components will be integrated not only in cars, but also in industrial and process automation and in home and building automation.

These are technological influences. As already explained above, economic influences include the demand for saving energy and costs in services. Automation will provide solutions by collecting information about all processes, by building networks of an increasing number of sensors. Facility management is a good example: it costs a lot of logistic organization and money to collect all the data needed. How many devices are in a building, in a hospital, or in a plant? Managers want to have this information up-to-date and at a low price. Concepts for such solutions based on fieldbus systems and specific sensors are an area of intensive research.

3.1.8
Outlook and Conclusion

As discussed above, the number of FANs will increase in a dramatic way in the near future, and there will be thousands of networked nodes in all kind of applications. It is becoming more and more common to take biological systems as a model for innovations or for efforts to improve existing features. Material science, avionics, and other disciplines have successfully borrowed concepts from nature. To be able to do this, we can compare different forms of life, eg, ameba, bugs, and human beings. If we look at energy consumption, housing, and information processing, it is interesting to see how nature changes the underlying principles, ranging from very simple organisms to highly flexible systems such as a human being. Of course, the energy consumption and the housing of an ameba and a bug are optimal for their life, but, on the whole, the system of the human being is the most flexible and the one with the highest performance.

Let us now try to concentrate on information processing: the nervous system. The eyes are only one component of the human 'I/O system'. We have thousands of other sensors to perceive the reality. Similarly, cameras are not enough to collect relevant information, eg, for smart houses. Also, we do not have only a brain to control the inside and outside processes, but also an enormous amount of peripheral nerves that do already some information preprocessing. Similarly, a PC or IPC is also not enough to control physical processes.

We have a perception of muscles, to collect information about the position and state of our arms, legs, and face. We have an internal perception to control many processes in our body including our breathing. And we have an external perception to feel objects, temperature, and even the wind. In a technical system, the fieldbus takes the role of the peripheral nerves. So, the principle of collecting information through large numbers of sensors seems to have all characteristics of an optimal technology.

3.1.9
References

1 TÖPFER, H., KRIESEL, W., *Regelungstech. Praxis* **24** (1982) 336.
2 WANSER, K., *Automatisierungstech. Praxis* **27** (1985) 237.
3 Many philosophers and psychologists refuse to speak of 'intelligence' in this context. However, we define technical intelligence in the area of digital chip design as the ability of an electronic circuit to store or manipulate data. So the term intelligence is not to be compared with the intelligence of an animal or human being.
4 This was the dramatic and most important evolutionary step in telecommunication systems, when digital connections were introduced and replaced analog systems.
5 KOOMEN, C. J., *The Design of Communicating Systems*; Dordrecht: Kluwer, 1991.
6 TANENBAUM, A. S., *Computer Networks*, 2nd edn.; Englewood Cliffs, NJ: Prentice-Hall, 1989.
7 TANENBAUM, A. S., *Distributed Operating Systems*, Englewood Cliffs, NJ: Prentice-Hall, 1994.

8 *Auswahl von Feldbussystemen durch Bewertung ihrer Leistungseigenschaften für verschiedene Anwendungsbereiche*, VDI/VDE 3687, VDI/VDE-Richtlinien; Frankfurt: Verein Deutscher Ingenieure – Verband Deutscher Elektrotechniker; ICS 35.240.50, 1998.

9 KOPETZ, H., *Design Principles for Distributed Embedded Applications;*, Dordrecht: Kluwer, 1997.

10 DIETRICH, D., LOY, D., SCHWEINZER, H.-J., *LON-Technologie, Verteilte Systeme in der Anwendung*, 2. Aufl.; Heidelberg: Hüthig, 1999.

11 SINHA, P. K., *Distributed Operating Systems*; IEEE, New York, 1997.

12 SAUTER, T., WOLLSCHLAEGER, M., *it + ti* **42** (2000) 7.

13 GÜRTLER, G. H., in: *Feldbustechnik in Forschung, Entwicklung und Anwendung*, DIETRICH, D., SCHWEINZER, H. (eds.), Springer, 1997, p. 2.

14 THOMESSE, J.-P., LEON CHAVEZ, M., in: *Fieldbus Technology*, DIETRICH, D., NEUMANN, P., SCHWEINZER, H. (eds.), Berlin: Springer, 1999, p. 2.

15 DIETRICH, D., SAUTER, T., in: *Proceedings of WFCS 2000*; Porto: IEEE, 2000, p. 343.

16 DIETRICH, D., KASTNER, W., SAUTER, T., *EIB Gebäudebussystem*; Heidelberg: Hüthig, 2000.

17 SCHWEINZER, H., *VENUS-Vienna Embedded Networking Utility Suite*; Loytec Electronics, www.loytec.com, Wien, October 1999.

18 Motorola, Austin (TX), *LonWorks Technology Device Data*, DL159, Rev. 4, Q4/97, 1997.

19 LonMark, *Application Layer Interoperability Guidelines*, Version 3; LonMark Interoperability Association, 1996.

20 BAUER, A., SOUCEK, S., in: *Proceedings of LonWorld*, Orlando (FL), 2000.

21 *A Data Communication Protocol for Building Automation and Control Networks*, ANSI/ASHRAE Standard 135-1995; ASHRAE, December 1995.

22 BENDER, J., NEWMAN, M., *BACnet/IP Tutorial;* www.bacnet.org/Tutorial/BACnetIP/default.html.

23 *Data Communication for HVAC Application – Automation Net – Part 2: EIBnet*, ENV 13321-2:2000; Brussels: CEN, February 2000.

24 SAUTER, T., PALENSKY, P., *e & i* **17** (2000) 314.

25 In the classical definition Intranet is a LAN using the Internet protocol. Today, especially FAN user groups have taken over this expression and take the original pure meaning of this word: Intranets are all networks inside of a building or a house. To simplify matters we also follow this idea.

26 BANGEMANN, TH., DÜBNER, R., NEUMANN, A., in: *Fieldbus Technology*, DIETRICH, D., NEUMANN, P., SCHWEINZER, H. (eds.); Berlin: Springer, 1999, p. 180.

27 RÜPING, S., KLUGMANN, H., GERDES, K.-H., MIRBACH, S., in: *Fieldbus Technology*, DIETRICH, D., NEUMANN, P., SCHWEINZER, H. (eds.); Berlin: Springer, 1999, p. 240.

28 NEUMANN, P., IWANITZ, F., in: *Proceedings of WFCS'97*; IEEE, Piscataway (NJ), 1997, p. 247.

29 KNIZAK, M., KUNES, M., MANNINGER, M., SAUTER, T., in: *Proceedings of WFCS'97*; IEEE, Piscataway (NJ), 1997, p. 223.

30 DIETRICH, C., in: *Fieldbus Technology*, DIETRICH, D., NEUMANN, P., SCHWEINZER, H. (eds.); Berlin: Springer, 1999, p. 90.

31 DÖBRICH, U., NOURY, P., in: *Fieldbus Technology*, DIETRICH, D., NEUMANN, P., SCHWEINZER, H. (eds.); Berlin: Springer, 1999, p. 414.

32 QUADE, J., in: *Feldbustechnik in Forschung, Entwicklung und Anwendung* DIETRICH, D., SCHWEINZER, H. (eds.); Berlin: Springer, 1997, p. 12.

33 RAUSCHER, TH., FISCHER, P., KABITZSCH, K., *it + ti* **42** (2000) 38.

34 PALENSKY, P., in: *Proceedings of ETFA'99*; Barcelona: IEEE, 1999, p. 917.

35 PALENSKY, P., SAUTER, T., in: *Proceedings of WFCS 2000*; Porto: IEEE, 2000, p. 27.

36 DIETRICH, D., *it + ti* **42** (2000) 5.

37 DIETRICH, D., NEUMANN, P., SCHWEINZER, H., *Fieldbus Technology – System Integration, Networking, and Engineering*; Berlin: Springer, 1999.

3.2
Wireless In-building Networks
MIKE BARNARD, *Philips Research Center, Redhill, Surrey, UK*

3.2.1
Introduction

For a building to be 'intelligent' one might expect that a 'central nervous system' would be required to connect together the various sensors and actuators within the structure. To continue this medical analogy further would imply that the building be wired with 'nerves' of wire or optical fiber, but of course this is not necessarily true as the option almost always exists to use a wireless rather than a wired link.

In this section, the features or peculiarities of wired and wireless links to and from sensors are compared, followed by overviews of existing and emerging wireless standards and products. Throughout the main focus will be on low-power/low-bandwidth networks suitable for sensors, telemetry, etc.

3.2.2
Network Characteristics

To understand how and when a wireless network might be appropriate, it is first necessary to consider the fundamental characteristics of wireless networks and compare these with the alternative of a wired network. In this section the differences between wired and wireless networks are therefore considered, along with the typical requirements that different types of sensors might make upon such networks. In doing this a range of sensor types are assumed from environmental sensors (eg, temperature or humidity sensors) with simple requirements to complex surveillance sensors (eg, CCTV cameras) with obviously much greater requirements.

3.2.2.1
Wired vs. Wireless?

The main thrust of this section is to establish the capabilities and limitations of wireless systems in the reader's mind, so in principle enabling in any situation an answer to be given to the question of whether a wired or a wireless network is more appropriate. However, in many practical cases a more relevant question is whether or not to wire, especially in an existing building. This is not exactly the same question since many so-called 'no new wires' solutions [1, 2] rely on the reuse of existing wiring such as mains cabling or telephone wiring.

In general, the benefits of using a wireless system can be summarized as one of two possibilities. Either the primary benefit of a wireless approach is that it allows mobility of someone or something connected to the in-building network or it allows for easier installation of an essentially fixed network. One has only to consider the increasingly widespread use of cordless telephones within buildings to see how the benefit of mobility in the telephone function is appreciated and desired. On the other hand, the variety of home security systems now available which link the sensors wirelessly together is a good example of how the installation of such a system can be made much easier when there are no wires.

Conversely, wireless networks are by no means ubiquitous and a number of benefits can also be claimed for wired systems. For example, wired systems are almost always lower in cost than their wireless analogues, at least if wiring installation costs are ignored. Second, it is generally much easier to achieve a given capacity in terms of bandwidth (the number of bits per second that can be carried) in a wired network. Third, the 'quality' of a wired network, in terms of its reliability, immunity to interference, and even its security, is inherently higher than that of a wireless system. Lastly, there is the small matter of power. If a sensor is connected by a wire, then that wire can be used to carry electrical power if required to the sensor. Without the wire, each sensor has to be battery powered.

So in general terms there is always a trade-off. Either the requirement for mobility is strong enough or the 'cost' advantage (in terms of money, time, etc.) is large enough that a wireless solution is better or not, in which case the wired solution is obviously to be preferred.

3.2.2.1.1 A Full Range of Choices

The choice, of course, is not as simple as between wired or wireless, as there are many options for both. In the wired case a basic distinction may be drawn between newly installed wiring and the so-called 'no new wires' solutions. In the latter case the wires that are reused are normally either the mains wiring [3] or existing telephone connections [4]. Similarly for wireless, there is a basic choice between infrared (IR) and radiofrequency (RF) connections.

IR technology is already very familiar in its use within remote controls. As many people are familiar with the behavior of such devices, it should be easy to appreciate that IR has a major limitation in that it requires a 'line of sight' between the two ends of the link. This has at least the effect of confining any IR transmission within one room, in the circumstances either a useful feature or a very awkward restriction. The main benefits of this limitation are that IR frequencies are unregulated and therefore freely available for use, and the security and immunity to outside interference are inherently higher than for RF. However, for most in-building applications there is an assumption that the network will cover many if not all rooms and therefore some other method of moving signals between rooms needs to be found.

RF, on the other hand, will normally pass through walls (depending ultimately upon the material and thickness of the wall and the frequency of the RF signal), which makes it the natural choice for systems where movement between rooms and/or wireless inter-room communication are required. However, RF also has a downside in that the frequencies are regulated, with only certain frequency bands available for use, and external RF signals can interfere with the system unless suitable system design considerations are followed.

3.2.2.1.2 A Compromise
Of course, there does not have to be a straight decision as to which technology to use, as many systems successfully combine more than one type of technology. Indeed, good modern network design allows such mixing of technologies very easily. The so-called Internet Protocol (IP) [5] upon which the Internet is based is primarily a defined method of sending messages over many different networks, and is thus a very prominent example of how such mixes of technology can be made. Indeed, many of the proposed in-building networks, such as universal plug and play (UPnP) [6] or LonWorks [7], are specifically designed to allow a mixture of technologies in the network links.

In the case of IR it would in fact be necessary to mix technologies in order to make inter-room links, and even in the case of RF to cover a whole building may require several 'cells', each normally connected together by some kind of wiring. Hence it can be seen that mixed technology or heterogeneous networks are not only possible but in fact might be necessary or highly desirable. In this way wireless technology is used for those links where mobility is required or where new installation of cables is difficult, whilst wires are used on fixed links where trunking routes are easily available.

3.2.2.2
Sensor Network Requirements

The characteristics of the network, whether wired or wireless, depend greatly upon the characteristic of the sensors themselves. Some sensors, eg, a thermometer or thermostat, are relatively undemanding, producing a minimal amount of information at a slow rate. Other sensors, such as CCTV cameras, produce many complete images per second, which can be very demanding on the network. In between these two extremes come remote controllers, which like the simple sensors will have a very low net data rate but which nevertheless are demanding in terms of the maximum permitted latency in response. In other words, if an operator pushes a button on a remote control then the system should respond very quickly.

In general, the demands that a sensor places upon the network are basically set by the amount of information the sensor produces, how far that information needs to be sent around the building, and how fast it needs to be delivered. In addition, the aggregation of all the sensors in a building place additional demands upon a wireless network in terms of the number of device addresses that must be accommodated, the number of separate wireless networks that are within interference range of each other, and so on. Finally, there are a number of integrity demands that an intelligent building network will place upon a wireless network, in terms of an acceptably low probability of messages failing to reach their destination, of message errors occurring without detection, and of device identifications becoming corrupted.

The information flow rate is normally measured as bits (or kilobits, megabits, etc.) per second (b/s, kb/s, Mb/s, etc.), whilst the distance that the information needs to be sent is of course measured in meters. Any network technology will have upper ratings for the available bandwidth and range, whether this is point to point range for a wireless system or tolerable length of cable for a wired system. The response time, or latency, of a system is usefully measured in milliseconds (ms) and can become an issue with both wired and wireless systems, although the situation is normally more critical in the wireless case as extra delays can result from the efforts required to protect messages from corruption during transmission.

3.2.2.2.1 Range

To understand the mechanism by which range limitations occur in a wireless system, it is perhaps helpful to consider briefly how radio signals propagate. Generally the most easily perceived range limitation arises from the fact that the further that a radio receiver is from the transmitter the weaker is the received signal. Anyone with experience of a normal broadcast radio signal will be readily aware of the apparently increasing amount of noise associated with a weakening signal from a radio station. In a digital system this noise will tend to corrupt the received signal data, which can be prevented for a while by applying various error protection or error correct schemes. Sooner or later however the received signal-to-noise ratio (SNR) becomes so low that the radio link breaks down.

Other effects also conspire to limit the effective range of a wireless signal, particularly two known as shadowing and multipath. In the case of shadowing, an object or objects which are opaque to the radio signal block (or at least attenuate) the signal and so reduce the effective range. In the case of multipath, what typically happens to a radio signal is that in addition to being partially absorbed by objects in its path it is also partially reflected. These reflected signals can travel by many different routes to end up at the receiver, which can lead to a mechanism known as destructive interference. This means that the received signal will appear to be very much weaker in certain places that might be expected owing to shadowing or straight forward distance losses.

Fortunately, all of these effects are well understood and can be modeled. Indeed, a number of computer simulation packages are available which can predict in great detail how a radio system will behave in a given building, eg, the WISE

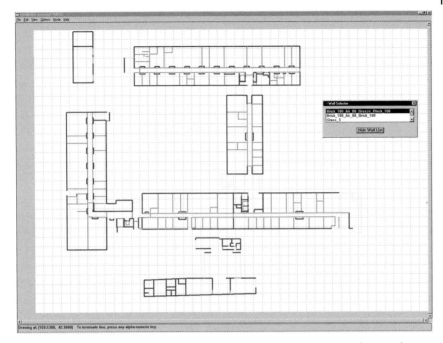

Fig. 3.2-1 Radio propagation prediction tool screenshot showing (color) classification of construction materials

tool from Bell Laboratories [8] and the PlaceBase tool from Ascom [9]. These packages generally require a plan of the building to be entered into the computer, either in electronic form (generally as a file in Drawing Interchange Format (DXF)) or as a scanned image. The user then overlays information on this plan about the building materials used to construct the various walls, and finally a computation is performed to determine both the distribution of the radio signal strength and the likely magnitude of any multipath effects. Typical screenshots from such a package are included in Figures 3.2-1 and 3.2-2, showing, respectively, the building plan with building material information superimposed on all the walls and calculated signal strength from one radio transmitter.

Additional information such as the local multipath conditions can also be determined by such tools, enabling in principle a wireless system planner to dimension a radio system accurately prior to installation.

As mentioned above, interference between neighboring radio systems can also be an issue, and of course these same software tools can also be used to predict levels of interference in a given setting.

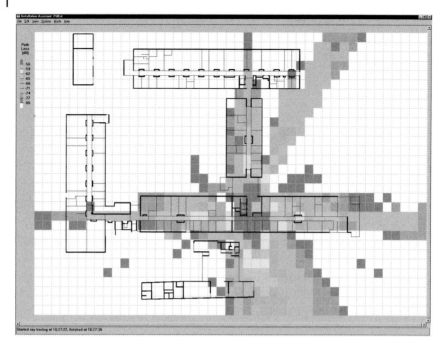

Fig. 3.2-2 Radio propagation prediction tool screenshot showing predicted signal strength of signal from one transmitter

3.2.2.2.2 Speed

Two main factors limit the speed at which information can flow through a wireless network. The first is basically governed by the amount of radio spectrum available to contain the signal, together with the amount and type of error protection applied to the signal, which limits the bit rate of the signal transmitted. The second factor is the way in which the timing of the radio system is arranged, which limits the speed of response of the system to even the smallest amount of data.

The bandwidth of a radio signal is measured in hertz (Hz) and is limited within any radio system in order to keep the transmitted signals inside the allocated spectrum. Within this bandwidth for a digital system a given number of bits per second can be transmitted. The exact ratio between the bit rate and the bandwidth depends upon the type of modulation used for the radio signal, and can range from very much below 1 for so-called spread spectrum systems to 5–6 for very high-order modulation schemes. Whilst the details of this relationship between bit rate and bandwidth are of limited importance to the overall network designer, the type of modulation scheme used can affect how neighboring radio systems interfere with each other. This is discussed further below in the section on co-existence.

The bit rate is, of course, critical in situations where a continuous stream of data, eg, from a CCTV camera, has to be transmitted. In this case (assuming a digitized signal!) the information rate is known, to which needs to be added an overhead data rate to allow for error protection schemes. This additional overhead generally varies from 50% to 100% of the original bit rate, depending upon the level of error protection required. The resulting bit rate must, of course, be lower than the capacity in the radio link for a system to perform properly.

In addition to a straightforward consideration of the bandwidth of the radio link, the speed of response of the system is also limited by the timing scheme used. Such timing schemes occur for two main reasons, the first being to organize the 'traffic flow' of information between many nodes in a network and the second being to allow battery-operated equipment to power down for a large proportion of the time, thus extending the battery life.

Typically, a radio system will have a defined frame structure, which essentially means that there is a periodic cycling of the various activities that are required to keep the network running. So, for example, Bluetooth [10] has defined time slots of 625 μs which are available alternately for the 'master' and any of the 'slaves' within the network to send data. In principle then one might expect a delay of no more than 1.25 ms before a single packet of data can be sent. However, this simple consideration ignores the fact that several of the available timeslots may already be pre-allocated by the system to other units on the network, meaning that a wait of many timeslots might ensue before a particular packet of information is sent. Furthermore, since communication flow in Bluetooth (and in many other radio systems) is only between the master and a slave device, then any packet requiring ultimately to go from one slave device to another has to make two hops, the first from the originating slave device to the master and the second from the master to the destination slave. In principle this could mean twice the delay since each hop must wait until a free timeslot is available. Of course, if an acknowledgement is required for the packet sent, then yet further delays might result.

A far more serious delay can occur if one or more of the radio nodes is powered down. This is typically the case for battery-operated equipment, as this is a way to save a large amount of power and so extend the lifetime of the batteries. The duration of the sleep periods can vary enormously from one system to another, but figures anywhere between 1 s and 1 min are not uncommon. However, response delays for the case where an operator pushes a button should ideally be below 100 ms so in a situation where this is likely to be an issue the sleep period of any wireless devices will need to be adjusted to take this into account, and balanced against the potential power saving that could be made.

3.2.2.2.3 Reliability

In any radio system errors will occur in the data bits that are transmitted over the air. Since one of the main causes of these errors is the ever-present electrical noise that occurs in radio channels, it is easy to appreciate that this error rate can be reduced to fairly low levels in most conditions by increasing substantially the amount of power transmitted, so that the received signal is very much stronger than the noise. However, it is rarely desirable to transmit this much power since this wastes energy from what might be a limited battery supply and also it reduces the use that other systems may make of the nearby radio spectrum, which is a limited resource. Also, turning up the transmitted power in situations where there are severe multipath effects causes little improvement since the transmitted signal is interfering with itself. Thus, for all practical digital radio systems the occurrence of errors is taken for granted and the system is designed to cope with them. In fact, the 'raw' error rate in a radio channel can easily be as high as one bit in 100, so the error protection required is very significant.

Error protection is implemented using the twin approaches of error detection and error correction. These in turn are achieved in two basic ways, both of which involve sending extra information.

The first basic scheme involves sending packets of data to which extra bits called a checksum are added. At the receiving side the data is used to generate a local copy of the checksum which is compared with the received checksum. If the two agree then the packet is acknowledged by sending back a short signal to the transmitter. If the two checksums do not agree, however, then one or more errors have occurred in transmission and so either a non-acknowledgement signal is sent or no acknowledgement signal is sent. Either way the same data is transmitted again and the checksum compared as before. This is known generically as Automatic Repeat reQuest (ARQ).

The second scheme involves using a predefined coding scheme to translate from the given data packet to a much longer data packet which is then transmitted. On the receiving side a decoding process is used to recover the original data bits. To be of any use the coding scheme used must be such that the decoding can recover the sent bits even though many errors have occurred in the coded data which was transmitted. There are many well known coding schemes such as Reed-Solomon coding or Bose-Chaudhuri-Hocquenghem (BCH) coding which are commonly used to improve the error performance of radio systems [11]. In practice, a combination of coding and ARQ is often used since the ability of coding to detect errors far exceeds their ability to correct them.

These basic approaches to error protection, or more complex versions of the same basic approaches, can have a very profound effect on the error rate actually suffered by the data being transmitted. Although error rates of one in a hundred may be common on the radio channel itself, the protected data may well be suffering much fewer than one in a million, which makes the radio system substantially more useful. Of course, extra layers of protection may be added still to give exceptionally high levels of error protection, although always at a cost of extra complexity and normally extra delay.

3.2.2.2.4 Co-existence

Traditionally, radio signals were kept separated by allocating to each a different frequency. Most people are of course familiar with this concept since a radio set must be 'tuned' to the frequency of whichever radio station the listener wants to hear. However, there is only so much usable radio spectrum so sooner or later it becomes inevitable that radio systems must share the same spectrum. Even so, if the two systems are some distance apart this is not a great problem since the signals from one are very weak by the time they reach the other.

A more pressing problem occurs when two or more radio systems are in close proximity. In this case there are three basic ways in which these radio systems may co-exist, each of which involves dividing up the available spectrum in some way. The easiest is for each system to only use part of the allocated spectrum at any one time (known as Frequency Division Multiple Access or FDMA). Another is for each system to only use the allocated spectrum for part of the time (known as Time Division Multiple Access or TDMA) and the third is for each radio system to use special and unique codes on their transmissions (known as Code Division Multiple Access or CDMA). These last two terms are very common, for example, in cellular telephony systems; indeed, two of the cellular standards used in the USA are commonly referred to as 'TDMA' and 'CDMA' after the fundamental scheme that each uses.

In practice, for the smaller radio system that might typically be used within a building a mixture of FDMA and TDMA is often used. Most new local area radio systems such as IEEE 802.11 [12] and Bluetooth [10] send data in discrete packets that occur only at certain times and which are restricted to a small frequency band within the overall allocated band in which they operate. In fact, both of these systems will normally 'hop' frequencies so that each subsequent packet is sent at a different frequency in an apparently random sequence. In this way several systems may operate in close proximity with the probability of a 'collision' at any given moment quite low. Of course, collisions do occur, which add to the noise-induced errors encountered by the data. However, these collisions are normally well spaced by careful selection of the 'random' frequency-hopping patterns and the error protection is sufficient to maintain a good link.

3.2.2.2.5 Positioning

One requirement that may exist for a wireless sensor network is that of being able to determine the current position of a sensor within the building. Typically this might be in order to track a valuable item, or person, around the premises. Alternatively, there may be a requirement to correlate measurements taken by a mobile sensor with the position of the sensor at the time of each measurement.

Two basic positioning techniques can be employed by wireless systems. The first is to use a set of low-power beacons whose signals can only be detected when a receiver is physically close to the beacon (in the same room, for example). A more sophisticated version of this approach is to also measure the power of the received beacon signal and use this to estimate the range from the beacon to the receiver. In the extreme of this case the power of the beacons can be turned up so that signals from several beacons can be received in any given location. Provided that the signal from each beacon can be correctly identified and its strength measured, then a predetermined (measured or calculated) database can be used to turn any particular combination of beacon signal powers into a unique position inside the building. Simple IR beacon systems have been used on many occasions, ranging from automatically controlling a commentary given to visitors in a museum based upon their current location to a system designed to call up a users computer details wherever he or she approached a computer terminal.

Alternatively, the distance between a beacon and a moving receiver can be determined by measuring the time taken for the signal to travel from one device to the other. This can be done in principle because radio waves are electromagnetic waves, another manifestation of which is light waves, and like light waves, radio waves travel at the (constant) speed of light. In practice, of course, within a building the distance between any beacon and the moving receiver will be short and so the time taken for the radio wave to make the journey will be very short. To make measurements of the required accuracy either a two-way measurement has to be made or some sort of differential measurement is made between the signals from two beacons. Either way, the possibility of a clock error between the two radio units is eliminated and accurate measurements can be taken. This ranging method is also used in the well-known Global Positioning System (GPS) [13], which uses satellites above the earth as its beacons. However, GPS signals at the earth's surface are not very strong, and very often are unable to penetrate sufficiently the walls or roof of a building to be of much use in determining positions indoors.

3.2.3
Existing and Emerging Standards

The field of wireless networks is growing rapidly and as a result there is a proliferation of new proprietary and standard solutions becoming available. In general, there are many aspects to each of these solutions, which for simplicity can be categorized into three groups. The first of these is the design of the wireless network itself (which in itself can, of course, be broken down into many more detailed facets), the second is the way in which the network is used and controlled, and the third is the frequency bands which the system uses. The first two are, of course, common to any network, wired or wireless, whereas the spectrum issue is a peculiarly wireless issue.

A typical pattern for any particular application area is that a number of companies will produce solutions which are at least to a degree proprietary, whilst simultaneously the industry at large tries to reach agreement on a standard. Once an application area becomes established then there is a great likelihood that such a standardized solution will come to prevail, although not always. A good example of this is the wireless Local Area Network (LAN) market, where under the auspices of the IEEE many companies laboured to agree a standard. Whilst this was still under discussion, however, several products were brought to the market, all of necessity proprietary since the standard was not finalized. Once the standards work was finished, however, then a lot more products started to appear which adhered to the standard, and gradually the market shifted towards 'compliant' products. It should be noted, however, that this pattern can only occur with products which use the 'unlicensed' radio spectrum. With the licensed spectrum, there is far greater control from the regulatory authorities, which tends to favor a delay of product introduction until the standard is complete.

3.2.3.1
Network Standards

As was hinted at above, when considering the wireless network various further degrees of functional categorization are often used. In principle, this is often based upon the ISO/OSI seven-layer model [14], which embodies the notion of an application making use of a protocol stack to communicate to another application. This protocol stack is further defined so that the 'upper' layers are concerned with the overall behavior of the network and the 'lower' layers are concerned with the behavior of individual links within the network. An important principle within this arrangement is so-called peer-to-peer working. This means that an application at the top of the stack will effectively behave as if it is directly communicating with a remote (peer) application, although in fact it is doing so via the complexities of the network protocol upon which it rests. Similarly, the upper protocol layers will appear to behave as if they are directly communicating with their peers in other network nodes, even though they are actually doing so via the lower layers to which they are connected. In fact, only the lowest layer literally communicates via the physical wire or via the radio link.

The importance of these distinctions is the notion of independence between the various layers, so that the lower layers of the network can be one of a number of wired or wireless possibilities (and of mixed type even inside one network), whilst a unified upper layer resides on top of this. This is shown in Figure 3.2-3, where two links are connected together to make a connection between the two applications at either side. It can be seen from this diagram that only the lowest (physical) layer actually communicates directly with its peer at the other side of the local link. The data link layers also operate locally in order to control the flow of data on the physical links and ensure adequate link performance. Above these layers the network layer ensures the correct routing of the information (trivial in this diagram but not so in a real network). Above these three basic lower layers then the protocol effectively works from end to end, ensuring signal integrity. In actuality, all the data flow down the originating stack, across the first physical link, up to the intermediate network layer, back down to the next physical link, and then all the way up the receiving stack.

Having established this differentiation in layers of the available standards the rest of this section is devoted to the network or 'upper' layers. Table 3.2-1 shows the main standards that exist at the time of writing, although as was emphasized above this is a dynamic area and this list would be expected to change over time. Since these network standards are in general able to work over any type of link, they become immediately relevant to the subject in hand since wireless networks will make use of these standards as well as of the lower layer wireless links.

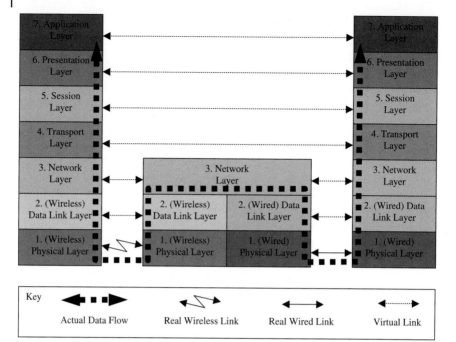

Fig. 3.2-3 Illustration of peer-to-peer working at various levels of the protocol stack, showing different levels operating locally or across the network, allowing a mixture of wired and wireless links

Clearly, a number of these standards have much more capability than is required for a simple sensor system, but with the growing trend towards interconnecting different networks together it might well be the case that a highly configured network is required for one particular function and then also serves all other easier interconnection needs.

3.2.3.2
Wired Links

There are, of course, a number of standards defined for wired networks, and whilst this is not the main thrust of this section there is still some relevance in considering them briefly, since they have tended to dictate the form of a number of the in-building network standards above. In this context it can become necessary for a wireless link to mimic in some way the behavior of the wired link in order for the network to be able to accommodate a wireless connection. A list of the most common wired standards is shown in Table 3.2-2.

Tab. 3.2-1 In-building network standards

Standard	Description
CEBus [15] (Consumer Electronics Bus)	Defines a Common Application Language (CAL) within a 'home plug and play' specification, providing equipment manufacturers with a uniform way of producing products that will work together over multiple media and communications protocols
EHS [16] (European Home System)	Defines a complete network system which supports all domestic functions, in a modular, easily extendible and automatically configurable way based upon the OSI reference model
HAVI [17] (Home Audio Video Interoperability)	Defines certain software elements common to all (particularly audio and video) appliances on the network to ensure that the software elements of different appliances will work together
HES [18] (Home Electronic System)	Intends to specify hardware and software that enable a manufacturer to offer one version of a product for connection to a variety of home automation networks, also to networks for command, control, and communications in commercial and mixed-use buildings
JINI [19]	Provides simple mechanisms which enable devices to form an 'impromptu' community – a community put together without any planning, installation, or human intervention. Each device provides services that other devices in the community may use, and also provides its own interfaces to ensure reliability and compatibility
LonWorks [7]	A distributed network with no central controller which consists of: 'Local Control Nodes' (which control the operation of attached sensors and/or actuators) along with 'Supervisory Nodes' (which collect and log data from local control nodes, or coordinate the operation of one or more local control nodes), 'Network Services Tools' (which install, configure, etc., the network) and 'Routers' (which provide transparent connectivity between two LonWorks network channels)
UPnP [6] (Universal Plug and Play)	A distributed, open networking architecture that leverages TCP/IP and the Web to allow seamless proximity networking in addition to control and data transfer among networked devices in home, office, small businesses, and commercial building environments. There is support for zero-configuration networking and automatic discovery whereby a device can dynamically join a network

Tab. 3.2-2 Wired standards

Standard	Description
EIA232 [20] (RS232)	Developed as RS232 in the early 1960s as a common interface standard for digital data exchange between a centrally located mainframe computer and a remote computer terminal. Since that time the standard has been modified three times, resulting in the EIA232E standard introduced in 1991. Today, virtually all current serial interfaces are EIA232-like in their signal voltages, protocols, and connectors
USB [21]	USB was designed to simplify connections to a PC by (i) replacing all the different kinds of serial and parallel port connectors with one standardized plug and port combination, (ii) eliminating the requirement to adjust internal PC settings, (iii) allowing 'hot-swapping' so that peripherals can be attached or removed without restarting the PC, and (iv) facilitating the connection of many peripherals at one time by supporting a daisy-chained network. In practice, the central hub of a USB network need not be a PC, and the 'peripherals' could be any type of building sensor, complex or not
Ethernet [12] (IEEE 802.3)	Developed originally by Xerox (and subsequently by Xerox, DEC, and Intel), Ethernet is currently the most widely installed Local Area Network (LAN) technology. An Ethernet LAN typically uses coaxial cable or special grades of twisted pair wires to provide transmission speeds up to 10 Mbps. Devices are connected to the cable and compete for access using a Carrier Sense Multiple Access with Collision Detection (CSMA/CD) protocol
Token Ring [12] (IEEE 802.5)	Based on the original IBM Token Ring protocol, this is a LAN in which all devices are connected in a ring or star topology and a bit- or token-passing scheme is used in order to prevent the collision of data between two devices that want to send messages at the same time. After Ethernet this is the second most widely used LAN protocol and currently provides for data transfer rates of either 4 or 16 Mbps
X10 [3]	X10 communicates between transmitters and receivers by sending and receiving signals consisting of short rf bursts over the mains wiring. This system assumes that all units are connected to the same mains wiring infrastructure and thus has no defined method of supporting a network layer. Rather, there is a complete set of codes defined to control a standard set of units

Tab. 3.2-2 continued

Standard	Description
HomePNA [4]	This is designed to facilitate the easy installation of an Ethernet-compatible network within a building by using the existing phone wire infrastructure. This potentially makes every phone jack in the building into a port on the network as well as a phone extension. The technology requires the same PC software drivers as existing Ethernet cards, and uses Frequency Division Multiplexing (FDM) to handle simultaneously existing and emerging telephone services along with networked data traffic
P1394 [22]	The IEEE High Performance Serial Bus (P1394, sometimes known as Firewire) is a serial digital bus connector for interconnecting domestic units such as computers, camcorders, VCRs, printers, TVs, and digital cameras. It is particularly targeted at transferring video or still images from a camera or camcorder to a printer, PC, or television, with no image degradation

3.2.3.3
Wireless Links

There are two dominating factors in the standardization of wireless (RF) links. The first of these relates to which frequency band(s) the link will use, and second to how well the equipment will co-exist or even interoperate with other wireless links in the neighborhood. The next two sub-sections examine each of these issues in turn, before attention is briefly turned to the use of IR.

3.2.3.3.1 **Frequencies**

Usage of the radio spectrum, that is the range of available radiofrequencies, is closely controlled by governments around the world. In the main this is because of the ease with which radio systems could cause interference with each other if they were operated in an uncontrolled manner. The practice is then to divide the radio spectrum into very many frequency bands, each having a defined frequency range, and defining who can use each band and how.

Globally RF allocations are agreed at meetings held by the Radio Communication sector [23] of the International Telecommunication Union, which is in itself an agency of the United Nations. Further details and refinements to these spectrum allocations are then made by regional and governmental bodies, such as the Federal Communications Commission (FCC) [24] in the USA and the European Radiocommunications Office [25] based in Denmark. In the majority of these frequency bands individual licenses are granted by the government(s) concerned to permitted users of the band. However, in some bands unlicensed use is permitted, subject to radio equipment meeting certain criteria in terms of the type and power of the radio signal transmitted. Given the complexities and cost of obtaining radio licenses, it is most likely that in-building radio networks will operate within one of these unlicensed bands.

The impact of choosing one or other of the available unlicensed frequency bands is to dictate:

- the bandwidth of the spectrum available which sets an upper limit on the achievable bit rate of the system;
- the number of other users who are operating in the band, which then gives a probability that interference will be encountered;
- the amount by which the radio signal is attenuated by, for example, the walls in the building, which has the effect of setting a range limit on the system;
- the cost of the radio equipment.

In general, the higher the frequency of the band, the greater is the available bandwidth, the smaller is the number of other users, the greater is the attenuation through walls, and the greater is the cost of the radio equipment.

3.2.3.3.2 'ISM' Standards

One particular type of frequency allocation has been made in various ways around the world for industrial, scientific, and medical (ISM) uses. In principle, this was originally intended as spectrum where various effects of radio signals other than communication could be exploited. Perhaps the best known example of such usage is microwave ovens (microwaves being a particular frequency range of radio waves), which use radio waves within an ISM band to cause a heating effect. However, one of the key characteristics of these ISM bands is that they are lightly regulated; for example, no licence is normally required to operate a radio at that frequency. Another key feature is that, particularly at 2.45 MHz, there is a virtually worldwide ISM allocation. Thus, if a radio communication system is built to work in the ISM band, it may be used without a license and in almost any country. So, despite the obvious drawback of interfering radio sources such as microwave ovens, the ISM bands are increasingly being used as to accommodate the type of local radio system that would be deployed within a building. Several well-defined radio standards have been developed to operate in these ISM bands, the most prominent being listed in Table 3.2-3.

3.2.3.3.3 IR Standards

As has been noted already, one key feature of IR is that it does not pass through building walls. Another,particularly relevant after considering available RF frequency allocations, is that IR has an enormous spectrum that is not regulated at all. These two facts are, of course, not entirely unrelated, in that because it is difficult to interfere with a neighbor's IR system no regulation is deemed necessary. For a

Tab. 3.2-3 ISM standards

Standard	Description
IEEE 802.11 [12]	The IEEE 802.11 defines PHYsical (PHY) and Medium Access (MAC) layer protocols for wireless transmission of data using a variety of options in two ISM bands and also one IR option. The physical layer data rate varies from 1 to > 20 Mb/s, depending on the options chosen. The MAC layer specification is very similar to the IEEE 802.3 Ethernet wired line standard, making the inclusion of wireless links into an otherwise wired Ethernet system very easy
Home RF [26]	The Home Radio Frequency (Home RF) Working Group was formed to establish an open industry specification for unlicensed RF digital communications for PCs and consumer devices in and around the home, resulting in the SWAP-CA (Shared Wireless Access Protocol – Cordless Access) specification. It differs from 802.11 with a more relaxed PHY layer, the inclusion of beacons to manage isochronous traffic, a simplified protocol enabling IP data at up to 2 Mb/s, and also direct support of cordless telephony. The design focus is for a range to cover the typical home and garden (range of > 50 m indoors) with up to 127 devices supported per network
Firefly [27]	Firefly aims at supporting a class of devices, differing from Home RF SWAP-CA by allowing for lower data rate communications at a cost point and power consumption curve substantially below existing SWAP-CA implementations. Such devices include PC peripherals, gaming peripherals, home security units, home automation devices, home control units, and toys. This system has a total bit rate of 128 kbps
Bluetooth [10]	Bluetooth was developed by an alliance of mobile communications and mobile computing companies to define a short-range communications standard which will allow networks of handheld computing terminals and mobile terminals to communicate and exchange data at ranges of up to 10 m

similar reason until recently no universal standards had been created for IR systems. Instead, each manufacturer of IR equipment (most commonly remote controls) had developed their own system, protocols, etc. More recently 'universal' remote controls have been available on the market that can control consumer electronics equipment from most if not all manufacturers, but such products are achieved by a growth in the capabilities of the remote controls, not a convergence of any of the standards.

However, in the last few years one general standard known as IrDA [28] has emerged, aimed mainly at the remote connection of PCs. In this case, of course, there is a distinct benefit in having a well-defined standard, in that different PCs can quickly be connected together. This is described in Table 3.2-4.

Tab. 3.2-4 IR standards

Standard	Description
IrDA [28] (Infrared Data Association)	Defines an interoperable, low-cost IR data interconnection standard that supports a walk-up point-to-point user model, usable with a broad range of appliances, including computing and communications devices
Remote control (various)	Most CE manufacturers use their own proprietary standards for remote controls which have very many predefined codes for required control events. Multistandard controls are now available

3.2.4
Existing and Emerging Wireless Products

There are so many wireless products on the market, and new products are emerging so rapidly at present, that any list of wireless products given here would very rapidly go out of date. Nevertheless, a brief overview of the various types of wireless product available is given just to complete this section.

3.2.4.1
Remote Controls

The IR remote control used with consumer electronics equipment such as TV, Hi-Fi and VCR is almost ubiquitous. However, remote controls are now appearing for a number of other uses, particularly controlling lighting, power switches, security systems, and other control and monitor functions.

3.2.4.2
Security and Telemetry

For many years now, intruder detection systems based on wireless links have been available for easy installation within a home environment. More recently items such as wireless cameras have also been appearing, which connect easily to a TV, VCR or PC. For all of these products the key benefit of wirelessness is the ease of installation, and the downside is the reliance upon batteries in the remote sensors and cameras.

3.2.4.3
Data Networks

A number of PC manufacturers and other vendors now offer products which enable a wireless link to be established either between PCs in a home environment or into a corporate LAN in the workplace. In the main these products are based upon either the IEEE 802.11 or Home RF SWAP-CA protocols. Many of these products have a form factor which allows them to be fitted into the PCMCIA slot available on nearly all laptop PCs, thus allowing these machines to roam around a building whilst staying connected to the network.

3.2.5
References

1 *Multiple Home PCs Create Need for Home Networks*; Intel Press Release, 12 November 1998, archived at URL: http://www.intel.com/pressroom/archive/releases/hn111298.htm.

2 *Microsoft and ITRAN Communications Join to Create a No New Wires Solution in the Home*; Microsoft Press Release, 9 February 2000, archived at URL: http://www.microsoft.com/PressPass/press/2000/feb00/itranpr.asp.

3 *X10 Powerline Carrier (PLC) Technology*; available at URL: http://www.x10.com/support/technology1.htm, or further details from X10 Wireless Technology, Inc., 15200 52nd Ave South, Seattle, WA 98188, USA, E-mail: info@mail.x10.com, URL: http://www.x10.com/homepage.htm.

4 *The Home Phoneline Networking Alliance: Simple, High-speed Ethernet Technology for the Home*; available at URL: http://www.homepna.org/.

5 *Internet Protocol*, RFP 791; available from The Internet Engineering Task Force: IETF Secretariat, c/o Corporation for National Research Initiatives, 1895 Preston White Drive, Suite 100, Preston, VA 20191-5434, USA, URL: http://www.ietf.org/.

6 Further details from URL: http://www.upnp.org/, or from Microsoft Corporation, One Microsoft Way, Redmond, WA 98052-6399, USA, URL: http://www.microsoft.com/.

7 Further details from Echelon Corporation, 4015 Miranda Ave., Palo Alto, CA 94304, USA, E-mail: lonworks@echelon.com, or URLs: http://www.echelon.com/ and http://www.lonmark.org/.

8 FORTUNE, S.J., GAY, D.M., KERNIGHAN, B.W., LANDRON, O., VALENZUELA, R.A., WRIGHT, M.H., *IEEE Comput. Sci. Eng.* **2** (1) (1995) 58–68, or see URL: http://cm.bell-labs.com/cm/cs/what/wise/paper-toc.html.

9 Further details from Ascom Systec AG, Applicable Research and Technology, Gewerbepark, CH-5506 Mägenwil, Switzerland, URL: http://www.ascom.com/art/htmldocs/placebrochure.html.

10 Specification of the Bluetooth System, published by the Bluetooth SIG, available at URL: http://www.bluetooth.com.

11 See, for example, STEELE, R., HANZO, L. (eds.); Mobile Radio Communications, 2nd edn.; New York: Wiley, 1999.

12 Further details from The Institute of Electrical and Electronics Engineers Standards Association, at IEEE Customer Service, 445 Hoes Lane, P.O. Box 1331, Piscataway, NJ 08855-1331, USA, URL: www.standards.ieee.org/catalog.

13 *GPS SPS Signal Specification*, 2nd edn.; 1995, available from the US Coast Guard Navigation Center, USCG NAVCEN, 7323 Telegraph Road, Alexandria, VA 22315, USA, URL: http://www.navcen.uscg.mil/ or http://www.nismirror.com.

14 *Information Technology – Open Systems Interconnection – Basic Reference Model: The Basic Model*, ISO/IEC 7498-1; 1994, available from International Organization for Standardization (ISO), 1 rue de Varembé, Case Postale 56, CH-1211 Geneva 20, Switzerland, E-mail: central@iso.ch, URL: http://www.iso.ch.

15 Further details from CEBus Industry Council, Inc., 2 Wisconsin Circle, Suite 700, Chevy Chase, MD 20815, USA, E-

mail: cebus-staff@cebus.org, URL: http://www.cebus.org/.

16 Further details from European Home Systems Association, Excelsiorlaan 11, Bus 1, B-1930 Zaventem, Belgium, URL: http://www.ehsa.com/.

17 Further details from URL: http://www.havi.org/.

18 Further details directly from IEC/ISO JTC1 SC25 Working Group 1 at URL: http://www.metrolink.com/sc25wg1/, or indirectly from International Electrotechnical Commission, 3 rue de Varembé, P.O. Box 131, CH-1211 Geneva 20, Switzerland, E-mail info@iec.ch, or International Organization for Standardization (ISO), 1 rue de Varembé, Case Postale 56, CH-1211 Geneva 20, Switzerland, E-mail: central@iso.ch, URL: http://www.iso.ch.

19 Further details from The JINI Community, URL: http://www.jini.org/.

20 Further details from Electronic Industries Association, Corporate Engineering Department, 2500 Wilson Boulevard, Arlington, VA 22201, USA, URL: http://www.eia.org, or Global Engineering Documents, Suite 400, 1990 M St., N.W. Washington, DC 20036, USA, URL: http://global.ihs.com.

21 Further details from USB Implementers Forum Inc., E-mail: admin@usb.org, URL: http://www.usb.org.

22 Further details from 1394 Trade Association Office, Regency Plaza, Suite 350, 2350 Mission College Boulevard, Santa Clara, CA 95054-1552, USA, URL: www.1394ta.org, or see ref. [12].

23 Further details from the Internation Telecommunication Union, Place des Nations, CH-1211 Geneva 20, Switzerland, E-mail: brmail@itu.int, URL: http://www.itu.int/ITU-R/index.html.

24 Further details from the Federal Communications Commission, 445 12th St. S.W., Washington DC 20554, USA, URL: http://www.fcc.gov/.

25 Further details from the European Radiocommunications Office, Midtermolen 1, DK 2100 Copenhagen, Denmark, E-mail: ero@ero.dk, URL: http://www.ero.dk/.

26 Further details from the Home RF Working Group 5440 SW Westgate Drive, Suite 217, Portland, OR 97221, USA, E-mail info@homerf.org, URL: www.homerf.org.

27 Further details from the FireFlyWorking Group, URL: http://www.firefly-rf.org/.

28 Further details from The Infrared Data Association, P.O. Box 3883, Walnut Creek, CA 94598, USA, E-mail: info@irda.org, URL: http://www.irda.org.

3.3
Sensor Systems in Modern High-rise Elevators

ENRICO MARCHESI, AYMAN HAMDY, RENÉ KUNZ, *Schindler Elevators & Escalators, Ebikon/Luzern, Switzerland*

3.3.1
Elevator System – Overview

This section describes the application of sensor systems in modern elevators of Schindler Elevator and Escalator Company. One focus is on the so-called shaft information system which is deployed for control purposes of the elevator car on its trip along the shaft and positioning at the landings as well as for safety purposes in order to guarantee a safe and reliable operation. The shaft information system is nowadays an interesting field for the application of new sensor technologies. An appropriate section gives an overview of possible solutions and concludes with Schindler's choice for this task. The final section describes a technology referred to as active ride control. It is an example of a state-of-the-art mechatronic system that relies heavily on fast and accurate sensors.

3.3.1.1
Functional Description

Despite the emergence of several innovative elevator designs in the past few years, an electric traction elevator is commonly arranged as follows: a car that contains the passengers, a counterweight the function of which is to reduce the power required to move the car, a set of ropes connecting the car to the counterweight, a traction machine driving the rope and thus the car up and down, an electric interface between the main power available in the building and the traction machine, and the control software establishing the elevator logic.

It should be noted that an alternative to the electric traction elevator is the hydraulic elevator where the traction machine and ropes are replaced by an oil piston powered via an oil pump.

Figure 3.3-1 shows an overview of the main mechanical parts of a modern high-rise elevator.

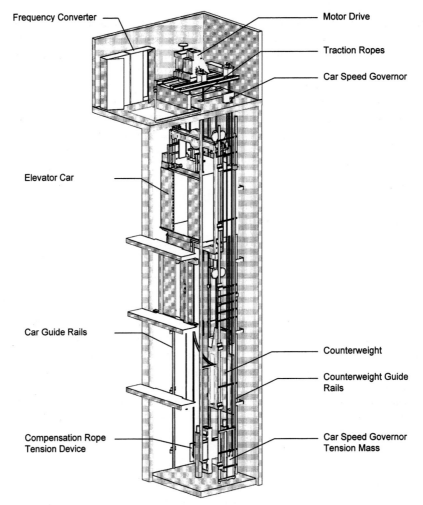

Fig. 3.3-1 Elevator system

3.3.1.2
Sensor Applications in Elevators

A modern traction elevator deploys a great many sensor systems. Functionally, four areas are of special interest: the shaft information system, the door sensorics, cabin load measurement, and the safety circuit. Because of its highly complex functionality and very narrow conditions for deployment, the shaft information system is the most challenging and interesting area. Therefore, the shaft information system was picked as one of the main topics of this article. It is described in Section 3.3.2. Only a short overview will here be given on the safety circuit, load measurement, and the door sensorics.

3.3.1.2.1 Elevator Safety Chain

Since elevators are a transportation system for people, very stringent demands are put on safety issues. One backbone of elevator safety is the safety circuit. It consists basically of a large number of contact switches. These mechanical switches are mounted on shaft and cabin doors, buffers in the shaft pit, tensioning devices for ropes, and various other mechanical systems whose failure can lead to or indicate an unsafe condition of the elevator. The switches are connected in series by a simple voltage circuit and the final signal is provided to the elevator control to be checked at a high sampling rate. If one of the systems monitored by safety switches fails and opens the switch, the safety chain circuit is interrupted. The consequence is an immediate emergency brake of the elevator car that can only be released by instructed service personnel.

3.3.1.2.2 Load Measurement Sensor

A load measurement sensor is used to determine the load inside the car before a new elevator trip starts. This information is important for calculating the torque needed to prevent the car from uncontrolled motion upon the release of the drive brake and also to calculate the trajectory of the motor nominal torque during the trip. The load is measured using gravity forces acting on the cabin floor. For this purpose four cantilever beams that contain strain gages carry the platform on which the passengers stand. The output of the strain gages is processed by an analog circuit before being digitized and sent to the elevator controls in the machine room via a local operating network (LON) bus. It should be noted that the load measurement system is expensive since it necessitates special mechanical construction of the cabin floor. The electronic circuitry required is sensitive, needs calibration, and is also fairly expensive. However, options for alternative load measurement systems are not easy to find.

3.3.1.2.3 Door Sensorics

The primary goal of the door sensorics is to avoid accidents and possible injuries caused by closing automatic doors. The standard solution for this purpose is light barrier systems with various complexity, depending on the elevator type. Simple low-rise elevators may have only one or two light barriers where in expensive state-of-the-art installations the whole door opening is surveyed by a light curtain. These light curtains ensure that even very small objects (eg, a dog leash) are detected. In addition, some of these systems feature a lobby surveillance system consisting of proximity sensors. These sensors detect people moving towards an open or just closing elevator door. Consequently, the doors are reversed or the closing is delayed in order to allow the passenger to enter the elevator car safely.

If, despite the light barrier, an object or a person gets jammed between closing door wings, the doors are equipped with so-called safety edges that will detect this object on contact. These sensors again can be simple mechanical switches or inductive contact sensors in more expensive elevator installations. Finally, the closing force of the doors is measured and monitored. If a person persists in blocking the doors for some reason, the closing will first be reversed based on the output of the light barrier or the safety edge. After a certain number of attempts the door will start to close anyhow. In case of a sensor error of the light barrier and the safety edge, the door can now close and the elevator can take up service. If an object is in between the doors the door will continue closing with a limited force, pushing objects out of the way or reminding persons to clear the door area.

3.3.2
Shaft Information System

The shaft information system is a collection of sensors of different types that is needed by the elevator system for numerous control and safety functions. Currently, Schindler's shaft information system for high-rise elevators is composed of the various standard modules as described in the following sections.

3.3.2.1
Control Sensorics

3.3.2.1.1 Incremental Encoder on the Traction Sheave
An incremental encoder placed on the shaft of the traction sheave is used to measure its angular position. The position value can be read by the computer and differentiated digitally to determine the angular speed. The following advantages of incremental encoders led them to dominate in this industrial application:

- the output is the exact digital description of the position and there is no transformation error as with the case of sensors with analog output followed by analog-to-digital conversion;
- the accuracy and repeatability of the sensor's output are equal to its sensitivity and resolution;
- the two pulse trains represent a digital output of the sensor that is immune against noise and can be transferred to the processing circuits over long distances without loss of the bandwidth;
- the sensor and the processing circuits are obtainable at very reasonable prices;
- they provide position as well as speed measurements (using digital differentiation, to be implemented in the software).

However, the following disadvantages remain:

- the coupling of the sensor to the rotating shaft required to measure its speed is a major source of errors;
- the higher the resolution, the more sensitive the sensor will be to shocks and vibrations;
- the position measurement system needs initialization as the sensor cannot measure the absolute position.

The most important specification of incremental encoders is the resolution as it determines the accuracy of the measured position. The so-called field-oriented vector control of induction motors needs the motor angular position for the calculation of the coordinate transformations as part of the controller algorithm. This places conditions on the resolution of the sensor and its mounting accuracy. The sensor resolution and the sampling frequency of the controller determine the quality of the speed obtained by differentiating the position. For a given sensor resolution increasing the sampling frequency of the controller will result in much noisier speed measurements owing to the effect of smaller time intervals on the numerical differentiation.

3.3.2.1.2 Incremental Encoder on the Governor Roller

The governor roller is part of a system that prevents the elevator car from falling in case of drive shaft or rope breakage. The over-speed governor is a wheel mounted in the machine room and is driven by a steel rope that forms a closed loop starting and ending on the car frame. The rope drives the governor roller as well as the tensioning pulley used to keep the rope in tension (Figure 3.3-1). The connection point of the rope to the car frame is a lever mechanism that will activate two identical brakes from each side of the car to grip on the two guide rails.

Mounted on the shaft of the governor roller is another incremental encoder (IG200, see Figure 3.3-3). The difference between this sensor and the incremental encoder of the traction sheave is only in resolution. The fact that it is used for position control only and not for speed control puts milder conditions on the resolution.

3.3.2.1.3 Car Optical Switches

The position measured by the incremental encoder on the governor roller is not accurate as it is influenced by rope slip, rope elasticity, building shrinkage, and building expansion due to temperature changes. To correct for slip errors, additional position measurements indicating the position of the car relative to the landing stops are used (Figure 3.3-2). Two optical switches are integrated in a unit that is mounted on the car, while metal sheet vanes are fixed to the shaft at every landing position. As the elevator passes a certain landing position, the vane will activate the two optical switches successively. The output of the switches is used to correct the position calculated by the governor encoder according to a lookup table determined during a learning trip. The learning trip is done during the start-up of the system by letting the elevator travel along the whole shaft length while registering all sensor outputs. A constant low speed during the learning trip is important to minimize the slip of the governor roller.

The realization of the position correction is done by the tacho interface (TIF application-specific integrated circuit (ASIC)) (Figure 3.3-3). This circuit contains a digital filter and a multiplier for the pulses of the incremental encoder, a 24 bit up/down counter, freeze registers, and control logic. The pulses are multiplied by

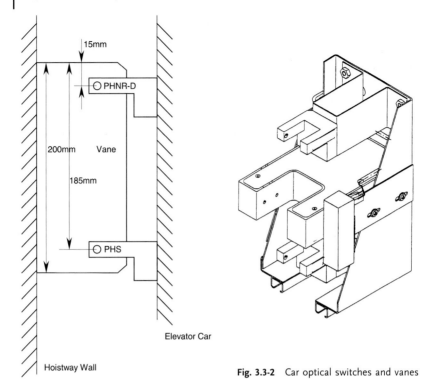

Fig. 3.3-2 Car optical switches and vanes

four to increase the numerical accuracy and then used to clock the up/down counter. When the infrared beam of a landing zone detector is interrupted, the content of the up/down counter is stored in the corresponding freeze register. The drive controller reads the freeze registers and compares their contents with the position values determined during the learning trip to calculate the correct position within ±0.5 mm.

3.3.2.1.4 Normal Terminal Stopping Device

To determine the upper and lower shaft ends, a bistable reed switch is used as normal terminal stopping device. It is integrated in the same sensor module with the two car optical switches (light barriers 1 and 2, Figure 3.3-2). Magnets are positioned in the shaft 2 m away from the highest and the lowest landing positions in order to operate the switch. During the learning trip the output of the reed switch is used by the drive controller to identify the first and last landing positions in the shaft. This is necessary since the incremental encoders and the car optical switches do not provide absolute position information. During normal operation the output of this switch is used as a position safety check. In some countries this switch is also used as a safety device. In such applications it defines the last possible break point to stop the car at the bottom or top landing, within regular operation.

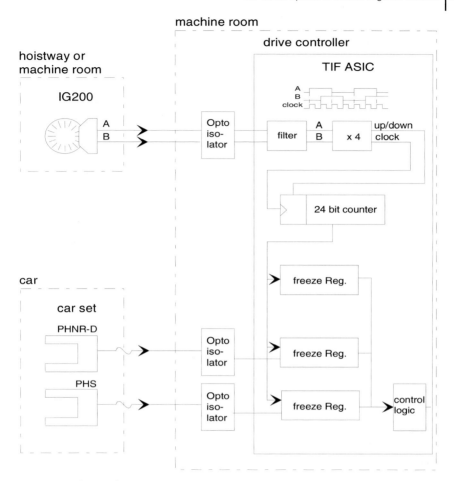

Fig. 3.3-3 Tacho interface (TIF ASIC)

3.3.2.2
Safety Sensorics

3.3.2.2.1 **Final Terminal Stopping Device**
A mechanical switch is used as a final terminal stopping device in case the normal terminal stopping device fails. A ramp activates the switch when the car passed the final landing position. Consequently, the mechanical switch interrupts the elevator safety circuit and the car is stopped.

3.3.2.2.2 **Slowdown Monitoring**
In the hoistway pit buffers are installed in case an elevator passes the final bottom landing and continues its trip despite all other safety devices. The buffers are designed to absorb the motion energy of the car and bring it to a safe stop with a phy-

siologically acceptable deceleration. Since the motion energy is a function of speed the buffer size can be reduced if the impact can only happen at a reduced speed. Therefore, the slowdown monitoring system is deployed to monitor the car speed near the highest and the lowest terminal zones of the hoistway. It has two sets of vanes mounted in each of these two zones. These vanes will interrupt the beam of an additional optical switch mounted on the car. Knowing the distance between the vanes, the time between two successive interruptions is a measure of the car speed. If this time is less than a specified limit, the speed is too high and the sensor circuit interrupts the safety circuit to stop the car. The closer the vanes are to the hoistway end, the nearer they are mounted to each other. This will decrease the allowed speed when the car comes nearer to the shaft end to assure that the car is being stopped (see Figure 3.3-4).

The optical sensor is designed as a failsafe device with two redundant channels. Each channel has a micro-controller to process the output of the following three infrared sensors (see Figure 3.3-5):

- sensor A to check if the vane was passed or just toggled;
- sensor B to measure the time while the beam is interrupted by a vane;

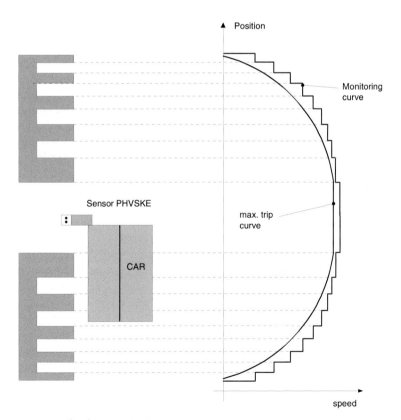

Fig. 3.3-4 Slowdown monitoring

Fig. 3.3-5 Emergency terminal speed-limiting device

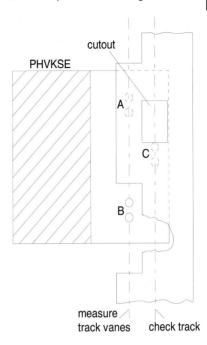

- sensor C faces the check track to ensure that the sensors A and B are correctly aligned with the measuring track.

A hardware counter is used for time measurement. The counter starts when the beam of sensor B is interrupted and stops when the vane has been passed. The state of beam A is used to qualify the measurement: if the vane was passed, the measurement is valid. Valid measurements of the two redundant channels must be the same within a certain tolerance. A comparison circuit is responsible for this check.

3.3.2.2.3 Door Zone Detection

Under regular operating conditions the opening of the car doors will interrupt the safety circuit and cause the car to stop. Another optical switch is used to override the door safety contacts used to monitor the locked position of car doors. Door override allows for pre-opening of the doors during the landing approach slightly before the car has reached the final landing position. In this way the overall travel time for the passengers will be reduced. This function is also needed to re-level the car with opened doors when load changes cause a substantial deviation from the landing position. The door zone detection is integrated in the same car module as the two car optical switches used for final positioning and the normal terminal stopping device.

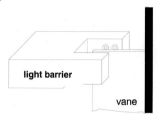

Fig. 3.3-6 Door zone detection sensor

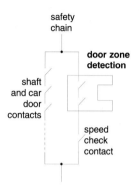

Fig. 3.3-7 Door override circuit

The infrared beams of the two redundant optical sensors will be interrupted by a vane within the allowed door overriding zone (Figure 3.3-6). Such a vane is mounted at each shaft door. A control circuit monitors the equal status of the two channels and activates an output relay. The sensor circuit has an input to enable the device. If the device is not enabled, the relay contacts will remain open. The purpose of the enabling is to avoid switching the relays when passing vanes other than the vane of the destination floor. In addition to the optical sensor, the door override function utilizes a speed check device to assure that the doors will be only opened when the elevator speed is slow enough (Figure 3.3-7).

3.3.2.3
Comments on Currently Used Sensors

As can be deduced from the above description, infrared optical switches dominate the applications in the shaft information system. They are used for landing zone detection, slowdown monitoring, and door overriding devices. Even the incremental encoders of the traction sheave and the governor roller may be considered as a special case of infrared optical switches. The reasons for this wide spread are the reliability, cost, and simplicity. Magnetic sensors fulfill the same requirements. However, they are not as accurate as optical switches as they show inherent hysteresis. Mechanical switches such as the final terminal stopping devices and the switches used to detect door openings are also widely used for the same reasons.

It is also obvious that the so-termed shaft information system consists of several independent subsystems. The question of whether the current shaft information can be replaced by a simpler and cheaper system is justified. This leads to the question of why the current system needs so many components to meet some relatively simple needs. One answer lies in the fact that the current system does not provide a continuous absolute position information of the car within the hoistway. If this kind of information was available, such a system would meet some important needs for a replacement. Further needs would include the design as a safety system as well as aspects of signal quality and robustness. The signal quality is of specific interest for the elevator motion control system that relies on the two incremental encoders described above. High-rise elevators in particular are facing control problems that are addressed with new control algorithms as described in the next section.

3.3.3
Present Developments for High-rise Elevators: New Shaft Information System

3.3.3.1
The Conflict of High-rise Traction Elevators

Although elevators with extremely long rises represent a small portion of the present worldwide elevator market, they continue to be of great importance for elevator producers. This importance is mainly due to the fact that the quality of these elevators is considered as a measure of the capability of the elevator producer. In many cases the winning of a certain contract with many elevator types depends on the readiness of the elevator supplier to provide a good solution for the high-rise segment of the contract.

The conventional elevator with traction ropes will remain the only economically realizable answer to the demands in the high-rise segment, until this technology is totally replaced (eg, with self-propelled elevators with no traction ropes). A major problem facing high-rise traction elevators results from the elasticity of the traction ropes. The longer the ropes are, the less rigid they will be. With decreasing rope rigidity, the resonance pattern of the elevator mechanical system wanders to a lower frequency domain. As a result, it will not be possible to regard the mechanical system within the controller bandwidth as a single rigid body. A simple dynamic model of the high-rise traction elevator system is shown in Figure 3.3-8, representing the ropes as springs. According to this model the elevator car will encounter low-frequency vibrations in the vertical direction (lower than 1 Hz), making the positioning time unacceptably long, and also affecting the ride comfort.

The maximum load on the traction ropes is given by the weight of the whole system including the weight of the ropes themselves. The number of traction ropes required for a given rope safety factor IZ will increase massively with the elevator travel height HQ as shown by the curves in Figure 3.3-9. In order to travel in shafts higher than 300 m using an acceptable number of ropes, the rope safety factor must be reduced to ≤ 12. For super high-rise elevators with a reduced

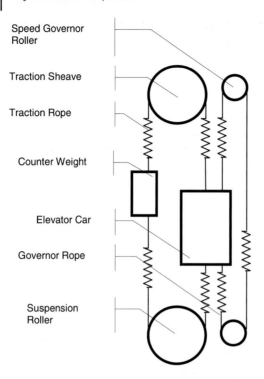

Fig. 3.3-8 Dynamic model of the high-rise elevator system

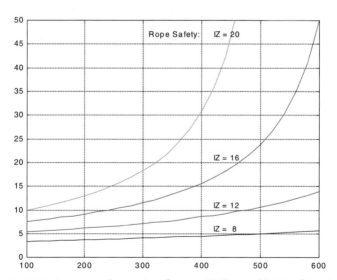

Fig. 3.3-9 Increase in the number of ropes with the travel height of the elevator

rope safety factor the ropes become very elastic so that the problem of the elevator car oscillations will become very serious.

3.3.3.2
New Challenges of Motion Control

A model-based motion controller that takes the whole system dynamics including the elevator car, the traction ropes, and the counterweight into consideration should be able to substantially reduce the vertical oscillations of the elevator car. In this case the control system will have a higher bandwidth resulting in a faster and more comfortable positioning of the car. However, this task is not so easy owing to the following main difficulties:

1. The dynamics of the mechanical part of the system depend strongly on the vertical position of the elevator car and the counterweight. Modeling and simulation studies have shown that the dependence of the system dynamics on the car vertical position cannot be neglected when designing a model-based controller.
2. A new measurement system of the car position and speed along the shaft is necessary, so that new controller concepts will be possible.

A new motion controller will be sophisticated compared with the present type. It needs to determine the dynamic model of the plant as a function of the car position along the shaft. The determination of such a model for a certain plant needs measurements to be processed by a specially developed identification algorithm. An on-line adaptation of the controller to the plant model is also needed and this will result in a nonlinear controller scheme. The new controller will definitely need the measurement of the car position and velocity in the shaft in addition to the already available measurements.

3.3.3.3
Specifications of the New Shaft Information System

3.3.3.3.1 Need for New Shaft Information Systems
With the requirements of new high-rise elevators, the current shaft information systems are reaching the limits of their applicability. From a controller point of view the present position and speed measurement systems do not meet performance requirements in terms of sensor accuracy and bandwidth. Furthermore, the cost and complexity of present shaft information systems increase significantly when moving towards high-rise and super high-rise applications. A main goal of the ongoing research and development efforts in the elevator industry is therefore the conception of a shaft information system that exceeds current systems in terms of measuring performance, reduces complexity and cost, and allows for easy scale-up while meeting all safety requirements.

3.3.3.3.2 Requirements of the Motion Controller

The new motion controller sets the highest requirements on measurement accuracy and speed. Within the complete travel distance of 500 m or even more, position must be resolved to within as little as 0.5 mm. In order to compensate for the dynamic effects produced by rope elasticity, it was determined that an acceptable motion control system should provide an open loop bandwidth of 20 Hz. This value applies for the complete control loop including the sensors, the data transmission, and the computer controller. Special attention must be given to the data link between the sensors, eg, on the elevator car and the controller in the machine room. In high-rise elevators the cable distance between these two systems can be as long as several hundred meters. A slow data link within the control loop produces a time delay and this introduces a phase shift in the loop. For real-time control applications this will limit the controller performance to a great extent. Generally, a parallel data link is not feasible for this application. As for serial communication, the transfer rate must achieve a certain speed in order to keep the time delay within acceptable limits. This will require transfer rates as high as 1 Mbit/s. Unfortunately, current standard serial buses with copper wire transmission do not allow for such high transfer rate over the distances indicated above. A custom development of a high-speed serial data transmission system is then needed. Replacement of the copper wire as transmission medium with high-bandwidth links such as optical cables will also help to attain higher transfer rates while applying standard protocols.

When selecting position sensors for the motion control of the car it is important to distinguish between the absolute and the relative positions. The absolute position of the car is the distance of the car from the end of the hoistway while the relative position indicates its position relative to the landings. Absolute and relative car positions may vary over time owing to the settling of the building structure after its completion and also owing to temperature effects. A shaft information system must be able to deliver the absolute position for control purposes in addition to the relative position needed to position the car properly at each landing.

3.3.3.3.3 Requirements of the Safety Functions

For high-rise elevators a variety of safety functions is required, including the need for speed and position measurements. In contrast to the sensorics needed to control the elevator car movement, where high precision and speed of data transmission are the highest goals, the requirements for fulfilling the safety functions are rather focused on the (fail)safe design and operation of the sensor electronics.

The present shaft information system utilizes a lot of measurement modules that are partially mounted on the car and partially mounted in the shaft. This adds to the system complexity and increases the installation costs to a significant level. The requirements on the new shaft information system are to reduce the number of independent subsystems by combining the possible safety functions in new intelligent modules. The total number of modules in the shaft information system is then minimized and this will reduce both the hardware system costs and the installation costs. This target is in principle possible as all functions of safety and motion control are satisfied with a reliable and accurate position and speed measurement of the car in the shaft. Since the functioning and operation of the shaft information system has to comply with very stringent safety codes, all system components need to be designed as redundant and failsafe systems.

3.3.3.3.4 Reliability

Unlike in other industrial sensor applications, the safety requirements for elevators specify redundancy and a failsafe design in order to guarantee the safe functioning in all operational states. Hence, a two-channel sensor unit, for example, is used to increase safety, but it could also be used to increase the reliability of the overall elevator system. Depending on the reaction of the system to a difference in the two channels or the redundant systems, the function can be made more safe or more reliable:

- When focusing on safety, a difference in the two comparing systems will cause the overall system to fall into a safe state, which is normally an emergency stop, that leads to entrapment of passengers, but prevents injury.
- With the same amount of sensor and electronic expenditure, it would also be possible to use the parallelism of two systems/channels to ensure at least one part is functioning. A difference in the two channels would therefore lead to the overruling of the less reliable channel in order to continue to provide the necessary sensor signals/functions. This would enable the system to function properly, even if there was a fault in one part of the sensor system.

Since safety functions are a major requirement in elevator sensorics, it is not reasonable to double the already redundant/failsafe safety systems again in order to increase the reliability. Therefore, there are high requirements on the design, manufacturing, installation, and maintenance of the elevator sensorics to achieve high reliability with as little complexity/expenditure as possible. Another source of errors and thus call backs is the installation process. The higher the number of parts to be mounted and adjusted in the shaft, the more delicate the sensor system is in terms of reliability.

Since the elevator is being operated in a rough environment, the building shaft, often exposed to dirt, dust, humidity, temperature differences, EMC, etc., the sensorics have to be extremely robust.

3.3.3.3.5 Economic Considerations

Since the elevator market is under a high cost pressure, the economic aspects may not be disregarded when evaluating new sensor technologies for high-rise elevators. Whereas the cost for the main sensor apparatus with most of today's shaft sensor systems is the same for any high-rise elevator, installation time and material costs rise considerably with the travel height of the elevator because of the large amount of material needed to be mounted in the shaft. Thus cost dependance on travel height should be minimal.

3.3.3.4
Candidate Sensors

The selection of a sensor system to measure the position and speed for an elevator car seems to be a trivial problem at first glance. However, this is not the case. Problems arise not only from the sensor itself in terms of measurement accuracy and noise, but also from a large set of boundary conditions such as described in Section 3.3.3.3, eg, safety issues, mounting considerations, environmental conditions, and data transmission. All these conditions apply together and narrow the choice of suitable sensor systems considerably. Considering safety issues, a system with high accuracy but low failure tolerance is of no use for this application. Installation is also of high importance since complex systems with numerous sensors on the car may require installation times of several days performed by highly skilled personnel, and this adds considerably to the total system cost. System functionality must be guaranteed in fairly adverse conditions comprising temperature ranges from −10 up to 60 °C and relative humidity up to 100% with possible dust and smoke in the hoistway. Under all these conditions the system must operate in a reliable way without the need for extensive or expensive maintenance. Therefore, preference goes to a modular system with simple and cheap standard modules. In the following subsections, an overview is given of numerous sensor systems that appear promising for application in elevator shaft information systems. A combination of several systems is also conceivable.

3.3.3.4.1 Opto-mechanical Sensors:
Incremental Encoders with Friction Wheels: Odometer

As explained above, current elevator control systems rely on incremental encoders mounted on the traction sheave and the speed governor for position measurements. The main problems with these sensors are the mechanical slip and the elasticity of the ropes. In order to measure the actual car position using an incremental encoder, the sensor must be mounted on the car itself. The translation of the linear motion into circular motion can be performed using a friction wheel that rolls along the rail with the encoder mounted on its axle as shown in Figure 3.3-10. With this system the error due to rope elasticity can be eliminated. Thanks to the wide availability of incremental encoders as standard industrial sensors for rotation, the cost of such a system is fairly low. The utilization of a multi-turn encoder instead of an incremental encoder allows for an absolute position measurement. The resolution of commercially available encoder systems easily meets and exceeds the requirements for elevator position control. With an appropriate electrical driver the incremental encoder signals can be transmitted for distances over a maximum distance of 600 m without the need for a data transmission link. A high bandwidth is then assured.

However, some disadvantages arise from this solution. First, the problem of mechanical slip persists. In order to minimize this slip a careful design of the wheel support and a suitable choice of the wheel material are of great importance.

Fig. 3.3-10 Friction wheel and incremental encoder (odometer)

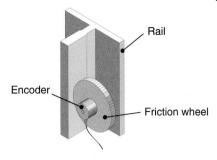

Nevertheless, slip cannot be completely eliminated. Consequently, another position information system is needed to allow for slip compensation as it is the case in the current solution.

3.3.3.4.2 Opto-mechanical Sensors: Cable Extension Position Transducer

Many industrial applications rely on cable extension position transducers for linear position measurements. These commercially available devices use a flexible cable and a spring-loaded spool for linear position and speed measurements. The cable would be attached to the measurement point on the elevator car while the unit would be mounted on the shaft floor or ceiling. The actual measurement is done by a rotary encoder device mounted on the spool as shown in Figure 3.3-11. These products offer an interesting off-the-shelf solution for linear position mea-

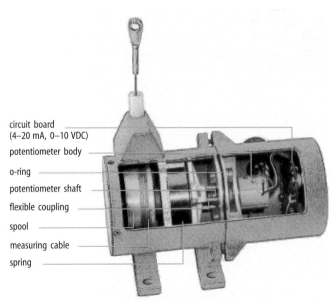

Fig. 3.3-11 Cable extension position transducer. Courtesy of Celesco Transducer Products Inc.

surement. The position information is accurate and quickly delivered to the controller thanks to the short transmission distance. For high-rise applications questions arise concerning the mechanical aspects of the spanned wire. On going to wires of several hundred meters the elasticity starts to play a role, causing measurement errors similar to the case with the current incremental encoders. Car vibrations and air disturbances will easily excite horizontal oscillations of the wire, causing dynamic measurement errors. The longer the wire, the more severe these problems will be. On the other hand, these systems are robust and very tolerant against adverse conditions such as moisture, dust, and smoke in the hoistway. The position information of this sensor will be absolute, ie, relative to the shaft end. Therefore, a second system is needed to give reference to the landing positions.

3.3.3.4.3 Opto-mechanical Sensors: Contact-less Linear Incremental Encoders and Marking Tape

Linear encoders are based on the same principle as rotary encoders. An optical sensor detects the position information marked on a carrier, such as a glass ruler or a tape (Figure 3.3-12). This technology is widely used today where highly accurate linear position information is needed, as for example in tooling machines. Position can be coded as either incremental or absolute information. If the information carrier can be mounted on the building structure, this system is able to deliver the absolute position relative to the shaft end and also give reference to the landing positions. No additional reference system for the landing zones is then needed. As a shaft information system this sensor offers many advantages. Commercially available systems may have resolutions as sensitive as 0.1 µm but the overall measuring distance is generally far too short for an elevator shaft. The cost of an information carrier of the desired length and resolution is expected to be high. The installation of the information carrier in the shaft is a difficult and time-consuming task. Meanwhile, mounting the sensor on the car adds to the system complexity since these systems are very sensitive to distance variations be-

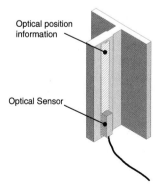

Fig. 3.3-12 Contactless linear incremental encoder

tween the sensor and the information carrier. Horizontal movements of the car beyond the system's tolerance will necessitate an expensive suspension and guiding mechanism for the sensor. Since this is an open optical system it is prone to become spoiled by the dust, grease, and water that are present in an elevator hoistway.

3.3.3.4.4 Optical Sensors: Laser and Infrared

Today distance can be measured using laser or infrared distometers. These sensors emit laser or infrared light to be reflected by a solid object on the measurement point. The measuring device receives the reflected rays, determine the runtime, and calculate the distance from this information. Products based on this principle are widely used today in applications ranging from standard industry to sensitive military equipment. The technology is well established and a large choice of complete systems is commercially available. As a shaft information system laser and infrared distometers yield a considerable set of advantages over other sensor types. A significant cost reduction can be achieved through the easy and straightforward installation procedure. The sensor head is mounted in the machine room with no work needed in the hoistway except mounting the reflector on the car (Figure 3.3-13). The measuring system is located next to the elevator controls and this allows fast and easy data transmission.

The biggest drawback of optical distometers when used for elevator shaft information originates from safety considerations. In emergency situations involving

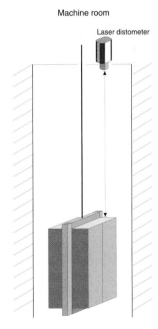

Fig. 3.3-13 Optical sensor

smoke and heavy dust in the hoistway, immediate passenger evacuation is essential. In these situations optical distometers are completely blind and a fully functional backup system must be available. Meanwhile, commercially available optical distometers are designed for applications with much less stringent dynamic requirements as in the case of the elevator motion controller. Consequently, their sampling rate and the resulting bandwidth is too low for this application. Furthermore, measurement accuracy and resolution do not meet the stated requirements, either.

3.3.3.4.5 Optical Sensors: Charge-coupled Device (CCD) Sensor

An interesting proposal to solve the problem of position and velocity measurements in elevator shafts employs a CCD sensor. This component is widely used today in image scanners. It may be considered as a simple optical camera having a number of optical measuring cells arranged in line. The system makes use of the random surface pattern of any part in the hoistway, which in our case is the guide rail (Figure 3.3-14). During a learning trip the optical surface pattern along the complete travel distance is scanned and stored. With this information it is possible to determine the position of the car at any point in the hoistway by comparing the on-line scanned image with the previously stored information.

This idea is interesting since such a system needs no shaft installations. However, there are too many open questions at the present time and further research is still needed. In particular, the necessary computing power must be determined and an appropriate software package needs to be developed. At this point it is still questionable if the position and speed information can be generated with the required bandwidth at acceptable prices. It is nevertheless expected that these problems will be overcome with the continuous increase in the available computing power. Concerning the cost of such a system, detailed analyses are still required. Finally, there is the soiling problem that all open optical systems have in common.

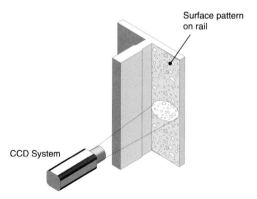

Fig. 3.3-14 CCD sensor

Fig. 3.3-15 Optical sensors and correlation algorithm

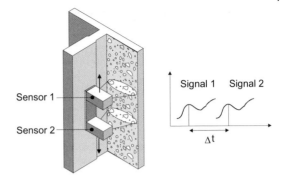

3.3.3.4.6 Optical Sensors: Optical Sensors and Correlation Algorithm

This optical measurement system also makes use of the surface pattern on the guide rails. It uses two light-emitting diodes (LEDs) as light sources and two photoresistors as receivers for the reflected rays (Figure 3.3-15). The time-continuous output signals from the receivers are sampled at a sufficiently high frequency to be processed by a real-time computer. While the car is moving the two signals will look identical but with a time delay between them. Using correlation methods the phase shift between the two signals can be calculated and the car speed is then easily determined. The car position can also be obtained either by integrating the speed measurements or by comparing the on-line measurements with data stored during the learning trip. Although the proposal looks attractive, a lot of research and development are still needed to reach an operable state. Meanwhile, the problems that can arise are similar to the case with the system utilizing the CCD sensor. At this point it is too early to draw definitive conclusions about the applicability in elevator systems.

3.3.3.4.7 Acoustic Sensors: Runtime Measurement of Sound Waves in a Wire

Currently, elevator positioning systems based on acoustic principles are available as commercial products. Such a system is composed of a steel wire vertically suspended in the hoistway so that it will pass through a coil mounted on the elevator car (Figure 3.3-16). The coil induces sound waves in the wire at sampling rates up to 1 kHz. These sound waves are registered by receiver units at each end of the wire at the top and bottom of the hoistway. Knowing the speed of sound in the steel wire, a simple runtime measurement of the induced sound waves is performed and the car position can be calculated.

This system certainly has a high market potential. According to the manufacturers, the operating principle allows position resolution in the millimeter range. The system is robust and insensitive to smoke or heavy dust. Because of the contactless operation there is virtually no wear. Material costs are relatively low and mechanical installation is easy. Tests with these systems have indicated the need to improve the electrical installation and calibration processes, still laborious and

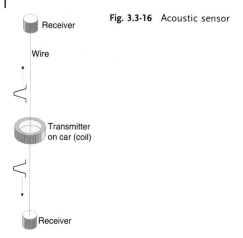

Fig. 3.3-16 Acoustic sensor

time consuming. For low- and medium-rise plants this system is fairly competitive and represents a good alternative to conventional systems. For high-rise installations the application of this system is restricted. Above a certain height mechanical problems with the spanned wire are expected. The 1 kHz sampling can be heard and is a nuisance. Furthermore, the system cannot meat the bandwidth requirements imposed by the motion controller.

3.3.3.4.8 Magnetic Sensors: Magnetic Band and Reader

The layout of a magnetic sensing system for position measurement in the elevator shaft is similar to a linear encoder system. The main difference is that the sensor principle is based on the Hall effect, a phenomenon that allows the detection and quantification of magnetic fields. Many disadvantages of the optical encoder system no longer exist in the case of the magnetic sensor. First, a magnetic system is completely insensitive to dust, oil, smoke, or other factors that may cause an optical system to fail. In addition, the price of Hall sensors is very affordable. They are available as OEM (original equipment manufacturer) parts from almost any semiconductor producer at prices starting from US$ 1 or less per piece. As with the case of optical incremental encoders, the sensor output signal can be transferred to the processor over long distances without the need for a special data transmission link. The magnetic system allows for both incremental and absolute coding with high resolution. The simplest system will have one sensor and

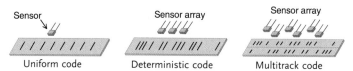

Fig. 3.3-17 Magnetic sensors: band and reader

Fig. 3.3-18 Magnetic tape coding using a template

one track carrying uniform linear magnetic information to yield an incremental measurement system (Figure 3.3-17). For an absolute system more than one sensor is used while the information is coded in a deterministic way such that any position of the sensor head over the whole travel distance will result in an unique absolute signal signature. The use of several parallel information tracks permits virtually any desired resolution for any travel distance while keeping the material cost affordable.

The most critical cost factor of this system will most likely be the process of coding the information on to the carrier. The finer the resolution desired, the higher the cost will be. The material to be used as information carrier could be a magnetic tape selected from the production line of any of the several available manufacturers. These tapes are either self-adhesive or may be glued to the rails, making the installation relatively fast and easy. Coding of the information on to the tape can be performed in different ways. The industry standard process is to code the magnetic information directly on to the tape. The resolution will be sufficient, especially when using several parallel tracks. However, process costs should not be neglected, as they will increase with higher resolution and longer distances. A cheaper way to do the coding may be the use of a permanently magnetized tape to be covered with a punched foil of non-magnetizable material (eg, aluminum). A Hall sensor can now detect the punches in the tape, and hence read the coded position information. This coding method allows both incremental and deterministic code inscription. Several parallel tracks to increase the resolution are also possible (Figure 3.3-18).

3.3.3.5
Conclusion for Future Shaft Information Systems

The wide range of sensor systems in the industry provides a large selection and many possibilities to solve the above problems with high-rise elevators, such as building movement, rough environment in the shaft, transmission to the control unit, safety requirements, etc. However, since the requirements are so manifold, the major focus must be the reduction of complexity of the sensor system in order to guarantee a safe and reliable/robust shaft information system. Depending on where a manufacturer sets its priorities, different sensor systems may be evaluated as the best for the individual requirements.

Linear optical systems are promising in a variety of application variants, but have to be questioned because of their sensitivity to soiling. Systems based on odometers are preferred because of their robustness and simplicity, but face the problem of slip in their drive unit and of high costs if high resolution is needed. Magnetic systems also show a very robust characteristic. Yet, mechanical align-

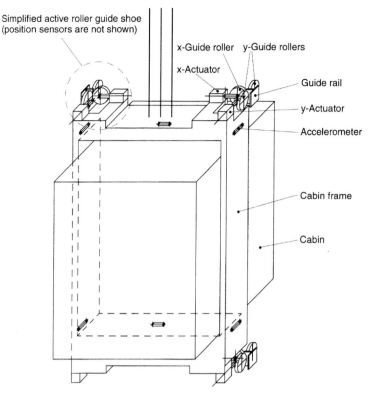

Fig. 3.3-19 The active damping system

3.3.4.4
Controller Scheme of the Active Damping System

The primary goal of the controller is to suppress the vibrations in the frequency range between 1 and 20 Hz using acceleration feedback. The secondary goal of the controller is to correct the orientation of the car in the presence of asymmetric cabin load using position feedback. This is needed to optimize the car position and angle relative to the two guide rails in order to provide sufficient actuation distance for each actuator. The controlled system therefore has two distinct feedback loops for the acceleration and the position. Both of these loops are multi-input, multi-output and are designed according to their own requirements for the bandwidth, loop shape, and performance (Figure 3.3-22).

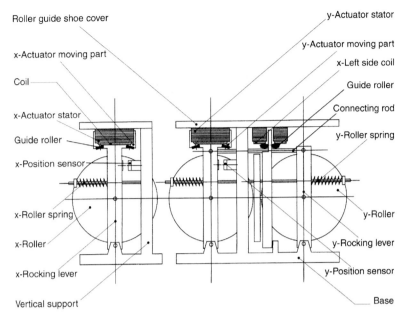

Fig. 3.3-20 The active roller guide shoe

3.3.4.5
Sensor Specifications for the Active Damping System

3.3.4.5.1 **Accelerometers**

The main difficulty with the accelerometer is its price. Inertial sensors are fairly expensive components, especially for a measurement range of ± 1 g. Recent developments in the area of micro-machining allowed the production of suitable sensors at reasonable prices. Owing to their low cost, continuously improved characteristics, and compactness, micro-machined piezoaccelerometers are now dominating the market. In all possible types of these accelerometers the displacement of a micro-machined mass relative to a casing is taken as a measure of the acceleration. The mass is connected to the casing through a cantilever beam with certain elasticity and the movement takes place in a damping environment. However, the accelerometers differ according to their measurement principles as follow:

1. in piezoelectric accelerometers, the electric charge accumulated between the mass and the casing is measured;
2. a piezoresistive accelerometer contains a strain gage on one of the cantilever surfaces to measure the deflection as a function of the acceleration using a resistive bridge;
3. the piezocapacitive accelerometer measures the distance between the mass and the casing by determining the capacitance between the two surfaces.

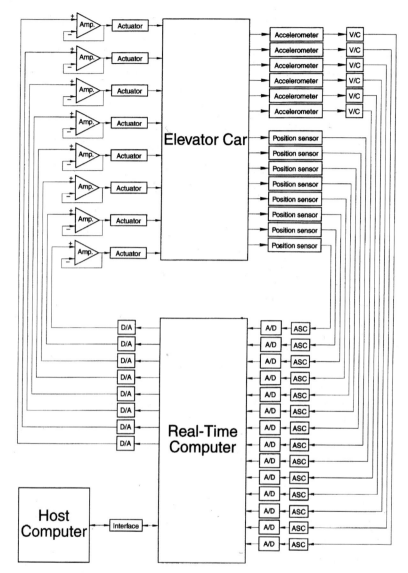

Fig. 3.3-21 Hardware scheme of the active damping system

The piezoelectric accelerometer was excluded from the evaluation for two main reasons: first, it is not capable of measuring accelerations below a certain frequency, and second, the measured signal is interrupted in cases of overload and sometimes upon start-up. The other two types are suitable for active damping systems, with the main difference that the second type needs temperature compensation while the third type is the most accurate. The first version of the active system was equipped with piezoresistive accelerometers, as they were the only available ones fulfilling the

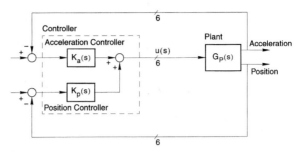

Fig. 3.3-22 Controller structure for the active damping system

specifications at an acceptable price. Over the past 2 years, the technological development in micro-machining together with the widespread use of accelerometers in mass-produced systems such as automobile suspensions, wheel anti-lock systems, and air bags encouraged manufacturers to produce low-cost piezocapacitive accelerometers. Piezocapacitive accelerometers are now offered at a lower price than piezoresistive accelerometers, while achieving a better performance.

The key specification of the accelerometer for the active damping system is its measurement range. Considering the accelerometers offered on the market it is usually easier to find devices with higher measurement range, such as ±20 g or more, than with lower measurement ranges, such as ±1 g or less. The active damping system is required to reduce the peak-to-peak acceleration signals to below 7 mg. According to experience with high-rise elevators, the maximum acceleration measured before the installation of the active system ranges from 40 to 100 mg depending on the conditions of operation and the quality of the rails. In addition to the acceleration resulting from the vibrations, the accelerometer measures a certain component of the Earth's gravitation. just 1° of tilt is responsible for about 17 mg of additional acceleration. With a few degrees of tilt it is reasonable to assume that the absolute value of the maximum measured signal will be less than 200 mg. Since high amplification of the signal would increase the noise level and deteriorate the signal-to-noise ratio, it is not advisable to amplify the measured acceleration signal too much. The ideal measurement range of the accelerometer will then be ±0.5 g.

Another important feature of the accelerometer is its bandwidth. The acceleration controller is required to achieve performance in the frequency range between 1 and 20 Hz. For the stability of the system it is necessary to have good measurements up to 60 Hz with very little phase error. The accelerometer selected can be modeled as a second-order mass/spring system with a damped natural frequency of about 450 Hz and a damping factor of about 0.4. Significant variations of these two parameters from one sensor to another are common but the phase and the magnitude errors remain small in the relevant frequency range.

3.3.4.5.2 Position Sensor

As shown in the constructional sketch in Figure 3.3-20, the position sensor measures the distance between the cabin frame and the rocking lever. To avoid mechanical wear, it is advantageous to use a contactless sensor. For this reason an inductive position sensor was chosen. It contains a coil activated by a high-frequency oscillating supply (typically 120 kHz), so that a metallic object located in the vicinity of the sensor will absorb part of the oscillator system energy. The resulting attenuation in the amplitude of the coil voltage is a measure of the distance between the sensor and the metallic object. An electronic circuit produces a signal proportional to this attenuation, linearizes it, and amplifies it. A first-order low-pass filter is then used to reduce the noise level of the output.

The bandwidth of the position control loop will be far below that of the acceleration control. For this reason, the bandwidth of the position sensor is allowed to be correspondingly lower than that of the accelerometer. Consequently, the noise level of the output may then be kept at a very low level. The bandwidth of the sensor chosen is 80 Hz and its noise level is about 1.6 µm peak-to-peak. The measurement range of the sensor is 7 mm (from 4 to 11 mm). The maximum linearity error is ±3% of full-scale and the temperature error is less than ±5% of the same value.

3.3.5
Conclusions and Outlook

Traction elevators have been working on the same basic operation principle for more than 140 years, and many of the subsystems and components involved have not changed significanty during this time. These systems are tuned to the extent of their technical capability – quantum leaps cannot be expected without new concepts and the deployment of new technologies. Todays deregulation will push major innovations throughout the industry. This applies also to sensors in elevators. Most existing sensor systems used for control and safety purposes were developed years or even decades ago. Further developments during the time since were mainly directed towards the goal of further cost reduction.

The rapid developments in the area of microelectronics and industrial automation in recent years have now opened up new opportunities. This has spurred the interest in new sensor systems for elevator applications that go beyond the functionality of the traditional systems. The development trends are:

- increased precision for advanced control purposes;
- integrated circuit design for signal preconditioning and safety purposes;
- 'intelligent' sensor systems that allow a more differentiated reaction to events;
- deployment of modern bus solutions for communication.

Especially the use of modern bus systems from the area of industrial automation now offers completely new possibilities. Sensors and the communication links between them function as intelligent entities. The higher data bandwidth of modern buses allow for much more information content to be transferred. A good example is the safety circuit. The voltage circuit is replaced by an industrial bus. Consequently, the simple mechanical switches used on doors and other places can now be intelligent sensor units that offer far more than the simple open/closed information. In the case of a door, additional information can include the exact closing status of the door (how far closed, how fast closing), as well as status information of the sensor unit.

Competitive cost still remains one of the primary targets for components of modern elevators. Further boundary conditions apply for sensors as shown in this article:

- *Safety:* Elevators move people. Safety is therefore a design goal that cannot be tampered with. Failsafe design and inherent redundancy are essential factors required by codes and standards.
- *Robustness:* Elevator shafts are a harsh environment. Relative humidity close to 100% and temperatures from as low as below freezing to as high as 60 °C must be withstood without any problems. Vibrations, shocks, and heavy soiling complicate the task even more.
- *Modularity:* Ideally, sensor solutions can be used from simple two-story home elevators to high-rise systems of several hundred meters. Higher rises and increased functionality can added by standardized modules that match clearly defined interfaces.
- *Installation:* Easy installation and configuration is of high importance. Extensive fine tuning and complicated setup procedures are not acceptable.

Generally, it is believed that the trend of intelligent automation systems will grow stronger in the coming years. The rapid development and the achievements in the IT world are now moving into the area of industrial automation. The actual outcome of this trend is hard to predict, but the new systems to come will be far more than some simple contacts and light barriers that are somehow interconnected. Industrial automation and thus sensor systems on its backbone will see a revolution just as the IT world is right now – and so will modern elevator systems further develop towards sophisticated traffic management systems.

3.4
Sensing Chair and Floor Using Distributed Contact Sensors

HONG Z. TAN, *Haptic Interface Research Laboratory, Purdue University, West Lafayette, IN, USA*
ALEX PENTLAND, *The Media Laboratory, Massachusetts Institute of Technology MIT, Cambridge, MA, USA*
LYNNE A. SLIVOVSKY, *Haptic Interface Research Laboratory, Purdue University, West Lafayette, IN, USA*

3.4.1
Introduction

As computing becomes more ubiquitous and distributed, there is a growing need for the computing environment to become more aware of the people present and activities that take place around it. In a futuristic intelligent office building, every piece of furniture and building structure becomes a perceptual user interface – it sees, hears, and feels its surrounding as well as the people around it. This can be accomplished by providing objects with sensory mechanisms similar to our own – camera for vision, microphone for hearing, and pressure sensors for touch. Numerous systems have been developed that explore the idea of perceptual intelligence [1–8]. Among these, very few employ touch-based sensory information. Our Sensing Chair and Sensing Floor projects are conceptualized to explore the use of *distributed* pressure information, from sensors that are analogous to artificial skin, to achieve perceptual intelligence.

Perceptual intelligence for a building cannot be achieved by merely collecting and displaying sensory information, such as most webcams do. Perceptual intelligence results from an understanding of what the sensory data reveal about the state of the environment and people. The key research problem to be addressed with the Sensing Chair and Sensing Floor, therefore, is the automatic processing and interpretation of touch sensor information, and the modeling of user behavior leading to such sensory data. We envision tomorrow's buildings where all objects are outfitted with a layer of artificial skin (for example, a sensing chair, a sensing floor, a sensing file folder). We expect the algorithms and behavior models that we develop with the Sensing Chair and Sensing Floor to be extensible to large-scale distributed haptic (touch-based) sensing and interpretation.

To enable a chair to sense and interpret its occupant's actions, pressure distribution sensors are surface-mounted on the seatpan and backrest of a Sensing Chair. Work on the Sensing Chair draws upon current advances in computer vision, pattern recognition and stochastic modeling, taking advantage of the similarity between pressure-distribution maps and gray-level images. To enable a floor to sense and estimate the positions of its occupants, force-sensing resistors are placed under the corners of floor panels that make up a suspended floor structure. The sensor readings are then combined and compared with a threshold to determine whether the floor panel is occupied.

The successful implementation of a Sensing Chair and a Sensing Floor will impact many areas including ergonomics (by monitoring a person's sitting posture and giving feedback when necessary), multimodal human-computer interface research (by providing new haptic systems that can be integrated with other state-of-the-art user interfaces for multimodal interaction), intelligent environment (by creating interfaces that can feel their environment with contact sensors), universal access (by empowering people with limited sensory and motor capabilities with assistive interfaces), and safety of automobile operation (by augmenting a driver's seat with sensors that can automatically regulate airbag deployment force).

3.4.2
Related Work

Many systems have been developed around the structure of a chair. The British Telecom SmartSpace, for example, is a concept personal working environment of the future, built around a swivel chair (http://www.bt.com/innovation/exhibition/smartspace/index.htm). It is equipped with a horizontal LCD touchscreen, video projection, and 3D sound space. In contrast, the goal of our Sensing Chair is to achieve information extraction by instrumenting the chair itself.

BCAM International (Melville, NY, USA) has developed a recliner with pneumatically controlled air bladders placed near the surface of the recliner that can be inflated to 'hug' and support the occupant's body. This technology, called the 'intelligent surface', has recently been implemented in United Airline's Connoisseur Class seats [9, 10]. It should be pointed out that the air bladder activation patterns are based on ergonomic considerations, rather than on the needs of its occupant. The Sensing Chair can provide the needed intelligence to such mechanisms so that surface distribution can be altered in response to the real-time pressure distributions in the chair in an ergonomically beneficial manner.

Pressure distribution sensors have been widely used for the evaluation of weight-supporting surfaces in shoes, chairs, and beds. Examples of shoe studies include the assessment of seven types of shoes with regard to their ability to reduce peak pressure during walking for leprosy patients [11], the evaluation of the generalizability of in-shoe peak pressure measures with data collected from numerous subjects over a period of time using two calibration schemes [12], and the validation of the use of total contact casts for healing plantar neuropathic ulcerations through reduction of pressure over the ulcer [13]. Studies of seats include the development of a measurement protocol and analysis technique for assessing pressure distribution in office chairs [14], the use of body pressure distribution measures as part of a series of tests for assessing comfort associated with five automobile seats [15], and an interesting review of how objective pressure measures can lead to improved aircrew seating with more evenly distributed pressure patterns, thereby potentially improving a pilot's task performance by reducing or eliminating pain endured during high-acceleration maneuvers of the aircraft [16]. Examples of bed studies include an investigation of support surface pressure and reactive hyperemia (the physiological response to pressure) in the older population [17], and a recent development of body posture estimation system for sleepers based on pressure distribution measures and a human skeletal model [18].

Our Sensing Chair and Sensing Floor projects are similar to the last study cited [18] in that we focus on the automatic processing and interpretation of contact sensor information, whereas the other studies rely on expert analysis of pressure distribution measures. Of particular importance is the development of real-time systems that can be used to drive other processes such as a sitting posture monitoring system for persons with chronic lower-back pain, or for the prevention of such ailments.

3.4.3
The Sensing Chair System

3.4.3.1
Overview

The long-term goal of the Sensing Chair project is to model the sitting postures of the person occupying the Sensing Chair (Figure 3.4-1). As shown in Figure 3.4-2, the Sensing Chair project is further divided into the two components of Static Posture Classification (identification of steady-state sitting postures), and Dynamic Posture Tracking (continuous tracking of steady-state as well as transitional sitting postures). In each case, we start with a single-user system and proceed to a multi-user system. Our ultimate aim is a robust, real-time, and user-independent sitting posture tracking system.

Fig. 3.4-1 The sensing chair

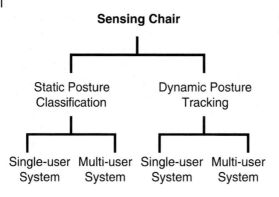

Fig. 3.4-2 Overview of the sensing chair project

Given the similarity between a pressure map and an 8-bit gray-scale image (see Figure 3.4-3), it is speculated that pattern recognition algorithms developed for computer vision would be applicable to the interpretation of sitting postures from pressure distribution data. There are two major approaches to object representation and recognition in computer vision: model-based (eg, [19]) and appearance-based (eg, [20]). The latter is considered more applicable since the concept of object model does not apply directly to pressure maps. Appearance-based modeling and object recognition involves the two steps of training and recognition. First, a set of training images is obtained. The technique of principal components analysis (PCA, also known as 'eigenspace methods', 'eigen-decomposition', or 'Karhu-

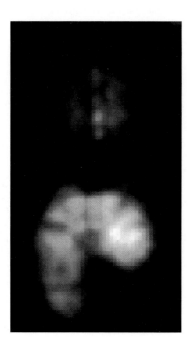

Fig. 3.4-3 A full pressure map for the posture 'left leg crossed'. See text for details

nen-Loeve expansion') [21] is often applied to the training data to obtain a low-dimensional representation (eigenspace) for the training and test images. Recognition of a test object is performed by projecting a test image on to the eigenspace and comparing the distance between the test image and image models derived from training. This is the approach that we have taken for static posture classification. Our work is based primarily on the computer face recognition work conducted at the MIT Media Lab [22, 23].

3.4.3.2
The Sensor

Our Sensing Chair is equipped with a commercially available pressure distribution sensor called the Body Pressure Measurement System (BPMS) manufactured by Tekscan (South Boston, MA, USA). The office chair shown in Figure 3.4-1 is fitted with two sensor sheets (hidden inside the protective pouches) on the seatpan and the backrest. Each sensor sheet is an ultra-thin (0.10 mm) flexible printed circuit. The sensing units are arranged in 42 rows and 48 columns with an equal inter-element spacing of 1.016 cm. Each sensing point acts as a variable resistor in an electrical circuit – its resistance changes in inverse proportion to the pressure applied. This resistance is then converted to an 8-bit (0–255) digital value that corresponds to a pressure range of 0–4 psi. Each sensor sheet is attached to a multiplexing unit called a Handle (see Figure 3.4-1). The Handles for both sensor sheets can then be connected to a PC interface board that occupies a 16-bit ISA expansion slot of a personal computer.

The Tekscan BPMS system comes with a software environment where pressure distribution data from multiple Handles can be displayed in real time as color-coded two-dimensional maps. It also provides numerous functionalities including the recording of pressure maps as 'movies' for later viewing or analysis. The current version runs in a Windows 95/98 environment. This software is useful for checking the integrity of the sensor sheets and for gaining intuition about the structure of pressure distribution associated with various sitting postures.

The Sensing Chair project requires real-time access to pressure distribution data collected from the two sensor sheets. For that purpose, Tekscan provides a simple hardware interface API (application program interface). The Tekscan API is a 32-bit static library developed for the Microsoft Visual C++ 6.0 environment. It provides functions that enable a user, who may not have extensive knowledge of how to control and interface to the sensor hardware, to perform tasks such as initializing the sensor sheets, checking sensor parameters (number of rows and columns, and total number of sensor sheets), and capturing a frame of pressure distribution data in a buffer. The throughput sampling rate supported by the API can be up to 127 Hz.

The Tekscan BPMS system has been selected for several reasons. First, the inter-element resolution of the sensor sheets is 1.016 cm. This resolution is considered to be very high so as not to become the bottleneck of the performance of the Sensing Chair system. Another reason for using a sensor system with a higher

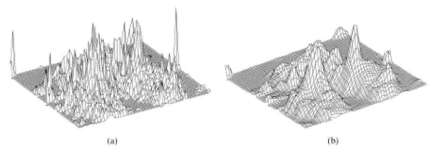

Fig. 3.4-4 3D views of a pressure map for the posture 'seated upright', (a) before and (b) after smoothing

than necessary resolution has to do with the integrity of the data: raw pressure readings can be very noisy (see spikes in Figure 3.4-4a). Local averaging smooths raw data (see Figure 3.4-4b) at the cost of reduced sensor resolution. This reduction in resolution is compensated for by the use of a sensor system with high resolution. Second, the sensor sheets are very thin (0.10 mm in thickness). The flexibility of the sheets makes it possible for them to conform to the shape of a chair. Third, the Tekscan pressure measurement systems are widely used by major research and industry laboratories including the Natick Army Research Laboratory (for the design of army boots) and Steelcase and Herman Miller (for chair evaluation). This enables us readily to compare our findings with those of other researchers. Finally, an important reason for our selection of the BPMS system is Tekscan's willingness to provide an API that has enabled us to access and process pressure distribution data in real time.

There are several known problems associated with the Tekscan BPMS system. Like any resistive sensors, the Tekscan sensors suffer from nonlinearity, nonuniformity, hysteresis, drift, temperature sensitivity, and limited sensor life. The noise introduced by these characteristics turned out to be manageable. Another problem with the sensor sheets is that they are designed for flat surfaces. When a person sits on a chair, the sensor sheet interfacing the person and the chair surface can bend, thereby introducing additional noise to sensor readings that are dependent on both the individual and the sitting postures. Repeated bending results in cracks in certain parts of the sensor sheets. When this happens, the entire sensor sheet needs to be replaced. An additional limitation of the Tekscan sensor is that it can only measure the pressure component that is perpendicular to the sensing elements. One can easily imagine how knowledge of pressure components that are tangential to chair surfaces can be useful in determining an occupant's sitting postures. Finally, the high cost of the Tekscan BPMS system has restricted its use to a handful of research systems.

3.4.3.3
Preprocessing of Pressure Data

The image shown in Figure 3.4-3 is a full pressure map for the static sitting posture of Left Leg Crossed (after noise removal). The top and bottom halves of the pressure map correspond to the pressure distribution on the backrest and seatpan, respectively. To understand the orientation of the pressure map, imagine standing in front of the chair and its occupant and unfolding the chair so that the backrest and the seatpan lie in the same plane. Therefore, the top, bottom, left, and right sides of the pressure map shown in Figure 3.4-3 correspond to the shoulder area, knee area, and right and left sides of the occupant, respectively. The size of each full pressure map is 84-by-48 (two sheets of 42-by-48 maps), or equivalently, 4032 sensels (sensing units).

As mentioned earlier, the raw pressure distribution map is typically noisy (spikes in Figure 3.4-4a). The noise is removed by convolving the pressure map with a 3-by-3 smoothing kernel:

$$\frac{1}{7}\begin{bmatrix} 0.5 & 1 & 0.5 \\ 1 & 1 & 1 \\ 0.5 & 1 & 0.5 \end{bmatrix}$$

The smoothed pressure map (Figure 3.4-4b) contains pressure artifacts (at the top left and right corners of the image) due to the corners of pressure sensor sheets being wrapped around the chair. Since these artifacts are common to all pressure maps, their removal is not necessary for the real-time posture tracking system that we have developed. Finally, although the sensor sheets can be calibrated to display pressure readings in psi or other standard units, the raw digital data are used since we are only interested in the *relative* pressure distribution on the chair surfaces. The raw pressure readings are normalized, separately for the seatpan and the backrest maps. The rest of the discussion on the Sensing Chair system assumes that all pressure maps have gone through the above-mentioned preprocessing procedures.

3.4.3.4
Static Sitting Posture Classification

To date, we have developed both a single-user [24, 25] and a multi-user [26] Static Posture Classification System (see Figure 3.4-2). For the multi-user system, a Static Posture Database has been collected on 30 subjects (15 females and 15 males) for 10 sitting postures. The subjects were selected with the goal of covering a wide distribution of anthropometric measurements. The ranges of subject's height, weight, and age were 152–191 cm, 45.5–118.2 kg, and 18–60 years, respectively. Each subject contributed five pressure distribution samples per posture. There are therefore a total of 150 training samples per posture, and the training database consists of a total of 1500 pressure distribution maps.

The postures contained in the Static Posture Database are (1) seated upright, (2) leaning forward, (3) leaning left, (4) leaning right, (5) right leg crossed, (6) left leg crossed, (7) leaning left with right leg crossed, (8) leaning right with left leg crossed, (9) leaning back, and (10) slouching. These postures are considered to be representative of the typical sitting postures that can be found in an office environment [27].

For each of the 10 postures, the 150 training samples are used to calculate the eigenspace for that posture (called an 'eigenposture space'). During classification, a new pressure distribution map is first tested for 'empty seat' by comparing the sum of all pixel values with a preset threshold. Once a pressure map has passed this initial test, it is projected on to the 10 eigenposture spaces. The posture label associated with the eigenposture space that best represents the new pressure map is then assigned to the new map. A more detailed description of our posture classification algorithm can be found in [26].

3.4.3.5
Performance Evaluation

The accuracy of our multi-user Static Posture Classification system was evaluated with additional pressure maps collected from two groups of subjects. First, an additional 200 pressure distribution maps were collected from 20 of the 30 subjects who contributed to the Static Posture Database (one sample per posture per subject). These pressure maps were then labeled with respect to their corresponding postures by the Static Posture Classification system. Figure 3.4-5 shows the classification accuracy in terms of percent-correct scores averaged over postures, as a function of the number of eigenvectors that are used in the classification algorithm (see [26] for details). As expected, overall classification accuracy increases as a function of the dimension of eigenposture space. The curve in Figure 3.4-5 shows a knee point at 15 eigenvectors with a corresponding average accuracy of 96.0%.

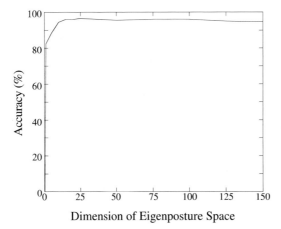

Fig. 3.4-5 Classification accuracy for 'familiar' subjects

Classification accuracy by posture (averaged across subjects) ranges from 90.3% (for posture 'leaning back') to 99.8% (for 'slouching'). The system is also able to discern among postures that have very similar pressure distribution maps (eg, 95.2% for 'leaning left', 95.1% for 'right leg crossed', and 93.5% for 'leaning left with right leg crossed').

Second, a total of 400 pressure distribution maps were collected from eight new subjects (five samples per posture per subject) who did not contribute to the Static Posture Database. The ranges of subject's height and weight were 160–191 cm and 65.9–93.6 kg, respectively. These anthropometric values are within those represented in the Static Posture Database. For these 'new' subjects that the system has never 'felt' before, the average classification accuracy at 15 eigenvectors dropped from 96.0 to 78.8%. In an effort to locate the sources of error, we examined the posture labels associated with not only the eigenposture space that best represents the test pressure map, but also those with the next two closest eigenposture spaces. The classification accuracies that can be potentially achieved if the correct posture label is associated with the first three closest eigenposture spaces turned out to be 99.0 and 97.5% for 'familiar' and 'new' subjects, respectively.

The execution time for the classification subroutine as a function of the number of eigenvectors used was also measured, with source codes that have yet to be optimized for speed. This is an important parameter for any real-time application of our system. The average classification time for 5, 10, 15, and 20 eigenvectors is 62.1, 107.8, 168.1, and 241.0 ms, respectively. The corresponding average classification accuracy (for 'familiar' subjects) is 88.5, 94.5, 96.0, and 96.0%, respectively. In view of these measurements, 15 (out of 150) eigenvectors corresponding to the 15 largest eigenvalues are used for our current version of the multi-user Static Posture Classification system.

3.4.4
The Sensing Floor System

3.4.4.1
Overview

The goal of the Sensing Floor project was to track single or multiple people by instrumenting floor panels. It takes advantage of a suspended floor structure where the weight of each floor panel is supported along its edges by aluminum railings (Figure 3.4-6). Force-sensing resistors (FSRs) are placed between the floor panels and their supporting structures. Readings from FSRs are then combined and compared to a threshold to determine whether a floor panel supports weight. While the overall concept is straightforward, sensor stability and accuracy turned out to be the limiting factors for the success of the Sensing Floor project.

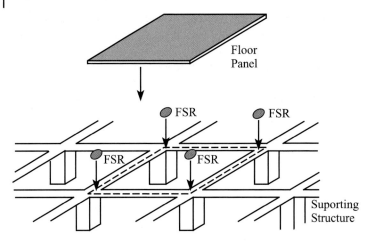

Fig. 3.4-6 An illustration of the sensing floor components. See text for details

3.4.4.2
The Sensor

The floor sensors are based on the FSRs manufactured by Interlink Electronics (Camarillo, CA, USA) (Part No. 402) [28]. Each sensor has a circular active sensing area that is 12.7 mm in diameter. It is interfaced with the floor panel and its supporting structure via rubber pads. Four sensors are used for each floor panel at its four corners. Each FSR is connected (as a variable resistor) to a measuring resistor in a simple force-to-voltage conversion configuration. The analog voltage level is then sent to a computer and digitized as an 8-bit integer.

The FSRs were selected for their ease of use and relatively low cost. The FSR User's Guide clearly states that 'FSRs are not suitable for precision measurements' and 'only qualitative results are generally obtainable' [28]. It was anticipated, however, that if a 2-bit accuracy per FSR channel could be achieved, then it would be possible to detect and track movements on the floor panels. In practice, we had considerable difficulties with sensor drift, thermal sensitivity, and hysteresis. It was difficult to compensate for random sensor variations as FSR readings would at times fluctuate as much as 50% in idle condition (that is, with no one standing or walking on the floor panels).

3.4.4.3
Data Processing

Real-time data acquisition is accomplished with a PC board that can support 64 analog inputs simultaneously at a sampling rate of up to 500 kHz (Part No. AT-MIO-64E-3, National Instruments). On-line data processing is performed with the LabVIEW software (National Instruments) that features a graphical-based programming environment. The initial implementation of the sensing floor included a total of 16^2 ft^2 floor panels (as limited by the 64-channel data acquisition system). Readings from the four sensors corresponding to the same floor panel are summed and compared with an empirically determined threshold. If the total force reading exceeds the threshold, an icon representing the floor panel will change its color from green to black on the computer screen. When sensor readings are stable, the system can correctly track a person walking across the active floor area.

3.4.5
The Future

We have described two systems based on distributed contact sensors. The Static Posture Classification system is based on a Sensing Chair that monitors the pressure distribution patterns on its surfaces in real time. Future work is aimed towards a Dynamic Posture Tracking system that continuously tracks not only steady-state (static) but also transitional (dynamic) sitting postures. It is expected that techniques such as hidden Markov modeling (HMM) commonly used for speech recognition can be successfully applied to sitting posture tracking. A robust posture tracking system to be completed in the near future will support many exciting applications such as a sitting posture monitoring system for ergonomics, and automatic adjustment of airbag deployment forces for automobiles.

The Sensing Floor system is based on floor panels instrumented with force-sensing resistors. Although we have demonstrated that it is possible to track people on the floor in real-time, we have had considerable difficulties with unstable readings from FSRs. Sensors that are more stable and accurate are needed in the future. It is conceivable that force-sensing units can be manufactured as an integral part of the floor structure to enable universal access to force distribution data on an active floor.

3.4.6
Acknowledgments

Part of this work was presented at the First Workshop on Perceptual User Interfaces and at the Eighth International Symposium on Haptic Interfaces for Virtual Environment and Teleoperator Systems. The first two authors were partly supported by British Telecom, Things That Think consortium at the MIT Media Laboratory, and Steelcase NA. The first and last authors were partly supported by a National Science Foundation Faculty Early Career Development Award under Grant No. 9984991-IIS.

3.4.7 References

1 COEN, M.H., *IEEE Intell. Syst.*, March/April (1999) 8. Vol. 14, no. 2.
2 FLANAGAN, J.L., *IEEE Intell. Syst.*, March/April (1999) 16. Vol. 14, no. 2.
3 FRANKLIN, D., *IEEE Intell. Syst.*, Sept./Oct. (1999) 2. Vol. 14, no. 5.
4 MOZER, M.C., *IEEE Intell. Syst.*, March/April (1999) 11. Vol. 14, no. 2.
5 PENTLAND, A.P., *Sci. Am.* **274** (1996) 68.
6 TORRANCE, M.C., in: *Proceedings of CHI'95 Research Symposium*, Denver, CO; Association for Computing Machinery, New York, NY, USA, 1995.
7 TURK, M., (ed.), *Proceedings of the Workshop on Perceptual User Interfaces (PUI'97)*, Santa Barbara, CA, PUI Workshop, Banff, Alberta, Canada, October 19–21, 1997.
8 TURK, M. (ed.), *Proceedings of the Workshop on Perceptual User Interfaces (PUI'98)*, Santa Barbara, CA, PUI Workshop, San Francisco, CA, November 4–6, 1998.
9 COY, P., *Business Week*, Nov. 4 (1996) 199.
10 ROACH, M., *Discover*, March (1998) 74. Vol. 19, no. 3.
11 BIRKE, J.A., FOTO, J.G., DEEPAK, S., WATSON, J., *Lepra Revi.* **65** (1994) 262.
12 MUELLER, M.J., STRUBE, M.J., *Clin. Biomech.* **11** (1996) 159.
13 CONTI, S.F., MARTIN, R.L., CHAYTOR, E.R., HUGHES, C., LUTTRELL, L., *Foot Ankle Int.* **17** (1996) 464.
14 REED, M., GRANT, C., *Development of a Measurement Protocol and Analysis Techniques for Assessment of Body Pressure Distributions on Office Chairs*; Technical Report, Ann Arbor, MI: University of Michigan, 1993.
15 HUGHES, E.C., SHEN, W., VERTIZ, A., *SAE Tech. Pap. Ser.* (1998) 980658, pp. 105–115.
16 COHEN, D., *J. Aviat. Space Environ. Med.* **69** (1998) 410.
17 BARNETT, R.I., SHELTON, F.E., *J. Prev. Healing: Adv. Wound Care*, vol. 10, no. 7 (1997) 21.
18 HARADA, T., MORI, T., NISHIDA, Y., YOSHIMI, T., SATO, T., in: *Proceedings of the 1999 IEEE International Conference on Robotics and Automation*, May, Detroit, MI; IEEE (The Institute of Electrical and Electronics Engineers) Piscataway, NI, USA (1999) pp. 968–975.
19 BROOKS, R.A., *IEEE Trans. Pattern Anal. Machine Intell.* **5** (1983) 140.
20 MURASE, H., NAYAR, S.K., *Int. J. Comput. Vision*, **14** (1995) 5.
21 FUKUNAGA, K., *Introduction to Statistical Pattern Recognition*, 2nd edn.; New York: Academic Press, 1990.
22 PENTLAND, A., MOGHADDAM, B., STARNER, T., in: *Proceedings of IEEE Conference on Computer Vision and Pattern Recognition*; IEEE (The Institute of Electrical and Electronics Engineers) Piscataway, NI, USA (1994) pp. 84–91.
23 TURK, M., PENTLAND, A., *J. Cogn. Neurosci.* **3** (1991) 71.
24 TAN, H.Z., in: *Proceedings of the ASME Dynamic Systems and Control Division*, Vol. 67, Olgac, N. (ed.); New York: American Society of Mechanical Engineers, 1999, pp. 313–317.
25 TAN, H.Z., LU, I., PENTLAND, A., in: *Proceedings of the Workshop on Perceptual User Interfaces (PUI'97)*; Turk, M. (ed.), Banff, Alberta, Canada, October 19–21, 1997, pp. 56–57. Santa Barbara, CA, PUI Workshop
26 SLIVOVSKY, L.A., TAN, H.Z., *A Real-time Sitting Posture Tracking System*; Purdue University Technical Report, TR-ECE 00-1, West Lafayette, IN: School of Electrical and Computer Engineering, Purdue University, 2000.
27 LUEDER, R., NORO, K., *Hard Facts about Soft Machines: the Ergonomics of Seating*; Bristol, PA: Taylor & Francis, 1994.
28 *FSR integration Guide and Evaluation Parts Catalog*; http://www.interlinkelec.com/, Interlink Electronics, Camarillo, CA, USA, 2000.

4 Safety and Security

4.1
Life Safety and Security Systems

MARC THUILLARD, *Siemens, Cerberus, Männedorf, Switzerland*
PETER RYSER, *Swiss Federal Institute of Technology, EPFL, Institut de Production Microtechnique, Lausanne, Switzerland*
GUSTAV PFISTER, *Siemens, Cerberus, Männedorf, Switzerland*

4.1.1
Introduction

This chapter describes intelligent sensors in life safety and security systems. "Life safety" includes fire alarm (4.1.2), gas alarm (4.1.3), extinguishing (4.1.6) and evacuation (4.1.6) systems. "Security" refers to intruder alarm (4.1.4), access control (4.1.5) and CCTV (4.1.4, 4.1.5) systems.

Since the dawn of time it has been a basic need of all living organisms to feel comfortable and safe. While the basic dangers arising from the four basic elements of science from Ancient Greece, namely earth, water, air, and fire, still represent formidable threats in today's life, it is man himself who has contributed to extend substantially the threat potential. Indeed, man has felt threatened to a large degree by his own kind. One has only to think of wars, tribal feuds, raiding parties, piracy, breaks-ins, robbery, arseny, and terrorism. The list could go on!

It was, of course, recognized very early that a timely mobilization of defensive forces by early warning could reduce or even eliminate the potential damage. In ancient times watchmen played a key role in these measures (watchmen on fire towers, smoke signaling, etc.). With time man involved technology to assist and eventually replace him from the early warning duty. A most prominent invention in Antiquity was the key which was used separate unwanted from wanted access and which, of course, has evolved into the modern access control systems of today.

With the advent of electricity the first alarm warning systems were introduced. With all kinds of ingenious inventions one wanted to detect dangerous situations as early as possible. Most of the technology used contacts which triggered an alarm clock when open or closed, for instance by a burglar stepping on or breaking through a wire strung along the perimeter of the premises to be protected. To detect fire incidents the ideas span from using canaries in a cage which fall on to a contact when gassed by fire fumes (this was probably the first bionic application proposed for sensing danger phenomena) to bimetallic cantilever sensor designs which close a contact when exposed to a slowly rising temperature [1].

The development of sensors has a formidable benchmark in the animal world. Indeed, animals have developed sensorics to spot danger (and food) which even today outperform the achievements of modern technology. Many of the physical principles which form the basis of sensor development have been perfected by Nature. Examples are the use of ultrasonic waves by bats, infrared sensing by rattlesnakes and the outstanding sense of smell which is on the molecular level for many animals.

In this section we want to discuss the technologies and physical principles of sensors designed for early warning of dangerous situations in buildings such as fire incidents, toxic or explosive gas concentrations and unwanted intrusion into the premises. Therefore, we discuss fire sensing, intrusion sensing, gas sensing and access sensing. By definition we call a sensor the device which transforms the fingerprint from the danger indication into a signal which can be analyzed. The detector has the sensor integrated and processes the sensor information.

The physical and chemical principles underlying modern sensorics are with few exceptions well understood. The main challenge in detector design is not so much the detection of the danger phenomenon – this can be achieved with relatively simple electronics and mechanical designs – but rather the identification and suppression of deceptive inputs. In most cases danger and deceptive phenomena are very similar and may even be of the same origin. A high-quality smoke detector has to differentiate between aerosols and smoke particles of a true fire source from dust particles which in a light-scattering detector produce the same effect. Radiation from the sun heating up, for instance, a desk top may generate temperature fluctuations (Schlieren) which can be picked up by a passive infrared motion detector and wrongly be taken for an intruder. Reflections from sunlight can generate deceptive signals in flame detectors. Deceptive sources also exist for seismic detectors used to identify assaults on vaults, ultrasound and microwave detectors used in intruder alarm systems, linear beam smoke detectors, and gas detectors. In access control it is the main challenge to identify those people who have the right of access at any given moment (day, time, duration) and reject the others. This must be done with a high level of confidence.

It is the challenge of the engineer to design detectors which have a high detection intelligence. This is a loose term referring to the sensitivity of catching real danger situations in due time and the rejection probability of deceptive phenomena. This must be achieved under the demanding economic requirement of low detector cost and under the often difficult environmental conditions which are dictated by the application (humidity, temperature, electromagnetic interference). The detector design involves all disciplines – mechanics, electronics and software (signal processing) – which must be optimized. We will give examples and in particular address recent developments in signal processing.

Although the main theme of this book is intelligent sensors, we found it appropriate not to limit this section to sensorics alone – it will, however, be the dominant part by far – but to extend the discussion and include aspects of the alarm system and in particular of the emergency handling systems which are activated by the sensors described. We therefore spend a few paragraphs on voice evacuation systems, extinguishing systems, and alarm-receiving centers. Thus the entire loop from the detection of a dangerous situation to its elimination is closed.

The contents of this section are largely based on the vast knowledge built up in our company which for years had its focus on the development of life safety and security systems, and in fact with the invention of the ionization detector in the 1940s pioneered this industry [2]. We would like to thank our company colleagues who have made valuable contributions and suggestions, specifically R. Beckers (alarm-receiving centers), U. Krienen and M. Schumacher (identification sensing), M. Müller (extinguishing systems), and H. Kupfer (voice evacuation systems).

4.1.2
Fire Sensing

4.1.2.1
Fire Physics, Smoke Aerosols, Gases, and Flames

Characteristic signatures of smoldering fires are the emission of smoke aerosols and gases which are complemented by radiative emission and a temperature rise once the smoldering fire has broken open. Fire detectors use any of one these physical effects either alone or in combination to signal a fire incident at an early stage in order to prevent or limit possible damage to lives and property. With the evolution of the trend-setting technologies also fire and smoke detection has undergone substantial changes and improvements.

For over 50 years, smoke detectors were predominantly of the ionization type. Smoke is detected by measuring the modification of the conductivity of an ionization chamber in the presence of smoke aerosols. Within the last 10 years photoelectric smoke detectors which measure the scattering of light from the smoke particles in an optical chamber have begun to replace ionization detectors.

Fire detectors are not only limited to these two principles. Light absorption is used in linear beam detectors. Also fire detection can rely on the detection of the thermal radiation of a fire. Flame detectors measure the flame radiation in the infrared or ultraviolet region. Thermal detectors give an alarm when a significant increase in temperature is recorded, for instance with a thermistor or by registering modifications of the optical properties of an optical fiber. Many other effects have been thought of to detect a fire: quartz microbalances, variations of sound propagation speed, and fluctuations of pressure, optical refraction, or temperature are just a few effects that have been tested. Even the cracking noises of special additives to wall, paint have been investigated. However, none of these principles, although successfully tested in the laboratory, was developed into a high-volume product of any commercial impact. Indeed, the reliability, manufacturing cost, and detection performance of today's fire and smoke detectors set a high benchmark for the commercialization of new detection methods.

The response of a detector to a fire depends on its detection principle and on the type of smoke. Figure 4.1-1 compares the sensitivity of a light-scattering smoke detector with that of an ionization detector as a function of particle size. Ionization detectors respond better than optical smoke detectors to small particles, whereas the reverse is true for large particles.

The response of a detector depends also on its exact design. For a light-scattering smoke detector the sensitivity to different smoke types is related, for instance, to the

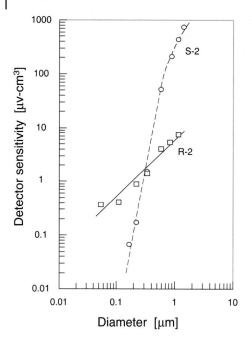

Fig. 4.1-1 Detector sensitivity of two smoke detectors: ionization principle (R-2) and light scattering (S-2). After Bukowski et al. [3]

angle between the primary axis of light beam impinging on the aerosol cloud and the axis of the detector which measures the scattered light emerging from the cloud.

Figure 4.1-2 compares the qualitative responses of four types of detectors to different fires. The two optical detectors respond better than ionization detectors to smoldering fires, whereas the ionization detector is superior for small aerosol particles typical of open fires. By combining an optical principle with thermal sensors, a homogeneous response to all fires can be achieved.

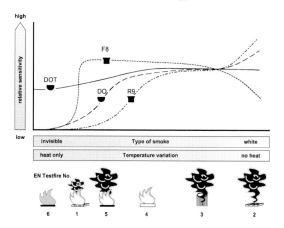

Fig. 4.1-2 Qualitative responses of different smoke detectors for different fires. F8, ionization detector; R9 and DO, photoelectric (light-scattering) detectors with different scattering angles; DOT, photoelectric detector (=DO) with additional temperature sensor (multisensor principle). The fire numbers refer to the test fires as defined in the European standard EN 54-7. After Pfister [4]

'Smoke' detectors which sense the combustion gas from burning material have recently been introduced. At the present time, however, no combustion gas detector approaches the detection performance, reliability, and robustness of standard smoke detectors. Smoke detectors are guaranteed to work properly from −30 to 75 °C and have a lifetime of more than 10 years. Both criteria cannot be met with the gas sensors which are proposed today. However, significant research and development effort is being put into new technologies for smoke gas sensing and significant advances in the development of reliable products are expected in the near future.

In unwanted fires, smoke may result from different combustion mechanisms such as pyrolysis, flaming, or smoldering. Pyrolysis results from the heating of a material by an external heat source. The difference between pyrolysis and smoldering is that in smoldering the combustion process is self-sustaining. While most materials can be pyrolyzed, only a relatively small number of materials are able to smolder. These materials include a number of very common materials such as paper, some foams, and wood.

The physical characteristics of smoke and the temperature rise can be related to some extent to the type of fire. Pyrolysis and smoldering generally lead to large, light-colored smoke droplets containing a substantial amount of water. By comparison flaming often produces smoke containing a large amount of invisible small and 'visible' black aerosol soot. Smoldering and pyrolysis temperatures are generally much lower than flame temperatures in a flaming process.

Light scattering depends on the particle size and number and on the particle's refractive index. The computation of light scattering and absorption of small spheres of known size and distribution is a classical problem in optics. The application of Mie and Rayleigh theory is generally sufficient to predict light scattering from computer simulations. The characterization of smoke from optical measurements is a very difficult inverse problem. In their book on light absorption and scattering, Bohren and Huffman [5] compare the problem of particle characterization from optical measurements to the determination of an extinct animal from its fossilized foot traces! The computation of light scattering under real conditions is often complicated by the fact that smoke particles may agglomerate into complex shapes. Such an effect is typical of smoke aging. Smoke aging is characterized by the coagulation of smoke particles resulting into a decrease of the particle density at fixed mass [6].

Scattering and absorption depend very significantly on the wavelength and polarization of the light source. Figure 4.1-3 shows the computation of light scattering for a sphere of known refractive index. A number of simple rules can be extracted from this example:

Large particles
- Large particles have a scattering cross-section with a number of lobes. The scattering of light depends very much on the scattering angle.
- Most of the light is scattered in the forward direction.

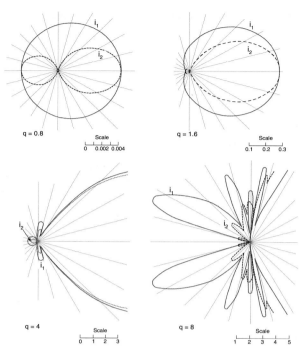

Fig. 4.1-3 Scattering of linearly polarized light of wavelength λ by a sphere of radius r ($n=1.25$, $q=2\pi r/\lambda$). Red, light scattered with perpendicular polarization; blue, horizontal polarization. The incident light is along the horizontal axis from the left. The scattering angle is measured counter-clockwise from the direction of the incident beam. The particle size increases from the top left to the bottom right. After Born and Wolf [7]

Small particles

- Light scattering depends strongly on polarization.
- The ratio of forward to backward scattering is lower than for large particles.

These simple rules have important consequences on the design of an optical scattering smoke detector. For instance, more light is scattered into a small angle (forward scattering) than into a large angle (backward scattering). Depending on the aerosol, the ratio between forward to backward scattering can be as high as 50. Consequently, the electronics to amplify the signal in a detector which uses a backward scattering design are more complex. The difference between forward and backward scattering is illustrated in Figure 4.1-4.

The design of an optical detector depends on the desired response to the different types of smoke. The choice of the scattering angle determines to a large extent the relative response of the detector to open fires and smoldering fires. A detector with an optical detector using a large scattering angle (backward scattering) is more sensitive to small particles.

The scattering of light by small particles is generally described on the basis of Mie and Rayleigh equations. The scattering by very small smoke particles can be

Fig. 4.1-4 Sketch showing the difference between forward and backward scattering

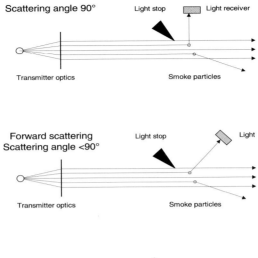

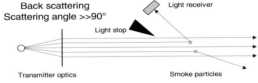

described with Rayleigh theory, whereas large particles are often within the range of validity of Mie theory. The geometric mean diameter of smoke particles is generally in the transition region between the Rayleigh and the Mie domains. The particle size distribution of smoke aerosols depends on the type of fire and the smoldering/burning material. Open fires have a large proportion of particles below 0.3 µm, whereas in smoldering fires the particles are larger. A typical particle size distribution is shown in Figure 4.1-5.

4.1.2.1.1 Rayleigh Scattering

Rayleigh theory [9] describes the light scattering of small particles, when the particle radius r is significantly smaller than the wavelength λ ($2\pi r/\lambda < 1$). Table 4.1-1 summarizes the main results of Rayleigh theory.

A number of relevant remarks can be done. Light scattered by a particle is proportional to the sixth power of the radius. In other words, the scattering cross section decreases rapidly for particles smaller than the wavelength (the S-2 data in Figure 4.1-1 confirm this result). Figure 4.1-6 shows the degree of polarization of light scattered in the Rayleigh domain. Light scattered at 90° is linearly polarized. This effect can be used to design detectors which are more sensitive to small particles [9, 9a].

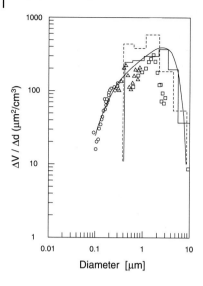

Fig. 4.1-5 Example of volume size distribution of smoke. After Mulholland [8]

Tab. 4.1-1 Equations describing the optical properties of particles [1]

Extinction	Scattering + absorption
Absorption	$Q = (2\pi r/\lambda) \times 4 \times \text{Im}\left(\dfrac{m^2 - 1}{m^2 + 2}\right)$
Scattering	$I_s = \dfrac{8\pi^4 n r^6}{\lambda^4 d^2} \left|\dfrac{m^2 - 1}{m^2 + 2}\right|(1 + \cos^2\theta) I_0$
Polarization	$P = \dfrac{\sin^2\theta}{1 - \cos^2\theta}$

1) Q = Absorption cross section; I_0 = incoming light intensity; I_s = scattered light intensity at distance d of a particle of radius r and refractive index n; θ = scattering angle, $m = n_{\text{air}}/n$.

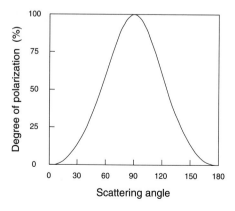

Fig. 4.1-6 Degree of polarization of light scattered in the Rayleigh domain

4.1.2.1.2 Mie Theory of Scattering

The Mie domain [10] is characterized by rapid oscillations of the scattering cross section as a function of the refractive index, wavelength, scattering angle, and particle radius. The distribution of smoke particles is generally fairly broad. In real fires, the aerosol particles are not monodisperse and the scattering lobes are generally partially averaged out. Nevertheless, a qualitative difference does exist between large white particles and small black particles. Figure 4.1-7 shows the ratio between the scattering cross-sectional area of an open fire and a cotton wick fire as a function of angle. The black/white ratio is much higher at large scattering angles. In practice, this means that a detector using backscattering is more sensitive to small black particles than a detector using forward scattering. The main disadvantage of backward scattering is the small scattering cross section. This necessitates a more energetic light pulse or a higher amplification of signal on the receiver.

4.1.2.1.3 Absorption

Figure 4.1-8 shows the absorption cross sections of water. In the region dominated by scattering (0.5–5 µm^{-1}), the absorption cross section decreases very rapidly. The Mie domain is characterized by oscillations. For very large particles, the absorption cross section tends to 2 (the so-called extinction paradox).

The extinction cross section decreases less rapidly than the scattering cross section. A point extinction detector is consequently comparatively more sensitive than a light-scattering smoke detector to small absorbing smoke particles. Figure 4.1-9 compares the response profile of a point extinction detector with that of a scattering detector. The extinction point detector is more sensitive to open wood fires or liquid heptane than to smoldering fires.

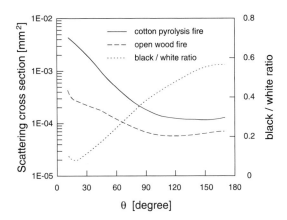

Fig. 4.1-7 Scattering cross-sectional area as a function of the scattering angle θ for two test fires: TF1 (open wood fire) and TF3 (cotton wick). After Ryser [11]

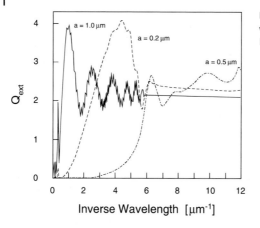

Fig. 4.1-8 Extinction efficiencies for water droplets of different size a. After Bohren et al. [5]

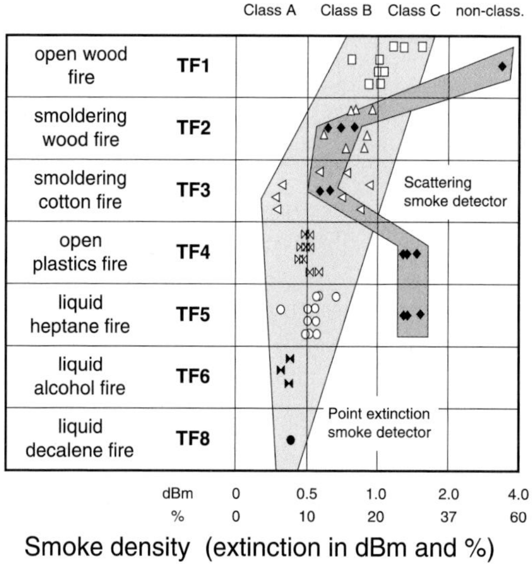

Fig. 4.1-9 Response of a point extinction smoke detector compared with that of a scattering smoke detector to EN54 test fires. After Müller et al. [12]

4.1.2.2
Smoke Sensing Principles

4.1.2.2.1 Ionization

The basic principle of an ionization detector can be simply explained [13] (Figure 4.1-10). The measuring chamber is delimited by two electrodes and electrically isolating parts. A small radioactive source, generally ^{241}Am, emits α-particles. The air contained in a small measuring chamber is ionized (mostly singly charged small ions of both polarities are produced). A small voltage (typically 5 V) is applied between the electrodes. In the absence of smoke, a small current of the order of 10–20 pA is recorded. In the presence of smoke in the chamber, a significant portion of the ions attach to the large smoke particles. The smoke particles act as recom-

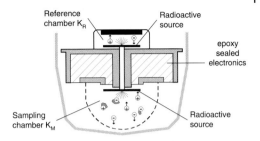

Fig. 4.1-10 Ionization chamber

bination centers for the small ions. The drift velocity of the large smoke particles is orders of magnitude smaller than that for small ions. The global effect is a decrease in the ion current.

The ionization chamber has been optimized over five decades. Ionization chambers are pressure and temperature dependent. By careful design of the chamber, the dependence on temperature and pressure can be kept small. The main optimization parameters are the energy of the radioactive source, that determines the range of the α-particles, the chamber geometry, and the applied voltage. For a given ionization chamber, a small ion lifetime is related to the applied field in the chamber. At high applied fields, ions drift rapidly in the chamber and a large current is measured. Almost all ions produced reach the electrode. In this range, the ionization chamber behaves as a current source. At low fields, the current is small but the sensitivity to smoke is large. The electric field on a measuring chamber is chosen so as to have a large enough ionization current and a good sensitivity to smoke. The chamber geometry is best optimized, when the right balance between unipolar and bipolar domains is found (the bipolar domain corresponds to the region in which ions are created. The unipolar domain corresponds to the region in which ions of one polarity drift under the influence of the electric field, but no small ion is generated by the radioactive source).

The current-voltage relationship of an ionization chamber not only on the aerosol concentration in the chamber volume but also on the cleanliness of the radioactive source surface. Examples are shown in Fig. 4.1-11. The different I–V characteristics can be used to distinguish the origin of the current decrease. Specifically the dominant contaminant water condensation can be distinguished from aerosols [15].

The measurement of small pA currents requires a high-ohmic resistor. In the one-chamber principle, a resistor (typically 20 GΩ) is taken as reference element. In the two-chamber principle, a second chamber operated in the saturation regime is taken as reference. The reference chamber is impermeable to smoke and has the function of a current source.

The modeling of a smoke chamber is difficult and goes beyond the scope of this section. A few simple equations can be used to explain the main property of an ionization detector, namely that ionization smoke detectors detect open fires very well (small particles are produced in an open fire).

Let us consider the equilibrium concentration n of small ions. For simplification, we assume that ions of both charges have the same density n, the same elec-

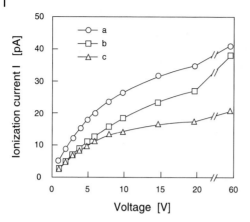

Fig. 4.1-11 Current-voltage relationship of an ionization chamber. The voltage dependence is shown in the case of (a) clean air, (b) smoke, and (c) water condensation on the radioactive source. The measurement of the chamber at two different voltages permits one to differentiate between a decrease in the conductivity due to smoke and coverage of the radioactive source (eg, water). After Thuillard [15]

tric mobility μ (in reality the mobility ratio μ^+/μ^- varies with humidity between 1.1 and 1.5 [14]) and that the average charge on large ions is small. The concentration n can be estimated by solving the equation

$$I = an^2 + 3\pi E r^2 n N \mu \tag{4.1-1}$$

with I the number per unit volume of ions created per unit time, a the recombination rate of small ions, N the number per unit volume of smoke particles of radius r and E the electric field.

One sees that the attachment of small ions on an uncharged large particle is to a first approximation proportional to the square of the radius of the particle.

Equation (4.1-1) permits us to understand why ionization detectors detect small particles better than large particles. Consider a fixed mass m and a monodisperse aerosol of radius r. The number of small ions that become attached to smoke particles is roughly proportional to $1/r$ (attachment rate ~ number of particles (m/r^3) × attachment rate on a particle (r^2)). One concludes that, for a given mass of smoke, small particles are much better detected than large particles.

4.1.2.2.2 Optical Light Scattering
General Principles

An optical smoke detector measures the light scattered by smoke aerosols and particles. A light-emitting diode sends a light pulse at regular intervals. The light scattered by the particles is measured with a receiver photocell positioned at an angle to the optical axis of the incident light beam. The scattered light is generally focused on to the receiver with a lens. Despite the fact that the general principle is simple, a closer examination of the detector shows a number of interesting aspects. Let us look first at the electronics.

Electronics

Smoke detection systems are vital security systems that must work in all kind of applications and environmental conditions which in some cases can be extremely harsh (temperature and humidity variations, electromagnetic interference, mechanical vibration, dust, and dirt). The detection of minute-scattered light signals under this broad range of environmental conditions therefore places a heavy demand on the design of the electronic circuitry (the intensity of the scattered light at the alarm point is about one hundredth of the intensity of the primary light beam). Furthermore, the power consumption of the detectors must be kept as low as possible because in case of a mains power failure the detectors must be powered from batteries for a prescribed number of hours or days. The importance of the low power requirement becomes obvious when considering that in larger industrial installations the number of detectors can be as high as several thousand. Fortunately, smoke detection can benefit from the development of low power electronics in other industries which require battery-operated products (mobile telephones, laptops, watches).

In most commercial photoelectric detectors used today the light-emitting diode (LED) emits in the near-infrared region at 880 nm (IRED). Recently studies with LEDs emitting in the red and blue part of the visible spectrum were published [11]. Since the scattering efficiency is particle and wavelength dependent, one expects that using different wavelengths detectors can be developed with enhanced application-specific detection properties (broad band detection and/or enhanced detection of specific aerosols). Whether such detectors will be commercialized is dependent on component costs (a sensitive parameter in high-volume products), lifetime (>10 years), power consumption, and emitted light intensity.

The LED operates in either the pulse or the burst mode. In the pulse mode, a short ($\sim$ 150 µs) energetic pulse is sent, whereas in the burst mode a pulse train is emitted. The advantages and inconveniences of both modes are still subject to discussion. High-performance photocells and amplification circuits in complementary metal oxide semiconductor (CMOS) technology working in the range of nanoampères have been developed to measure and process the low level of the intensity of the scattered light [16]. The circuit converts the signal from the receiver photocell into a voltage which is directed into an A/D converter. In order to cancel possible errors due to amplifier offsets or leakage current, three distinctive measurements are performed. The first and last measurements are carried out without any light pulse and the infrared emitter is switched on only during the second measurement. The filter integrated in the ASIC (Application Specific Integrated Circuit) performs the function shown in Figure 4.1-12.

Besides high-performing electronics and optics, there are a number of features that are essential to a smoke detector, namely the labyrinth, the protective grid,

Fig. 4.1-12 Output=M2−(M1+M3)/2

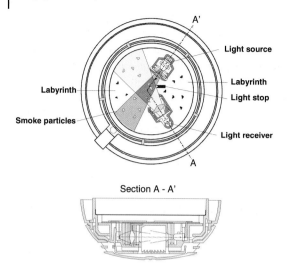

Fig. 4.1-13 Light-scattering smoke detector

the light stop, and the hood with its smoke entry slits (Figure 4.1-13). The light stop and the labyrinth have the functions of preventing light from coming directly to the receiver in the absence of smoke. The design of the labyrinth must be done with great care. Ray tracing programs and other optics programs are used to optimize the geometry of the chamber. Also, the quality of the plastic parts is of crucial importance. Bad quality of the plastics may lead to a large residual signal even in the absence of smoke. The intensity of the background signal represents a good indication of the detector state which can be used as a self-test of the detector. On the one hand, detectors with too large a background signal are rejected during production. On the other hand, the absence of the background signal is an indication that the detector has a problem (for instance, the IRED might not work properly). In this case, a 'trouble' signal is generated and the detector has to be replaced.

The protective grid prevents dust, soiling, or even insects from penetrating the optical chamber. Surprising as it may seem, many insects hatch their eggs in smoke detectors, finding a well protected environment. The metallic protective grid is also often part of the protection concept against electromagnetic perturbations. The choice of the right mesh for the protective grid is a crucial point that determines not only the robustness of the detector against soiling or electromagnetic perturbations, but also the penetration of the smoke. Too small a mesh may result into bad penetration of the smoke into the chamber.

The shape of the cover hood determines, together with the labyrinth and the protective grid, the ease with which smoke gets into the detector. The development of a new detector necessitates a long testing phase to determine optimal profiles for smoke penetration. Computational fluid dynamics programs and wind or water channel experiments complete intensive testing in fire rooms.

Optical smoke detectors are about to replace the ionization smoke detector in most markets. There are a number of reasons that have led to this change. First, optical smoke detectors have benefitted greatly from the development of electronics. The reliability of vital components in optical smoke detectors, such as the LED, has increased tremendously. Also, the appearance of low-cost microprocessors and ASICS have permitted new functions to be built for self-testing, calibration, or drift compensation. These algorithms have increased significantly the reliability of light-scattering smoke detectors. Furthermore, environmental concerns connected with the recycling of the radioactive source has facilitated the change from the well-proven ionization technology to the more modern technology of light scattering.

High-sensitivity Systems
A number of high-sensitivity detection systems such as aspiration systems use optical detectors. Different systems are also commercialized. A laser particle counter is implemented in some systems (VESDA), while some other systems use scattering smoke detectors in aspiration systems. Air is aspirated in 'pipes', then filtered to remove dust. A high-sensitivity smoke detector evaluates the smoke level. Such very sensitive systems are typically used for protecting valuable objects, such as computer rooms.

4.1.2.2.3 Optical Extinction
The measurement of light extinction (Figure 4.1-14) plays a central role in fire detection. Light extinction is related to visibility. Extensive tests were conducted to determine the visibility of light-emitting signs through smoke. Although the visibility of an object can be defined and measured with scientific instruments, tests have been conducted with persons to evaluate the vision of fire fighters as a function of smoke levels. Factors such as eye adaptation are better captured through a test with human subjects. The level of smoke at which the signs were still clearly visible to the human eyes was found to correlate fairly well with the extinction factor. Light extinction has become the standard to characterize optically smoke in different standards, such as the European standard EN 54-7 for optical smoke detectors.

There are two types of optical detectors working on the extinction principle, the linear beam detector and the point extinction detector. These two detectors differ only in the length of the optical path. For the linear beam detector this length is

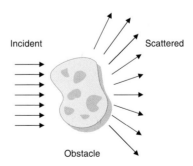

Fig. 4.1-14 Light extinction is the sum of light scattering and absorption

typically meters (5–100 m) whereas for the point detector it is centimeters (5–15 cm). Since extinction depends on light absorption and scattering, this detection principle allows the design of detectors with sensitivity to a broad range of fire aerosols and particles. The linear beam has been commercialized since the beginning of the development of early-warning fire systems. Although it would be desirable for detection properties, it has so far not been possible to develop a commercially acceptable high-volume point extinction detector product. Only recently have the first designs been shown which bear the potential for successful commercialization. The reason is simple to understand: at the alarm point the extinction is approximately 3%/m, ie, the intensity of the light beam is reduced to 97%/m. While this decrease can easily be measured with a pathlength of the order of meters, this measurement becomes a technical masterpiece if the pathlength is of the order of 10 cm. Furthermore, a reliable measurement has to be made under all conditions to which smoke detectors are subjected (wide ranges of environmental parameters, power consumption and product costs). Different approaches have been tested to overcome this difficulty using a reference path or correlation methods [17].

The ratio between light scattering and absorption depends strongly on the kind of smoke. White particles absorb light much less than soot particles. According to Grosshandler [18, 19], 'absorption can account for 75% of the extinction for smoke produced from flaming fires, while scattering can account for 80% or higher of the extinction for smoke produced from smoldering or pyrolyzing materials'.

Beam detectors are typically used in large building halls (industrial production sites, shopping malls, warehouses, hotel lobbies, etc.).

An energetic light pulse is emitted by a LED, a lamp or a laser [20]. Light extinction is measured by comparing the light intensity at the receiver with a reference value. The receiver is placed on the other side or in the same housing as the emitter. In the latter case, light is sent back to the detector by a retro-reflector. For short distances, retro-reflecting foils are used. The foil is made of small prisms or spheres embedded in a matrix. For long distance, typically 50–100 m, a single high-quality retro-reflector is necessary (Figure 4.1-15). Retro-reflectors should not be confused with mirrors. Retro-reflectors send the light back along the same axis as the incoming beam. In a mirror, small variations of the mirror position do lead to considerable changes of the light pathway.

Fig. 4.1-15 Linear beam detector, together with two examples of retro-reflectors (high-quality prisms and retro-reflector foils)

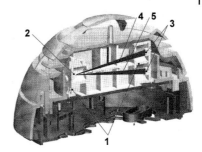

Fig. 4.1-16 Prototype of a point extinction detector. 1 = Aluminum optical body; 2 = LED; 3 = receiver cells; 4 = measuring path; 5 = reference path. After Müller et al. [12]

Linear beam detectors using retro-reflectors present a number of advantages in comparison with the separate emitter-receiver configuration. Retro-reflectors are very insensitive to vibrations or winding of the building walls on which they are mounted. Also, having the emitter and the receiver in the same housing permits one to improve the robustness of the detection electronics. The receiver is switched on only as a light pulse is about to be sent. This makes the detector immune to false alarms caused, for instance, by ambient light. In the separate emitter–receiver configuration, the switching of neon light may fool the receiver (if the neon light pulse is of lower intensity than the normal pulse, it may result into a false alarm!).

Beam detectors using the fluctuation of the refractive index of heated air to detect open fires have appeared on the market. The mixing of hot and cold air leads to large local gradients of the refractive index. This leads to rapid fluctuations of the apparent extinction coefficient. Such an effect can be visualized in the hot summer above a road as the landscape seems to vibrate. The main problem with such an approach is to distinguish variations due to a real fire from air mixing due to a high-temperature working process or a climatization.

The miniaturization of the linear beam detector to the size of a smoke detector is a very difficult problem, as the absorption of smoke diminishes rapidly with the beam length. Point extinction detectors (Figure 4.1-16) require very precise electronics. The response of a point detector to smoke is slightly different from that of a linear beam detector, in particular for smoke with large particles. The difference is due to the fact that most of the forward scattered light on a long pathway does not reach the receiver. In a point extinction detector, some of the forward light will still reach the detector.

4.1.2.3
Heat/Temperature-sensing Principles

Heat detectors are more restricted in their application as smoke detectors as they rely only on temperature measurements. They are typically used in locations in which a large amount of smoke, vapor, or dust may be present during regular operations and in which a rapid temperature rise might be expected in case of a fire.

4.1.2.3.1 Thermoelectric Detectors

Figure 4.1-17 shows the construction of a thermoelectric heat detector. The air temperature is compared with the detector case temperature. Case and air temperatures are measured with two precision NTC temperature sensors. The two measurements are electronically corrected to compensate for different effects (eg, the thermal capacity of the sensing element).

According to the European standard EN 54-5 heat detectors are tested for their response fixed temperatures and for different rates of temperature rise. Based on these results, the detectors are assigned to different classes.

4.1.2.3.2 Fiber Laser

Different methods using fiber optics as sensors have been developed in recent years (Figure 4.1-18). These sensors are based on several physical effects. Recently systems based on the inelastic scattering of light by optical fibers have been commercialized. The Fibro-Laser II (Figure 4.1-19) of Siemens Building Technologies (Cerberus Division) uses Raman scattering, whereas other systems measure the Brillouin effect. For research purposes, the Brillouin system is superior to Raman. The LASBI system from the Metrology Institute in Lausanne, that uses stimulated Brillouin, has been successfully implemented to control the temperature of concrete in large construction works and to record the temperature of Lake Geneva along a few kilometers [21]. The Raman system is more economical. The main advantage of fiber systems in comparison with point heat detectors is the economy of scale. The price of a fiber laser does not increase much with the fiber length, while the price of heat detection is roughly proportional to the size of the installation. Fiber laser systems are typically applied in fire warning systems for tunnels.

The Raman effect arises when the incident light excites molecules in the fiber. The interaction between the molecules and light results into some light being scattered. Most of this scattered light is at the same wavelength as the incident light. Some light is also scattered at a different wavelength. This inelastically scattered light is called Raman light.

The energy difference between the incident light (E_i) and the Raman-scattered light (E_s) is equal to the energy involved in changing the molecule's vibrational state. This energy difference is called the Raman shift:

$$E_v = E_i - E_s \tag{4.1-2}$$

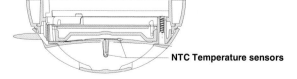

Fig. 4.1-17 Thermoelectric detector

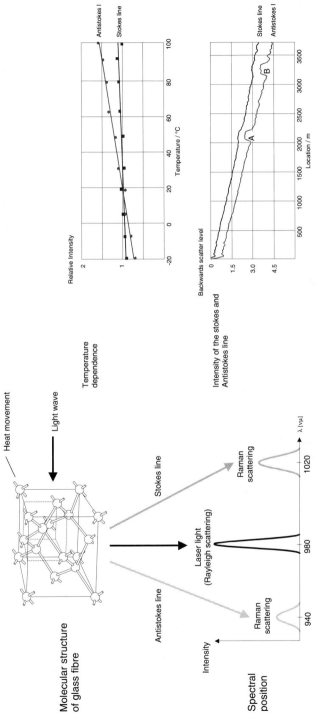

Fig. 4.1-18 The temperature of a fiber can be evaluated by comparing the Stokes and anti-Stokes components of the light scattered locally by a fiber

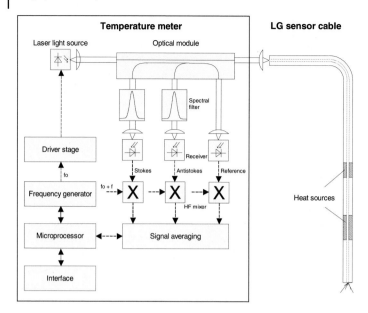

Fig. 4.1-19 With the commercial Fibro-Laser system a temperature rise along the fiber can be localized to a resolution of about 1.50 m. After Notz [22] and with the courtesy of Felten & Guillaume

The Raman backscattered light contains two spectral components: the Stokes and the anti-Stokes line. The Stoke line, corresponding to the higher wavelength, is almost independent of the temperature, whereas the anti-Stokes line is temperature dependent. The comparison of the intensities of two lines permits the temperature to be estimated.

Optical frequency domain reflectometry (OFDR) permits one to analyze the signal and to determine automatically the position of the temperature change. A spatial resolution of 1.5 m can be obtained with such a system (a compromise between detection time and resolution must be found as detection time increases with the resolution!).

4.1.2.3.3 **Temperature Fluctuation Detector**
The release of energy by a fire results in the heating of the surrounding air. The warm air mixes with the colder air around, resulting in large temperature fluctuations which can be detected by a pyroelectric sensor (Figure 4.1-20). Methods were proposed by Kaiser et al. [23] to differentiate between fluctuations caused by a fire from those resulting from air mixing caused, for instance, by the sudden opening of a window.

Fig. 4.1-20 Construction of a sensor for detecting the temperature fluctuations resulting from the large heat release by a fire. After Kaiser et al. [23]

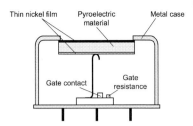

4.1.2.4
Flame-sensing Principles

Flame detectors are based on the detection of infrared (IR) or ultraviolet (UV) radiation by a flame. Flame detectors detect open fires very rapidly. Flame detectors are faster than smoke detectors. In a smoke detector, some time is necessary for the aerosol to be transported from the fire to the detector. This is not the case for a flame detector as the light radiated by the flame is detected almost instantaneously. Flame detectors do not react to non-flaming fires and their use is therefore limited to a number of important applications (high bay, protection of chemical plants, areas with high risks for hydrocarbon fires, etc.). In many critical applications, fire warning is guaranteed by a combination of a flame detector with another type of detection principle, such as a linear beam detector. Flame detectors using different sensor combinations are on the market (eg, two IR sensors, three IR sensors, UV-IR, IR-visible). Generally, IR flame detectors are well suited to detect hydrocarbon fires or more generally open fires with carbon-based combustibles (most liquid fires, wood, most foams, and plastics). UV flame detectors (Figure 4.1-21) are better suited for detecting metal or hydrogen fires, ie, materials that do not contain carbon.

IR flame detectors use the CO_2 radiation at 4.3 µm to detect a flame which can be measured with pyroelectric materials (see also Section 4.1.4.2.1). In a pyroelectric sensor the incoming light is absorbed by a thin surface layer resulting in a temperature difference between the two faces of the pyroelectric platelet. This in turn leads to a change in the electric polarization of the material which can be measured as a voltage across the pyroelectric element. Figure 4.1-22 shows an example of a three-wavelength flame detector. The first sensor measures the strong radiation at 4.3 µm that is generated during the combustion of carbonaceous materials. The other two sensors record the radiation at two different wavelengths in order to differentiate flame radiation from potential deceiving phenomena. Figure 4.1-23 shows that the spectra of flames, sun radiation, or blackbody radiation have some characteristic features that can be used in signal processing. The reliability of flame detectors depends to a large extent on the quality of the algorithms analyzing the different signals. Important new developments in this area are explained in Section 4.1.7.

328 | *4 Safety and Security*

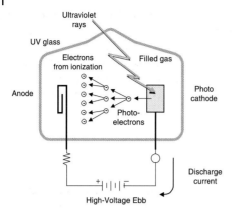

Fig. 4.1-21 UV flame detectors measure flame radiation with phototubes

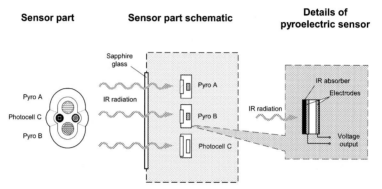

Fig. 4.1-22 Flame detectors use a number of sensors to discriminate a real fire from blackbody or sun radiation

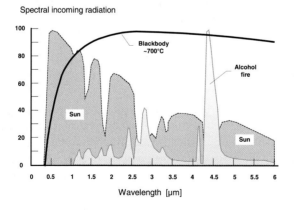

Fig. 4.1-23 Relative spectral incoming radiation on to a flame detector for blackbody, sun, and alcohol radiation

4.1.2.5
Multicriteria/Multisensor Detectors

In recent years, a number detectors which combine different sensors for smoke, ionization, temperature and gases have been tested or commercialized [4, 24–32]. Also detectors which use one sensor principle only but apply different criteria for the evaluation of the signal were proposed and commercialized, eg, light-scattering smoke detectors with more than one measuring angle or ionization detectors which exploit the non-linearities of ionization chambers [15].

Multisensor detectors are defined in the European standard EN 54-1 as detectors which respond to more than one phenomenon of fire. Fire phenomena are defined, for instance, as smoke, heat, flame, or gas. Hence, under the definition of EN 54-1, only the first group of detectors mentioned above are detectors with multisensor technology. The second group would be referred to as multicriteria detectors. Obviously one can also design detectors which use both multisensor and multicriteria technologies.

It is obvious that reliable fire detection is better achieved with different sensor types as one sensor alone cannot uniformly capture the broad spectrum of all fire signatures. In addition, by using multisensor/multicriteria technology the immunity to false alarms can be significantly improved when compared with one-sensor signal evaluation.

The photoelectric smoke sensor is especially sensitive to the large, white aerosol particles (of the order of 1 µm) which emanate from smoldering fires. By the same token it is sensitive to dust particles which are often the cause of false alarms. The photoelectric smoke sensor is, however, fairly insensitive to the small, black aerosol particles (of the order of 0.1 µm) produced by open fires (this imbalance could partially be compensated for by choosing a large scattering angle of $>120°$ but the consequence of this is a lower signal level by about one order of magnitude and hence a higher electronic amplification, which in turn reduces robustness and increases hardware cost (see also Section 4.1.2.2.2).

Consequently, in order to detect smoldering and open fires with a photoelectric smoke sensor, the sensitivity of the sensor has to be increased to catch the small, dark aerosols of the open fire. This results in an unnecessarily high sensitivity with respect to the large, white aerosols from the smoldering fire and also to dust particles. While much can be achieved by proper design of the light-scattering chamber (robustness to dust) and signal evaluation (multicriteria), a much better improvement is achieved by adding temperature as the second sensing input to the detector. The sensitivity of the photoelectric smoke sensor can now be reduced to levels necessary to catch smoldering fires while the temperature sensor is used to detect the temperature increase associated with open fires. The signal level of the temperature sensor can be used, for instance, to control the sensitivity of a smoke sensor, ie, the high sensitivity is established only when necessary. As a result we have a robust smoke detector which detects open and smoldering fires with about equal sensitivity.

Tab. 4.1-2 Qualitative response sensitivity of a photoelectric smoke detector, a heat detector and a detector which combines both sensing principles [1]

	Light scattering	Temperature	Multisensor
Open fire	•	•••	•••
Smoldering fire	•••	•	•••
Dust	•••	•	•

1) •••, Highly sensitive; •, not sensitive. After Pfister [4]

Qualitative responses of different detectors are given in Table 4.1-2 and a multisensor detector is shown in Figure 4.1-24.

In a practical example (Siemens Building Technologies, Cerberus Division) the more sophisticated technique of 'neural nets' was used for the combination of the photoelectric smoke and the temperature. The neural net is parameterized and hence the detection characteristic can be changed and adjusted to the specific requirements of an application. A range of algorithms for three different applications were developed: (a) life safety, (b) equipment and machinery, and (c) property. The net itself is multilayered with a single output port at which the output signal is mapped three levels. The highest level 3 initiates the alarm procedure while the lower levels 1 and 2 are used to indicate, respectively, a possible application problem and a 'maybe' alarm situation (prewarning) (for a more complete description, see [4] and Section 4.1.7).

In Figures 4.1-25 and 4.1-26 the response of a light scattering/thermal multisensor detector is compared with the response of detectors using one sensing principle only (optical, ionization) for the test fires heptane (open fire) and smoldering wood.

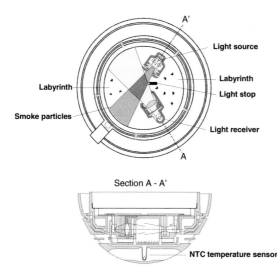

Fig. 4.1-24 PolyRex multisensor detector

Fig. 4.1-25 Open heptane fire. Upper part: temperature (1, right abscissa) and smoke (2, left abscissa) evolution with time after igniting the fire. Lower part: signal output from a standard photoelectric smoke detector (5), a standard ionization detector (6), and the PolyRex multisensor detector equipped with different application-specific algorithms, 'life safety' (3) and 'property' (4). Note that for the multisensor detector only the danger level (3) triggers the external alarm. Levels 1 and 2 are used for internal information (application problem warning). The standard photoelectric detector is a light-scattering detector with a forward scattering optical arrangement. PolyRex© is a registered trade mark of Siemens Building Technologies, Cerberus Division

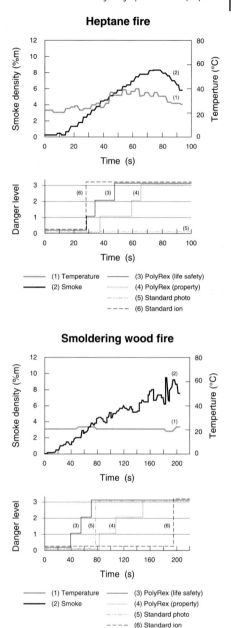

Fig. 4.1-26 Smoldering wood fire. Details as in Figure 4.1-25

Note the significant difference in the response of the standard photoelectric and ion detectors to the two fire cases. For the heptane fire the standard photoelectric detector does not trigger the alarm on the time-scale of the experiment whereas the standard ion detector is the first detector to trigger the alarm. For the smoldering fire, however, it is the standard ion detector which barely reaches the alarm threshold. By contrast for both parameter sets the multisensor detectors trigger the alarm well within the experimental time.

4.1.2.6
System Concepts

In most *residential* applications the fire/smoke detectors are battery operated and contain a signaling device (alarm horn). The number of detectors installed is typically less than five. These detectors are autonomous, ie, they have no communication link to a central control unit. In case of fire, the alarm horn of one or more detectors is activated. Because the signaling device is placed in the detector housing the origin of the fire can be easily localized. However, there is no transmission of the alarm signal to an invention force such as the fire brigade, nor is there a possibility of remote and recorded acknowledgement or resetting of the alarming devices. This type of 'fire alarm system' is called *single station* detection.

In *commercial* applications, the number of installed detectors in a fire alarm system depends on the type and size of the building and typically ranges from a few tens in a small hotel to more than a few thousand in larger buildings or, in some cases, even to more than 10000, eg, in high-rise buildings. These detectors communicate with one or more central control units by wired or wireless means, ie, detectors and control panels form a fire alarm system. From the control panels the system can be operated (switch on/off, reset, acknowledge, change of sensitivity, etc.) and the proper functioning of the system is supervised. The remote transmission to call the intervention forces or maintenance is centrally initiated and controlled.

The evolution of system concepts since the 1950s when the first electronic early-warning fire alarm systems began to appear on the market followed the general trend of the electronics industry. The first systems employed what is known today as conventional or collective technology. The fire/smoke detectors are connected to a two-wire stubline on which all detectors are equal, ie, cannot be localized at the control panel. Regulations limit the number of detectors per stubline to 32. Depending upon the detector circuit employed the alarming detector either increases its current draw from typically 0.1 to 45 mA or it pulls down the line voltage from typically 24 to 7 V. The change of the line electrical characteristics is detected by the control panel which indicates a 'line alarm'. The alarming detector may be localized via an optional alarm indicator lamp connected to the detector circuit. To do so the occupant or the fire brigade has to walk along the alarming detector line. Trouble conditions of line devices are also detected as a line condition. Again, the troubling device has to be found on the detector line by physical examination.

The conventional systems (Figure 4.1-27) are still in use today for smaller and noncritical applications as they have low product, installation, and commissioning, costs.

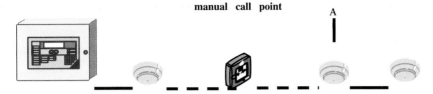

Fig. 4.1-27 Conventional/collective system concept. Up to 32 detectors are connected on a two-wire stubline. Alarms of individual detectors are signaled as line alarm, either line voltage or line current change. Localization of the alarm site by optional alarm indicator lamp integrated into or connected to the detector (A)

The conventional system concept has obvious limitations. In the late 1970s new systems appeared on the market which began to exploit the rapidly advancing microelectronics components (ASICS, microcontrollers, microprocessors). The key was the availability of components with low power consumption, high enough performance and a cost which could be justified for products of the fire alarm industry.

The system concepts introduced stepwise were (i) addressable, (ii) analogue and in the early 1990s (iii) interactive. Unlike with the conventional technology, the detectors in the modern systems now communicate with the control panel with an increasing flow of information exchange. The location of the detector can now be displayed at the control panel, usually as text information but increasingly as graphical information (floor plan). The detectors are now wired on a two-wire detector loop in numbers up to the allowed regulatory limit which depending upon the country is 128 or higher. Isolator circuits integrated into the detectors or connected to the wire loop as separate devices guarantee that a single short or opening of the loop does not reduce the detection capability of the fire alarm system (no loss of detector communication). The system concept is illustrated in Figure 4.1-28.

In *addressable* systems the detectors transmit its status (alarm, trouble) to the panel. In *analogue* systems the actual time-dependent value of the sensor is sent to the control panel which processes the signal and decides on further actions (alarm yes/no, etc.). In *interactive* systems the microcontroller in the detector performs the signal processing and the decision alarm yes/no is taken by the detector. The control panel analyses the status of different detectors at different locations. The latter feature is important if, for instance, the control panel initiates the release of an extinguishing agent. Detailed information on the detector status, trouble diagnostics, sensitivity, and relevant maintenance information can be made available either at the control panel or at a remote service organization (Central Monitoring Station, CMS).

Although today with a few exceptions all systems are wired, it is expected that in the near future hybrid systems of wired and wireless installations will become state of the art. Much progress has recently been made in the development of reliable wireless communication technology which can be employed to connect detectors to control panels. In Europe a frequency band (864 MHz) is being reserved for exclusive use of life safety and security devices.

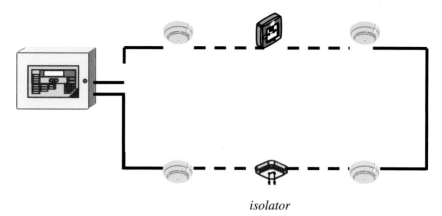

Fig. 4.1-28 Addressable/analogue/interactive system concept. Detectors and line devices are connected on a two-wire loop. There are up to 128 and more detectors per loop. Isolators 'isolate' single loop failures (short, opening)

Hybrid systems (Figure 4.1-29) will become important for retrofit installations, exhibitions, and the protection of architecturally sensitive buildings which do not allow pulling of wires to the detectors.

4.1.2.7
Application Concepts and Criteria

The application of the different detector and system principles demands detailed expertise. Also codes of practice and application rules which are often country specific or even specific to the demands of local authorities must be taken into consideration, such as the distance between detectors or the number of detectors per square meter (all of which depend on the detection principle, ceiling height, etc.). The application spectrum is indeed very broad and includes not only enclosed environments (buildings, transportation vehicles such as airplanes, trains, or ships) but also open spaces such as the supervision of tunnels and industrial fields. With the more recent systems one not only has a choice of the detector principle (optical, thermal, beam, flame, gas, multisensor detectors) but also of the detection characteristics which can be adjusted via the parameterization of the detection algorithms. Furthermore, one has to decide with the customer whether a conventional system or the more advanced addressable or interactive system will be the best solution for the particular needs.

Important parameters to consider in designing a fire alarm system for a specific application are (i) possible sources of a fire incident, (ii) material exposed to the fire, (iii) humans and property to be protected by early warning, (iv) possible measures of extinguishing a fire (sprinkler, foam, water mist, inert or even reactive

Fig. 4.1-29 Hybrid fire alarm system. Wired and wireless devices are connected to one control panel

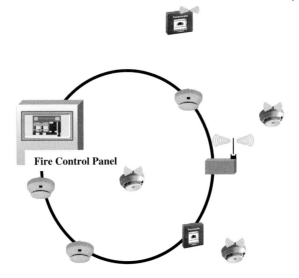

gases), and (v) environmental causes which could lead to the triggering of a false alarm (dust, electromagnetic interferences, humidity, temperature fluctuations, drafts, etc.). As an example, for a small hotel a conventional system with light-scattering detectors is often sufficient whereas for airplane hangars or automobile plants intelligent systems which include all different detection principles may be necessary to guarantee optimal protection. Of course, the fire alarm system must be considered as a member of an integrated protection concept including extinguishing, public announcement systems, building construction means and organizational measures (emergency concepts, fire brigade, etc.).

4.1.2.8
Trends

The fire alarm industry is not a leader in the development of basic technologies. Rather, it wants to be a leader in the application of novel technological developments if they have proven to be stable and reliable. Looking into the future, the following major technology developments and trends will impact the further development of fire alarm systems: (i) continuing increase in performance at lower prices of microelectronic components; (ii) microsystems and sensor technologies (silicon-machined sensor designs, gas sensors, fiber optics, video processing, sensor arrays, etc.); (iii) open system architectures, standardization of SW and HW platforms, communications and components (OPC, WinCE®, WinCC®, SW libraries, component ware, etc.); (iv) Web/Internet, multimedia/communication technologies (voice, video, wireless).

Proper use of these technological developments and trends will help in the development of the next generation of fire alarm systems with substantial added customer benefits. Specifically, one may expect: (i) further improvements in detection intelligence through use of multisensor technologies and advanced signal-processing methods; gas sensing, optical, and microsystem technologies are expected to play an important role; (ii) decentralized system architectures with intelligent peripheries providing more flexible user-specific applications and reduced cabling through digital and wireless communications; (iii) remote services including software up-/download as key for maintenance, troubleshooting, commissioning, and software update processes, custom-specific applications, and extended customer services; (iv) easier to understand and better accepted user interfaces for fire brigades, system operators, and service personnel; (v) more precise information for fire-combating forces (maybe future fire brigades will wear a kind of 'fireman's cyber helmet'); (vi) integration of fire alarm systems into total building protection concepts (the key phrase is performance-based codes (PBCs)); (vii) integration with other security systems (intruder alarm, access control, CCTV, gas); (viii) integration with building management services such as heating, ventilation, air conditioning, lighting control, process control, and intruder detection systems.

4.1.2.9
Standards

EN 54 is a series of standards developed by CEN TC72 for fire alarm systems. These apply to European countries. They are a set of basic standards. Individual countries may have additional requirements. All fire detection systems must be tested and certified by accredited test laboratories and certification bodies, respectively.

EMC Electromagnetic Compatibility
EN 61000-4-1	EMC Overview Over Immunity Tests
EN 61000-4-2	EMC ESD Electrostatic Discharge
EN 61000-4-3	EMC Radiated Radiofrequency Electromagnetic Field
EN 61000-4-4	EMC Electrical Fast Transients/Burst
EN 61000-4-5	EMC Surge
EN 61000-4-6	EMC Conducted Disturbances Induced by RF Fields
EN 50081-1	EC Generic Emission Standard

Environment
EN 60068-2-1	Test A: Cold
EN 60068-2-2	Test B: Dry Heat
EN 60068-2-3	Test Ca: Damp Heat (Endurance)
EN 60068-2-6	Test Fc: Vibration (Sinusoidal)
EN 60068-2-27	Test Ea: Shock
EN 60068-2-30	Test Db: Damp Heat (Cyclic)
EN 60068-2-75	Test Eh: Impact (Hammer)

EN 54-1	Introduction (Published)
EN 54-2	Control and Indicating Equipment (Final Voting)

EN 54-3	Alarm Devices (Draft)
EN 54-4	Power Supplies Equipment (Final Voting)
EN 54-5	Heat Detectors (Final Voting)
EN 54-6	Rate of Rise Detectors (Withdrawn)
EN 54-7	Smoke Detectors (Final Voting)
EN 54-8	High Temperature Detectors (Withdrawn)
EN 54-9	Fire Tests (Withdrawn)
EN 54-10	Flame Detectors (Final Voting)
EN 54-11	Manual Call Points (Draft)
EN 54-12	Beam (Draft)
EN 54-13	System Requirements (Draft)
EN 54-14	Planning, Design, Installation (Draft)
EN 54-15	Multisensor Detectors (Draft in ISO)
VDE 0833-2	Alarm Systems for Fire, Intrusion and Hold-up
VdS 2489	Guidelines for Automatic Fire Detection Systems
VdS 2496	Guidelines for Control of Extinguishing Systems
VKF	Fire Protection Guidelines

In the Americas the following standards are applied:

UL 268	Smoke Detectors for Fire Protective Signaling Systems (only for USA/Canada)
UL 864	Control Units for Fire Protective Signaling Systems
ULC-S527-99	Standard for Control Units for Fire Alarm Systems
FM	Relates to NFPA 72
NFPA 72	National Fire Alarm Code (Detectors) (only USA/Canada)
NF S 61-950	Fire Detection Equipment
NF S 61-936	Fire Safety Systems

In Australia the relevant standards are:

AS 1603.1	Heat Detectors
AS 1603.2	Point Type Smoke Detectors
AS 1603.4	Automatic Fire Detection and Alarm Systems
AS/NZS 4428.n	Fire Detection, Warning and Intercom Systems – Indicating Equipment (Draft)
AS/NZS 1670.n	Fire Detection, Warning and Intercom Systems – Design, Installation (Draft)

In Asia Pacific, both EN and UL standards are used. Some countries have additional requirements.

4.1.3
Gas Sensing

For many industrial and residential applications gas detection systems are an integral component of a danger management system in order to protect life from toxic gases, eg, from the work process, or prevent explosions, eg, from combustible gases used in heating. Accordingly, gas detection systems are designed for detecting either combustible, explosive, or toxic gases.

The first early-warning systems were well known from the mining industries where canaries monitored toxic gas concentrations in the air. They suffered more rapidly than humans from the toxic effects. Due to regulations gas detection technology evolved in the last 30 years to commercial, industrial, and residential installations.

4.1.3.1
Toxic and Combustible and Explosive Gases

4.1.3.1.1 Toxic Gases

Toxic gases have to be very selectively monitored at very low concentrations which are usually expressed by the volume concentration in air in ppm (parts per million) or by the mass concentration in mg/m^3. The toxicity of a gas is expressed by the threshold limit value (TLV) which is the maximum allowable concentration to which a human can be exposed during an 8-h working day.

The TLV is sometimes also called the maximum allowable concentration (MAC). To express the importance of the dose of a toxic gas, the time-weighted average (TWA) over an 8-h day is computed. Another important value often indicated for toxic substances is the level immediately dangerous to life or health (IDLH), which determines the value where no irreversible effects occur under the condition that an escape from the dangerous area within 30 min is possible.

Some TLV values are given in Table 4.1-3.

Tab. 4.1-3 Threshold limit values (TLVs) for some toxic gases in Switzerland (the TLV may change from country to country)

Substance	Formula	TVL (ppm)
Ammonia	NH_3	20
Carbon monoxide	CO	30
Ethylene oxide	C_2H_4O	1
Hydrogen chloride	HCl	5
Hydrogen cyanide	HCN	10
Hydrogen sulfide	H_2S	10
Nitric oxide	NO	25
Nitrogen dioxide	NO_2	3
Sulfur dioxide	SO_2	2

4.1.3.1.2 Combustible/Explosive Gases

Combustible/explosive gases mixed with air are only explosive within a specific concentration range which is defined by the lower explosion limit (LEL) and the upper explosion limit (UEL). If a mixture of combustible medium with air is too lean, it does not ignite: its condition is below the LEL. If the mixture is too rich, it also cannot be ignited; the concentration is above the UEL. The range between these two levels, which is specific for every gas or vapor, is dangerous. Therefore, the gas sensors used for explosion protection detect concentrations below 100% LEL. Usual alarm levels are around 60% LEL. The LEL concentration is in most cases a few volume percent. Usually the measurements are expressed as a percentage of the LEL. Note that gas sensors for most toxic gases have to be more sensitive by orders of magnitude (volume ppm versus percent).

To ignite a fire, three prerequisites are necessary, namely combustible matter, oxygen, and an ignition source which is necessary to provide the starting energy. If the concentration range of combustion matter (such as the common hydrocarbon gases methane, propane, and butane) is in the range between the LEL and UEL, the velocity of oxidation is increased until the reaction is called an explosion (Figure 4.1-30).

All areas in which explosive gases or vapors may be present are classified according to the probability and duration of their occurrence, ie,

- *zone 0*: area in which a potentially explosive atmosphere is present either continuously or for extended periods (inside containers, reaction vessels);
- *zone 1*: area in which a potentially explosive atmosphere will occur periodically or occasionally (immediate neighborhood of zone 0 sites);
- *zone 2*: area in which a potentially explosive atmosphere is expected only rarely and for short periods <2 h.

The gas sensors must be in a special robust spray and waterproof housing and are approved by Ex designations. In Europe, this is according to CENELEC EN 50014/18 EEx d IIC T6, where d means pressure-tight or flameproof housing

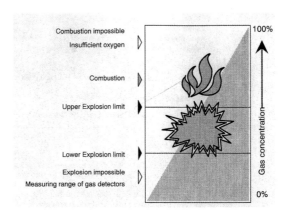

Fig. 4.1-30 Schematic representation of explosion and combustion ranges versus oxygen concentration

Fig. 4.1-31 Industrial gas detector housing

(which can resist the pressure if an explosion occurs inside the housing and will therefore not transmit the explosion to the outside atmosphere). An industrial gas detector housing is shown in Figure 4.1-31.

4.1.3.2
Catalytic Devices (Pellistors)

Catalytic sensors are appropriate for detecting flammable gases and vapors in the LEL range. A pellistor (pellet-resistor) is a miniature calorimeter used to measure the heat of oxidation of combustible gases. Usually it consists of a platinum coil acting as heater and temperature-sensing material simultaneously, embedded in a bead of porous ceramic, impregnated with a catalyst, where the oxidation of the gas takes place. The measurement is made with a matched pair of elements: the active detector carries the catalyst, the compensating nonactive element is passivated so that oxidation of the gas does not occur. The elements are connected into a Wheatstone bridge circuit (Figure 4.1-32). The chemical reaction at the active catalytic element is exothermic, causing a temperature increase which in turn can be measured as a change in the resistance of the platinum wire. The resulting electrical imbalance of the bridge circuit is proportional to the gas concentration. To activate the reaction the beads are heated to an operating temperature of around 500 °C. The catalysts are made of precious metal such as palladium or platinum.

The exothermic reaction mechanism for the detection of methane (CH_4), for instance, is given by

$$CH_4 + O_2 \rightarrow 2H_2 + CO_2 \tag{4.1-3}$$

Recently, catalytic sensors manufactured from silicon with thin film deposition and micromachining techniques have been introduced. The sensitive element and the reference element are two identical heaters in form of thin platinum layers. These heaters are deposited on a 1 μm Si_3N_4/SiO_2 membrane. The catalyst is a palladium layer deposited on the sensitive heating resistance. The advantages of this sensor design are the relatively low power consumption (100 mW), small

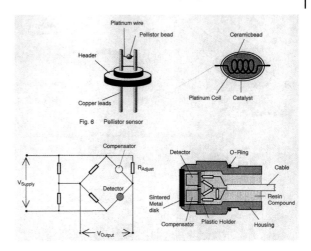

Fig. 4.1-32 Circuit diagram and assembly of a catalytic-type gas detector

size, reproducibility, and low cost due to batch production. The sensor chip is bonded on a TO39-type package and often supplied with a charcoal filter on top of the support [33] (Figure 4.1-33).

4.1.3.3
Photoacoustic Cells

The principle of photoacoustic gas detection (Figure 4.1-34) is based on the measurement of the pressure variation generated in a cavity by selective absorption of IR radiation. The IR source is provided by an incandescent lamp or a laser. The source is pulsed either by a mechanical chopper or by modulating the excitation current. The emitted IR radiation, after passing through a selective optical filter, is absorbed by the gas inside a cavity cell. The filter determines the selectivity of the gas sensor. The absorbed radiation increases the gas temperature. The resulting pressure change inside the cavity which has the same frequency as the incident IR radiation is measured with an acoustic microphone.

Gases that can be detected by the photoacoustic method are listed in Table 4.1-4.

Fig. 4.1-33 Catalytic sensor on TO 39 package micromachined in silicon for methane detection (Microsens MCGS-2102)

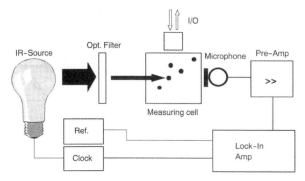

Fig. 4.1-34 Functional scheme of photoacoustic detector

Tab. 4.1-4 Infrared absorption bands used in photoacoustic gas detection

Gas	Absorption wavelength (μm)
CH_4, C_2H_6, C_3H_8, C_4H_{10}	3.4
CO_2	4.25
CO	4.7
Freon R-123	8.35
NH_3	10.45

Gas is either passively exchanged between the measurement cell and its environment by diffusion or actively by pumps and valves. The latter method is the ideal solution for multipoint measurement. The main application fields of photoacoustic cells are in-house air quality and carbon dioxide monitoring (Figure 4.1-35). Another important market is the leakage detection of freon and ammonia.

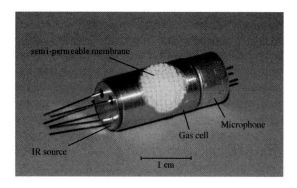

Fig. 4.1-35 Prototype of miniaturized photoacoustic carbon dioxide sensor cell. After Weber [34]

There are, of course, alternative optical methods for the measurement of gas concentrations such as conventional dispersive and nondispersive IR techniques and more recently the measurement of the gas-specific coloration of organic materials. None of these methods, however, is widely used in gas-sensing applications as described here owing due to cost, lifetime or size difficulties.

4.1.3.4
Electrochemical Cells

The simplest form of an electrochemical sensor is a diffusion-limited metal air battery comprising a semipermeable, fine-pored plastic membrane and two electrodes embedded in an electrolyte. If an electrochemically oxidizable (reducing gas) such as carbon monoxide is present in the ambient air, it will diffuse to the sensing electrode and cause a shift of its potential. Owing to the resulting potential difference between the sensing electrode and the counter electrode, a current will flow in the external circuit.

For a CO sensor the chemical reaction in the cell is

sensing electrode: $\quad 2CO + 2H_2O \rightarrow 2CO_2 + 4H^+ + 4e^-$ (oxidation) $\quad$ (4.1-4)

counter electrode: $\quad O_2 + 4H^+ + 4e^- \rightarrow 2H_2 \quad$ (reduction) $\quad$ (4.1-5)

overall cell reaction: $\quad 2CO + O_2 \rightarrow 2CO_2 \quad$ (4.1-6)

An oxygen supply from ambient air must be provided to the counter electrode to sustain the reaction. The above principle may be applied to measuring gases which can react electrochemically at the electrode. The resulting current is proportional to the gas concentration. For CO the current is in the region of 0.1 µA/ppm. Examples of commercially available electrochemical sensors are as follows:

- carbon monoxide (CO);
- sulfur dioxide (SO_2);
- ammonia (NH_3);
- nitrogen dioxide (NO_2);
- chlorine (Cl_2);
- hydrogen (H_2);
- hydrogen chloride (HCl);
- hydrogen sulfide (H_2S);
- phosphine (PH_3);
- silane (SiH_4);
- ethylene oxide (C_2H_4O);
- hydrogen cyanide (HCN).

To compensate for polarization of the counter-electrode, most sensors are equipped with three electrodes (Figure 4.1-36).

Electrochemical cells show a stable and linear signal, are very sensitive in the ppm range, have a very low influence of temperature, humidity and shock, and the power consumption is very low. The major drawbacks of electrochemical sensors are that for many applications their life time is too short (2–3 years) and the price is high compared with semiconductor or catalytic sensors.

Carbon monoxide electrochemical cell sensors are also used in combination with smoke detectors. These cells are made of an electrolyte containing sulfuric acid (H_2SO_4), filter material, a catalytic activated alloy as working electrode, and enclosed in a plastic housing with attached metal connection pins. Research work is going in progress on to improve the cells regarding lifetime and manufacturing cost.

A miniaturized electrochemical sensor is shown in Figure 4.1-37.

4.1.3.5
Metal Oxides

A solid-state metal oxide gas sensor is composed of sintered thick film (30 µm) or sputtered thin film of 1–2 µm tin oxide semiconductor which is heated to ca. 400 °C (Figure 4.1-38). In clean air oxygen, which traps free electrons, is absorbed on the SnO_2 surface, forming a potential barrier at the grain boundaries. This po-

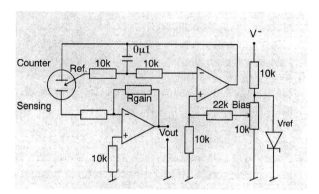

Fig. 4.1-36 Operating circuit of a three-electrode electrochemical sensor

Fig. 4.1-37 Miniaturized electrochemical sensor (Microsens)

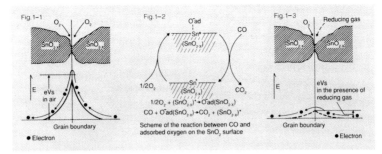

Fig. 4.1-38 Reaction mechanisms of semiconductor SnO_2 detection (Figaro Gas sensors Series 2000)

Tab. 4.1-5 Common gases and concentration ranges which can be detected with SnO_2 metal oxides

Gas	Detection range (ppm)	Application
Methane	500–10 000	Explosive, home security detectors
Hydrogen	30–1000	Explosive, industry process control
Ammonia	30–300	Toxic, leakage detection, refrigeration
Hydrogen sulfide	5–100	Toxic, industry process control
Carbon monoxide	30–1000	Toxic, CO monitoring, parking and residential

tential barrier restricts the electron mobility. When the sensor element is exposed to a reducing gas (combustible or toxic), eg, carbon monoxide, the SnO_2 surface is oxidized by the absorbed gas molecule. This lowers the potential barrier leading to an increase in the electrical conductivity in the semiconductor. The reaction to the gases (sensitivity, selectivity) depends on the operating temperature, the material composition, and the operating cycles.

The selectivity of SnO_2 gas sensors can be influenced by doping. Common gases and concentration ranges which can be detected are given in Table 4.1-5.

Most SnO_2 sensors are batch produced in large quantities. Either they are thick-film screen printed on an alumina-ceramic substrate or they are processed on silicon by micromachining and a thin-film deposition process (Figure 4.1-39).

The features of semiconductor sensors are long lifetime, good reliability, high sensitivity, quick response, high resistance to poisoning, low cost, low selectivity, nonlinear response, and high cross sensitivity to humidity and temperature. Using a special operating mode based on two heating temperatures, the often unwanted humidity influence can be minimized (Microsens).

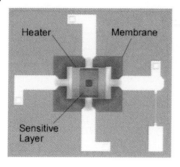

Fig. 4.1-39 Miniaturized gas sensor micromachined in silicon with a doped SnO_2 layer on top of a heater. The integrated heater resistance controls the sensitive layer temperature. The heater and SnO_2 layer are separated by a tin-SiO_2 isolation film. Thin-film sensors have lower heater power consumption than thick-film or bulk sensors (MicroChemical Systems [35])

4.1.3.6
Application Concepts and Criteria

The control unit of the gas detection system performs all functions required for operating the gas detectors. In residential home applications detectors are battery powered and contain a built-in piezoelectric buzzer. Most popular devices are sold to monitor carbon monoxide in kitchens. In commercial and industrial applications a number of gas detectors are connected in a gas detection system. These systems require several levels of alarm. The first level provides early warning of a developing hazard and notifies supervisory personnel to initiate corrective actions. The second warns personnel and automatically stops the process or the flow of gas, and also some action must be taken to control the hazard. The third level warns the operators of a detector malfunction such as loss of power to the system, communication errors, and self-supervision of the detectors.

In a gas detector installation the individual sensors have to be calibrated. The sensor's zero and response range must be checked in the field. Typically, clean, bottled air is used to set the zero; a known concentration of test gas must be used to set the range. Planning and commissioning of an installation have to be made very carefully. When positioning sensors, careful thought must be given not just to the density of the gas but also to the prevailing air flow. For instance, the ambient air in a parking garage is never still. The different wall, floor and ceiling temperatures keep the air in motion, causing the exhaust gases to be mixed with room air. Since CO is slightly lighter than air (under normal conditions of 0 °C and 1.013 bar, $CO = 1.25$ kg/m^3 and air $= 1.29$ kg/m^3), the movement of air is the greatest force in the dispersion of gas.

The points to be taken in consideration are:

- air movements;
- arrangement of the exhaust air ventilation;
- possible turbulence or dead zone distribution;
- diffusion with the room geometry;
- the installation height.

The detectors are often placed at nose height 1.60 m above the floor on columns or on walls. CO_2 sensors and other detectors to measure leakage are often placed 20–30 cm above the floor. The spacing of the detectors is between 25 and 30 m and 14–15 m from the walls. Temperature distribution may affect proper detection. If the air near the ceiling is much hotter than the rest of the room air, the ceiling air will have a lighter density. A 'thermal barrier' slows the diffusion of the carbon monoxide in order to delay or prevent detection. This thermal barrier is also well known in smoke detection from smoldering fires where air movement is slow, 10–20 cm/s.

4.1.3.7
Standards

EN 50014/18	EEx-d Approvals
EN 50054/57	Functionality Requirements
EN 50270	Electromagnetic Compatibility
EN 50271	Software Requirements.

4.1.4
Intrusion Sensing

Intrusion detection systems sense unwelcome intruders or break-ins by detecting physical quantities, such as pressure changes, moving sources of radiation, seismic effects, acoustic waves, interruption of light beams, Doppler shifts, breaking of wires, changes of wire capacitance, etc. The individual sensors transform the physical quantities into electrical signals, which are then processed and transmitted. Detection principles include manual detectors, typically a kind of push button which usually is silently and unnoticeably activated by the threatened person, eg, in the event of a hold-up or a break-in, and automatic detectors. These are divided into (i) passive and (ii) active detectors.

In passive detection, the sensors are merely receivers and register physical changes. Examples of this are passive infrared (PIR) motion detectors, seismic detectors, contact detectors, glass breakage detectors, and audio discriminators. Active detectors consist of transmitters and receivers, and compare the parameters of the signals transmitted and received. Examples of this are showcase surveillance (slow pressure modulation), motion detectors based on the Doppler effect (ultrasound/microwaves), IR light barriers, and electromagnetic field change detectors (perimeter protection).

4.1.4.1
Passive Sensing Principles

4.1.4.1.1 Passive Infrared (PIR) Detectors
PIRs are the main component of space surveillance in intrusion protection. In the early days of passive IR intruder detection, highly sensitive thermocouples and

bolometers were used. These sensors measured the temperature increase dT due to a person entering a room. Subsequently these sensing elements were replaced by the much more sensitive pyroelectric sensor elements which respond to temperature changes dT/dt.

People (and animals) are moving sources of IR radiation. In the case of people, with an average temperature of about 30 °C, maximum radiation in the IR range is at a wavelength of 10 μm. Since people can be distinguished from background radiation (room temperature), effective space surveillance can be achieved. This involves the use of an optical system which divides the space into a number of segments which are designed and optimized for specific applications, eg, intruder detection in rooms, in allyways, behind curtains, long- or short-distance detection, wide or narrow room angle detection, etc. Specific areas of the room can be blocked out, eg, 'pet allyways' to allow pets to move around without being detected and generating an unwanted alarm signal. The segmentation of the space is achieved by using either segmented parabolic mirror optics or Fresnel lens optics which focus an intruder entering such segment on to the pyroelectric element. The moving intruder thus moves in and out of the 'active' space segments which generates the necessary dT/dt at the pyroelectric element. A PIR and typical space segments are shown in Figures 4.1-40 and 4.1-41, respectively.

Figure 4.1-42 shows a pyroelectric IR sensor with two active elements with a typical size of 1×2 mm, made from lithium titanate (LiTaO$_3$), a field effect transistor (J-FET) and a high-ohm resistor. Sensor elements can also made from lead zirconate/titanate (PZT) instead of LiTaO$_3$ monocrystals. The sensor is sealed in an airtight enclosure. IR radiation passes through a silicon bandpass filter (8–12 μm) on to the absorber layers of platinum, gold black, or NiCr.

Optical filtering is one of the most important measure to eliminate risks of unwanted alarms due to external light. In order to suppress further the false signals from visible light sources, eg, sun light, additional optical filtering of the incoming light signal is necessary. Figure 4.1-43 shows the different filters typically used.

A pyroelectric IR sensor can be described with a simple model. The illumination function $S(t)$ of an intruder is determined by the flow of energy on the sen-

Fig. 4.1-40 Passive IR detector

Fig. 4.1-41 Room segmentation which is achieved using segmented mirror or Fresnel lens technology

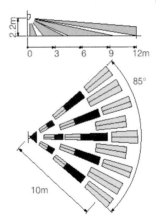

Fig. 4.1-42 Pyroelectric IR sensor with two sensing elements ('dual sensor')

sor elements through the filter and the optical system. In the following equations the temperature of the pyroelectric LiTaO$_3$ flakes is $\theta(t)$ and the voltage across the flakes forming a capacitor is $U(t)$:

$$\frac{d\theta}{dt} = \frac{S(t)}{dC_p} - \frac{\theta(t)}{\tau_{th}} \tag{4.1-7}$$

$$\frac{dU}{dt} = dp\frac{(\theta_1 - \theta_2)}{2\varepsilon_o\varepsilon_p} - \frac{U(t)}{\tau_{el}} \tag{4.1-8}$$

where τ_{el} is the electrical time constant, p is the pyroelectric constant, ε_p is the dielectric constant, d is the thickness and C_p is the specific heat of the material. The thermal time constant τ_{th} is the rate of heat dissipation of the flakes via air and the mounting support.

IR sensors can be described by a small set of primary characteristics, namely their size (pixels, flakes), sensitivity, signal, noise, and response time. From these, a set of additional characteristics, such as signal-to-noise ratio, noise-equivalent power, and normalized detectivity, can be derived.

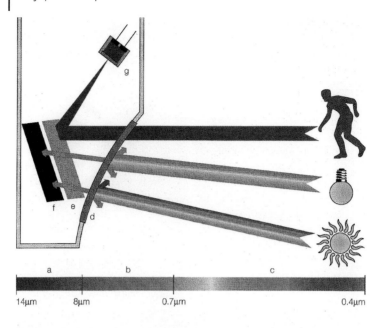

a) Typical infrared radiation emitted by human beings
b) Near infrared and
c) visible light, such as that occurring in lamplight and sunlight
d) The white light filter is virtually impermeable to interference radiation from lamps and sunlight
e) The special coating of the infrared mirror reflects only the typical infrared radiation emitted by human beings
f) The black base material absorbs the remaining residual interference radiation
g) Pyrosensor

Fig. 4.1-43 Optical filter processes in modern passive IR detectors g, pyroelectric sensor window (a coated silicon interference filter); d, white light filter (diffraction-type polyethylene loaded with tin oxide or diamond powder); e and f, indium tin oxide (ITO)-coated mirror; the IR radiation in the 8–14 μm range is reflected towards the pyrosensor whereas the coated film is transparent to all other wavelengths and absorbs them in the black base material

Sensitivity is the most frequently used property. It describes the response to radiation and is defined as the change in signal voltage ΔU per unit change in total incident power ΔP on the absorbing sensor area:

$$S = \frac{\Delta U}{\Delta P} \tag{4.1-9}$$

A typical detector has a sensitivity of 3000 V/W at 1 Hz. This value depends on the material, radiation wavelength, and incident angle.

The normalized detectivity D^* represents the signal quality:

$$D^* = \frac{\sqrt{\Delta f A}}{NEP} \tag{4.1-10}$$

It derives from the detectivity, which is the inverse of the noise-equivalent power (*NEP*) normalized at a bandwidth of 1 Hz. A and Δf are the sensor area and the bandwidth, respectively. For pyroelectric sensors using $LiTaO_3$ as active element, D^* is typically 0.5×10^8 cm $Hz^{-1/2}$ W^{-1}.

4.1.4.1.2 Trends in Infrared Sensor Technology (Arrays)

More than 10^8 pyroelectric sensors are produced annually worldwide. They are, for example, used for noncontact switching-on of light sources for convenience or deterrence. However, this technology reaches its limits as soon as multielement sensors (arrays) and sensors with directly integrated electronics for signal processing have to be developed. These so-called uncooled 'IR focal plane arrays' (IRF-PAs) are used in heat image cameras.

Uncooled IRFPAs are always two-stage transducers, with radiation absorption followed by conversion of the temperature increase into an electrical signal. Thermal IRFPAs measure the difference in temperature between the absorber surface and the environment.

The different types are listed in Table 4.1-6. All these types of arrays can be manufactured using silicon micromechanics.

4.1.4.1.2.1 Microbolometer

Bolometer arrays made from vanadium oxide (Vox) are commercially available and the most widespread technology for uncooled IRFPAs with high resolution (typically 320×240). These sensors are supplied together with interface electronics (NTSC/PAL standard video interface) for heat image cameras in the 8–14 µm IR range [36–38].

Nightdriver technology for vehicles also offers considerable potential for application in security systems (Raytheon, Lexington, MA, USA). So-called ferroelectrical microbolometers made from barium strontium titanate which are thermoelectrically temperature stabilized are now available.

The noise equivalent temperature difference (NETD) is a quantity for evaluating an IR system including its optics. Typical values for uncooled systems are between 50 and 100 mK for a pixel size of 50 µm.

Tab. 4.1-6 IR sensor types and physical effects used for detection

Type	Physical effect
(Micro)bolometer (dT)	Electrical resistance
Pyroelectric sensor (dT/dt)	Electrical polarization
Thermoelectric sensor (dT)	Seebeck voltage

4.1.4.1.2.2 Thermopiles

Thermopiles are attractive candidates for low-resolution IRFPAs (up to 64×64). Series connections of thermoelements can be achieved with photolithography and silicon micromechanics similar to microbolometers. Thin, bridge-like mechanical structures (cantilevers/membranes) are produced by selective etching, with one end attached to the absorber (warm contact) and the other end attached thermally to the silicon mass (cold contact). CMOS technology is very attractive for producing thermopiles, since the electronic amplifiers can be directly integrated [39] (Figure 4.1-44).

4.1.4.1.2.3 Pyroelectric Sensors

Arrays of 9×9 pixels or linear arrays for special applications (spectroscopy/military) are commercially available.

Pyroelectric arrays are currently produced in research laboratories (Figure 4.1-45). Thin PZT films are applied to undercut Si_3N_4/SiO_2 membranes by means of sol-gel or by sputtering. The optimum film thickness is in the region of 2 µm. The readout and amplifier electronics must be implemented on separate chips, so that the packaging is more complex than for CMOS thermopiles. In the laboratory approximately the same NETD is achieved for pyroelectric arrays as for bolometers. Pyroelectric sensor arrays can also be produced using thick-film technology. PZT ceramic pastes have been developed in quality grades (definition of response sensitivity) which come very close to those of $LiTaO_3$ monocrystals. The combination of functional layers with silicon is currently possible in the laboratory. Low-cost, high-volume industrial manufacture will certainly be possible within a few years [40, 41].

No technology has yet achieved the use of low-resolution (up to ca. 50×50 pixels) IRFPAs in motion detection. The manufacturing costs are significantly higher than those for established PIR technology. The additional information gained cannot be marketed as customer benefits. Further features such as locating and distinguishing between people and animals need to be added to the functional features of enhanced detection reliability to ensure success on the market.

Fig. 4.1-44 IR sensor thermopile 10×10 array with low-noise amplifier and multiplexer with CMOS 0.8 µm process

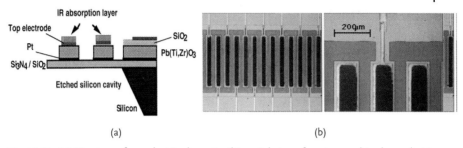

Fig. 4.1-45 (a) Structure of pyroelectric elements; (b) partial view of a micromachined pyroelectric array sensor

4.1.4.1.3 **Mirror and Fresnel Optics**

Most PIR detectors use Fresnel lenses (Figure 4.1-46). The Fresnel principle is based on the division of an optical system into small, partial lenses. For each surveillance zone, there is a partial lens which is subdivided into several lens segments. The IR radiation penetrates the lens with the pyrosensor located at its focal point. The design of the lens system requires approximately the same focal length for all surveillance zones. People in the near vicinity are therefore depicted on the sensor as disproportionately large, and those in the distant zones appear too small. A mouse moving through a zone not far away may generate the same signal as a human intruder further away.

This serious drawback is counteracted in high-quality PIR detectors by a system of reflecting mirrors. The optical system consists of curved mirror shells, comparable to parabolic mirrors, which divide the surveillance zone into different coverage areas (near, medium, and distant zones). The mirror is designed in such a way that a mirror segment with the appropriate focal length and size is available for each zone. The shape of the individual segments enables clearly defined boundaries to be formed between the individual coverage areas. The shape and the special coating minimize reflection and scattering losses. Response behavior can thus be optimized to the shape and size of the human body. The IR energy of small animals does not actuate an alarm.

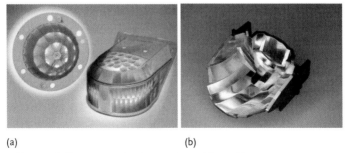

Fig. 4.1-46 (a) Fresnel lens and (b) segmented mirror optics and Fresnel lens

4.1.4.1.4 Cameras

A closed-circuit television (CCTV) system can be used in buildings for monitoring zones, objects, and events, for alarm verification in conjunction with intrusion/fire detection systems, or as a supplement to access control. A system consists of cameras, matrix switchers, video motion detectors, image storage systems, monitors, and image transmission units using public or private communications networks.

Modern cameras feature the following characteristics:

- 1/3 in CCD sensor with 752×582 pixels;
- remotely configurable via RS 485/232;
- set-up menu, character generator for date/time, camera number, etc.;
- remote power supply via the signal cable;
- switchable color/monochrome with poor lighting, IR capability (up to 1100 nm) for active illumination;
- sensitivity 0.1 lux monochrome, 1 lux color;
- DSP (digital signal processor) for image processing (lighting control, electronic shutter, and integrated motion detector).

These cameras provide excellent image quality with a high-quality lens system and are state-of-the-art for surveillance systems.

Low-cost cameras are increasingly being used for improved motion detection and alarm verification in conjunction with intrusion systems. These come from the consumer goods industry, eg, components of web cameras, or are developed for specific applications. A typical product is shown in Figure 4.1-47. This camera has a 1/3 in charge-coupled device (CCD) monochrome module camera, plastic lens system with integrated IR illumination, microphone, and loudspeaker for audiovisual alarm verification (see system trends).

In addition to the widespread CCD technology, image sensors using CMOS technology are increasingly being developed. Their advantages compared with conventional technology are:

- mainstream technology;
- high integration;
- low power consumption (3.3 or 5 V);
- high dynamics;
- low noise;
- active pixel sensing (APS);
- application-specific design, eg, motion or edge detection.

Image sensor development will continue to make considerable progress owing to the consumer products industry. Complete camera systems consisting of a single chip with housing and lens system which can be miniaturized to an exceptional degree are available on the market [42]. The combination of PIR and camera technology in intrusion detection promises excellent detection capability in the very near future.

Fig. 4.1-47 Detection and verification integrated in one device including camera, IR illumination, microphone, and loudspeaker

4.1.4.1.5 **Vibration/Seismic Microphones**

Breaking into a vault or an automatic teller machine (ATM), for instance, requires cutting out an opening in a steel enclosure. To do so one has to use steel cutting tools, eg, metal saws, drill, or oxygen torches. These break-in activities create mechanical oscillations which are transmitted as structure-borne sound. The frequency spectra of the mechanical oscillation of the various break-in methods are very different and span a very broad frequency range.

The elastic waves are propagated by the structure and can be monitored by an accelerometer of a seismic detector. In general, resonant piezoelectric accelerometers are used in most products on the market. The oscillations are picked up by a piezo ceramic sensor coupled to the structure via a steel plate or directly mounted. Other products are equipped with cantilever sensors made from a PZT bimorph (Figure 4.1-48). The resonant frequency of such a sensor is around 16 kHz, which is close to most useful noise generated by break-in tools.

Seismic detectors (Figure 4.1-49) operate as passive systems which analyze the frequency, amplitude, and duration of the incoming signal. The analysis of the sensor signal requires sophisticated algorithms and circuit designs because the true break-in signals overlap with signals from common oscillations which can be present in the steel structure, eg, from vehicle and railway traffic, elevator movement, or building vibrations which are transmitted to the structure to be protected.

The attacks on vaults illustrated in Figure 4.1-50 are well known. Correctly applied, seismic detectors detect all conceivable methods of attack on safes, strongroom walls and doors, modular vaults, etc.

Recently a new 2 g dual-axis accelerometer has appeared on the market which can serve as a multifunction tilt, vibration, and shock sensor. The accelerometer's output is passed through two parallel filters, a bandpass filter to extract shock/vibration information and a low-pass filter to extract tilt information. These sensors are made of surface micromachined polysilicon structures built on top of a silicon wafer. Polysilicon springs suspend the structure over the surface of the wafer and provide a resistance against acceleration forces. The sensors are complete measurement systems on a single monolithic integrated circuit (IC). Derived from automotive applications, micromachined devices in general are well suited for utilization in security systems [43].

Fig. 4.1-48 Bimorph accelerometer mounted on a TO5 header

Fig. 4.1-49 Seismic detector with bimorph sensor

4.1.4.1.6 **Contacts**

The opening of doors and windows can be monitored with microswitches and magnetic contacts. The magnetic contact consists of two units, a Reed switch and a permanent magnet. The units are mounted on the object to be monitored so that in the quiescent position they lie close together. In this position the Reed switch is closed owing to the effect of the magnetic field. If the permanent magnet is removed, the influence of the magnetic field decreases, so the switch opens and activates an alarm.

4.1.4.2
Active Sensing Principles

4.1.4.2.1 **Infrared Beam**

With IR light beams, invisible barriers can be set up which have to be crossed when attempting to break into a protected area. IR barriers are typically used outdoors along the perimeter surrounding the house or indoors across doors or hallways.

The infrared barrier (Figure 4.1-51) consists of a transmitter (LED) which transmits a narrowly focused IR beam (700–900 nm) and a receiver which collects and analyzes incoming IR energy. The emitted light is pulse-modulated for increased reliability in order to provide protection against extraneous light, especially in outdoor applications (sabotage, sun, fog, rain, birds, etc.).

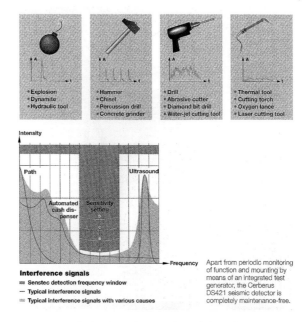

Fig. 4.1-50 Methods of attack on vaults and their characteristics

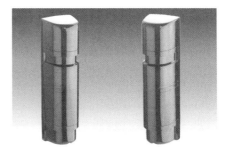

Fig. 4.1-51 Infrared barrier (transmitter/receiver pair)

4.1.4.2.2 Ultrasound

Ultrasonic motion detectors operate according to the Doppler principle (Figure 4.1-52). The frequency difference is evaluated between the transmitted and the received signal. Changes in frequency are caused by humans, animals, or objects within the sound field. The Doppler shift is proportional to the speed of movement of the target which is measured radially from the detector. The monitored coverage corresponds to a 50 m^2 area.

The ultrasonic detection principle is used in combination with a PIR detector and very seldom as stand-alone detection unit.

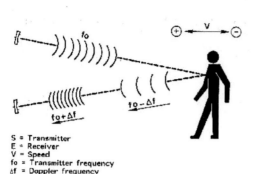

Fig. 4.1-52 Operating principle of an ultrasonic motion detector

S = Transmitter
E = Receiver
V = Speed
fo = Transmitter frequency
Δf = Doppler frequency

4.1.4.2.3 Microwaves

Like the ultrasonic motion detector, the microwave detection principle operates according to the Doppler effect. The microwave frequency is 9–10 or 24 GHz (K-band). Microwaves penetrate glass, wood, and other materials or are reflected on metallic objects. The correct application of microwave detectors is somewhat tricky. The principle is therefore mainly used as a second sensor in a PIR motion detector.

4.1.4.3
Multisensor Sensing

Multisensor movement detectors monitor space using two independent detection principles, ultrasonic and PIR, at the same time. Also well known is the combination of microwave and PIR techniques. The success factor of multisensor systems lies in the signal evaluation. PIR detection levels are in general high for tangential and medium for radial motion, whereas ultrasonic detection is low for tangential and high for radial motion.

The disturbance signal common to both principles are high frequency, line interference, and hot air turbulence. For ultrasonic detection there are additional influences such as cold air turbulence, airborne sound, vibrations, curtain movements, moving fan blades, etc. PIRs have additional unwanted signals from white light, hot spots, small animals, temperature variations, and mirror reflections.

It is obvious that only a very clever combination of both signal parameters will result in a high level of detection quality and protection against sabotage. For instance, one can evaluate the following criteria: frequency shift, distance and speed measurement for the ultrasonic and frequency, and amplitude for PIR. The multisensor detector has uniformly high sensitivity over the entire monitoring range (Figures 4.1-53 and 4.1-54).

Another example of multisensor detection is sophisticated glassbreak detectors. They rely on structure-borne sound detection by a piezoelectric sensor combined with the generated audible airborne sound wave detection in the 3–5 kHz range by a microphone. The signal evaluation also depends, similarly as described above, on clever multimode sensor fusion algorithms.

4.1.4.4
System Concepts

The detector line connects the individual detectors (data points) with the control unit. A distinction is made between two types of line transmission, as in fire detection technology, namely collective technology and addressed technology. In collective detector lines the locations of the data points cannot be identified; this is in contrast to addressed technology, where each detector has an identification address. The detection control unit incorporates the entire alarm setup in addition to the local human/machine interface.

Security against tampering is a core characteristic of an intrusion system. It calls for full-cover safeguards against tampering as well as monitoring of the detectors, the line, the control unit, the indicators, and remote transmission. Intrusion detection systems are certified by approval bodies, such as the Association of German Property Insurers (Verein Deutscher Sachversicherer, VdS), and have to display performance features described in standards (such as DIN VDE 0833).

There is a strong trend towards wireless transmission in intrusion detection systems. It is a requirement today for holdup contacts and switch-on/off devices to be wireless operated. Hybrid systems featuring conventional zone inputs and several receivers for radio detectors on a line extension will become established. The quality of the radio systems is determined primarily by the battery life of the detectors and the reliability and security against tampering of the radio transmis-

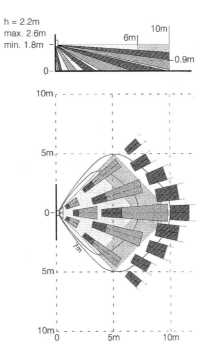

Fig. 4.1-53 Effective monitoring range with PIR zones and the shaded ultrasonic detection area

Fig. 4.1-54 Multisensor combining PIR and microwave detection technology

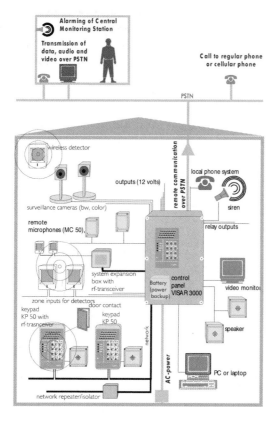

Fig. 4.1-55 Modern intrusion detection system. After Sysnova [44]

sion. Especially systems designed for residential premises are based increasingly on wireless transmission, since installation is very easy. Voice-controlled menus and the option of operation via mobile telephones are state-of-the-art. The messages can be forwarded as SMS (short message service) or via pager systems to the alarm and intervention organization (Fig. 4.1-55).

4.1.4.5
Trends

Over the past few years, the density of intrusion alarm systems has significantly increased worldwide owing to the rising crime rate. At the same time, the alarm mode has changed from roof sirens or yard lighting to alarm transmission to the police or alarm-receiving centers (ARC; see Section 4.1.6.3). The high nuisance alarm rate is a significant problem. Police statistics show that about 96% of all incoming messages are nuisance alarms, most of which have been triggered by incorrect human behavior.

In the near future, integrated security systems will be required which permit interactive alarm verification and systematic alarm responses. Experience in France and the UK, where audio alarm verification ('listen-in') is already well established, show that after verification the false alarm rate can be reduced by an order of magnitude, which means that the available intervention resources can be allocated more effectively.

The utilization of a full range of multimedia technologies in alarm verification has become economically viable. The security systems are supplemented with video sensors (cameras), microphones, and loudspeakers. When an alarm is triggered by one of the alarm sensors, pictures are taken at the event location and possibly at other important locations within the building. The resulting video data are compressed and stored. At the same time the integrated communicator establishes a connection to the alarm control system via the PSTN, ISDN or other communications networks and transmits the alarm event type. The operator is now able to listen for suspicious noises inside the building. He/she can subsequently request transmission of the pictures stored at the time the alarm was initiated and view them on a monitor. All audio and video data are stored by the ARC for several years and can be retrieved for review and also as evidence in case of legal disputes with customers or with the police.

The integration of multimode verification technologies could automate the procedures in the ARC, recognize noises, voices, motion, and faces, and assist the operator in decision-making processes.

4.1.4.6
Standards

The following standards have been published (for EMC standards and Environment, see Fire, 4.1.2.9):

Europe
EN 50130-4	Alarm Systems – EMC Tests for Immunity
EN 50130-5	Alarm Systems – Environmental Test Methods
EN 50131-1	Alarm Systems – Intrusion Systems – System Requirements
EN 50131-6	Alarm Systems – Intrusion Systems – Power Supply

Germany
VdS 2344	Tests and Homologation of Security Equipment, Components, and Systems including Passive Infrared
VdS 2312	Motion Detectors, Requirements
VdS 2326	Motion Detectors, Test Methods
VdS 2331	Seismic Detectors, Requirements
VdS 2227	General Requirements and Test Methods
VdS 2110	Environmental Test Requirements
VdS 2203	SW Requirements and Test Methods

AFNOR France
NF C 48-	Series Alarm Systems: Detection Intrusion

UL/ULC Americas
UL 639	Intrusion Detection Units
CAN/ULC-S306-M89	Intrusion Detection Units

F&P Denmark
SKAFOR 212	Intruder Alarm Systems

IMQ Italy
CEI 79-2	Intruder, Hold-up, and Anti-attack Alarm Systems; Particular Requirements for Apparatus

ANPI/NVBB; UPEA/BVVO Belgium
T.N. 120	General Requirements for Tests on Alarm Systems
C.T.K. 1.1	Requirements for Electronic Installations for Security Against Intrusion

SAA Australia
AS 2201.3	Intruder Alarm Systems: Detection Devices for Internal Use

LPCB United Kingdom
BS 7042	High Security Intruder Alarm Systems in Buildings
BS 4737-3 3.5	Ultrasonic Movement Detectors
BS 4737-3 3.6	Acoustic Detectors
BS 4737-3 3.7	Passive Infrared Detectors
BS 4737-3 3.10	Vibration Detectors
LPS 1167	Ultrasonic Doppler Detectors
LPS 1168	Microwave Doppler Detectors
LPS 1169	Passive Infrared Detectors

4.1.5
Identification Sensing

Unlike with intruder systems, the purpose of which is the detection of any intrusion into protected premises, access control systems are designed to allow the entry of people who have the right of access and to reject those who have not. This necessitates identification of the people wanting access. Although in many cases, eg, controlling the access to companies, a guard at the entrance door decides on the right of access, a range of identification technologies have been developed which allow the control of access to premises without human intervention. They range from conventional PIN identification known from card-based money withdrawal from automatic teller machines (ATMs) to more modern chip-based access control cards to biometric identification principles such as face, voice, or fingerprint recognition. The most commonly used principles are described in the following.

4.1.5.1
PIN Code

The personal identification number (PIN) is a user's personal code, known only to him/her and thus allocated to him/her. PIN codes can be used in simple door-locking systems as simple substitutes for identification cards or keys.

The PIN usually consists of 4–6 digits. Longer PINs would certainly be desirable for security reasons. As PINs are easy to use and for many applications offer adequate security, they have become widely accepted for all kinds of applications such as telephones, gasoline stations, access to the operation of intruder panels, etc. They are easily allocated, renewed, and changed.

For many applications the PIN is used in combination with an identification card (eg, a credit card). In this case possession of the card and knowledge of the PIN are complementary, creating a unit with a higher degree of security than either of the individual components.

A further functional feature can be combined with a PIN: a stress code or silent alarm. If a unique digit, different for each holder, is added to the PIN, notification of a threat can be given by entering a different digit (eg, the operation of an intruder panel). Access is then granted, but preparations are made to respond to the intrusion without actuating an immediately recognizable alarm and thus without directly exposing the person concerned to further danger.

So-called encrypted PIN pads, which effectively prevent onlookers from seeing PIN inputs, are now coming into more widespread use. For computer applications which need secure access, eg, to connect to corporate networks or banks, a PASSCODE is required which consists of a sequence of the PIN code and a code which is read from a SecurID card (individual six-digit random number synchronized with the account displayed on a card LCD).

4.1.5.2
Reading Methods for Identification Cards

Identification cards are used in order to record automatically the data on the card and link them with the user (personalization). A wide variety of reading methods have been developed for a wide range of applications. Application criteria are security aspects in the transaction between card and reader, ambient conditions, range, etc.

In addition to the methods described in detail below, a number of additional reading principles are in use: (i) inductive cards (copper foil with holes for the data); (ii) IR cards (coded data in different formats and reading modes, masked by IR-permeable foils/materials); (iii) bar code cards (different bar codes, exposed or masked); (iv) Wiegand cards (small metal pins on the card: the Wiegand effect is utilized to read their layout as a code).

4.1.5.2.1 Magnetic Strip

Identification cards with a magnetic strip were one of the first technologies developed for automatically reading and writing card data. The data are read out by sliding the magnetic strip through a reading head. The magnetic strip is divided into three tracks; their format is described in ISO 7811.

The storage capacity of the magnetic strip is only small (ca. 1000 bits), but this is sufficient to store the target data volume for many relevant applications. The information on the individual tracks can be read off with a simple reading device. The card data can be coded/encrypted using standard routines. This presupposes that all users – system operators, reading units – use the same method of reading and writing.

The main field of application for magnetic strip cards is bank/credit cards; for access control the use of magnetic strips has in the meantime dropped back into second place.

The main advantage of the magnetic strip card is its low cost. On the negative side is the limited protection offered against abuse of the written data. A wide variety of methods for ensuring the authenticity of the data have failed, also owing to the higher costs involved.

4.1.5.2.2 Smart Cards

The use of cards incorporating a chip, an electronic microcomponent, has only become technically and economically feasible with advances in miniaturization. These cards are often referred to as 'smart cards' without specifying their functional features or technology in detail.

A distinction can be made on the basis of reading methods: (i) contact-related, where the data are read out via contact surfaces, and (ii) non-contact reading. A further breakdown can be made on the basis of the relevant storage and access functions: unique codes, memory cards, and processor cards.

For economic reasons, 'passive' cards are used in widespread applications; this means that the electric power required for processing the information on the card is transmitted with the first 'contact' and stored in capacitors on the chip. The power is sufficient to perform the transaction. In the case of 'active' cards, electric power is supplied by batteries in the card; their operating life is currently no more than 5 years.

Smart cards have a wide range of memory capacities for recording volumes of data. These range from several hundred bytes to several kilobytes. There are various security mechanisms for protecting the data; the approaches used range from simple write protection, through protection by a PIN code, to authentication procedures. This severely limits the opportunities for the abuse of card data or makes this impossible. If authentication procedures are used even the production of a duplicate is doomed to failure, since the secret codes are unknown, ie, they are issued by an independent third party.

4.1.5.2.2.1 Contact Cards

Contact-related cards are divided into two categories: memory cards and microcontrol-based cards. The feature common to both is the 6- or 8-pole, gold-colored contact surface visible in the top left-hand corner on the front of the card.

The card is supplied with power via the contact surface when it is inserted. The other contact surfaces are intended for data interchange between card and reader. The use of each contact surface and data interchange with the card is standardized and described in the separate sections of ISO 7816.

4.1.5.2.2.2 Memory Cards

Memory cards consist essentially of (i) contact interface, with synchronous communication, (ii) control logic for controlling the processes in the card, usually a state machine, eg, communication, memory access, security logic, encryption, and (iii) memory (EEPROM).

Communication with memory cards takes place via synchronous interfaces in two- and three-wire technology or via a I2C-bus interface. The EEPROM (electrically erasable programmable read only memory) varies in size from 416 bits to several kilobytes. The memory has a linear structure. The entire memory can be accessed at any time for reading. In certain types of chip, access to a second page requires the input of a password or PIN. Access protection when writing the memory ranges from no protection through simple write protection, to writing after input of a password.

Memory cards are widely used for telephone cards and health insurance cards.

4.1.5.2.2.3 Microprocessor Cards

Microprocessor or processor cards consist essentially of (i) contact interface with asynchronous communication, (ii) processor for controlling the processes in the card (CPU), eg, communication, memory access, security logic, encryption, as stipulated by the operating system, reset logic, (iii) card operating system which provides the basic functional features of the card and is securely stored in the card's ROM, and (iv) memory (random access memory (RAM) and EEPROM).

Communication between reader and processor card takes place via the asynchronous interfaces. The CPU is the processor card's computing unit. In common with any computer, the processor card can only be used to a very limited extent without the card operating system. The card operating system organizes all the functions and capabilities of a processor card in a standardized and thus uniform operating mode for each operator.

The RAM is a temporary working memory. The EEPROM varies in size from 4 to 32 kbyte. Allocation of the memory areas is not linear. The memory must be structured according to its intended use. The separate storage areas – files – are allocated together with access protection and provided with an area identifier (file ID).

Access to the individual files in the memory always takes place via the CPU in combination with the card operating system. Direct access to the memory from the interface is not possible. This 'filtered' access enables a very varied range of security mechanisms to be used: from 'read only' to 'read with prior PIN input' or 'write only after authentication' – proof that the card is genuine. Each file can be secured individually against access.

By analogy the processor card can be compared with a PC by substituting (i) the contact interface for keyboard, monitor connection, power supply, (ii) CPU for PC processor, (iii) card operating for PC operating system, eg, DOS system, (iv) RAM for PC working memory, and (v) EEPROM for PC hard disk or floppy disk.

Communication with a processor card takes place, as in the case of the DOS PC operating system, by means of defined commands: create file, select directory/file, read data, etc.

The organization of the EEPROM on the processor card is entirely comparable to that of the hard disk: all files are laid down in the main directory or a sub-directory. Access protection is comparable rather with the principles of a UNIX system. Some processor cards are also equipped with a crypto processor in order to process longer security codes.

In order to be able to use a processor card, all issues concerning functional features, structures, and access protection must be clarified beforehand. This can only be done uniformly throughout the system by relevant specialists. The data structure arrived at is tested for attacks on data integrity and can only then be used for the production of user cards, ie, personalization. This procedure provides an appropriate guarantee against unauthorized abuse or duplication.

Familiar applications for processor cards are the cash card, eg, issued by the ZKA (Central Bankers' Association) in Germany, and cards for computer or network access with, for example, cryptographic capabilities (RAS, remote access system).

4.1.5.2.2.4 Non-contact Cards

Non-contact identification cards function without making any contact, even for power supplies. These cards generally manage without batteries. The necessary power is transmitted to the identification card via the coupled resonant circuit of the card's rf interface. The energy source is the carrier frequency of the reading system (Fig. 4.1-56).

Non-contact identification cards consist essentially of (i) rf interface for power supply and data interchange, (ii) control logic for controlling the processes in the card, usually a state machine, eg, communication, memory access, security logic, encryption, and (iii) memory as ROM for the unique code and EEPROM for the read/write area.

The frequency used governs the relevant parameters such as transmission rate and read/write range for the type of system:

125 kHz ca. 2–4 kbaud
 Read only and read/write
 Ranges: reading, ca. 10, 50, to 70 cm
  writing, 5–7 cm

13.56 MHz ca. 50–100 kbaud
 Read/write
 Ranges: reading, ca. 3, 10, to 30 cm
  writing, 3–5 cm

2.54 GHz Read only
 Ranges of several meters up to 100 m, depending on the environment

In principle, distinctions are made between the following types of card:

Read only Unique serial number as a unique code
OTP Programmable unique code (one-time programmable chip)
Read/write A memory area for writing and reading is available in addition to the unique serial number
Password Access to data in the read/write area is protected by a password
Key Access to data in the read/write area is protected by one or more pairs of keys
Cryptography Data transmission at the air interface between reader and card is secured cryptographically (eg, challenge-response method)

Contact-related and non-contact cards are similar as regards functional features and capabilities. The existing non-contact systems are all proprietary.

The main applications of non-contact smart cards at present, ticketing and labeling, impose specific standards in respect of functional features. Estimated sales of non-contact transponders are between 200 and 350 million per year. They are used with the following aspects in mind: production and transport logistics, identification during use, and security against abuse. Standardization, as in the case of

Fig. 4.1-56 Antenna coil of the reader generates a strong, high-frequency, electromagnetic field. Propagation occurs in both directions relative to the antenna coil in the shape of a flat beam. The antenna coil of the transponder (identification card with chip and antenna coil) couples (inductive coupling) on penetrating the transmission field of the reader with the transmission frequency. This generates a voltage which supplies the chip with power

contact-related smart cards, is urgently necessary so that users can choose between different manufacturers.

4.1.5.2.2.5 Hands-free

Hands-free is the term used to describe the ability to pass through a barrier with a card which does not have to be held in the hand, ie, while retaining the card in a pocket or in a wallet. This presupposes non-contact reading capability over longer distances.

A hands-free range has to be ca. 60 cm. In all other respects the structure and function correspond to that of the non-contact card.

4.1.5.3
Biometric Reading Principles

Certain unique human features are used for automatic identification by means of an electronic device.

Among the main problems and challenges of biometrics are 'failure to recognize unauthorized persons' or 'rejection of authorized persons'. These two curves, representing false accept rates (FARs), and false reject rates (FRRs), define the reliability of a system. This cannot be 100%, since it is comparing human beings and not black with white. Its reliability, however, can be improved in combination with other methods, including biometrics.

Two possible methods define the systems used:

(i) *Identification.* From a large number of recorded reference patterns of individuals, the one which agrees most closely with the current comparison is defined. In time-consuming comparison with a correspondingly large number of reference patterns, pure biometric identification can take some time. It is used when it is certain that a comparison with all patterns can be completed within a pre-specified period, eg, 2–4 s, or when pre-identification is not possible it may be possible to search for pre-specified individuals from an observation position.

(ii) Verification. The user gives a prior pointer to a reference pattern by entering a PIN or by means of an identification card, and the system verifies this claim. Entering an allocation number means that in this case a comparison with only one or a few patterns is necessary, which is practical in the case of time-consuming comparisons. In the case of very large volumes of data a pre-selection is made, so that not all patterns are compared.

4.1.5.3.1 **Fingerprints**
The unique nature of fingerprints is well known to everyone. The fingerprints taken from each individual are stored as reference data of the recorded characteristics of the finger pattern minutiae. The minutiae are recorded via special optical sensors or capacitive sensors and evaluated in accordance with criminological principles. The features identified are compressed by means of appropriate algorithms and stored. When a comparison is performed the new pattern is recorded, processed accordingly and compared with the stored pattern. This process of recording, optimizing, and comparing is performed via appropriate special modules (DSPs, ASICs) or via the firmware of the devices. Reproduction of an image from the stored print of the reference data memory is not possible, since the data concentrate only on certain distinguishing features. A nexample is shown in Figure 4.1-57. These devices are easy to operate, and detailed instruction is not necessary. An example is shown in Figure 4.1-57.

Intentional abuse by means of deceptive impressions, eg, assault, terror, is countered by 'live recognition' which means one wants to identify 'live' from 'dead' imprints. This is performed on the basis of a wide variety of technical facilities, such as color error recognition, pulse recognition, oxygen measurement, etc.

The retina of the eye is a unique identification means similar to the fingerprint. Indeed, systems with eye retina recognition have been designed for access control. A human wanting access has to look on to an optical device which records the retina pattern and compares them with the stored data. As with fingerprints, PIN code entry can be used to shorten the process time and improve the FARs and FRRs.

Fig. 4.1-57 Fingertip sensor from Infineon with a complete processing unit (Courtesy Siemens AG, Switzerland)

4.1.5.3.2 Facial Comparison/Recognition

Facial recognition is a perfectly normal process for human beings; we recognize the faces of individuals known to us without any great difficulty, but it has only become possible to achieve this by technical means since the availability of high-performance computers.

Several distinctive techniques are used:

(i) *Facial metrics technology* relies on the measurement of specific facial features (eg, the distance between the inside corners of the eyes, the distance between the outside corners of the eyes and the outside corners of the mouth, etc.) and the relationship between these measurements.

(ii) *Neural networks* employ algorithms to determine the similarity of the unique global features of live versus reference faces, using as much of the facial image as possible. An incorrect vote, ie, a false match, prompts the matching algorithm to modify the weight it gives to certain facial features. This method should lead to a higher ability to identify faces in difficult conditions such as changed illumination.

(iii) *Eigenface* is a technology which utilizes two-dimensional, global grayscale images representing distinctive characteristics of a facial image. It has been postulated that every face can be assigned a 'degree of fit' to approximately 150 eigenfaces; further, only the template eigenfaces with the 40 highest 'degree of fit' scores are necessary to reconstruct a face with over 99% accuracy. Variations of the eigenface method are often used as the basis of other face recognition methods [45].

(iv) *Local feature analysis (LFA)* is the most popular face recognition technology. This technology is related to Eigenface, but is more capable of accommodating changes in appearance or facial aspect LFA, which can be summarized as an 'irreducible set of building elements', utilizes dozens of features from different regions of the face, and also incorporates the relative location of these features.

There are several products on the market running with the computing power of a modern PC [46–48]. The face recognition technology is typically used in the following ways:

(i) *Identification* (one to many) determines the user's identity and computes the degree of overlap between the face in a picture taken at the access point and those associated with known individuals stored in the database.

(ii) *Verification* (one to one) is a typical scenario in access control systems where the user claims the identity by means of his/her card or PIN code.

(iii) *Monitoring* uses face detection and face recognition capabilities to follow the presence and position of a person in the field of view.

(iv) *Surveillance* identifies faces anywhere in the field of view, tracks them, and crops them out of the scene.

Compared with card readers, face recognition technology is relatively expensive, as each sensor (camera) needs the processing power of a modern PC. The high cost and the difficulties in application techniques (illumination control passive/active, cooperation of the user, etc.) are major hurdles with this emerging technology for successful entry in the mass market of access control in buildings.

The problem of abuse is especially acute in the case of facial comparison, since a three-dimensional face can only be recorded in two dimensions by a camera. A good photograph of a person could confuse a system in this case and allow the photograph to be accepted. For this reason, facial comparison is combined with additional functions to ensure authenticity, ie, the presence of a living person, such as eyelid movement, lip movement, voice recognition, etc.

Recording and comparison for facial comparison systems have to be practiced. The procedure in front of the camera must be rehearsed in order to obtain optimum reference data and ensure high reliability of the system.

4.1.5.3.3 Voice Recognition

Speaker verification is a part physical, part behavioral biometric reading method that analyzes patterns in speech of each individual. Speech recognition is not a biometric method and should not be confused with speaker verification. Speech recognition involves recognizing words as they are spoken and does not identify the speaker.

There are two major technologies in speaker verification:

(i) *Text-dependent* systems require a speaker to say a specific set of numbers or words. Text-dependent methods are usually based on template-matching techniques. In this approach, the input is represented by a sequence of feature vectors, generally short-term spectral feature vectors [49].

(ii) *Text-independent* systems create voiceprints from unconstrained speech and do not require a speaker to say a specific set of numbers or words. One of the most successful text-independent recognition methods is based on vector quantization (VQ). In this method, VQ codebooks consisting of a small number of representative feature vectors are used as an efficient means of characterizing speaker-specific features [50].

4.1.5.4
Concepts for Automatic Processing of Card Data

An access control system can range from a very simple, single-reader approach to a complex installation with control unit, branches, and operator consoles with several hundred controllers and card readers. The requirements imposed on the system-related functions of the individual units and the overall system connected to them therefore differ widely.

The basic precondition for recording card data automatically is always the availability of an appropriate card reader. This device incorporates the technological counterpart of the card and ensures correct reading or writing of the card or the file carrying the relevant information at the location of the reader.

The reader is a computerized device which prepares the data taken from the card for processing and transmits them to a higher level system or also performs processing steps directly in the reader. Processing capacity, communication facilities, and power supplies are important features of a device which also has to satisfy a whole range of environmental requirements.

In line with the use of a PIN pad, a single-reader approach calls for storage of the accepted identification card numbers in the reader. A comparison of the identity number with the authorization in the memory produces the decision.

In the case of larger and more complex systems, it is impossible to enter all authorized cards individually into large numbers of readers. The user-friendly requirements of simple handling and distribution of the parameterized data call for a central processor with an appropriate access system.

The functional features of a system can differ according to the application. Basic functions are (i) parameterization of the functional features, (ii) recording of the authorized personnel or authorized individuals, (iii) authorization profiles of the individuals (who can gain access when and where), (iv) parameterization of the peripherals used, and (v) downloading data, and saving data. System functions are verifying access authorization with regard to time and place, addition of a PIN code, anti-passback, double-entry barrier.

All data must, of course, be available for subsequent verification; all inputs are saved in logbooks. Recording and saving of movement data are mandatory. In certain situations these data need additional protection so that personnel profiles are not drawn up.

In larger systems covering several locations, the relevant data must be transmitted from the central system to the branch systems. The individual authorizations can be administered centrally or locally. A clear concept is necessary here so that data consistency is guaranteed in the system as a whole.

The individual system components can be interconnected via a wide variety of serial connections or via a local area network (LAN). Party line connections based on RS485, modem connections, and other network concepts – ISDN, X 25 – have become established for bridging long distances and using LAN networks. Data interchange via LAN networks must take place in encrypted form when public LANs are used for effective prevention of abuse, contamination by viruses, or attacks on data content.

In an overall concept for a global access control system, supplementary systems are required to perform individual tasks outside the access control system but in accordance with the processing rules in order to utilize the functions of the access control system. They are, for instance, (i) recording personnel data and producing personnel identification cards, (ii) recording and managing visitors, (iii) attendance panel displaying personnel present and how they can be reached, and (iv) connection of systems such as tank data collection, time recording and processing, canteen cash registers and automatic dispensing machines use the identification card as a means of identification for debiting the transactions to the relevant holders and their accounts.

4.1.5.5
Trends

The rapid increase in the use of proximity cards in all areas of security systems is leading to further developments in this field. The focus here is on issues of data volumes and reading distance. Through the use of new technologies and with a view to multiple functions, increasing efforts are being made to combine several applications on only one card. With the security aspect in mind, a combination of non-contact and contact-related smart card technology is being called for in some of these applications, with access to the same, common database on the chip. This will be the so-called dual-interface card, a highly integrated microprocessor card.

The combined use of an identification card inside buildings will increasingly lead to the use of a card for a wide variety of applications, such as access to the building, activating/deactivating the intrusion detection system, and activating and deactivating building services management system functions such as the control of lighting and heating in certain zones.

An important aspect with regard to wide-ranging rf identification systems is asset tracking. Objects, eg, people or items of office equipment, are provided with tags which are recognized on their passage through security zones. People – employees, guests, suppliers, etc. – or objects are tracked along their selected route and can be guided accordingly.

An employee can take his/her laptop with him/her only if the recorded ID tag of the laptop matches his/her access authorization, his/her personnel identification card, otherwise an alarm is actuated. The unauthorized removal of items will be prevented. Linking people with the items of value allocated to them, such as cars, laptops, etc., will give employees freedom of movement throughout all areas of the company site. These technical facilities can also be used for the surveillance of people and property of any kind. For example, babies can be protected against unauthorized removal from hospital and substitution can be prevented.

The functions related to the latest card technologies are applications in multiple application areas such as CityCards, RegionCards, and Campus- or University-Cards, in which authorizations for services, access to buildings/zones, wallet functions for canteens and copy shops, and access to computer systems are stored with matriculation data on a card. These authorizations can be granted by issue or acquired for cash. Further fields of application are Clubcards for achieving incentive targets, booking and attending events of all kinds, combined with authorized access to certain information on the Internet regarding these club activities, such as 'Event information; for insiders only', VIP cards.

Most recently smart cards which "guarantee" secure identification of the user and secure access to the equipment (i.e. mobile phones, PC, cash dispensers, etc.) and the transmission media (internet) have been announced (Acter). These cards combine on the size of a normal credit card: GPS (global positioning system) for localization of the user (including date, time), FP (finger print) for user identification, BT (bluetooth industry standard for wireless communication) for connectivity to BT compatible equipment.

4.1.5.6
Standards

ISO 14443	Proximity cards with ranges of up to 10 cm for ticketing applications; ÖPNV, CityCards, etc.
ISO 15693	Vicinity cards with ranges of ca. 100–150 cm for labeling applications, airline baggage, parcel services, etc.
EN 50133-x	Access Control Systems.

4.1.6
Emergency Handling

The alarm detection system has done its job when it has detected a true danger situation and transmitted this information to human intervention forces and/or emergency handling units. These now take over with the task to limit or even prevent the potential damage to human lives and property. While human intervention is in most cases required (fire brigade, police forces, private organizations) technical systems are put in place to support such actions. Although the main focus of this book is on sensor technologies it is worthwhile to spend a few pages describing the principles and actions of the most popular interventions systems in order to complete the cycle which is initiated by an alarm event.

4.1.6.1
Voice Evacuation Systems

Voice evacuation systems are intended to inform endangered people and help the evacuation in an emergency situation. In North America, Asia Pacific and China, ie, in countries with typically large building complexes (sky scrapers, campus-style housing), voice evacuation is often a mandatory and regulated functional part of fire alarm systems. They cover typically up to 35% of all installations. In other parts of the world, specifically in Europe, where buildings tend to be much smaller and regulations for evacuation are less demanding, voice evacuation is a non-mandatory optional additional function. However, also in these countries there is growing interest in voice evacuation systems and in several European countries local regulations have already been published or are being prepared.

A voice evacuation system consists of one or several controller units which drive loudspeaker systems distributed in the building. In larger installations the number of loudspeakers can be in the thousands and even surpass the number of smoke detectors. The evacuation systems can either be integrated with the fire alarm system or act as a self-contained system which is initiated by the fire alarm. Often voice evacuation systems have additional functions such as paging or provision of background music.

In case of an emergency the system activates its preprogrammed evacuation sequence. All functions of the evacuation system not relevant to the emergency handling are automatically switched off. In a high-rise building, for instance, the sin-

Fig. 4.1-58 Head end of fire and voice communication system

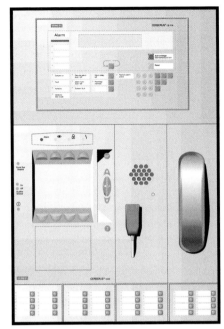

gle-channel voice system alerts the occupants on the fire floor and the floor above and below with a prerecorded (standardized) evacuation tone or with a voice message such as, 'May I have your attention please, may I have your attention please. A fire has been reported in the building, please evacuate using the nearest stairway, do not use the elevators'. With the more complex multichannel voice systems, additional alert and evacuation information can be sent to specific floors and locations in the building.

There is always the possibility that the fire officer in command of the evacuation wants to override manually the preprogrammed evacuation sequence because he wants to start or stop the prerecorded alarm messages, select more appropriate prerecorded alarm messages, switch on or off of selected speaker zones or broadcast live voice messages via the emergency microphone. Manual controls can be provided at the central head end of the voice system and also at specified remote head ends or transponders distributed in the building (Fig. 4.1-58).

As buildings become more complex and bigger, voice evacuation systems will become an increasingly important additional component in fire alarm systems. For economic reasons the voice evacuation function will be integrated into the fire alarm system, ie, fire and voice share the same central processing unit (CPU) and control and indicating equipment (CIE). System structures are arranged to increase immunity to failure with decentralized intelligence.

Digital audio technology offers new possibilities to extend the applications of voice evacuation systems. Through digital signal processing and transmission it is possible today to operate simultaneously several audio channels for transmission of different signals to various loudspeaker sectors. Digital amplifiers are considerably more efficient and yet smaller and cost less than analog amplifiers. The first digital, eight-channel systems are already on the market. It is also conceivable, in the future, that in an emergency CCTV video data will be transmitted via the data lines of fire detection and voice evacuation systems for a better assessment of the local situation.

4.1.6.1.1 **Standards**

UL 864	Control units for fire protective signaling systems
UL 1711	Amplifier for fire protective signaling systems
NFPA 72	Installation, Maintenance and Use of Protective Signaling Systems
CAN/ULC-S525	Audible Signal Appliances
CAN/ULC-S527-M87	Standard for Fire Control Units for Fire Alarm Systems
IEC/EN 60849	Sound Systems for Emergency Purposes
BS 5839	Fire Detection and Alarm Systems for Buildings, Part 8 – Code of Practice for the Design, Installation and Servicing of Voice Alarm Systems
DIN EN 457	Safety of Machinery, Auditory Danger Signals, General Requirements, Design and Testing (ISO 7731: 1986 modified)
DIN EN 842	Safety of Machinery, Visual Danger Signals, General Requirements, Design and Testing
DIN 33404	Gefahrensignale für Arbeitsstätten, Part 3 – Akkustische Gefahrensignale, Einheitliches Notsignal, Sicherheitstechnische Anforderungen, Prüfung
EN 60849	Sound Systems for Emergency Purposes

4.1.6.2
Fire Extinguishing Systems

The fire alarm system detects a fire at a very early stage in its development. In most cases the fire is still in its smoldering phase, ie, the damage done is very limited and the extinguishing of the fire is most effective. Therefore, in many applications fire extinguishing systems are put in place as integral part of the overall fire protection means. The extinguishing systems are automatically triggered by the fire alarm system or as with voice evacuation (Section 4.1.6.1) the automatic trigger can be manually overridden.

Fire extinguishing systems can be segmented according to (i) the extinguishing agent used which is either wet (water) or dry (gas), (ii) the type of application which can be full room protection or protection of room segments or objectives (eg, EDP (Electronic Data Processing) equipment), and (iii) protection objective (eg, limit fire spread, limit objective damage, protect lives).

The basic extinguishing effect sought is the reduction of the flame temperature to the region of 500 °C where the oxidation reduction reaction process cannot be maintained any longer, therefore leading to fire extinguishing. For wet systems the dominating flame cooling effect is direct water evaporation. For dry systems the cooling is indirectly achieved by reducing the oxygen transport to the flame due to either the reduction of the remaining volumetric oxygen concentration after discharge of the gaseous extinguishing agent and/or by imposing an oxygen transport barrier on the fire surface.

In most cases the basic protection objective for wet systems is the limitation of the fire spread (eg, sprinkler systems) whereas dry systems which are usually automatically triggered by the (early warning) fire alarm system focus on extinguishing the fire at an early stage in order to avoid damage to highly valuable equipment or loss of operational business (shutdown costs). In recent years, automatic wet extinguishing systems such as pre-action sprinklers, water mist, and spray systems have entered the market as complementary to the dry extinguishing agents.

In the case of full room protection, the complete sector is flooded with the extinguishing agent whereas object protection only covers a part of a sector or a specific object, eg, a machine or a cabinet.

Sprinkler systems were used in the early 1900s and at 60% still represent the major part of the global extinguishing market. Automatic dry extinguishing systems achieved booming business in the 1960s and 1970s with the introduction of Halon 1301 as the dominating extinguishing agent. Halon 1301 has excellent extinguishing properties (no chemical damage to property and harmless to living beings). This together with the facts that Halon systems are very easy to engineer, are readuky available and are inexpensive further accelerated the Halon boom and with it the overall dry extinguishing market. In certain countries Halon 1301 was complemented by CO_2 as a dry agent.

The Montreal Protocol of 1987 banned Halon 1301 and new agents had to be developed as replacements.

4.1.6.2.1 Wet Extinguishing Systems

Typical sprinkler systems are activated by radiation due to a fire. If the temperature at the sprinkler head reaches around 70 °C, a tiny glass pipe at the sprinkler nozzle melts, leading to the activation of the system since the pipes are filled with water. By using different types of sprinkler heads the activation temperature and therefore the reaction time can be modified. In the event of fire, usually a couple of sprinkler heads are activated and therefore the spread of the fire and the production of smoke and toxic fire gases are reduced significantly, enabling the fire brigade to come to the fire for extinguishing without needing special protective equipment. In most cases sprinkler systems automatically alert the fire brigade upon activation. In addition to conventional sprinkler systems, pre-action sprinkler and so-called water spray or water mist systems are becoming more and more popular. As a common difference to sprinklers, these systems are linked to a fire alarm system and are activated after having a confirmed alarm. Whereas pre-action sprinklers are mostly used for building protection, the major part of water spray and water mist applications is for object protection. The difference between water mist and water spray systems derives from the droplet diameter which is produced at the nozzle. Water spray systems operate at lower pressure regimes of about 5–10 bar and therefore produce larger droplets (100–400 µm), whereas water mist systems operate at higher pressure regimes up to 100 bar, producing very tiny droplets (10–100 µm). Both systems are based either on pumps or on pressurized gas cylinders used as a propellant to distribute the water at the nozzle.

Finally, foam systems belong to the group of wet extinguishing systems. Foam systems form a physical barrier which also reduces the transport of oxygen to the fire. In addition, the foam has a cooling effect.

4.1.6.2.2 Dry Extinguishing Systems

Today, all alternatives to Halon 1301 require more agent to protect against the same risk, which makes the design of new systems and upgrading of installed Halon systems a complex issue. Alternative installations require more specific engineering know-how.

The alternatives to Halon 1301 can be separated into the groups (i) chemical gases and (ii) natural gases.

Chemical alternatives to Halon belong to the chemical group of halofluorocarbons (HFCs) and in their chemical behavior and extinguishing performance are very close to Halon 1301, but have an ODP (ozone depletion potential) value of zero and therefore have no negative effect on the stratospheric ozone layer. The ODP value of zero is realized by replacing the critical components of Halons, chlorine and bromine, which led to the destruction of the stratospheric ozone layer by fluorine.

The group of HFCs used for fire protection is today dominated by FM200, which is the dry extinguishing agent closest to Halon 1301 regarding design concentration and required storage area. FE13 and CEA 410 also belong to this group. HFCs are stored as liquids either suppressed with nitrogen used as a propellant or under their own vapor pressure.

Natural gases are used in pure form or pre-mixed, stored as a gas under a pressure of 150–200 bar, except for carbon dioxide, which is stored as a liquid under its own vapor pressure.

Chemical alternatives perform very similarly to Halon 1301. Lower design concentrations are required and liquid discharge of the agent enables a short discharge time of only 10 s. Owing to the extinguishing effect of natural gases, which is the reduction of the volumetric oxygen concentration, more gaseous extinguishing agent is needed. Because natural gases are usually discharged as a gas, the discharge time is 60 s. Only carbon dioxide discharges in two phases into the protected area.

Since the extinguishing effect of chemical gases takes place locally in the flame zone, such agents are very fast acting. Owing to high temperature, the large molecules are split into smaller compounds, leading to a volume increase in the flame zone which reduces the oxygen mass transfer to the flame. From one mole of agent (ie, a volume of 22.4 L), several moles are formed close to the flame. Some of the oxygen is also consumed during the splitting of the chemical agent. Therefore, the extinguishing effect of chemical agents is comparable to that of natural gases with the only difference that the extinguishing mechanism acts locally at the flame.

However, as with natural gases, proper fire suppression with chemical gases also requires perfect homogenization of the agent in the protected area as rapidly as possible. Toxic by-products during fire suppression with chemical agents can only be reduced to a negligible level if the correct design concentration is provided homogeneously in the sector.

4.1.6.2.3 Applications
Typical applications are given in Table 4.1-7.

4.1.6.2.4 Standards
ISO 14520
PrEN 12094-1   Extinguishing Systems (Draft)

4.1.6.3
Alarm Receiving Centers

Both voice evacuation and extinguishing systems are emergency handling systems which can run automatically without human intervention, and many emergency situations can be handled in this way. Since the alarm signal is first transmitted to the building-responsible security officer, human intervention may be possible, eg, to put out a fire in its incipient stage before the alarm is transmitted externally to the fire brigade. Still in many danger situations the intervention of professional forces maybe be necessary (large, out-of-control fires, hold-ups, etc.).

The transmission of the alarm signal to alert external intervention forces is in many countries regulated by local standards. In case of a fire in some countries, for instance, it is mandatory to alert the fire brigade directly if the alarm signal is not acknowledged and the alarm systems not reset within a specified period of time. In other countries the external alarm first has to be transmitted to an alarm-receiving center (ARC) which itself will call the fire brigade or other intervention forces only under certain conditions which were agreed upon with the customer. The same holds true for security systems in which case the police department accepts the alarm and initiates an intervention only if certain conditions are met. The reason for this 'reluctance' to accept an alarm by official intervention forces (fire brigade, police) is the relatively high percentage of false alarms, ie, alarms which were initiated by the system maybe due to a deceptive phenomenon (an alarm-pretending situation), a wrong operation of the system by the local personnel or a technical fault. Fortunately, this troublesome false alarm situation could be significantly improved in recent years by much advanced processing of alarm

Tab. 4.1-7 Typical applications of wet and dry extinguishing systems

Type of system	Typical applications
Wet extinguishing systems	
Sprinkler	Building protection, stocks of goods
Pre-action sprinkler	Building protection, industrial applications
Water spray, water mist	Turbines, engines, cable ducks
Foam	Stock of chemical goods, hangars
Dry extinguishing systems	
FM200	EDP room, telecommunication, archives
FE13	Low-temperature storage
CEA410	EDP room, telecommunication, archives
Nitrogen	EDP room, telecommunication, archives
Argon	EDP room, telecommunication, archives
Carbon dioxide	Transformer stations
Inergen (52% N_2 – 40% Ar – 8% CO_2)	EDP room, telecommunication, archives
Argonite (50% N_2 – 50% Ar)	EDP room, telecommunication, archives

signals (see Section 4.1.7), alarm verification procedures, better application of the systems, and better education and training of the personnel handling the systems [51]. Indeed, there are systems on the market today which guarantee that a transmitted alarm is a genuine alarm, in which case the manufacturer takes responsibility and covers the expenses incurred in case of an unjustified excursion of the intervention forces.

The ARC has many responsibilities. The ARC will make sure that the right type of intervention force arrives as quickly as possible at the location which generated the alarm. In case of a fire alarm it is the fire brigade, but when an intrusion is detected it is the police or private guards which have to arrive at the scene as soon as possible.

Besides handling messages transmitted by the alarm systems, an ARC performs the following tasks: (i) supervision of the communication link to all connected systems, (ii) handling of technical trouble conditions reported from the alarm system, (iii) verification of alarm conditions, and (iv) remote operation of the alarm system.

Alarm systems communicate with the ARC through a variety of different wide area networks (WANs) (Figure 4.1-59). The most common used network is the public switched telephone network (PSTN).

In general there are two types of communication links between a life safety and security system and an ARC: (i) a permanent connection and (ii) a temporary connection.

A permanent connection is used exclusively by the life and safety system and the ARC. This kind of connection is chosen when the risks monitored by the life and safety system are very high. As an alternative to the exclusive connection between the system and the ARC, temporary connections are used which have constant com-

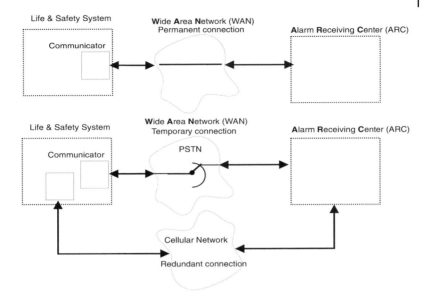

Fig. 4.1-59 Communication paths between alarm systems and ARCs

munication with the ARC via a nonexclusive network (eg, X25). In any case, the ARC knows instantly when an alarm occurs or communication is not possible.

A temporary connection shares the communication link to the ARC with other communication devices in the premises (eg, telephone, fax, PC) and in the ARC with other alarm systems which also report into the ARC. When an alarm system uses a temporary connection, it sends a test message in predefined intervals to the ARC. The ARC uses this test message to supervise the communication with the system. In the case of a missing test message the ARC will act as agreed with the system owner (eg, inform the technical service). However, between the test message transmissions the ARC does not know whether the alarm system can transmit alarms. In order to reduce the risk of having no communication, many alarm systems have a redundant communication link with the ARC via a different network (eg, PSTN and GSM).

As there are normally thousands of alarm systems relying on a single ARC, every part of the latter must be optimized for uninterrupted operation. This affects the ARC's premises and the equipment used.

With a few exceptions, which will be discussed later, an ARC receives the information from the alarm systems in a digital form in one of its receivers (Fig. 4.1-60). There are many different protocols in use for the transmission of alarms. However, the data received in the ARC consist typically of (i) the account number, which identifies the life and safety system in the ARC, (ii) the type of alarm, which indicates whether its a fire, an emergency, or a technical alarm, and (iii) a sensor or area descriptor, which helps to identify more precisely the location of the alarm in the premises.

When the message is received and understood by the receiver, it acknowledges the reception to the alarm system. Each of the receivers handles a number of network connections (eg, telephone lines). At the minimum there are two connections in order have a back-up connection. Also, there is a second back-up receiver. Because messages arrive in different protocols at the ARC, the receiver converts the incoming messages and forwards the data to a computer. For the case when the link to the computer fails, the receiver often has additional annunciators such as an LC display or a printer and additional operating elements integrated, which allow the ARC to continue the service but of course at a lower level of efficiency.

The computer uses the data sent from the alarm receiver(s) to fetch information from its database about the client, and with operator assistance decides what to do in case of this particular alarm. The database information is then forwarded to the next available operator. Certain messages such as a routine test call are not forwarded to the operator. They are handled independently by the computer.

It is the operator's duty to perform all actions that the computer lists on the screen, such call the fire brigade to the premises or call the owner and inform him/her about the event. Having done this, the operator acknowledges the alarm and is ready to work with the next event.

There is a second computer, which mirrors the first computer and takes over if the latter fails. In order to have a complete back-up for each technical part of the ARC equipment, there is usually a power generator which takes over in case the primary power supply (mains) fails.

Alarm verification is provided as an additional service by a number of ARCs. This service enhances the overall performance of the alarm systems as it greatly reduces the number of false interventions. Generally there are two types of alarm verification: (i) automatic alarm verification and (ii) human alarm verification. In the automatic verification the management software, for example, tracks the sequence of the incoming alarm messages. So only after a second alarm from another sensor in the same system is reported, does the computer notify the operator. With human verification the operator verifies the cause of the alarm by using the transmission of camera, microphone, and speaker signals which are installed in the alarm system to look, listen into the premises, or to speak with people on-site. The multimedia functionality – handling of audio and video – of the ARC is also used to provide service for elderly people, elevator emergency calls, or automatic helpdesk connection.

Improvements in the foreseeable future include (i) a higher degree of automatic alarm management – the computer calls the fire brigade, not the operator, (ii) more and better multimedia handling – replacing guard tours with automatic video transmissions, and (ii) migration and partial integration in the Internet – expanding the palette of service offerings toward the owner of the life, safety and security system.

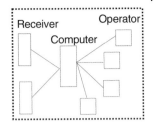

Fig. 4.1-60 Set-up of the ARC

4.1.6.3.1 Standards

The following standards are published:

EN 50136-1-1	Alarm Transmission Systems and Equipment System General Requirements for Alarm Transmission System
EN 50136-1-2	Alarm Transmission Systems and Equipment Requirements for Systems Using Dedicated Alarm Paths
EN 50136-1-3	Alarm Transmission Systems and Equipment Requirements for Systems with Digital Communicators Using the Public Switched Telephone Network
EN 50136-1-4	Alarm Transmission Systems and Equipment Requirements for Systems with Voice Communicators Using the Public Switched Telephone Network
EN 50136-2-1	Alarm Transmission Systems and Equipment General Requirements for Alarm Transmission Equipment
EN 50136-2-2	Alarm Transmission Systems and Equipment Requirements for Equipment Used in Systems Using Dedicated Alarm Paths
EN 50136-2-3	Alarm Transmission Systems and Equipment Requirements for Equipment Used in Systems with Digital Communicators
EN 50136-2-4	Alarm Transmission Systems and Equipment Requirements for Equipment Used in Systems with Voice Communicators.

4.1.7
Signal Processing

For almost three decades, fire detectors used only analogue electronics. When the sensor reached a threshold value, it triggered a comparator circuit in the detector. The possibility of bringing computational intelligence to the detectors was very limited and the level of unwanted alarms due to deceptive phenomena relative to genuine alarms was too high. In the late 1980s and 1990s, the main challenge to the smoke and intruder detector industry was the reduction of the number of false alarms [30, 52, 53]. With the appearance of low-price ASICS and microprocessors, it was suddenly possible to analyze the measured signals with mathematical algorithms and to propose improved methods for checking the electronics with so-called self-test procedures. This led to many new ideas. Multisensor fire and intruder detectors became a central issue. A few years later, the first positive effects of this small 'revolution' were clearly visible. Alarm statistics taken in major cities in Switzerland showed a rapid improvement of the situation. With fires, the number of false alarms per detector was reduced from 0.95 per 100 detectors per year in 1990 to 0.69 in 1995. A new spectacular improvement followed the introduction of multisensor/multicriteria. The false alarm rate for these types of fire and smoke detectors was reduced to about 0.05, an improvement by a factor about 20! With intruder alarm detection similar improvements could be achieved. In fact, in intrusion detection signal processing and multisensor technology was introduced some years before the fire industry followed the same technology trend.

Signal processing for fire and smoke detection was first done in the central CPU, ie, the detectors sent the actual sensor value to the CPU. Only later did the price performance of microprocessors allow the decentralization of the signal processing into the detector, ie, to the location 'where the action takes place'. In some countries, eg, France, fire alarm systems with central processing and alarm decision are not accepted for approval by the official certification bodies. Intruder systems did not go through the same technological evolution but bypassed the central signal processing phase. (The response time for intruder systems must be shorter (<1 s for assault) than for fire alarm systems (<10 s for alarm annunciation), which did not allow a centralization of the signal processing.)

It is not only that algorithms and multisensor detectors alone resulted in such an improvement. Parallel to efforts in signal processing, much work was done to improve the intrinsic quality of the sensors. The sensitivity level to electromagnetic perturbations was reduced by a factor of five over the same period of time. Dust and soiling resistance of light-scattering smoke detectors, for instance, was significantly improved through constructive refinements. The transmission of the detector state to the control unit was made more secure with the introduction of digital communication protocols. Organizational procedures such as the day/night operational mode for fire alarm systems also contributed to improving the situation.

Most manufacturers or research institutes have developed proprietary algorithms [4, 54–58]. Specifically, fire and smoke detection has become an active field for the application of soft computing techniques and artificial intelligence [14, 59–65]. Specific soft computing methods such as fuzzy wavelets and fuzzy wavelet networks have been invented [66].

In the following we will limit the discussion to a number of intelligent methods that have been used during the development of several smoke detectors. They show the importance of signal processing for both the development of algorithms and their implementation in detectors. The methods developed are fairly general and can be adopted for other danger detection disciplines also. Finally, the results of applying signal processing methods to two actual examples, the linear beam detector (Section 4.1.2.2.3) and the flame detector (Section 4.1.2.4) are discussed.

4.1.7.1
Intelligent Development Methods

Experience shows that the better are the sensors, the more efficient is signal processing. In other words, it is wrong to believe that signal processing can be substituted for sound and robust sensor design. Good algorithms can only be developed if a deep knowledge of the sensor physics and representative data are at hand. Knowledge acquisition and management become an essential part of the development work. Computer-assisted methods based on fuzzy logic have been implemented during the development of several products. In this section, we give a short introduction to these methods.

The development of a new product requires past experiences with previous product generations to be combined with possibilities offered by new technologies or ideas. The available knowledge must also be completed by selected tests, qualification, and intensive field testing. The sum of the information contained in the acquired data during in-house tests and field testing together with implicit and explicit knowledge should guarantee that the development work will lead to products that not only work to satisfaction under all the specified conditions but also show an improvement with respect to the previous product generation. In general this improvement relates to (i) the detection intelligence which describes the capability of the detector to identify real danger situations at a specified early time and block out deceptive phenomena, eg, cigarette smoke in the case of a smoke detector, and (ii) the cost/performance ratio. It is essential to collect data very thoroughly. Qualification and field testing should begin at a very early stage in the development work. Field testing with a few prototypes can help in catching important design errors in an early stage, when it is easy and not expensive to correct them. Fine tuning and final qualification must be done with a large amount of detectors in a large number of representative locations to evaluate the robustness and the quality of the detectors with good statistics. Note that the detectors have a lifetime typically exceeding 10 years and that fire incidents are a rare event which the detector has to catch at any time during its life. The reliability and detection intelligence are therefore of utmost importance and must be placed at the top of the priority list in the development plan.

Learning starts from a databank containing knowledge on fires and deceiving phenomena, in the form of signal recordings, obtained through field testing and laboratory work (Figure 4.1-61). An alarm surface separating the alarm from the non-alarm conditions is extracted from data for fire and non-fire situations. The alarm surface is put under the form of a number of fuzzy rules.

Learning methods combining fuzzy logic and wavelet theory (fuzzy-wavelet) have been developed to facilitate the extraction of knowledge from the databank and also to simplify the comparison of experts knowledge (Figure 4.1-62).

A considerable advantage of such a working process is that it takes advantage of the available computing power, while always keeping humans in the loop. The human experts are not run over by the computer. Since, at the end, the information is in a linguistic form, controlling of the computer results is always possible. The

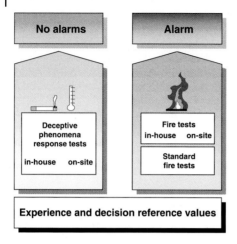

Fig. 4.1-61 The development of intelligent fire detectors is carried out by extracting information from databanks containing a large amount of data from measurements of fire and non-fire situations

checking of the results by human experts is indeed essential, as one should never forget that the results furnished by the computer can only be as good as the data supplied to it.

One of the most interesting stages in the development process corresponds to the fusion of the knowledge extracted from data to the already existing expert knowledge. The compatibility between the different experts must be checked prior to knowledge fusion. The comparison of expert knowledge can be supported by computer-assisted methods. It relies on a geometrical interpretation of knowledge, based on the fact that a series of linguistic fuzzy rules can always be expressed as surfaces (in n-dimension surfaces for n input variables). In this geometrical approach information processing from different sources corresponds to comparing surfaces (or hypersurfaces), studying how they complete each other and how they overlap. The compatibility between fuzzy rules generated by human operators or from data can be assessed by comparing the hypersurfaces.

Suppose that we have two experts who express linguistically their knowledge on a certain process under the form of a number of fuzzy rules. If the two experts have knowledge of different parts of the process, their knowledge may complete each other's, as shown in Figure 4.1-63. This corresponds mathematically to fus-

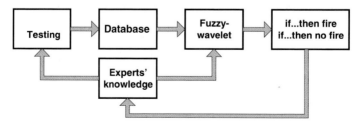

Fig. 4.1-62 Field testing and qualification are two essential processes in development. Fuzzy-based methods have been developed to extract the relevant information from a databank

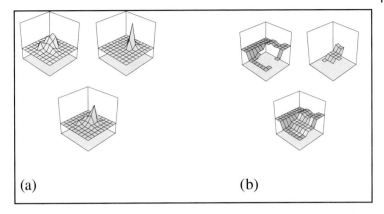

Fig. 4.1-63 (a) The knowledge of two experts can be analyzed by a deformable template method using as input the fuzzy-wavelet coefficients describing the rules. (b) The partial knowledge of two experts can be fused together using approximation techniques

ing two different surfaces into a single surface. If the two experts have knowledge of the same part of the process, their information might partially contradict. In this case, the computer can make proposals to reconcile the contradictory information. This is done by deforming the surfaces corresponding to the different expert knowledge with a minimum of pre-defined operations, such that the surfaces become comparable [65]. Each deformation results in a penalty and the surface minimizing the total penalty is proposed as a compromise solution. The deformed surface can then be translated into linguistic expressions that are submitted to the experts. The process can be iterated until an agreement between the experts is reached.

4.1.7.2
Application of Multi-resolution and Fuzzy Logic to Fire Detection

Smoke detectors have suffered over many years from the so-called 'false alarms' problem [15]. A spectacular improvement was made in recent years, owing to the combined influence of more reliable components, better alarm organization, and the appearance of high-quality microprocessors in the low-price segment. Signal processing is playing an increasing role in smoke detectors as more sophisticated algorithms are being developed. Fuzzy logic has played a central role in fire detection [53]. The acceptance of fuzzy logic has been substantial after the first successful implementation of fuzzy logic in fire detectors. At the beginning, fuzzy logic was used mostly in classifiers, but with time it became to be used to describe the alarm surface in multisensor detectors. We present below two commercial products that have benefitted significantly from fuzzy logic: a linear beam detector and a flame detector.

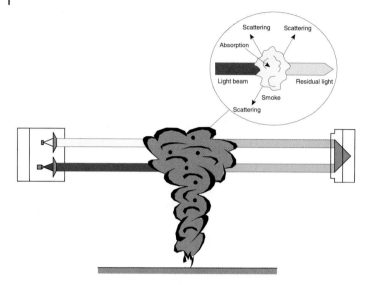

Fig. 4.1-64 A linear beam detector is a smoke detector in which light attenuation is used as the alarm criterion

4.1.7.2.1 Linear Beam Detector

The basic operation principle is shown in Figure 4.1-64. An energetic pulse is emitted by a LED and travels a distance between typically 5 and 150 m. The light is reflected back to the detector by a high-quality retro-reflector. The detector will go into alarm if a certain smoke level is in the beam path.

In order to prevent false alarms, fuzzy algorithms were developed that are capable of distinguishing the signature of non-fire (signal intermission due to a moving object, eg, a bird or a truck, signal attenuation due to the Schlieren effect resulting from mixing of cold and hot air, etc.) and fire events. In a first stage, a number of features are extracted from the signal. These features are combined with a number of fuzzy rules (Figure 4.1-65). These rules serve two purposes, first to furnish a differentiated diagnosis of potential problems to the operator, and second to modify the alarm parameters depending on the diagnosis [67].

The algorithms were developed with a neuro-fuzzy method, using a multi-resolution Kohonen type of network for signal classification. Data were collected in extensive field testing and fire testing. Data were classified using a constructive multi-resolutional approach. First, the network was optimized with only two membership functions per variable using a Kohonen network to determine the best partition and a simulating annealing method to optimize the shapes of the membership functions (Figure 4.1-66). The rules were then validated and the data corresponding to the validated rules were removed. The procedure was repeated by splitting the membership functions into two new membership functions to refine the fuzzy rules.

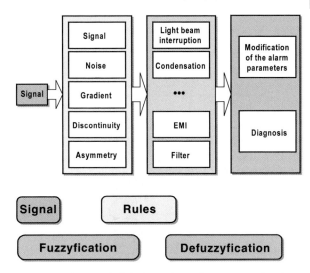

Fig. 4.1-65 The signal of the linear beam detector is analyzed on-line by a system containing a number of fuzzy rules

4.1.7.2.2 **Flame Detector**

Wavelet theory can be combined into an efficient spectral analysis method. We discuss here an example in some detail to illustrate the power of the method. Flame detectors use pyroelectric sensors to record in the IR domain the radiation emitted by flames. Flame radiation is measured at three different wavelengths (Figure 4.1-67).

Flame detectors must be sensitive enough to detect small fires, but must not be 'fooled' by deceptive phenomena such as sun radiation and strong lamps. The ratio between the sun and flame radiation is much lower in the IR than in the visible range. For that reason, the radiation is measured in the IR domain and not in the visible. Even in the IR region sun radiation is typically much larger than the radiation from the smallest fires one wants to detect. By chance a hydrocarbon fire is characterized by two features. First the flame pulsates and second the flame emits strongly around 4.3 μm, the emission line of CO_2 (Figure 4.1-23). The ratio between the signal at the different wavelengths and the spectral analysis of the flame fluctuations can be used to characterize a true fire [68, 69]. The spectral analysis is carried out by online by a method combining fuzzy logic with wavelet analysis (Figure 4.1-68).

Under laboratory conditions, a hydrocarbon fire pulsates at a very regular frequency in the range between 0.5 and 13 Hz. The larger the fire, the smaller is the pulsation frequency. A very simple law describes the pulsation frequency: the pulsation frequency is inversely proportional to the square root of the fire diameter. Amazingly, the pulsation frequency is to a first approximation almost independent of the fuel. This can be explained by the fact that a purely hydrodynamic instability causes the flame to oscillate. This instability is caused by the density dif-

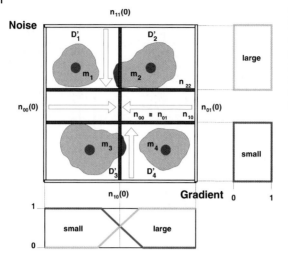

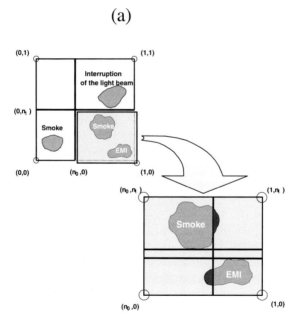

Fig. 4.1-66 (a) A Kohonen-based neuro-fuzzy system was developed and implemented during the development of the rule-based fuzzy system. (b) A multi-resolutional approach permits new membership functions to be added iteratively

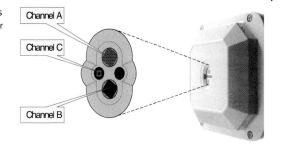

Fig. 4.1-67 A flame detector records the radiation of a flame at a number of wavelengths. The signal is analyzed with a number of rules in the spectral domain and in the frequency domain

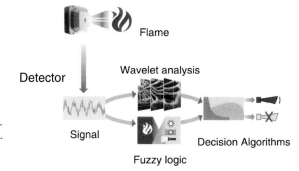

Fig. 4.1-68 Spectral analysis, feature extractions, and classifications are made by combining fuzzy logic and wavelet analysis

ference between the very hot air and the surrounding air flow, giving rise to an unstable gravity wave. We have suggested that the regular flame pulsation results from a resonant effect, that takes place when the wavelength of the gravity wave is in a simple ratio to the fire diameter.

In real-world applications, the regular flame pulsation may be easily perturbed by different influences. A window may be opened and an air current will destroy the regular flame pulsation. We found that, under those circumstances, flame pulsation still had some typical features. In order to understand why, we carried out a number of controlled experiments with oscillating membranes. With such experiments, we were able to show that external perturbations coupled parametrically to a flame. Flame pulsation can be fairly well modeled by a self-excited van der Pol oscillator with parametric coupling [68]:

$$X'' + \omega_0^2 X + a(X^2 - K)X' = F(\omega, t)X \qquad (4.1\text{-}11)$$

with $F(\omega, t)$ describes the perturbation, a and K are constants, ω_0 is the natural pulsation and X is the average flame radiation.

Figure 4.1-69 compares the pulsation of a flame excited with an oscillating membrane with the van der Pol model. At low excitation, the flame couples to the membrane oscillating at a frequency of the order of the natural flame pulsation frequency. As the amplitude of the excitation is increased, a bifurcation to the subharmonic takes place. The flame begins to pulsate at exactly half the excitation frequency. The comparison of many experiments with very different excitations furnished a qualitative understanding of real-world situations. The final outcome of this research was a catalog of the possible flame fingerprints.

The exploitation of these fundamental research results required an efficient spectral analysis method to recognize the different fingerprints. Short-time Fourier transforms in combination with a classifier were considered. The necessary power was too high for the microcontroller at our disposal (the main limitation is given by the electric power required for computation, not the computing power of the microcontroller).

An alternative to Fourier transformation was furnished by wavelet theory. Wavelet decomposition can be carried out by using a cascade of filters. A filter is associated with each level of resolution of the wavelet decomposition. For orthogonal wavelets, the energy conservation relation holds. It follows that the low-pass and high-pass filters corresponding to the two decomposition filters for the first level of decomposition fulfill the power complementarity condition, as illustrated in Figure 4.1-70:

$$\sum_{m=1,\ldots,p} |T_m(\omega)|^2 + |T_{\text{low}}(\omega)|^2 = 1 \tag{4.1-12}$$

In a wavelet decomposition, the signal after low-pass filtering and attenuation is filtered with the same two filters. This corresponds to splitting the low-frequency filter band into two new bands.

At any decomposition level, the corresponding filters fulfill the power complementarity condition. The filter transmission functions in Figure 4.1-70 can be interpreted as fuzzy variables, eg, 'low frequency' or 'very high frequency'.

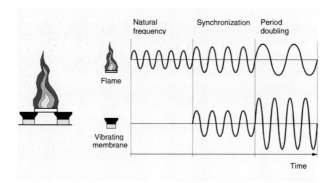

Fig. 4.1-69 a

Fig. 4.1-69 Fundamental research on the physics and dynamics of flames led to a new model of flame pulsation. In an experiment, a flame was excited with an oscillating membrane. Depending on the amplitude of the oscillation, the flame synchronized with the excitation frequency or period doubling was measured. The experiment can be modeled qualitatively very well with a forced van der Pol model

EXPERIMENT

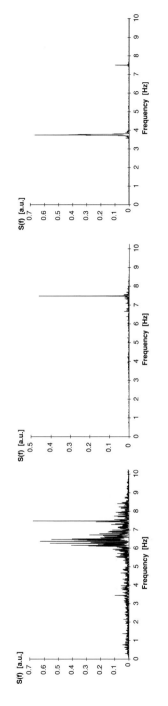

MODEL

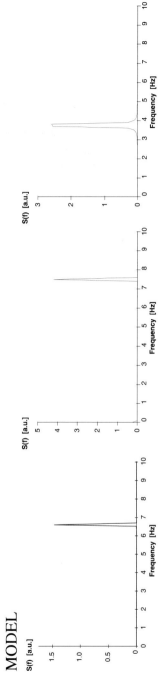

Fig. 4.1-69 b

erties by feature extraction and the subsequent classification by statistical or neuronal methods is a discipline which has attracted a great number of researchers and has generated many good results. Here the human visual system itself has served as an example and motivation for the development of pattern recognition systems: to detect a well-known face in a crowd takes only a split moment. The technology has been well developed within the last few years and, eg, as with the biometric feature fingerprint, supplies a fine selectivity allowing one to discriminate even between twins. The systems of today profit not least from the many years of experience in criminology.

Second, there are obvious advantages that biometric methods have: the biometric feature cannot be forgotten or lost. If, using the classical access control mechanisms, just one employee loses a token (key), the exchange of the whole system becomes inevitable in a security-sensitive case. This problem is eliminated by using a biometric key.

Also, deliberate handing over of an access permit for unauthorized access to a high-security domain by support of an (administrative) internal accomplice is, in general, excluded. Here biometry seems to provide a higher security level.

Biometry addresses a basic problem with which everyone is confronted: the password flood! If the classical token-based access control is supplemented by personal identification number (PIN)-based mechanisms, the number of PINs and passwords one uses regularly or, what is worse, irregularly not infrequently may exceed 20.

Therefore, great hopes are attached to biometry. This following contribution describes the mechanisms of biometric access control and, in addition to the advantages mentioned above, also examines the technical limitations of these methods.

4.2.2
Access Control

Being part of an organizational and logical security concept for the protection of valuable or secret goods or information, access control can be understood as a technical measure to regulate the physical access to these values. The access control process can be subdivided into the following steps [1]:

- *identification* for connecting a person to his/her identity or pseudonymity;
- *authentication* as proof of a correctly executed connection between person and pretended identity;
- *protocolling* of the actions executed in the system for recording and, if necessary, for an analysis of security-relevant occurrences.

Authentication as a crucial step of access control checks if the pretended identity definitely belongs to this person. This check can be done by different authentication mechanisms, in the simplest case by interrogating special information such as a PIN or a password known only by this person. Knowledge-based authentication can easily be implemented and is straight forward to apply, but shows the drawbacks mentioned in the Introduction: the knowledge or a recording medium (memo paper) can get lost. Moreover, especially very short passwords are not safe from (bruce-force) attacks systematically checking through all combinations possible.

Alternatively, authentication can be achieved through a permit, ie, when a person presents a personalized token, eg, a magnetic or chip card. Often, both methods are combined: a token and, additionally, a special knowledge is checked. Here, a transmission of knowledge and/or token – whether from indolence or carelessness – can, however, thwart a consequent concept of access control.

The third option is the proof of the connection by typical physiological or behavior-related characteristics of the person which are recognized and verified by a biometric method. The biometric properties considered can either be of a static nature, eg, fingerprints or hand geometry, or can describe a dynamic characteristic such as the voice or writing rhythm of the person. Biometric authentication does not have the special shortcomings related to the first two authentication methods mentioned above (loss, transmission, etc.). Nevertheless, even this method has drawbacks, which will be dealt with later.

4.2.3
Biometric Systems for Access Control

Access control systems realizing biometric authentication consist of several components which are commented as follows: being a crucial element, the *sensor* forms the interface between the user and the access control system. It records the biometric feature (eg, a scan of the finger-tip) and propagates the recorded pattern to the *processing unit* which generally generates a feature extraction in order to obtain a shorter and more selective pattern description. Furthermore, the processing unit performs the matching of the extracted properties with one or several reference profiles. Biometric systems can be shaped as a verification system or an identification system. In the case of a verification system the user indicates an identity for which a reference image exists or is provided as a template. If biometric systems are combined with token-based systems, the template can be stored on a chip card. For verification the recorded image is compared with the image on the template (1:1 comparison). Subject to the result of this comparison, access is allowed or rejected.

When an identification system is involved the recorded pattern is compared with possible and registered templates and the best match is identified. The difference between the presented pattern and the best match must, however, remain below a defined threshold value so that the authentication of the person and the tracing of the identity connected with the template can be performed with adequate statistical certainty. Identification systems present, at first sight, a greater user convenience; in practice, however, it often proves that the systems lose their practicability with respect to performance and recognition reliability already with medium-sized user groups (more than 30 persons). Another aspect is that for a $1:n$ comparison in the processing unit of the pattern with the n registered templates a secure data connection between the processing unit and the database must be established.

Before using a biometric access control system, an enrolement procedure is necessary in any case to record one or more templates of each user which are connected with his identity and serve as a reference model. These templates are recorded in a reference database or the chip card. Furthermore, the producer or the system administrator has to select the control parameters of the system for the definition of the threshold values for pattern comparison. Here the size of the user group should be taken into account.

The authentication systems frequently used today analyze the fingerprints of the user and derive properties from the minutiae of the image (Figure 4.2-1).

Such systems are interesting because of the low production cost and are mainly used for access control to computer systems, eg, integration in the keyboard for authentication at a computer workstation (Figures 4.2-2 and 4.2-3). In practice, eg, under Windows NT, the user provides his/her computer identification number and instead of a password a fingerprint is taken for authentication. In the near future providers plan to launch different comfort applications, eg, in the mobile device sector for unlocking of the mobile terminal. A use for the protection of high-security areas in buildings is imaginable if the sensors and systems feature an adequate trespassing security.

In physical access control, most systems are based on a picture or videostream evaluation and are integrated in an access control point. In addition to the picture taken from the face of the user (Figure 4.2-4), hybrid systems such as the system of DCS Co. additionally record and analyze the movements of the lips and the voice signal (Figure 4.2-5). The evaluation of the password in the additional infor-

Fig. 4.2-1 Record of a fingerprint

Fig. 4.2-2 Infineon FingerTIP as access control sensor. Chip size ca. 1×1.5 cm. Image courtesy of Infineon Technologies.

Fig. 4.2-3 Cherry keyboard with optical fingerprint sensor. Image courtesy of CHERRY GmbH

mation channel (audio) allows one to achieve stronger rules for acceptance and hence lower false acceptance rates (FARs).

Another alternative biometric feature which can be used for access control is the user's hand geometry. Such a system benefits from a quick evaluation of the geometry (Figure 4.2-6) mainly considering the hand contours. The user lays his/her hand

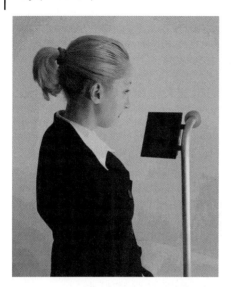

Fig. 4.2-4 ZN-Face, a face recognition system as physical access control point. Image courtesy of ZN GmbH

Fig. 4.2-5 BioID access control point with hybrid analysis of face, lip movement, and voice (spoken password). Image courtesy of DCS AG

on a red-lit glass surface safely mounted in a box (Figure 4.2-7). The contours of the hand are recorded and evaluated over a reflecting system by a frame grabber card.

Finally, an access control system belonging to the iris scanner type authenticates the user by a high-resolution shot of the iris (Figure 4.2-8). The system is well suited for the securing buildings since the camera is supported by two infrared radiators integrated in the box (Figure 4.2-9).

Fig. 4.2-6 Hand geometry of a user

Fig. 4.2-7 Access control by analysis of the hand geometry in the system of Dermalog Co.

4.2.4
Security of Biometric Systems

Biometry will only prevail in the future when a sufficient and reasonable degree of security is granted. Can the present generation of biometric systems be called secure? To answer this question we must take into account different aspects. These range from basic security needs of the end users in the context of data pro-

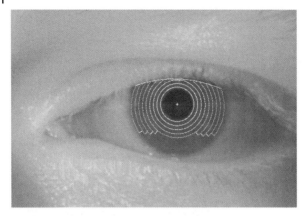

Fig. 4.2-8 High-resolution shot of an iris scanner

Fig. 4.2-9 Sensar Iris Scan camera with lateral illumination units

tection over the necessary quality assessment of a system on the basis of an investigation of the indices of the normal user statistics up to statements concerning the trespassing possibilities.

An adequate recognition of the user's legitimate interest in the protection of his/her personal data is one of the central tasks in the assessment of the consequences of technical developments. It has to be examined whether the storing of personal data, among which are undoubtedly the biometric properties of a user, is performed with due care and reduced to a minimum (concerning time and technique). Identification systems are by definition problematic with a view to data protection since a comparison is performed with the patterns of all users stored in a central (or distributed) database. A serious scenario would be an attack where user patterns are picked out. Compromised properties of a user are without any value and dangerous because biometric properties cannot be revoked – a seemingly trivial but altogether crucial difference from the classical PIN. The selection of different physiological properties that can be used as biometric basis (ten fingers, two eyes) may alleviate the problem but it does not solve it because the number of identifcation properties available is finite.

Verification systems offer, at least in the conception, a much higher level of data protection. For verifying a proclaimed user identity, it is in no way indispensible to perform a comparison between the sensor result and user pattern within the access control system. Alternative architectures can be considered instead [2] which use a (secure) personalized mobile terminal (Organizer, Palm, etc.) as a trustworthy assistant to do the comparison in a controlled environment of the user and only then authentically transmit the result to the authorizing system. Also, it is in principle not even necessary that the identity of the user itself is transmitted. Does this lead to the vision of an anonymous biometry allowing one to get an authorization and, at the same time, preventing unauthorized motion or interest profiles? The fact is that today's generation of systems cannot match this vision.

The statistical system performance of a biometric product is a describing measure for the expected reliability of a system. This is deduced from the false acceptance rate (FAR) and the false rejection rate (FRR) of a system on the basis of defined numbers of system users. For obvious reasons the rates are correlated and are subject to the selection of the system parameters and the size of the user group. In different trials [3, 4] it has been proved that the achievable rates in practice fall short of the rates indicated by the producers, which could result from interfering environmental circumstances (noise level, varying light conditions, etc.) compared with the laboratory conditions of the developers. Furthermore, a growing number of users within a group could cause worse rate results.

Another essential assessment criterion for biometric systems is the resistance to fraud, ie, the make-believe of a false identity in order to obtain a desired authorization.

Biometric systems are principally threatened by so-called replay attacks. These can be directed against different components whereby the necessary knowledge and the 'criminal energy' involved vary considerably. Considering the human-sensor interface we can see a problem especially for the static biometric systems, eg, face recognition, fingerprints. The effort necessary for the execution of such an attack can be rated low when it concerns public properties. Can a face recognition system yet be called a safe access control system when the presentation of a printed photograph to the sensor leads to a successful authentication? Fingerprints too are public biometric properties which, after being prepared, can help an attacker to succeed in a replay attack [5]. In these cases, only the development and integration of sharper live tests can afford a remedy. This is the necessary goal for the development of next-generation biometric identification systems. Dynamic properties such as lip movement, writing rhythm, etc., are in general less threatened by similar attacks. This is not true, however, for attacks towards the sensor-computer interface. Many systems must be called unsafe because it does not take much effort to record and replay the signal on the communication line between the sensor and the computer.

For a comprehensive assessment of the security of biometric systems, different (all) security levels must be considered: both the sensor and possible modifications of the sensor are subject to environmental influences, the algorithm for property extraction and classification, the physical security of the system, and other factors. Only when all levels are reliable can a system be assessed as secure. It is expedient to base an objective evaluation on well-defined evaluation criteria. Evaluation has to be redone with each new version of a biometric system if we expect an objective product assessment and not just an automatic transmission of a once gained credit to the next generation. This is an unsatisfactory situation for the end user in the acquisition stage, because the necessary comparative study is time and cost consuming and hardly practicable for the single user. Producer-independent evaluation centers serving the general public could deliver support.

4.2.5
Prospects

The future will show if in the next generation of biometric systems the above-mentioned problems will be solved. The discussion about the technical impact of biometry should also include the consideration of the possible implications that a stronger breakthrough of this technology will involve. One aspect that should not be neglected in any application is how to address non-users, ie, those persons who cannot deliver a certain property record or who have strongly varying properties so that they are regularly refused admittance. When an end user decides to embed the technology into the overall organizational security concept he/she should ensure that non-users get equal treatment. In the application scenario of access control it is unacceptable if access is generally refused to non-users or only admitted in the company of a security administrator. Alternative classical authentication methods for these persons must be included in the overall concept.

4.2.6
References

1 PETZEL, E., *Management der Informationssicherheit;* Weiden, 1996.
2 BLEUMER, G., in: *Multilateral Security in Communications;* New York: Addison Wesley, 1999, pp. 157–171.
3 *Vergleichende Untersuchung Biometrischer Identifikationssysteme;* http://www.igd.fhg.de/igd-a8/projects/biois.
4 *Ein interdisziplinärer Pilotversuch der Tele-Trust Arbeitsgruppe 6 zur Anwendung bioetrischer Identifikationsverfahren im Bankenbereich;* http://www.biotrust.de.
5 STEINHAUSER, A., in: *Die Datenschleuder;* #68/69, 1999.
6 DINKELBACH, W., HÜBNER, M., *Datenschutz Datensicherheit* 3 (1996).
7 DAVIES, D. W., PRICE, W. L., *Security for Computer Networks,* 2nd edn.; Chichester: Wiley, 1989.
8 FINNE, T., *Comput. Security* 15 (1996) 29.
9 National Biometric Test Center; http://www.engr.sjsu.edu/biometrics/publications_technology.html.

4.3
Smart Cameras for Intelligent Buildings

BEDRICH J. HOSTICKA, *Fraunhofer Institute of Microelectronic Circuits and Systems IMS, Duisburg, Germany*

4.3.1
Introduction

Modern electronics is invading all of aspects of human life – and buildings are certainly no exception. Although the use of electronics in residential buildings was proposed a long time ago, its extensive use has been rather sporadic so far [1]. However, now this is about to take place and promises to be a huge market. Nevertheless, there has been more use of electronics in nonresidential buildings and its use is on the increase [2].

But what can electronics really do in our buildings which has not yet been realized? The term that has been coined here with respect to buildings is that of an 'intelligent' building. Thus buildings can be made intelligent by employing electronics in order to provide significant improvements in the areas of fire protection, safety, security, comfort, emission and imission, illumination, transportation, power distribution, building automation, and communications.

One important electronic device used today in residential and nonresidential buildings is a video camera. It is usually used to carry out surveillance of critical locations for protection against threats to occupant safety, security, and health. Further, it often serves to provide remote video monitoring of building equipment, building structure, building contents, and access areas. Nevertheless, as we shall see, new applications are on the horizon, such as comfort control, energy management and conservation, and building automation. This will be possible by developing more sophisticated cameras that can serve not only as standard video cameras but also carry out additional tasks, eg, to determine occupancy state. But before we explore these new directions let us first have a look at the state of the art.

As mentioned above, today low-cost video cameras are widely used to provide remote surveillance of various locations by human guards. The main purpose so far has been to ensure security against intrusion. To keep the equipment costs low these video systems usually adhere to the TV standards used in public broadcasting such as NTSC or PAL (in contrast to public TV, however, these systems are closed and, therefore, they are called closed-circuit television (CCTV)). Nevertheless, the cost of TV equipment installation is still not negligible. Since the maintenance costs are usually not high, the major costs are related to the fact that the acquired or recorded images must be evaluated by human observers. In cases where security guards are on duty anyway (eg, surveillance in industrial plants, banks, museums, and railway stations), TV displays can be permanently viewed and evaluated by human observers, but this is subject to fatigue and reduced attention. Also, this requires 24 h duty and is therefore costly. For this reason, some security companies offer remote video monitoring at their monitoring centers where trained staff evaluate the acquired images. However, this calls for permanent or event-triggered video transmission of live images between the area under observation and the center. On the other hand, this transmission can use low data rates, but it still requires some sort of a communication link. Another possible solution is video recording but with no live images no immediate reaction is possible. The only benefit of archived video sequences is that they can be of benefit for a later investigation but they cannot help to prevent an intrusion.

There have been numerous attempts to develop image-processing algorithms for video images in order to yield reliable automated surveillance. However, under practical conditions this is much more difficult than it might appear at first sight. The very first problem is that of 'optical' dynamic range, ie, brightness range at the input that is severely limited in the case of todays' video cameras. In practice, illumination conditions in both indoor and outdoor scenes can differ greatly not only owing to varying daytime light conditions but also within a single scene owing to inhomogeneous illumination (eg, combination of direct sun or lamp and shadows). Commercially available CCD-based standard cameras deliver a dynamic range of only 50–70 dB, which is far below what is required. The limited dynamic range of these cameras thus severely limits their capability to capture all details in recorded scenes and thus hinders effective video-based surveillance coverage.

A fully automated video surveillance that would dispense with human observers requires a computer-aided scene analysis and understanding. However, this represents a formidable task. In order to achieve real-time image processing and to keep costs low, equipment manufacturers try to resort to the use of simple algorithms and hardware. However, simple algorithms are not robust enough and tend to fail frequently under practical conditions. A typical example is that of motion detection: this is an important indicator which signals temporal changes in an observed scene such as the arrival or departure of human intruders. Unfortunately, it also responds to nonhuman movements, eg, caused by animals. Similarly, movements of tree branches due to the wind and illumination changes caused by movements of clouds, shadows, or projected light beams have a similar effect. A robust real-time extraction of relevant motion (or suppression of irrelevant motion) still requires considerable computing effort and has eluded low-cost implementations in the past.

In the following sections we will show that the introduction of a new camera technology, namely complementary metal oxide semiconductor (CMOS)-based imaging, is going to change all this. We shall establish the concept of smart cameras, ie, cameras that possess built-in intelligence and realize enhanced imaging under critical illumination conditions and low-cost image processing. In this context we will distinguish four different application areas:

- video monitoring of remote locations for general surveillance;
- simple occupant sensing based on motion detection;
- advanced occupant sensing based on shape recognition; and
- biometrics for occupant identification and recognition and access control.

We will show how smart cameras can contribute powerful solutions to these four application areas. The key component to the development of smart cameras is going to be a solid-state imager based on CMOS technology. Unlike in charge-coupled device (CCD) imaging we shall not distinguish strictly in the following sections between 'cameras' and 'imagers'. As the CMOS imagers (or image sensors) almost always contain control, readout, and other key electronics, they no longer correspond to the classical CCD imagers which merely convert light into electrical signals. In fact, CMOS imagers can contain so much electronics that they become fully operational video cameras while containing just a single chip.

4.3.2
Technologies for Solid-state Imaging

The very first semiconductor imagers were so-called 'metal oxide semiconductor (MOS) imagers' based on an idea proposed by Weckler in 1967 [3]. Their images, however, were noisy and thus the CCD imagers which appeared somewhat later captured the market owing to their better image quality [4]. Nevertheless, the CCD imagers were also plagued by technical problems such as blooming and smearing and it took some time to solve these problems.

In the late 1980s and early 1990s, the advances in CMOS technology and circuit design reawakened the interest of the research community in MOS imaging, this time in CMOS (ie, complementary MOS) [5]. Although both imaging technologies, CCD and CMOS, rely on the light sensitivity of silicon, there are differences in fabrication sequences, processing parameters, and device structures. The main attraction of CMOS was, of course, that the CMOS technology had become widely used for mass volume fabrication of integrated circuits. This promised not only a drastic cost reduction but it was also surmized that CMOS technology would allow cointegration of highly complex analog and digital circuits together with imaging devices [6]. Actually, it turned out that black/white CMOS imagers featuring good image quality can be built with the standard CMOS technology as used for manufacturing standard integrated circuits. As expected, this has substantially reduced the costs with respect to CCD imagers because the availability of CCD technology is limited to only a small number of specialized manufacturers. Color CMOS imagers require the addition of color filters which is a technology essentially well known from CCD imaging. Hence an imaging technology has been reborn.

The co-existence of imaging devices and analog and digital circuits on a single CMOS chip implies that CMOS imaging has an edge over CCD imaging: circuits carrying out readout, amplification, image processing, storage, and A/D conversion can be now all be integrated on a CMOS chip together with imaging, control, and interfacing. This feature means that assembly and packaging costs can be also greatly reduced while reliability is increased in rugged environments. Also, this concept implies that a certain amount of 'computational' intelligence can be located on the imager chip – this is then often called a 'smart' imager.

In addition to pure imaging tasks, however, there are many more applications where CMOS imagers can be used. Examples involve 1D, 2D, and 3D ranging, edge detection, motion detection, and optical measurements (eg, velocity or angle) [7].

We have mentioned that CMOS technology basically exhibits the same light sensitivity as CCD technology based on the use of silicon. This is simply dictated by the underlying physics but we must not forget that a CMOS imager is capable of containing in situ circuits that allow manipulation of the photogenerated charge and the implementation of various forms of the photocharge transport to the interface. This is the main difference from the CCD imagers which are 'just' image-sensing devices with restricted types of photocharge transport. It is the combined capability of image sensing, on-chip intelligence, and versatility and flexibility of CMOS imagers that appeals to the users.

CMOS-imaging technology offers even more features that completely elude any comparison with CCDs. Thus CMOS imagers can be employed as 'dual'- or 'multiple'-use devices as they can be easily programmed, reconfigured, or multiplexed. For example, the same CMOS imager can be used as a motion detector or as a video camera if desired. The combination with motion detector could then be used for the realization of event-controlled video acquisition and/or transmission. Other applications are represented by combinations such as imager/brightness sensor (for lighting control), imager/sun sensor (for air-conditioning control), and imager with on-chip focus electronics.

4.3.3
Principles of CMOS Imaging

Basically, there are a variety of light-sensitive devices available with modern standard CMOS technology such as pn-diodes, photogate devices, parasitic bipolar transistors, and MOS transistors [8]. More or less they can all serve as imaging devices but they differ in detail with respect to sensitivity and noise. Nevertheless, all these devices exhibit sensitivity in the visible part of the light spectrum up to near-infrared (NIR) wavelengths, although practical use is limited up to about 950 nm.

The simplest imaging device is the pn-diode which can be formed, eg, either between a drain or a source diffusion and the substrate or a well, or a well and the substrate. Since all these structures are required for the generation of CMOS transistors anyway, the mutual compatibility of CMOS imaging and CMOS transistor functions is clearly established. This discussion can be readily extended to the other devices introduced above.

However, let us consider the pn-diode: photons incident to light generate electron-hole pairs in the depleted space-charge region of its pn-junction. This gives rise to photogenerated electrical charge flow. The conversion into a voltage can be realized either using a continuous-time diode loading, a transimpedance amplification, or a discrete-time charge integration. The load and the transimpedance can be both linear or nonlinear. The pn-diode (called now a photodiode) together with the readout circuit forms a 'picture element' or 'pixel'. In image area sensors the ratio of the light sensitive area to the total pixel area is called a fill factor. It is an important design criterion since a low fill factor implies a low pixel photoresponsivity and a wide spatial frequency response. The latter can induce undersampling of fine structures in images. This phenomenon, frequently referred to as 'aliasing', is visible as Moiré patterns in the images and can degrade image quality. The pixel spacing called 'pixel pitch' in practice determines the pixel count for a given lens size while the pixel count defines basically the spatial resolution.

Although there are a variety of other important design parameters, the discussion of which simply is beyond the scope of this contribution, we must mention one of the most important, namely the noise. Imagers can generate 'noisy' images due to statistical fluctuations of their properties in time and space: in practice we speak of 'random noise' and 'fixed-pattern noise (FPN)', respectively. The cumulative noise then determines other important parameters such as signal-to-noise ratio (SNR) and dynamic range (DR).

4.3.4
Examples of CMOS Imagers

A typical pixel circuit is shown in Figure 4.3-1 [9]. This is a so-called 'three-transistor' pixel circuit with a photodiode (PD). The photodiode capacitance C_D is periodically charged to the reference voltage V_{ref} using the reset device M1. The discharge voltage due to the integration of the photocurrent I_{photo} at C_D now corre-

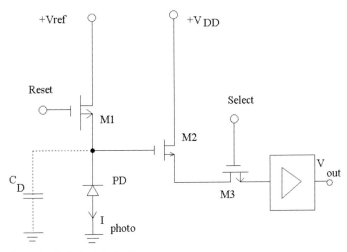

Fig. 4.3-1 Typical pixel circuit

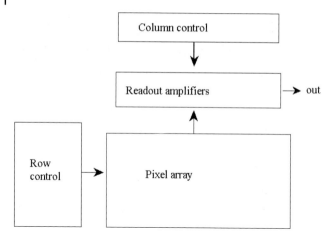

Fig. 4.3-2 Typical CMOS imager architecture

sponds to the irradiance measured by the pixel. It can be accessed by activating the select switch M3 and its value can be passed to the readout circuitry.

Typically an analog amplifier based on the switched-capacitor (SC) technique is used for the voltage readout of the pixel circuit. Of course, the pixel circuit in Figure 4.3-1 is not the only possible pixel construction as CMOS technology allows the realization of variety of different pixel structures.

An example of an architecture of a CMOS imager utilizing the pixel circuit in Figure 4.3-1 is shown in Figure 4.3-2. Note that the architecture is based on column-organized readout, ie, all pixel circuits belonging to the same column share the same readout amplifier located at the end of all columns. The access to the individual pixel circuits is then provided within each row in a cyclical manner.

An imager chip based on this principle is shown in Figure 4.3-3. It was fabricated with 1 µm standard CMOS technology featuring double metal, p-substrate, and n-wells. The imager can accommodate the CIF (common image format) format (ie, 360×288 pixels). The floor plan is depicted in Figure 4.3-4 and the technical data are given in Table 4.3-1. Note that it dissipates only 35 mW.

The column readout amplifiers employed in the CIF imager use a technique called correlated-double sampling (CDS) which is highly suitable for the reduction of low-frequency random noise and signal-independent FPN. The readout organization of the CIF imager implements inherently a rolling shutter with the integration time defined by the time delay between applying the reset and select signals. The imager also allows user-defined access to individual pixels or subregions at higher frame rates than the standard frame rate. Furthermore, the CIF imager is capable of capturing images at multiple integration times (ie, identical images acquired at different integration times).

Two examples of images captured using the CIF imager can be found in Figures 4.3-5 and 4.3-6.

Fig. 4.3-3 Chip photomicrograph of the CIF imager

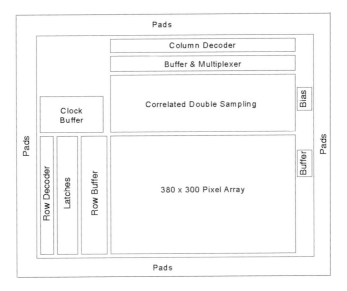

Fig. 4.3-4 Floor plan of the CIF imager

The CIF imager shown above exhibits an input (ie, 'optical') DR of about 60 dB at a single integration time. When using more integration times per frame the DR can be extended to more than 100 dB. This feature is very useful for many applications that require a high dynamic range, such as automotive cameras.

Automotive cameras must exhibit an extremely high input DR as road illumination conditions can vary greatly not only between the frames but also within a single frame owing to inhomogeneous scene illumination. Although typical for auto-

Tab. 4.3-1 Technical data for CIF imager

Readout organization	progressive scan
Total pixel count	380×300 pixels
Resolution (ie, pixel count to be read out)	max. 360×288 pixels
Shutter	rolling type
Fill factor	30%
Pixel pitch	$17 \times 17\ \mu m^2$
Chip area	$88\ mm^2$ in $1\ \mu m$ CMOS
Power supply voltage	5 V
Power dissipation	35 mW
Typical pixel clock rate	13.5 MHz
Max. pixel clock rate	20 MHz
Standard frame rate	100 Hz
Max. frame rate	140 Hz
Frame rate for 64×64 pixel subregion readout	1000 Hz
Input DR (at integration time of 30 ms)	60 dB
Max. SNR	52 dB
FPN	<2 mV
Min. detectable irradiance (at integration time of 30 ms)	$5 \times 10^{-5}\ W/m^2$

motive applications, such illumination conditions can also be found in a variety of other applications, eg, in buildings or building surroundings. A building perimeter or parking lot may pose exactly the same requirements as automotive applications.

A CMOS imager chip that has been developed specifically for automotive applications and that yields a high DR is shown in Figure 4.3-7 [10]. Unlike competing high dynamic range CMOS imagers that employ a logarithmic dependence on light irradiance, this imager uses a combination of multiple integration times with an automatic selection of the 'best' integration time and thus exhibits piece-

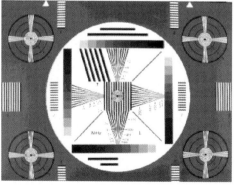

Fig. 4.3-5 Reproduced images from the CIF imager

Fig. 4.3-6 High-speed image sequences (1000 frames/s) of a rotating fan captured by the CIF imager

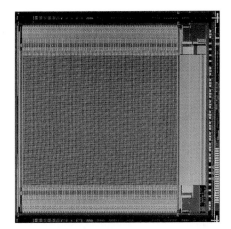

Fig. 4.3-7 Chip photomicrograph of the 256×256 pixel automotive imager

Fig. 4.3-8 Tunnel entrance

Fig. 4.3-9 Tunnel exit

Fig. 4.3-10 Laboratory scene. Left, image taken with CCD camera; right, image taken with CMOS camera

wise linear dependence. Hence it does not suffer from poor contrast and high noise typical of 'logarithmic' CMOS imagers. Besides multiple integration, the automotive imager utilizes an on-chip multigain readout control. This combination yields a 120 dB input DR unrivaled by other 'linear' imagers. The chip exhibits a resolution of 256×256 pixels while delivering 50 frames/s. Several scenes taken with the automotive camera can be observed in Figures 4.3-8–4.3-11. The images were prepared so that they fit the limited gray-level range of the displays.

Figures 4.3-8 and 4.3-9 show images taken from a passenger car before entering and leaving a highway tunnel on a bright sunny day. The images on the left were captured at 50 dB input DR and those on the right were taken at a DR of 120 dB using the same camera.

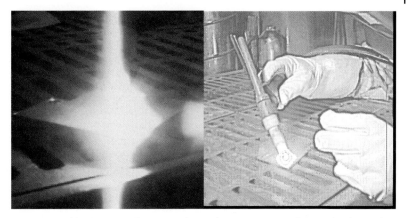

Fig. 4.3-11 Welding scene. Left, image taken with CCD camera; right, image taken with CMOS camera

Figures 4.3-10 and 4.3-11 depict a comparison between a standard CCD camera (on the left) and the automotive camera described above (on the right). Figure 4.3-10 features a laboratory scene with two high-intensity light sources and Figure 4.3-11 depicts a welding scene. Note the significant smearing effects in both examples in the case of the CCD image.

4.3.5
Simple CMOS Occupant Sensors Based on Motion Detection

So far our discussion has concentrated on CMOS imagers that compete with CCD imagers merely on the merits of cost and performance. In this section we discuss CMOS imagers that are suitable for occupant sensing and thus introduce a new feature class that goes beyond CCD imaging.

Occupant sensors are devices that respond to the presence and absence of people in monitored spaces. Classical occupant sensors rely on general motion detection without trying to localize the motion in space. These motion detectors use either ultrasonic sound, microwave, or passive infrared radiation (PIR) technologies [11, 12]. Such sensors are typically used in residential and commercial applications, eg, for intrusion detection in security systems, lighting or air-conditioning control for energy saving, or door entrapment detection.

Since all classical occupant sensors capture only global motion, these sensors can only detect the general presence of moving objects and not the absence of objects or people resting in the monitored space. Also, these sensors fail if additional features such as object localization, object identification, or object tracking are desired. To meet these requirements necessary for more advanced occupant sensing, not only a reasonable spatial resolution is required but also more complex image processing.

Now we shall consider the practical realization of optical motion detection suitable for occupant detection.

Motion detection using a stationary mounted camera with respect to the environment can be generally performed by the analysis of temporal changes in image sequences. This task can be tackled using various approaches such as detection of change in pixel intensities between consecutive frames, estimation of motion vector fields using local features, or differential estimation of motion vector fields. The simplest method is based on the analysis of intensity differences in consecutive frames and can be easily realized using time-recursive image processing. However, this technique is extremely sensitive to illumination variations caused by moving clouds or shadows, camera vibrations, and temporal noise occurring in the electronic components used. A possible in-pixel realization of the time-recursive algorithm is illustrated in Figure 4.3-12 [13]. The photodiode capacitance C_D is periodically reset using the switch M1 and then the discharge due to the photocurrent I_{photo} starts when the switch M1 is thrown open. The voltage across the photodiode (PD) can be easily accessed using the source follower M2 when the select switch M3 is turned on. It appears amplified at the output as $V_{out,I}$. Now, when the switch M4 is activated, a charge sharing between the capacitances C_D and C_R occurs. Since C_R is not reset it retains some charge from the previous period. This provides a suitable reference for evaluation whether there has been a change in pixel illumination or not. Note that the combination C_D, C_R, and M4 yields a simple temporal low-pass filter. Both the current and reference pixel values are available at the output as $V_{out,I}$ and $V_{out,R}$, respectively. Thus we obtain a currently acquired image and a reference image when accessing all pixels. Naturally, we do not have to implement the algorithm in the pixel as shown in Figure 4.3-12: an alternative is to use a standard imager and execute the image subtraction off-imager in the accompanying hardware.

The motion can be easily determined by pixelwise differencing the currently acquired and reference images [13]. All we need is to set a threshold and count number of pixel differences exceeding the threshold. The threshold can be chosen heuristically or adaptively using discriminant analysis based on evaluation of the pixel count and we can make a binary decision whether a motion has occurred or not.

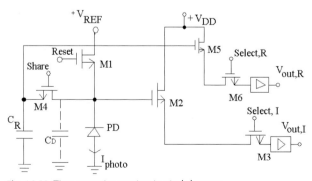

Fig. 4.3-12 Time-recursive motion in-pixel detector

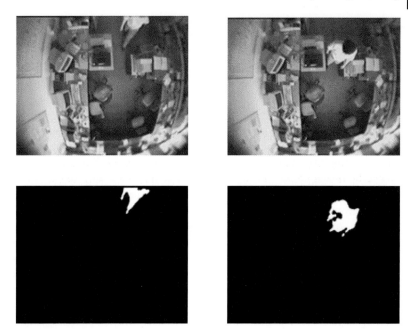

Fig. 4.3-13 Acquisition of images using a ceiling-mounted camera (top row) and occupancy detection (bottom row)

The computation of the difference images and the generation of the binary output signal can be easily done by an analog switched-capacitor (SC) circuit located at the periphery of the same chip.

Figure 4.3-13 shows images acquired by a ceiling-mounted camera in a laboratory (top row). The bottom row shows the difference images generated by a moving person and demonstrates the capability of this approach for occupancy detection. It is obvious that the passive system delivers reliable information about the occurrence of motion if the contrast between the background and the moving object is sufficient. This is true in most cases under daylight illumination conditions as extensive real-time tests have proved. Note that the presented method also allows the spatial localization of moving objects because the coordinates of moving objects are available. Nevertheless, the implementation of this feature requires additional hardware. The presented method can be also extended to include occupant tracking.

4.3.6
CMOS Imagers and Motion-based Occupant Sensors Using Active Illumination

The reliability of imagers and occupant sensors is crucial for successful applications in environments where the variations of the background illumination levels within the monitored scene exhibit an excessively high dynamic range. Thus, in both indoor and outdoor scenes the brightness can range from total darkness (<10 mlx) to very bright sunny days (>100 klx). Also, passive occupant sensor systems (ie, without active illumination) can be deceived by illumination variations, eg, caused by moving shadows, and they fail in total darkness. In cases where we wish to avoid this, it is mandatory to use active illumination. This can be solved by applying active illumination with a sufficiently high illumination intensity [13, 14]. If the reflected active illumination is higher than the background illumination for all reflectance factors, then the scene captured by the imager can be independent of the background illumination. To achieve this we can adopt an approach that relies on exposing the imager twice to the monitored scene. The first 'shot' is carried out using the active illumination (while still containing effects of the background illumination) and the second without, so that only effects of the background illumination are monitored. The subtraction of the two images yields a difference image without the effects of the background illumination. The time delay between the two images should be minimized so that motion effects are eliminated. Naturally, we can also apply the principle of active illumination to the occupant detection [14]. If we again employ the motion detector mentioned in the previous section and wish to apply the time-recursive approach, we have to repeat the just described procedure twice with a defined delay. This means, however, that we have to acquire four images in total as we need two 'shot' pairs now in order to obtain two difference images that have a defined delay in between. However, we can no longer realize the time-recursive algorithm in pixels, because the fill factor would become substantially degraded if we tried to accommodate all this processing in the pixel circuit. Hence we will have to read out all four images and process them off the imager.

Now let us consider the active illumination source. In order to maintain an eye-safe operation, the intensity of this source must remain below limits stipulated by the International Electrical Commission (IEC). While we can use continuous-wave sources in the visible range with CMOS imagers, there are more possibilities.

A useful way is to use sources emitting light in the NIR range where it is invisible to human eye but silicon is still light-intensive. Another intriguing idea is to pulse the active illumination source. As the eye-safety limit is basically defined by the energy received at the retina, we can increase the source power if sufficiently short pulses are used. This is a particular advantage of CMOS imagers as they can be easily equipped with on-chip synchronous shutters operating even below 50 ns. This feature allows the use of a synchronous imager shutter in combination with a pulsed active light source. Note that in principle it can be also applied to night vision.

An example of a CMOS pixel circuit with a shutter is shown in Figure 4.3-14 [15]. It has been successfully used in high frame rate CMOS imagers operating at frame rates exceeding 1000 frames/s. The shutter times could be varied between 30 ns and 150 µs. Each sensor pixel circuit consists of a photodiode formed by an n-diffusion in p-substrate. At the beginning of each exposure the reset and the shutter switches are closed synchronously in all pixels to charge the capacitance C_D of the photodiode and the storage capacitance C_S to the reference voltage V_{ref}.

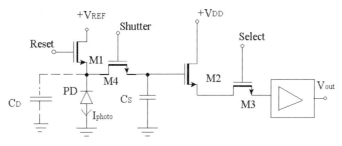

Fig. 4.3-14 Pixel circuit with on-chip shutter

After the reset switch has been thrown off the integration starts as the capacitances are discharged due to the photocurrent generated by the incident light. This process is stopped at C_S when the shutter switch M4 is thrown open. The resulting voltage stored at C_S can be accessed via the source follower M2 by activating the select switch M3.

Such a CMOS image sensor has been used to realize an active camera system [16]. For illumination purposes it contains an array of NIR diodes. In Figure 4.3-15 there are two examples of an application realized with the high frame rate camera system. The scene, a toy car on a laboratory table with instruments and windows in the background, is recorded once with and once without active IR illumination. The integration time has been set to 2 ms. The illumination device consists of a ring of NIR light-emitting diodes (LEDs) which is placed around the op-

Fig. 4.3-15 Laboratory scene with a toy car. The images on the left are illuminated additionally using NIR light, those in the middle contain only background illumination, and those on the right are the resulting difference images. The images in the top sequence were recorded without any IR filter, whereas the bottom sequence was acquired with an IR filter

tics and controlled by the camera system. Taking into account the spectral responsivity of silicon and the attenuation of the sunlight due to the atmosphere absorption, the wavelength of the NIR light has been chosen to $\lambda = 780$ nm. Additionally, in the bottom sequence an IR filter is used in order to obtain better daylight suppression. As can be seen, the shadow falling on the car from the left is visible in the original images but not in the resulting difference image. The sunlight in the windows has also been suppressed. In the top sequence it can be seen that saturation of the sensor (in the windows) in both original images is subtracted to zero in the difference image (black areas on the left side at the top). This problem has been solved by using the IR filter because the sensor then does not saturate due to the sunlight. The computation of the images with suppressed background illumination has been accomplished using an FPGA, which also generates the control signals for driving the high-speed sensor and a strobe signal to activate an LED array employed as an active illumination source.

4.3.7
Advanced CMOS Occupant Sensors Based on Shape Recognition

So far the presented CMOS occupant sensors have been based on motion detection. Their advantage was the combination with the standard CMOS imager. Hence such sensors can also provide standard images on demand, eg, after a motion has been detected. Nevertheless, they can be also used as CMOS imagers with attention detectors. In this section, however, we demonstrate another class of occupant sensors, namely sensors which are not based on motion but on the extension of range-finding principles. This leads to 3D imagers that can also evaluate shape of objects and thus can not only localize and track objects at all times but also acquire their shape. Note that this principle allows the realization of occupant sensors which permit the determination of an occupancy state irrespective of previous records, discrimination between objects, and shape-based recognition. This represents a major difference to the occupant sensors discussed in the previous section.

Several different approaches to the implementation of 3D imagers are feasible. One possible approach relies on employing 2D light-striping techniques and extending it to 3D by adding a scanning mirror. Another commonly used approach employs a time-of-flight method, ie, measurement of the time that emitted light pulses need to travel when reflected back [17]. Classical time-of-flight realizations are based on 3D laser scanning that involves a combination of lasers, single photodiodes, and rotating mirrors, but there is ongoing research on implementations that use photodiode arrays with on-chip electronics and do not require any mechanical scanning [18]. It is expected that this research will yield 3D imagers that can be used as low-cost shape-based occupant sensors.

4.3.8
Biometric Sensors

In this section we discuss briefly imager-based sensors that employ biometrics for automated occupant identification and recognition as well as for access control [19]. There are, of course, other sensing principles besides optical-based approaches that can also serve the same purpose. Thus badges with radiofrequency (rf) identification, capacitive sensor-based fingerprinting, or speaker recognition are also feasible, but optical methods, eg, face recognition, iris recognition, identity card or tag scanning, and optical fingerprinting, offer high reliability at low cost. Whereas the first two approaches require cameras with projection optics, ie, lenses, the second two dispense with lenses as they are contact-based. Figure 4.3-16 shows an example of a fingerprinting module (on the left) and a captured fingerprint.

4.3.9
Summary

In this contribution we have discussed smart cameras and their use in intelligent buildings. The underlying idea of the smart camera is that CMOS technology permits the realization of imaging, readout, control, image processing, and interfacing functions on a single silicon chip. Further, new functions are also feasible such as motion detection or shape recognition. This allows, eg, occupant sensing in addition to imaging. Thus the original concept of video surveillance used in buildings that was based on CCD cameras can be now extended to include occupancy sensing, localization, tracking, identification, and recognition.

Fig. 4.3-16 Fingerprint sensor (module developed by DELSY Electronic Components AG)

4.3.10
Acknowledgements

The author acknowledges useful discussions with his co-workers at the Fraunhofer Institute of Microelectronic Circuit and Systems in Duisburg, Germany. Supporting material was provided by A. Bussmann, J. Santos Conde, M. Hillebrand, C. Nitta, J. Huppertz, and N. Stevanovic.

4.3.11
References

1 FISCHETTI, M. A., HORGAN, J., WALLICH, P., *IEEE Spectrum* **22**, May (1985) 5.
2 SHER, N. C., in: *VLSI Electronics: Microstructure Science*, Vol. 9, EINSPRUCH, N. G. (ed.); New York: Academic Press, 1985, Ch. 10.
3 WECKLER, G., *IEEE J. Solid-State Circuits* **2** (1967) 65.
4 THEUWISSEN, A. J. P., *Solid-State Imaging with Charge-Coupled Devices*; Dordrecht: Kluwer, 1996.
5 FOSSUM, E. R., *Laser Focus World* June (1993) 83.
6 ACKLAND, B., DICKINSON, A., in: *Digest of Technical Papers at the IEEE International Solid-State Circuits Conference, San Francisco*; 1996, p. 22.
7 SCHANZ, M., BROCKHERDE, W., HAUSCHILD, R., HOSTICKA, B. J., in: *Proceedings of European Solid-State Circuits Conference, Southampton*; 1997, p. 236, Paris: Editions Fontières.
8 SCHANZ, M., BROCKHERDE, W., HAUSCHILD, R., HOSTICKA, B. J., SCHWARZ, M., *IEEE Trans. Electron Devices* **44** (1997) 1699.
9 FOSSUM, E. R., in: *Digest of Technical Papers at the IEEE International Electron Device Meeting, Washington, DC*; 1995, p. 17.
10 SCHANZ, M., NITTA, C., ECKART, T., HOSTICKA, B. J., WERTHEIMER, R., in: *IEEE Workshop on CCDs and Advanced Image Sensors, Nagano, Japan*; 1999, p. 17.
11 LYNNWORTH, L. C., in: *Sensors – A Comprehensive Survey*, Vol. 7, GÖPEL, W., HESSE, J., ZEMEL, J. N. (eds.); Weinheim: VCH, 1994, Ch. 7.
12 ROYER, M., MIDAVAINE, T., in: *Sensors – A Comprehensive Survey*, Vol. 6, GÖPEL, W., HESSE, J., ZEMEL, J. N. (eds.); Weinheim: VCH, 1992, Ch. 9.
13 TEUNER, A., HILLEBRAND, M., HOSTICKA, B. J., PARK, S.-B., SANTOS CONDE, J. E., STEVANOVIC, N., in: *Proceedings of the 10th International Conference on Image Analysis and Processing, Venice*; 1999, p. 1124, Los Alamitos: IEEE Computer Society.
14 PARK, S.-B., TEUNER, A., HOSTICKA, B. J., in: *Proceedings of IEEE at the International Conference on Image Processing, Chicago*; 1998, p. 967, Los Alamitos: IEEE Computer Society.
15 STEVANOVIC, N., HILLEBRAND, M., HOSTICKA, B. J., TEUNER, A., in: *Digest of Technical Papers IEEE International Solid-State Circuits Conference, San Francisco*, 2000, p. 804, J. H. Wuorinen, Castine (ME), USA.
16 GRASS, A., CARLEY, L. R., KANADE, T., *IEEE J. Solid-State Circuits* **26** (1991) 184.
17 WANGLER, R. J., OLSON, R. A., *Laser Focus World* July (1993) 105.
18 ZITTLAU, D., MENGEL, P., BOVERIE, S., *VDI Ber.* Nr. 1471 (1999) 213.
19 MILLER, B., *IEEE Spectrum* February (1994) 22.

4.4
Load Sensing for Improved Construction Site Safety
Peter L. Fuhr, *San José State University, San José, CA, USA*
Dryver R. Huston, *University of Vermont, Burlington, VT, USA*

4.4.1
Introduction

In April 1987, the L'Ambiance Plaza in Bridgeport, Connecticut, USA collapsed, killing 28 construction workers. This disaster was certainly tragic, but it is even more regrettable because it might have been prevented. The collapse of the plaza, which was 60% complete at the time, was blamed on the lack of temporary lateral bracing in the formwork. Unfortunately, this was not an isolated incident. In May 1994, two workers were killed and others were seriously injured when the shoring system being used in a refurbishment of bridge abutments on the SkyWay in Toronto, Ontario, Canada collapsed. Speculation as to the cause of the formwork failure is focusing on an improper footing for one key support member. On a broader time-scale, over the past 25 years, there have been more than 85 collapses of structures under construction that have been directly attributable to formwork failure [1].

In general, the primary causes of formwork failures are (1) excessive loads, (2) premature removal of forms or shores, and (3) inadequate lateral support for the shoring members. If a shoring system does not have sufficient strength and stability to carry the imposed loads, the result can be disastrous. Consequently, determining the load distribution during construction is essential in assessing the safety of the system [2]. A number of attempts have been made at developing computer models intended to evaluate the safety of a formwork structure [3–5]. However, an inherent limitation of these programs is that they cannot anticipate unusual or extreme circumstances that may occur at the site. Discussions with construction site supervisors reveals that many of the parameters used in the computer models do not accurately represent the reality of shoring support configurations. Furthermore, it is known that construction crews do not always follow the shoring system design. Hence a computer model may not correctly represent the loads and the resulting shoring load distribution present at the actual site [1].

A viable alternative to computer modeling is the real-time measurement of the actual formwork loads. This is accomplished by installing load cells in the shoring system. Load cells are available from a number of vendors, and are relatively inexpensive (in fact, the expense of the devices may be offset by a decrease in the cost of liability insurance). Data from several of the cells can be collected and analyzed on-site. The load information can then be used to alert the work crews to any hazardous situations, and allow sufficient time to correct the problem, and can be logged for load documentation and future shoring design purposes.

A potential use of the logged data would be in the development of wind-resistant shoring design codes. At the present, US Building codes offer little guidance for wind load design [6, 7]. A load cell data system would provide valuable information (such as wind speeds and direction versus the change in loads in the formwork), that can aid in formulating such guidelines.

The work described here investigates whether it was possible to detect (potentially dangerous) changes in load distribution of construction framework. Laboratory measurements have been taken in controlled situations, which included weak supports, soft footings, and uplifting of the supports, to ascertain the applicability of instrumented shoring at realistic construction venues. Extensions of this project into the 'real world' have included installation of load cells in the shoring of the following new constructions in the USA: (1) a seven-story courthouse in Burlington, Vermont (1995); (2) a maximum security prison in West Virginia (1996); (3) a parking structure in Burlington, Vermont (1996); (4) a four-story wooden/masonry building being renovated in Burlington, Vermont (1996–97); and (5) twin 26-story buildings currently under construction in Boston, Massachusetts.

While the prior laboratory and construction site data has provided valuable fundamental and practical information [8, 10], this article will focus on the twin tower activities due to a new type of shoring system used and the overall scale of the project.

4.4.2
Equipment and Data Processing

With the need for load and tilt measurements, each shoring member sensing 'platform' consists of a strain gage-based load cell, signal amplifiers, and a tilt sensor. In the wired case, this instrumentation is cabled to a data acquisition station which consists of additional signal amplifiers and conditioning circuitry and a Pentium-166 microcomputer with multichannel A/D capabilities. The load cells (Geokon Model 3000-300-3.5) are each capable of measuring loads ranging from 0 to 1 557 500 N. Multiconductor shielded electrical cables were constructed to attach the load cells to the signal amplifiers. The signal amplifiers are Vishay 2310 strain gage conditioning amplifiers. One of the key features of the amplifier is the adjustable calibrated gain. This allows the load cells to be sensitive and stable over a wide range of loads. The signal from the load cell is conditioned and amplified to a voltage between 0 and 5 V. The tilt sensors are from Applied Geomechanics, and allow for angular measurements ranging from -10 to $+10°$. An Analog Devices AD590 temperature sensor is also used to measure ambient temperature.

The overall data acquisition rate, in terms of cycling through each load and tilt channel and performing signal conditioning and assessment, is approximately 37 complete cycles per second. Specifically, it was determined that, at maximum throughput and processing, all channels would be sampled at a 10 kHz rate. Subsequent conditioning and interpretation take approximately 1.3 ms for the tilt sensors and for the load cells approximately 2.2 ms per channel plus an additional 16 ms for load history trend analysis. This gives a complete system response time of approximately 27 ms. We have compared this sensor system response time with actual shoring failure times that we have generated using our test assembly. We have found that the onset of collapse is indeed gradual whereas the collapse itself is swift. The sensor head components are robust and capable of quick response times. It is our belief that, given the nature of most collapses, this system provides very ample room for the capture of real-world failures.

4.4.2.1
Calibration

Calibration was performed by initially loading each of the load cells individually and then setting the gain on the Vishay amplifier. Each load cell was compressed slowly up to a maximum of 20 000 pounds (88 960 N) using a Tinius-Olsen mechanical press. This range was chosen based on the anticipated load levels to be found at various construction sites. The load cell voltage and the electrical output from the Tinius-Olsen's internal sensor were recorded. Once the load reached 20 000 pounds, the gain on the amplifier was set to provide a computer-displayed voltage reading of 5 V (±0.001 V). Any maximum load can be used by merely changing the gain on the amplifier. Hysteresis associated with each sensor is measured and minimized using computer-based signal conditioning and processing.

4.4.2.2
Sensor Head Configuration

The design concept led to the following realization: each shoring member to be instrumented will be outfitted with a small donut-like metallic attachment. This 'sensor head' is positioned between two pieces of interlocking shoring. The sensor head had the load cell positioned such that the shoring member's load is transferred to the load cell, where the load level is measured, and then mechanically transferred on to the lower shoring member.

From a measurement and identification standpoint, there are once again a number of issues that must be addressed in the design and fabrication of a tangible system. The 'alarm' feature of this system is, as previously stated, to measure the loads on each instrumented shoring member, extract the load distribution for the shoring members, then compare these values (and/or a time-tagged longitudinal record of the shoring load levels) with preset limits and observable trends to identify the onset of potentially hazardous conditions. Based on these comparative values, an alarm signal is generated. The alarm has two modes: in Mode 1, a simple high brightness light-emitting diode (LED) which is housed within the sensor head is illuminated when appropriate; in Mode 2 the LED is illuminated when appropriate and an audible warning signal is sounded. In each case the measurement values along with the onboard processor's determination of a warning status is conveyed via radio methods to a microcomputer which has been placed in proximity to a construction site official.

4.4.2.3
Wireless Communication Components

The difficulties presented in stretching cables in a construction site environment have led us to use a low-cost code division multiple access Spread Spectrum application-specific integrated circuit (ASIC) manufactured by Axonn Corporation. Devices and technologies currently exist which enable the designer to produce CDMA transmitters which are low cost, high performance and compact in size. The AX602 ASIC performs the processes and calculations required for chipping and data modulation, sleep counter and watchdog maintenance, and CRC calculations as well as rf circuit control, voltage multiplication, and low battery detection. When combined with a low-cost microcontroller, rf circuitry, and a small number of discrete components, a full performance direct sequence spread spectrum transmitter operating over the commercial temperature range is used to radio the information from the sensor platform to the data acquisition station.

Our current prototype instruments have exhibited a maximum range between sensor platform and base computer of approximately 800 ft (this value is our achieved distance in the presence of typical construction site-electromagnetic interference) and when within the metallic superstructure of an (in this case) old tenement housing. We envisioning being able to extending that range using the AX602 with an improved rf transmitter and embedded antenna.

4.4.3
Laboratory Work

The characteristics of the instrumentation and the operational procedures were initially evaluated in the laboratory. This included initially calibrating the load cells and physically configuring a signal conditioning and data acquisition system (and test protocols) which would be suitable for operation within the rough environment of a typical construction site. During the course of the work calibration experiments were conducted that demonstrated the use of the sensing system to detect a weak shoring member in a shoring system. Through these laboratory experiments of simulated potential failure mechanisms for shoring systems, it was determined that the instrumented shoring system is capable of differentiating between 'good' and 'bad' footing conditions and 'strong' and 'weak' shoring members through the monitoring and interpretation of the load distribution throughout the shoring. These findings have been reported previously [8–13].

Of particular importance is uplift occurring on shoring members. If this occurs with shoring on a construction site, the results can be catastrophic. If the shoring members were not connected together with cross-pieces and this occurred, one or more of the shoring members could collapse (as is believed to have happened at a major roadway bridge near Toronto, Ontario in May 1993 in which a number of workers were injured and killed).

An illustration of up-lift is depicted in Figure 4.4-1 where a beam is shown resting on supports 1, 2, and 3. The beam can be considered as an evenly distributed load (as shown in (a)). This situation is statically indeterminate because a portion of the total load being supported by 1, 2 and 3 cannot be calculated from static equilibrium alone. The situation changes when an additional load is applied between two of the supports such as between 1 and 2 (as shown in (b)). While the three supports may still be supporting the load, this fact cannot be verified using statics. As the load increases, the beam deflects (as shown in (c)), resulting in 1 and 2 supporting the entire load. The loads at 1 and 2 may now be calculated using statics and the deflection at 3 can be calculated using beam flexure theory.

The apparatus used for this laboratory experiment is shown in Figure 4.4-2. Three load cells were placed in a line with a single steel plate resting across the top of the load cells. A ball bearing was placed on the plate between loads cells 1 and 2 so as to produce a unidirectional moment-free point load that acts in the vertical direction. The load was applied with a Tinius-Olsen 267 000 N (60 000 lb) test frame.

As shown in Figure 4.4-3, as the applied load increased, each load cell experienced an increase in load. Once the load in each load cell reached approximately 37 800 N, the load in cell 3 decreased rapidly to zero. The beam was bent at this point with all of weight supported by cells 1 and 2. On a construction site, this situation can be particularly dangerous, since the unloaded (uplifted) shoring member may shift and become unstable. After a period of time, the applied load was then decreased, and load cell 3 then again measured an increase as shown in the graph. What is therefore apparent from these measurements is that the instrumented shoring system is capable of detecting uplift. When compared with the previously described experiments, it is apparent that uplift has a distinctly differ-

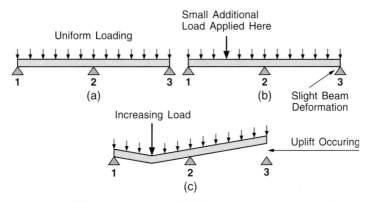

Fig. 4.4-1 Uplift occurs as an uneven load causes certain shoring members to assume more of the load while other members may feel little if any load

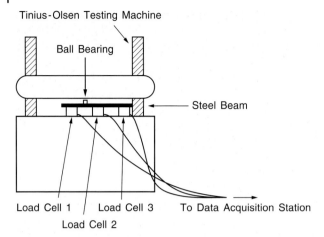

Fig. 4.4-2 Uplift simulation was performed in the laboratory using uneven loading by a press on to steel plates

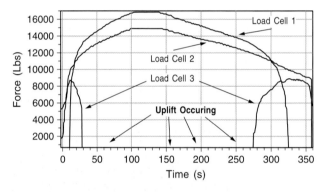

Fig. 4.4-3 Measurements taken during the simulated uplift failure tests show how certain load-bearing members must sustain higher loads, in the worst case loads that exceed the member's design criteria, while other shoring members have their loads diminish

ent 'load distribution signature' than for weak shoring members or weak footings. This situation may therefore be readily identified by the instrumented shoring system followed by an appropriate notification of the site engineers.

The mock shoring tests were designed to verify various collapse mechanisms for single- and multi-story structures. In essence they provided a controlled, laboratory environment where excessive loads and vertical and horizontal shearing forces could be applied to structures which possessed instrumented shoring members. In this way load and tilt data could be obtained as the structures were collapsed and subsequently compared with video and photographic records to determine alarm limits. The values obtained were then correlated with previously re-

Fig. 4.4-4 Data acquisition system for laboratory-based shoring measurements

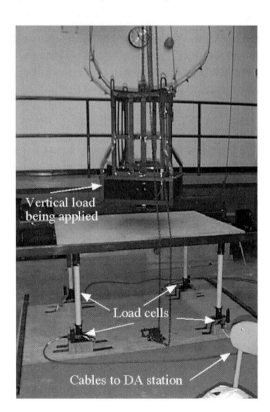

Fig. 4.4-5 The laboratory-based load-to-failure measurements used vertical loading while horizontal shearing loads were also applied

Fig. 4.4-6 This close-up of a single-stage shoring system shows the instrumented shoring members as well as the vertical load cage (weighing 365 pounds) being lowered on to the table via a crane

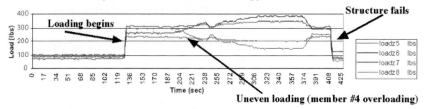

Fig. 4.4-7 This representative graph shows the temporal evolution of the loads and tilts varying as a vertical load increases while the horizontal shearing load of 275 pounds remains constant

ported measurements and theoretically obtained values to minimize uncertainty levels in structural failure alarm limits. Representative laboratory collapse setups are shown in Figures 4.4-4, 4.4-5 and 4.4-6. In this configuration vertical loads up to 1000 pounds were lowered via a crane assembly on to various locations along the structure while horizontal shearing loads of up to 600 pounds were applied. Load data and tilt measurements at two angles were simultaneously measured using the cabled and wireless measurement systems. A representative graph of this load-to-failure data is presented in Figure 4.4-7.

4.4.4
Uplift Monitoring

A final laboratory test simulated a condition referred to as 'uplift'. In these tests a two-story instrumented structure was designed and built. The structure was subjected to vertical loads which could be placed anywhere on the structure's top as well as variable horizontal side loads meant to simulate shearing loads. Load and tilt measurements were recorded with sensors cabled to the data acquisition microcomputer station. In these load-to-failure tests, the horizontal load level was fixed; in this particular instance a constant 275 pound shearing load was used, while the vertical load varied from 0 to 720 pounds. The vertical load was placed off-center purposefully to cause an uplifting condition.

As shown in Figure 4.4-7, as the vertical load was increased the two-story the measured loads increased until uplift began. At that time the load redistributed itself until certain shoring members began to fail causing an initial tilting of the structure (see Figure 4.4-8). Moments later the entire structure failed, as shown in Figure 4.4-9.

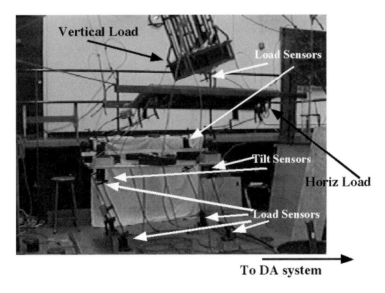

Fig. 4.4-8 As the two-story-instrumented structure collapses the loads continue to redistribute themselves unevenly among the shoring members still somewhat capable of supporting loads

Fig. 4.4-9 Moments later the entire structure fails. In this test condition the vertical loads as well as the structure's 'floors' are supported by numerous safety cables to prevent injury to the investigators

4.4.5
Field Measurements

As mentioned previously, field measurements were taken at five different sites. However, we wish to focus on the current effort, twin 26-story apartment towers with associated multilevel parking garages and a seven-story low-rise building in Boston, Massachusetts. This project, named the Museum Towers at North Point, is owned by Museum Towers LLC, and has Jung/Brannen Associates, serving as the architects, Weidlinger Associates serving as the structural engineering firm, and construction site logistics being handled by Beekman. Each of the $ 50–60M towers is to hold approximately 300 apartments which will rent in the $ 1500–2000/month range. This project constitutes the largest concrete-based building construction project under way in the USA.

For the past few years, particularly in the northeastern section of the USA, large buildings have been built of steel. However, such a steel structure would typically require 14–18 inches of floor separation whereas a concrete building requires a floor spacing of approximately 8 inches. When constrained to a maximum building height of approximately 230 feet, this means that 26 stories may be placed within the concrete building whereas only 23 or 24 floors could be placed into a steel structure. It is anticipated that this project's superstructure will be completed by late 1997.

4.4.5.1
Construction Site Specifics

As previously stated, this project involves the construction of two 26-story apartment buildings with an associated seven-story-attached low-rise apartment building and five-story parking garage. The Boston North Point region is located across from the Boston Museum of Science along the Charles River. Each building uses 8 inch-poured concrete slad floors (with rebar inforcements). A PERI shoring system is used with the shores on the outside portion (away from the internal central tower structure where the elevator shafts will eventually be) being placed on 9×9 ft center spacings. Eight load cells are used to measure loads from upper decks. This type of shoring requires that the poured upper level be carried by four floors of shoring. The concrete strength must reach a minimum of 1500 psi before the lowest level of shoring members may be removed. The required fully cured concrete strength is 4000 psi, a level that has been being achieved within 1 week; in fact the concrete is continuing to cure up to 6000 psi strength. A June 18 1997 photograph of the North Tower is shown in Figure 4.4-10. A closeup photograph of this building, more clearly showing the PERI shoring members, is shown in Figure 4.4-11.

Fig. 4.4-10 View of North Tower on June 18, 1997

Fig. 4.4-11 This closeup view of the North Tower more clearly shows the four levels of shoring members

Load cell readings are taken on the eight cells once per second (slow rate) or 10 times per second (fast rate). Fast recording is typically performed during the pouring cycle so that fast excitation and load variations may be recorded. After approximately 1 h after the end of the pour (a typical pour on this site lasts for 3 h, one entire floor at a time), the data recording reverts to the slow rate to look for longer term variations in the load distribution. An instrumented shoring member is shown in Figure 4.4-12.

4.4.5.2
Logistics of Field Site Work

The successful installation of test equipment and collection of data on an active construction site can be a complicated undertaking. Careful planning, cooperation, and coordination with the site engineer are of paramount importance. The goal was to get as much relevant data as possible without interfering with the construction.

The PERI shoring system used at the Twin Towers construction site requires that four levels of shoring be in place to support the upper floors. New floors are poured every 4–5 days. The lowest level of shoring cannot be removed until the concrete immediately above has reached a minimum of 1500 psi. Testing to see if the concrete has reached this hardness is performed using standard slump and cracking tests. In general it has been found that the concrete reaches 1500 psi within approximately 3 days; the expected performance criterion is 4000 psi, which has so far been achieved within approximately 1 week. This concrete has continued to harden and is currently reaching 6000 psi strengths.

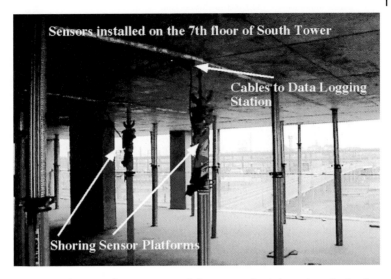

Fig. 4.4-12 This view of an instrumented shoring member is located on the South Tower's sixth floor

All of this concrete information implies that the lowest level PERI shoring members are being removed every 4–5 days. From our sensing standpoint, this means that we have sensors located on four levels with the lowest level varying every 4–5 days. Sensor installation follows the shoring installation. As the onsite construction crews have developed a better understanding of how to install optimally this type of shoring system, they are now able to remove shoring from one floor and reinstall it on another floor within approximately 45 min. This requires that we must be in close logistical contact with the shoring crew so that as they are installing the members to be instrumented we are ready for sensor installation. Once the members are in place, the cabling may be correctly placed.

From a logistical standpoint, two points are crucial for data collection and responsible activity at the construction site.

(1) Interference with the construction evolution is simply not acceptable (it is estimated that the labor costs for simply the concrete crew exceeds $ 3000/h). Close contact with the site supervisor must be maintained throughout the operation.

(2) To facilitate the quick installation and removal of testing equipment, everything was made manually portable. The equipment had to be arranged in a 'turn key' configuration, and the load cell-equipped shoring members must be interchangeable with standard PERI shoring posts. All equipment has to be rugged enough to endure rough handling as access to the site is difficult. Conditions on a construction site are not ideal for the implementation of computer-driven automated data acquisition, especially during the hot and humid summer months. In anticipation of this, we constructed a portable equipment container to protect the electronic gear from the construction site environment. This container was built using a high-density polymer waterproof case. An exhaust fan with a filtered in-

take port was installed to maintain a suitable temperature inside the box. A false bottom grating was included to keep the equipment out of the water in the unlikely event that the integrity of the box was compromised.

4.4.5.3
Site Data Acquisition

Sampling rates were initialized at ten samples per second. Since loading is dynamic during the first moments of the pour, a rapid sampling rate is necessary. Eight load cells and two tilt meters are sampled during this fast sampling period. Four hours after the pour the sampling rate is changed to once per second. Data are written onto a WORM optical drive.

Analysis takes two forms: in the first, the data are continually processed and correlated with known trends for shoring system failures. If a known failure trend is identified, alarm signals are issued. In the second form, the data are reviewed

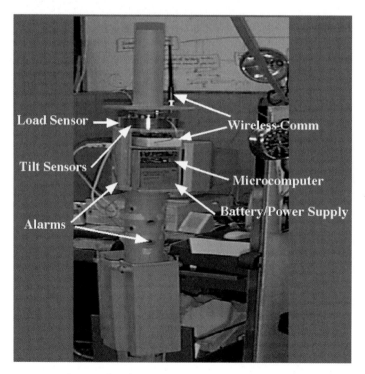

Fig. 4.4-13 The use of wireless communications and batteries for all power needs allows units to be quickly and easily installed throughout the construction site. This unit relies on spread spectrum communications in the 902–926 MHz band to achieve good signal fidelity in the presence of large of amounts of construction site-generated electromagnetic interference

Fig. 4.4-14 Street level view of the Museum Towers construction project on September 18, 1997

at daily-to-weekly intervals to examine loading histories and floor-by-floor comparisons of loading situations. Such data reviews are performed off-site.

4.4.6
Wireless Data Acquisition for Smart Shoring

One of the predicaments facing the real-world usage of such a shoring system is grounded in the worksite environment. In a conventional sense, power must be supplied to the load sensors and processing electronics and the data are carried to a microcomputer-data logging station, all via cables. The data logging system must be placed in a secure location typically in a more central location within the structure being built while the load sensors are attached to the predetermined shoring members which may be located on different floors potentially a long distance from the data logger. This, of course, necessitates running heavy and heavily shielded multiconductor cables around the site (see Figure 4.4-13). Such cables must be judicicously placed in lower traffic areas with the hope that they will not be broken or tampered with. Our experiences have shown that simply cabling the sensors is a laborious task, one that is compounded by the all too fre-

Fig. 4.4-15 View of the North Tower as seen from the top of the South Tower. The four floors of PERI shoring are readily visible

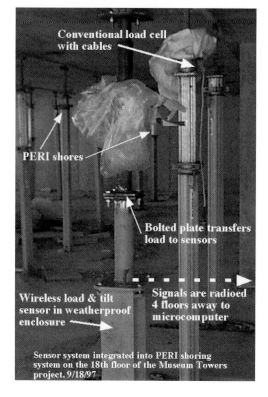

Fig. 4.4-16 View of the wireless load and tilt sensor installed on the 18th floor of the North Tower of the Museum Towers project

quent requirements to move the data logging station to a higher floor, which causes the cabling to be removed and re-run.

We have developed a battery/solar-powered standalone load, tilt, and temperature sensing system that directly attaches to the shoring member. The information is then transmitted via a wireless communication link to the data logging station. All cabling needs are removed, as are most requirements for almost any moving of the data logging station (in one case the station has remained in the main site trailer). While the electromagnetic environment associated with a construction site is 'noisy' (principally due to welding and large motors operating), our use of a spread spectrum wireless communication channel operating in the 902–926 MHz band minimizes data corruption and loss of signal integrity.

The wireless shoring sensing platform relies on a microcomputer (the Little Giant by Z-World) to perform routine data acquisition and general housekeeping of the platform (battery status, etc.). On-board processing of the load and tilt information is compared with preset limits which when exceeded causes alarm lights and buzzers to be activated on that particular platform, while, of course, this hazardous condition alarm message is radioed to the data logging system.

4.4.7
Field Use and Representative Data

Time-tagged data were recorded at this construction site using the wireless tilt and load monitoring system as well as the conventional cabled load cells. In each case, data were recorded at a centrally located microcomputer workstation. Specifically, a Pentium 166 class microcomputer was used to data log the analog readings. The maximum spacing between the conventional load cells and the workstation was three floors. Effectively, no matter where the workstation was located the stretching of multiple heavily shielded cables proved at best cumbersome and at worst potentially dangerous to the workers. Alternatively, the wireless battery pow-

```
Data Sets Found: 28
===================================================================================
Set#            Start Date      Average Tilt      Average Tilt  Average     Battery
                                (X-axis)          (Y-axis)      Load
0               12:37:55        16                81            1500        120
1               12:38:28        18                80            2255        116
2               01:16:00        11                83            1626        114
     .                .              .                 .             .  .
     .                .              .                 .             .  .
     .                .              .                 .             .  .
26              05:16:00        15                84            1056        107
27              05:56:00        14                86            1069        128
===================================================================================
09/19/97
09:46:48
End of Processing
===================================================================================
```

Fig. 4.4-17 The wireless unit's two-axis tilt readings are integrated with the vertical load readings, battery, and time information to provide a more complete picture of the performance of the vertical shoring members

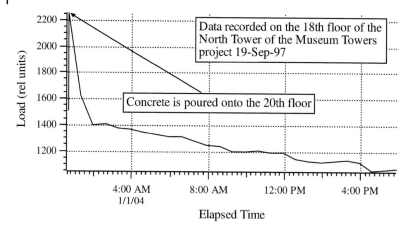

Fig. 4.4-18 The measured load for the PERI shoring member instrumented with the wireless unit is shown before, during, and after a large concrete pour occurs

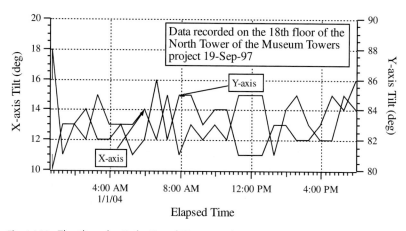

Fig. 4.4-19 The tilt angles in the X- and Y-axes are shown as a function of elapsed time. These measurements were made concurrent with those shown in Figures 4.4-18 and 4.4-20

ered system amptly demonstrated its inherent flexibility in being used in a field setting. The unit was easily disassembled from standing PERI shoring members, transported to another location within four stories (or approximately 600 ft) of the workstation, and then once again simply put into place on another shoring member. The Little Giant Z-class microprocessor within the wireless unit acquired readings from the tilt and load sensors, performed rudimentary signal processing and then radioed the readings to the workstation. There the information (data) was configured in the manner shown in Figure 4.4-17 for ease of subsequent processing. The workstation analyzed the data to look for the uplift and support member collapse data trends referred to in Section 4.4.2.

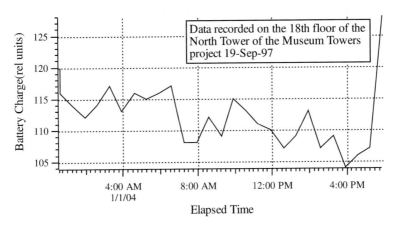

Fig. 4.4-20 The battery performance of the wireless unit plotted as a function of elapsed time

A representative graph of the wireless unit's readings for vertical load, two-axis tilt readings, and the battery, performance are shown in Figures 4.4-18, 4.4-19, and 4.4-20 as a function of elapsed time.

4.4.8
Conclusion

The purpose of the effort described here was to develop an instrumented smart shoring system and ascertain its applicability in providing load distribution information for construction site engineers. The smart shoring system was based around the use of four conventional load cells plus one unit equipped with a wireless data transfer system placed on various structurally important load-bearing support members of shoring configurations. Initial laboratory calibration measurements were made on these load cells. In a more controlled laboratory environment, the smart shoring system was studied to see its performance when presented with three potential failure mechanisms related to shoring systems: (1) weak shoring members, (2) soft fittings, and (3) uplift monitoring. In each case it was observed that such potential failure mechanisms are readily visible upon looking at the load distribution data presented by the instrumented shoring.

From a performance standpoint we feel that these measurements show that in principle instrumented shoring members may be used at actual construction sites to provide load information not previously available. No occurrences of the potentially hazardous shoring failure mechanisms investigated in the laboratory occurred at the site (or in the data). In addition, by using the instrumented shoring load measurements along with extrapolated time constants based on the measured data, it may be possible to determine the state of cure. Such information may be of enormous value to concrete engineers who rely on tried-and-true prac-

Fig. 4.4-21 The completed Museum Towers (as of February 1998)

tices such as cylinder tests to determine the state of the cure cycle. If accurate information regarding the level of cure of the concrete may be predicted via these measurements, or more importantly a prediction of when the concrete will be strong enough to support extra weight, then the project manager (or other cognizant individual) may save considerable costs by knowing when work can continue on to the next level (instead of just waiting), or can avoid potentially hazardous situations by delaying the construction process. Finally, yes, the Museum Towers are standing and are occupied (Figure 4.4-21).

4.4.9
References

1 Hadipriono, F.C., Wang, H.K., *J. Construct. Eng. Manag.* **112** (1986) 112–121.
2 Chen, W.F., Mossallam, K.H., *Concrete Buildings: Analysis for Safe Construction*, 1991.
3 Sbarounis, J.A., *Concr. Int. Des. Construct.* **6** (2) (1984) 70–77.
4 Liu, X.L., Chen, W.F., Bowman, M.D., *J. Struct. Eng.* **3** (1985) 1019–1036.

5 LIU, X. L., LEE, H. M., CHEN, W. F., *Concr. Int. Des. Construct.* (1988) 21–30.
6 RATAY, R. T., in: *Proceedings of the NSF/Wind Engineering Research Council Symposium on High Winds and Building Codes*, Kansas City, MO, November 2–4, 1987, pp. 431–436.
7 RATAY, R. T., *J. Aerospace Eng.* 2 (1989) 102–107.
8 AMBROSE, T. P., HUSTON, D. R., FUHR, P. L., DEVINO, E. A., WERNER, M. P., *J. Smart Mater. Struct.* 1 (3) (1994)
9 HUSTON, D. R., FUHR, P. L., AMBROSE, T. P., DEVINO, E. A., WERNER, M. P., *Proc. SPIE* 2191 (1994) 408–419.
10 HUSTON, D. R., FUHR, P. L., AMBROSE, T. P., in: *Proceedings of the 1996 ASCE Structures Congress*, Chicago, IL, April 1996.
11 ROSOWSKY, D., HUSTON, D., FUHR, P., CHEN, W.-F., *Concr. Int.* 16 (11) (1995) 21–25.
12 HUSTON, D. R., FUHR, P. L., ROSOWSKY, D. V., CHEN, W.-F., VAN LAAK, P., *Proc. SPIE* 2719 (1996) Paper 10.
13 HUSTON, D. R., FUHR, P. L., in: *Proceedings of the 1996 ASCE Structures Congress*, Chicago, IL, April 1996.

5 Maintenance and Facility Management

5.1
Maintenance Management in Industrial Installations
Jerry Kahn, *Siemens Energy & Automation, Peachtree City, GA, USA*

5.1.1
Introduction

The economic realities of today are forcing increased demands on the operation and maintenance of industrial facilities. Managers must continually seek out new ways to optimize their processes. The drivers are:

- increasing global competition;
- increased emphasis on up-time;
- decreasing operations and maintenance budgets;
- decreasing numbers of maintenance personnel;
- decreasing availability of in-house expertise, as a result of early retirements and attrition.

In the past, unscheduled outages to repair critical equipment resulted in lost income; now the result can also be lost market share. Repair and maintenance time must be minimized and optimized within production schedules to maximize plant availability. This must be accomplished with a maintenance work force reduced both in numbers and knowledge [1].

To meet these challenges, maintenance managers are relying more heavily on automatic data acquisition to provide the information needed to evaluate equipment states. New sensor technologies, for acquisition of equipment data, combined with new information technologies, to get the data to where it is needed, are becoming the backbone of the modern maintenance organization. Current equipment conditions can be readily evaluated and used to forecast future conditions using predictive maintenance techniques.

5.1.2
Predictive Maintenance and Condition Monitoring

Predictive maintenance (PdM) relies on the periodic monitoring and analysis of equipment conditions to predict future equipment performance [2]. The periodic monitoring is termed condition monitoring (CM). Various sensors (eg, pressure,

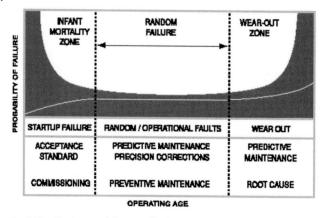

Fig. 5.1-1 Equipment failure profile (courtesy: Computational Systems Inc.)

temperature, vibration, flow) and fluid (eg, oil or air) samples are employed to obtain indications of system and equipment health, performance, and integrity. Trend analysis is employed to determine the optimum time when necessary maintenance must be performed in order to avoid equipment failure equipment. By continuously gathering data and assessing a piece of equipment's operating condition, progressive deterioration of components can be inferred without the necessity for equipment dismantling for physical inspection. A piece of equipment's failure profile, as characterized by the patterns of equipment reliability (Figure 5.1-1), can be modified [3]. Problems can be diagnosed and corrected early in the life cycle, and maintenance schedules can be better synchronized with plant operation.

The condition monitoring techniques used for detection include:

Tab. 5.1-1 Application of condition monitoring technologies

Applications	Vibration	Acoustic/ ultrasound	Lubricant	Infrared thermography	Process monitoring	Electrical testing	Sensory (visual)
Turbines	•	•	•	•	•		•
Generators	•	•	•	•	•	•	•
Motors	•	•	•	•	•	•	•
Pumps	•	•	•	•	•		•
Conveyors	•		•		•		•
Heat exchangers	•	•		•	•		•
Valves		•		•	•		•
Electrical systems				•	•	•	•

- vibration analysis;
- acoustic and ultrasonic monitoring;
- lubricant analysis;
- infrared thermography;
- process parameter monitoring;
- electric testing;
- sensory inspection.

The application of these technologies to the monitoring of industrial equipment is shown in Table 5.1-1. A brief survey of each these sensor-based monitoring technologies follows.

5.1.2.1
Vibration

Vibration is used to monitor and analyze displacement, velocity, and/or acceleration in rotating or reciprocating equipment (Figure 5.1-2). Such equipment includes steam and gas turbines, pumps, motors, compressors, rolling machines and mills, and gearboxes. Utilizing this technique, problems such as bad bearings, poor alignments, or improper balance can be identified. Each rotating or oscillating machine exhibits characteristic vibration frequencies, which are directly related to its geometry and operating speed. The impact of moving parts and imbalances causes shifts in the characteristic frequency spectra, which can be correlated to the defect type. This includes conditions such as:

- cracks, pits and roughness in bearings or rolling elements;
- imbalance in rotating parts;
- misalignment, eccentricity, and bends of shafts;
- misaligned or damaged teeth in gear boxes;
- deterioration of turbine wheels and pump impellers from erosion or corrosion.

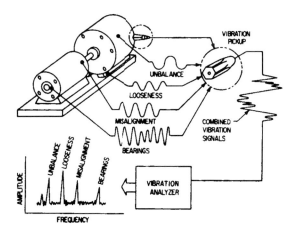

Fig. 5.1-2 The vibration monitoring process [4]

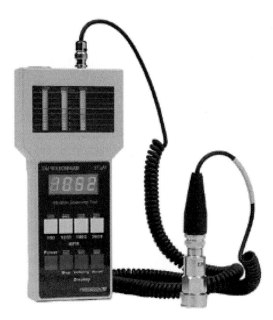

Fig. 5.1-3 The DLI Watchman ST-101 vibration screening tool (courtesy: Predict /DLI)

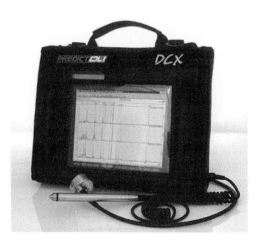

Fig. 5.1-4 The DLI Watchman DCX portable vibration analyzer (courtesy: Predict/DLI)

Good valid initial-test date is essential in order to establish a baseline for future monitoring.

A vibration analysis system usually consists of five basic parts: signal pickup(s), a signal-recording device, a signal analyzer, analysis software, and a computer for data analysis and storage. These basic parts can be configured as either a continuous or sampling on-line system, or a periodic analysis system using portable measurement and diagnostic equipment. Typical commercially available systems are shown in Figures 5.1-3 and 5.1-4.

5.1.2.2
Acoustic and Ultrasonic Monitoring

These techniques monitor energy emissions from sources. Acoustic monitoring covers the spectrum form about 20 Hz to 20 kHz; ultrasonic is used at higher ranges. Since most machines emit consistent sound patterns (sonic signatures) under normal operation, changes in these signatures can indicate a deteriorating state. The applications in predictive maintenance are as follows:

in the active (reflective) mode:
- measurements of material thickness;
- tracking rates of erosion and corrosion;

in the passive mode:
- leak detection;
- location of flow disturbances in process piping;
- coronal discharge;
- early stage bearing degradation.

Evaluating long-term trends in these signatures can also identify poor maintenance practices, such as improper bearing or seal installations.

5.1.2.3
Lubricant Analysis (Tribology)

Tribology is the science and technology of two interacting surfaces in relative motion and of related subjects and practices [5]. Historically, this 'science of lubrication' has been limited to following guidelines of the original equipment manufacturer. Machines were lubricated on set schedules using prescribed oils and greases, with little concern for optimizing either maintenance time intervals or the lubricant itself. Tribology analysis was slow and expensive. The advent in recent years of modern sensors and microprocessors has automated many of the analytical processes and with considerable cost reductions.

Several techniques are used for predictive maintenance:
- lubricating oil analysis;
- wear particle analysis;
- spectrographic analysis;
- ferrography.

Lubricant analysis is performed on a periodic basis to determine the condition of the liquid film that separates the rotating machine parts. Key parameters such as viscosity, acidity, oxidation, water content, and contaminant level give indications of lubricant breakdown or degradation. Wear particle analysis focuses on the particulates present in the lubricant sample. The size, shape, number and related characteristics of these particles provide indications of the internal condition of a machine. Since wear particles formed from different wear processes usually have different shapes, the type and location of a wear problem can be determined. Wear particle analysis can be augmented using spectrographic analysis to detect the presence and composition of trace elements in the particles. Ferrography uses a magnetic field to separate contaminants prior to visual, microscopic, or other analysis. It can be used for analysis of particles in the 10–100 µm range.

5.1.2.4
Infrared Thermography

Thermographic analysis of infrared images is used to detect abnormally hot or cold areas (or spots) on or around equipment. Since faulty electrical components will almost always generate heat before failing, thermography can be used as effective condition monitoring and predictive maintenance tool. The most commonly used instrument is the infrared video camera, which measures the infrared emissivity of the target object (Figure 5.1-5). Differences in surface temperature are highlighted on the produced images (Figure 5.1-6).

Fig. 5.1-5 Connector in normal light (courtesy: Logos Computing)

Fig. 5.1-6 Connector – infrared image (courtesy: Logos Computing)

5.1.2.5
Process Parameter Monitoring

A comprehensive predictive maintenance program must include monitoring of routine process parameters in order to identify and eliminate process inefficiencies. For example, vibration monitoring can be used to provide indications of a pump's mechanical condition, and infrared thermography can be used to analyze the condition of the motor that drives the pump. However, neither will indicate whether or not the pump is operating efficiently. A prescribed maintenance activity may fix a wearing bearing, but may not increase system efficiency as the root cause of the problem was outside the maintenance boundary. Only when the condition monitoring data are considered along with the process can the correct maintenance activity be taken to increase efficiency, quality, and production.

5.1.2.6
Electrical Testing

Evaluation of electric motors and equipment is critical to the predictive maintenance in industrial settings. Vibration (and thermography) can be used to identify and pinpoint problem spots in critical equipment. Motor current signature analysis (MCSA) can provide measurements of instantaneous load variations [6]. With this technology, all currents and voltages of a poly-phase motor are captured in both time and frequency domains. By comparing simultaneously obtained motor current signatures and mechanical vibration signatures, very early indications of wear, imbalance, and other degradations, which may have not been evident from the individual monitoring technique, become apparent. Long-term trends can identify both improper operating conditions and maintenance practices. Other techniques for electric apparatus condition monitoring are given in Table 5.1-2.

Since electrical motor failures can result from problems external to the motor, it is important to utilize motor circuit analysis techniques as well in electrical condition monitoring.

Intelligent motors are being developed by Status Technologies, a business unit of Computational Systems Incorporated, in conjunction with US Electric Motors

Tab. 5.1-2 Condition Monitoring and measurements for electrical equipment

Measurement technique	To detect
Resistance to ground testing	Insulation degradation
High-potential testing	Insulation integrity
Surge compression testing	Grounds and shorts
Motor current balance analysis	Impedance mismatch
Motor flux analysis	Rotor bar or stator faults
Normalized temperature analysis	Mechanical and stator faults
Coronal discharge monitoring	Arcing

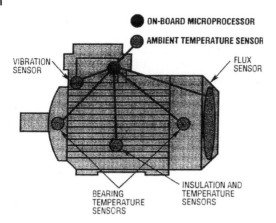

Fig. 5.1-7 Sensor placement on an intelligent motor

[7]. The motor has embedded sensors to monitor vibration, temperature, flux, voltage, and winding deterioration (Figure 5.1-7). Power comes from the main power supply to the motor. Data from the sensors are collected and stored on a central on-board processing unit, which is also used to record stars and stops, load profiles, and related lifetime data.

5.1.2.7
Sensory Inspection

Sensory inspection refers to the visual, listening, smelling, and touching capabilities of the human observer. These are the oldest condition monitoring techniques in existence, and ones that are difficult to mechanize. An experienced maintenance technician is often able to detect developing patterns that are precursors to failure, such as trends in color or texture changes, which are gages of deterioration. Periodic walkdown inspections are an invaluable supplement to a hardware sensor-based condition monitoring system.

5.1.3
Enhancing Condition Monitoring with Expert Systems

Many on-line and off-line sensors monitor a complex production line or industrial facility. In a large number of cases, both process and equipment condition data are needed to address the full range of potential failure modes for critical machinery. This requires an integrated condition assessment system. As equipment wears and degrades, the condition assessment system determines the current equipment condition utilizing a rule-based evaluation of all relevant sensor data. When this real-time data are combined with a knowledge base that includes historical equipment performance data and definitions of failure mode symptoms, logic algorithms can be applied to infer future equipment states. The result is an

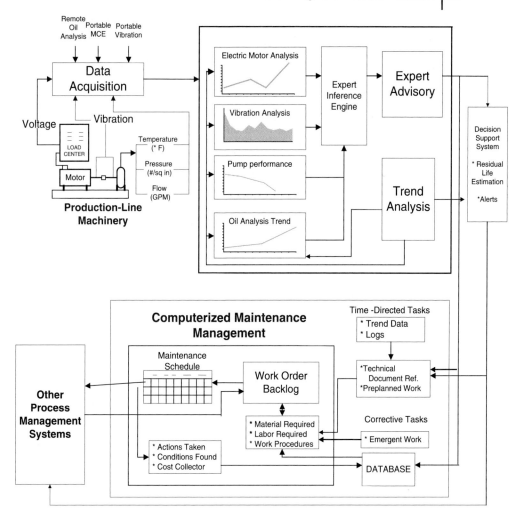

Fig. 5.1-8 Real-time expert system (courtesy: DEI Group)

expert system [8]. A flow schematic for such a real-time expert system is depicted in Figure 5.1-8. The core of the system is an expert inference engine, which deduces needs for maintenance and/or enhanced vigilance, and provides this advice to the maintenance engineer. If there is an indication of pending catastrophic failure, a request for maintenance can be automatically initiated to the computerized maintenance management system (CMMS).

CMMSs are used to track maintenance costs and equipment repairs. This is accomplished by monitoring equipment repair work orders, spare parts inventory, and preventive maintenance schedules. Management is supplied with the necessary information to keep track of maintenance budgets.

5.1.4
Integration with Plant Systems

Expert systems are being integrated into overall plant asset management (PAM) systems, which in turn are parts of broad plant information systems (Figure 5.1-9) [9]. By compiling and automatically assessing data from all aspects of operation, the PAM systems allow plant personnel to assess the impact of maintenance on production schedules. This can be done for individual product lines or for entire facilities. Operations and maintenance personnel are provided with the information necessary to make optimum asset use, risk-minimizing decisions for production and maintenance strategies. Gateways are available to provide information to other modules, such as plant automation systems (PAS), enterprise asset management (EAM) systems, and enterprise resource planning (ERP) systems.

For these integrated systems, there must be accurate detailed information on each equipment asset down to the sensor level. This is normally done in register modules, which contain key information such as:

- asset nameplate data and location;
- transducer type and location;
- transducer characteristics;
- post-processing algorithms.

The Machinery Information Open Systems Alliance (MIMOSA) currently publishes a universal set of open electronic exchange conventions, which unify ma-

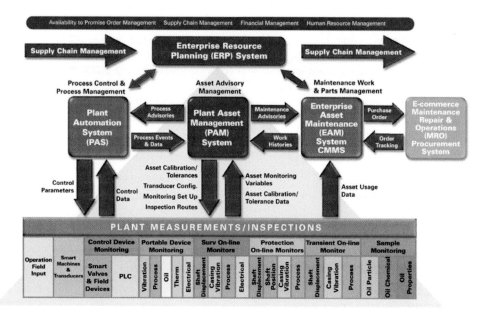

Fig. 5.1-9   Integrated plant information system (courtesy: Entek)

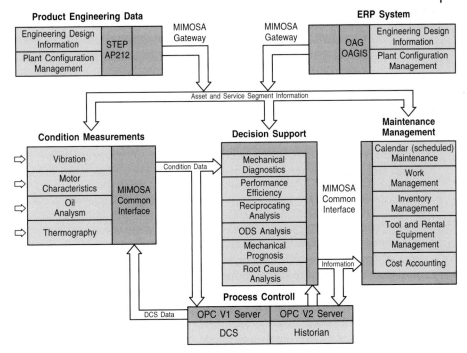

Fig. 5.1-10 MIMOSA open information exchange model (courtesy: MIMOSA)

chinery data taken from digital control systems, condition monitoring systems, computerized maintenance management systems, and operational logs in to order facilitate creation of comprehensive machine information management systems. A MIMOSA open system is one in which equipment asset software components are able to communicate and exchange data automatically without any proprietary or supplier-specific interface protocols.

Figure 5.1-10 depicts the current MIMOSA open systems information exchange model for product engineering data, ERP, control, condition monitoring, maintenance, and production systems [10].

5.1.5
Maintenance Management Methods

Predictive maintenance, condition monitoring, and sophisticated information systems are only effective when they are incorporated into a complete maintenance management program. The two most prominent programs are reliability-centered maintenance (RCM) and total productive maintenance (TPM).

5.1.5.1
Reliability-centered Maintenance (RCM)

Moubray defines RCM as 'a process used to determine the maintenance requirements of any physical asset in its operating context' [11]. RCM emphasizes cost effectiveness and relies heavily on the use of condition monitoring to establish equipment conditions and maintenance needs. The goal of RCM is to preserve the system or equipment function, not just to maintain operability. Statistical correlations are established for equipment, based on failure data, and probabilities of failure are established as a function of equipment age. When combined with equipment condition data from condition monitoring systems, an appropriate maintenance plan is established for each piece of equipment in an industrial setting.

Under the RCM concept, equipment maintenance falls into one of four categories:

- breakdown maintenance, or run to failure; this is usually applied to non-critical low-cost equipment, or where redundant systems are in place;
- preventive maintenance, based on periodic time-directed equipment maintenance (eg, lubrication);
- predictive maintenance based on condition monitoring; and
- proactive maintenance, which utilizes the techniques of root cause failure analysis (RCFA) and failure modes and effects analysis (FEMA) either to improve condition monitoring techniques such that impending failure can be predicted earlier, or to redesign the equipment to be less susceptible to failure.

The methodology of RCM involves systematic analysis and consideration of [12]:

- the functions of the system, subsystem and components;
- the failure modes of each function;
- the importance associated with each function and its failure; and
- identifying and prioritizing the preventive maintenance tasks which will reduce failure occurrences and are cost effective.

When used correctly, RCM results in significant improvements in maintenance effectiveness.

5.1.5.2
Total Productive Maintenance (TPM)

TPM builds upon the elements of RCM with the goal of maximizing equipment effectiveness at all times. Emphasis is placed upon overall quality, availability, life cycle costs, and continuous improvement of maintenance techniques and equipment design. The key element to a TPM program is total participation of employees at all levels within the organization, from top management to the operator on the factory floor. Employee involvement is facilitated by:

- instituting cross-functional work teams who work to reduce or eliminate downtime and quality, and to improve operability and maintainability; these teams are comprised of representatives from operations, maintenance, design, and other departments as required;
- empowering the teams to make routine operating and maintenance decisions; and
- providing a comprehensive training program in maintenance and operational skills.

There are eight pillars for implementation of TPM, the first five of which relate to production systems, and the last of which focus on management systems:

- utilization of project teams and small work group activities who focus on enhancing equipment efficiency (Kobetsu-Kaizen);
- establishment of a system whereby individual operators carry out routine inspections and maintenance of equipment (Jishu-Hozen);
- establishment of planned maintenance activities (corrective, preventive, and condition-based maintenance) by specialized maintenance personnel or by outsourcing;
- development of operation and maintenance skills (training);
- establishment of an initial-phase management system for equipment, which includes life-cycle costing and designing equipment for ease of maintenance;
- establishment of a quality maintenance system to create, maintain, and control the conditions for 'zero defect' equipment operation;
- establishment of efficiency improvement systems for administrative and supervisory departments (eg, purchasing, logistics, information technology);
- establishment of safety, health and environmental systems to maintain 'zero accidents' and 'zero pollution'.

Successful implementation of a TPM strategy also requires a disciplined approach to knowledge management. The effective capturing, development, and exchange of manufacturing knowledge are required to maintain and continuously improve a production line. This knowledge can be passed on via expert systems, product data libraries, and the employees themselves [13].

The effectiveness of a TPM program is measured by the overall equipment efficiency (OEE):

OEE = equipment availability × throughput rate (performance) × quality rate

Typical OEEs are under 60%. In well-run TPM programs, OEEs of over 85% have been achieved. For such operations, the Japanese Institute for Plant Maintenance (JIPM) bestows its 'Awards of Excellence' [14].

5.1.6
Future Directions in Maintenance Technology

The convergence of sensor and information technologies will soon be applied to maintenance management in several areas:

- wireless and smart sensors;
- human sensory sensors;
- e-maintenance on the World Wide Web.

5.1.6.1
Wireless and Smart Sensor Development

Wireless sensors will eliminate the need for the expensive cabling that is now required for real-time on-line condition monitoring. In fact, one such product is already available. Computational Systems is marketing the first fully integrated wireless sensor for use in harsh environments [15]. The RBM Consultant RF microanalyzer (Figures 5.1-11 and 5.1-12) monitors vibration and temperature, and utilizes the RF spread spectrum between 2.40 and 2.48 GHz to communicate with a transceiver up to 100 m away.

The use of smart-sensor networks has been in place for some time in the process monitoring area (Profibus, DeviceNet, etc.). However, the application to condition monitoring has lagged, primarily owing to the large volumes of data necessary to process real-time vibration or thermography data. Work is under way to develop a fourth-generation vibration sensor at Wilcoxon Research that will provide the necessary attributes of a smart sensor [16]:

- bi-directional command and data communication
- local digital processing and transmission;
- preprogrammed and user-defined algorithms;
- internal self-verification and diagnosis;
- on-board data and command storage.

Fig. 5.1-11 RBM consultant wireless rf microanalyzer (courtesy: Computational Systems Inc.)

Fig. 5.1-12 Internals of the wireless rf microanalyzer (courtesy: Computational Systems Inc.)

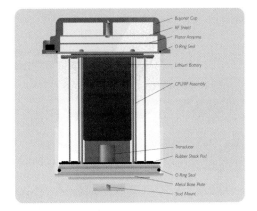

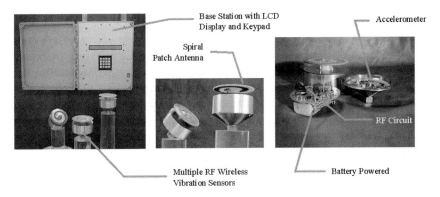

Fig. 5.1-13 RF smart sensor (courtesy: Wilcoxon Research)

The sensor currently under development is wireless, and utilizes a small internal battery (Figure 5.1-13).

5.1.6.2
Human Sensory Sensors

Until now, there has been no reliable way to automate the human sensory (hearing, vision, and smelling) methods of condition monitoring. However, with the recent introduction of the Human Interface Supervision System (HISS), such monitoring is becoming a reality [17]. The HISS technology has been developed jointly by Siemens, DaimlerChrysler Aerospace, Sennheiser Electronic, and BEB. HISS reproduces the functions of the human sense organs in a technical system. Information on movements, noises and smells is processed and analyzed by sensors that use self-learning logic (Figure 5.1-14).

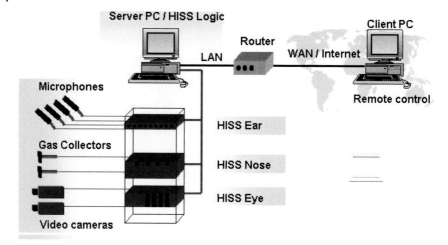

Fig. 5.1-14 HISS sensory system (courtesy: Siemens AG)

The HISS Seeing component consists of video cameras linked to an image-processing system, which carries out the image analysis. If a change occurs to the image, a message is passed on to the server. Image records can be made at adjustable sampling rates for future analysis and historical files. In the future, this technology could also be applied to infrared thermography.

The HISS Smell system relies on an array of sensors to monitor changes in concentrations of gas mixtures. The metal oxide semiconductor (MOS) sensors do not identify single substances, but rather recognize olfactory patterns, as does the human nose. An olfactory fingerprint is created, which can be analyzed and classified by modern pattern recognition techniques.

The HISS Audio (Hearing) component is based on the optical microphone technology developed by Sennheiser. Because of its small size and robust features, it can be employed in areas where direct human supervision is not possible, or in widely dispersed monitoring applications. The audio signal range (up to 24 kHz) encompasses that of the human year. As with the Seeing system, deviations from standard patterns are recognized and alerts are made via a link to remote servers. Neural fuzzy systems are utilized to distinguish and filter out ambient spurious noise.

Data are retrieved from these three sensor components and compiled by a HISS Logic model. Here comparisons are made with changes in process parameters to test the validity of fault signals from the sensing units. Event-related databases are continuously compiled, from which the system can be taught to learn new situations.

5.1.6.3
E-Maintenance via the World Wide Web

To date, critical machines have been monitored locally by on-line data collection systems. Even with expert automated fault diagnosis software, in-house experts are required to provide baseline information and interpret data. In the future, this task will be done off-site, utilizing a combination of smart machines sensors and external networks, such as the World Wide Web. A small staff of experts will be able to provide diagnostic services from a remote central service center, reducing on-site staff requirements. The company Predict/DLI, located in Bainbridge Island, Washington, USA has developed such a system called SmartMachine. The SmartMachine system utilizes three individual servers to provide information to networked clients [18]:

- a Data Server for spectral and process data;
- an Information Server to provide information to clients; and
- an OPC (OLE for Process Control) server to communicate with plant control and other systems.

The tools are built around Active-X and other Web technologies, so that the system size can range from a single industrial installation to a network of machines residing around the globe (Figure 5.1-15). Mimic displays provide the analyst with a graphical interface to monitor system performance, as well as machine condition information, and can allow actual remote controlling of monitoring systems.

5.1.7
Summary

Industrial maintenance in the twenty-first century will be governed by information exchange. Seamless exchange must occur between sensors, mobile devices,

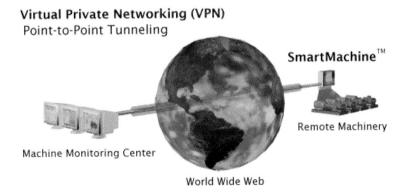

Fig. 5.1-15 Condition monitoring via the World Wide Web (courtesy: Predict/DLI)

computer networks, and via the World Wide Web to remote monitoring centers in order to maximize the benefits of condition monitoring and predictive maintenance. Sensor systems are becoming more integrated, more intelligent, more human-like in their sensing capabilities, and less reliant on wired connections. Real-time, comprehensive expert systems are needed to digest sensor data and provide maintenance management with quality knowledge to optimize asset productivity.

5.1.8
References

1 MARCHI, J. F., JR., in: *Proceedings of the Maintenance and Reliability Conference, Knoxville, May 10–12, 1999*; Knoxville, TN: University of Tennessee, 1999.
2 MOBLEY, R. K., in: *Maintenance Engineering Handbook*, Higgens, L. (ed.); New York: McGraw-Hill, 1995.
3 WISHNAFSKI, P. A., *The Road to RBM: Implementation of a Reliability-based Maintenance Program*; Knoxville, TN: Computational Systems, 1997.
4 NICHOLAS, J., YOUNG, K., *Predictive Maintenance*, 2nd edn.; Millersville, MD: Maintenance Quality Systems, LLC, 1999.
5 BHUSHAN, B., in: *The Engineering Handbook*, Dorf, R. (ed.); Boca Raton, FL: CRC Press, 1996.
6 JAMES, R., in: *Proceedings of the 11th Annual Predictive Maintenance Technology National Conference, Atlanta, November 15–18, 1999*; Minden, NV: SC Publishing, 1999.
7 BOWERS, S. V., in: *Technical Presentations from the CSI RBM University, 1999*; Knoxville, TN: Computational Systems, 1999.
8 FLOYD, C. A., paper presented at Main Tech South 1998, Houston, December 2–3, 1998.
9 BEVER, K., *Maintenance Technol.* July – Aug. (2000) 20.
10 WILLIAMSON, M. K., *Practicing Oil Anal.* Nov.–Dec. (1999) 7.
11 MOUBRAY, J., *Reliability-centered Maintenance*, 2nd edn.; New York: Industrial Press, 1997.
12 ZWINGELSTEIN, G., in: *Proceedings of the Maintenance and Reliability Conference, Knoxville, May 20–22, 1997*; Knoxville, TN: University of Tennessee, 1997.
13 DEWISSE, J., *Nat. Productivity Rev.* Summer (1999) 39.
14 SHIROSE, K., *Total Productive Maintenance – New Implementation Program in Fabrication and Assembly Industries*; Tokyo: Japan Institute of Plant Maintenance, 1996.
15 LINEHAN, D. J., in: *Technical Presentations from the CSI RBM University, 1999*. Knoxville, TN: Computational Systems, 1999.
16 BROOKS, T., *Machine, Plant Syst. Monitor* May–Jun. (1999) 15–18.
17 SIEMENS AG ATD – Advanced Sensor Technology. *ATD Insight* June 2000: 10–11.
18 LOFALL, D., *Diesel Gas Turbine Worldwide* Apr. (2000) 24–27.

5.2
WWFM – Worldwide Facility Management

ROLF REINEMA, *GMD Institute for Secure Telecooperation (GMD-SIT), Darmstadt, Germany*

5.2.1
Introduction

The advent of modern information and communication technology has changed work processes and contents of work significantly. Work can be and is being done at any place and at any time. Offices become merely places for face-to-face exchange. Workplaces are no longer set for a specific person but are flexibly assigned as required. Future work environments will be characterized by a high degree of diversity, flexibility, and dynamics.

Local work will be supplemented more and more by tele-cooperative collaboration, and cooperation within organizations will be supplemented by cooperation across organizations (Figure 5.2-1). The traditional local office work will go together with various forms of fixed and mobile tele-work. The work environments of the future have to constitute cooperative landscapes, in which information tech-

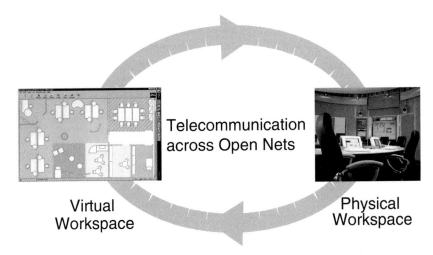

Fig. 5.2-1 Cooperative rooms

nology and physical work environments are highly integrated and can dynamically be reconfigured in response to alternating uses and different work situations [1].

There is more focus on meetings and teamwork today and offices are more and more becoming important meeting points, not just because of mobile and telework. Yet, the design of workspaces, especially of physical work environments in offices and buildings, has remained almost unchanged. Present work spaces are not very well prepared for teamwork, nor are they well equipped for integrating the IT infrastructure needed for work processes with the infrastructure needed for managing and operating the building (facility management).

To facilitate the evolution of new forms of work and organization at a higher level of quality, it is necessary to integrate modern information and communication technology (hardware, software, networks) with innovative spatial and physical structures. Challenges that architectural design faces include more flexible and adaptable building structures, dynamic reconfigurability of offices and work places, computer-based models for planning, operating, and lifetime maintenance of buildings.

This section introduces the need for worldwide facility management (WWFM), which is motivated by the realization that modern forms of work and organization require frequent and dynamic reconfiguration of physical work environments and easy adaptation of rooms to varying requirements. Part of the article describes the need for smart devices, which can be easily installed in any room and give access to a full set of services for that room. With the introduction of JINI (JAVA Intelligent Network Infrastructure), the development of a distributed system and therefore the usage of interconnected devices and services become more simplified. Integrated security features have to provide means for the secure identification and authentication of people, eg, entrance and access rights have to be defined by means of role-based access control.

5.2.2
Facility Management

In the course of fundamental technological and organizational changes at today's workplaces, new requirements on buildings and their facilities are being made. Modern information and communication technologies are strongly influencing work places and office structures. Status-afflicted assignment of office space is more and more being replaced by a more task-related usage. The dynamics of changing places of work require different structures and facilities and therefore ask for a greater flexibility in the building structure as well as the technical infrastructure.

The availability and usage of sophisticated telecommunication technologies liberates people from their place of work. The traditional form of office work will be substituted more and more by fixed and mobile tele-work in the future. People are working purely result-oriented. They are being freed from presence duties and status symbols. Their location is determined by their own initiative and self-organization. With respect to their needs they are changing freely between places, which allow concentrated work, offer access to information required, support team and project work, offer entertainment and relaxation, etc.

Present-day office buildings often house a larger number of companies and organizations, and thus must be capable of adapting the floor plan, technical facilities, and organization to changing uses and changing tenants on short notice and at no great expense. Owing to increasing economy pressure, the use of office space will become more diverse and intense in the future. Today, buildings are being managed with different systems on different hierarchical levels, which are often neither well coordinated among each other nor with respect to the needs of their users.

This makes it necessary that buildings take over services, which are currently rendered by traditional facility management systems, information systems, and various specialists. For the building and its management this means breaking with the traditional services and technologies. Hence integrated information and building management systems are being needed, which provide efficient and productive access and usage of the facilities within the provided physical workspaces. It is also required that these services be offered to users in an integrated and intuitively usable form.

Facility management (FM) is the practice of coordinating the physical workplace with the people and work of the organization. It integrates the principles of business administration, architecture, and the behavioral and engineering sciences. Although the facility management profession has been in existence for many years, only in recent times has it received worldwide recognition. Business entities have come to realize that maintaining a well-managed and highly efficient facility is critical to success.

The term facility management is often used to refer to a variety of activities ranging from control of facilities within individual rooms up to the management of buildings, business plants, real estate, etc. Here, the management of facilities in buildings and rooms is mainly being addressed, which is traditionally backed by (more or less centralized) systems and databases.

Within the COR R&D program at GMD, the WWFM project takes the above requirements into account as it focuses on the user, his/her work contexts and usage scenarios by providing a unified and worldwide accessible facility management platform based on open standards. Within the ongoing research work an approach from the side of data communication and data networks is being followed, such as the Internet/Intranet. It is based on the availability of ubiquitous universal networks (such as the Intranet/Internet) and inexpensive computing power (eg, provided by embedded systems) [2].

5.2.3
Worldwide Facility Management (WWFM)

The WWFM project integrates physical workspaces (rooms) and physical objects into virtual work environments. This is motivated by the realization that modern forms of work and organization require desk sharing by different users, frequent and dynamic reconfiguration of workplaces and easy adaptation of rooms to varying requirements. Offices need to be equipped with information and communication technology, devices to be located, rooms to be reserved, furnished, set up for a specific meeting, or re-set to a previous state.

The WWFM project develops strategies to link up buildings, which differ in construction, technical infrastructures, work, and organization forms, by means of unified, state-of-the-art technologies such that they can be organized and run worldwide like a single building. Moreover, WWFM investigates how users could implicitly take over typical functions of facility management themselves by organizing their personal work in the context of a virtual office.

The main objective is to make it possible to monitor, control, and manage both old and new buildings on a unified worldwide platform, irrespective of any particular local technology. WWFM creates the preconditions for locating, adjusting, and administering physical objects such as rooms, equipment, and other resources in a given work context. It implements a distributed application, which provides room services such as:

- control of light, heating, ventilation, air, and climate;
- communication facilities such as telephone, fax;
- Internet/Intranet access;
- reservation of rooms and required resources;
- localization of persons and equipment within rooms and buildings;
- organization of maintenance and house keeping;
- accounting and billing.

The first approach was to develop a device, which can be easily installed in any room and gives access to a full set of services for that room: the so-called *roomServer* (Figure 5.2-2). With roomServers, a facility management system would simply be a network of rooms, each of which has its own roomServer, linked together via the Intranet or Internet to form a worldwide facility management system. An expansion to include more rooms and resources can easily and inexpensively be done by adding roomServers and connecting them to the Intranet or Internet.

5.2.4
The RoomServer

In the old days, room control simply meant a couple of switches. Today's definition of room control encompasses a surplus of advanced pieces of environmental and

Fig. 5.2-2 The roomServer

multimedia equipment ranging from, eg, lighting control systems to VHS players, LCD projectors, and others. Of course, simply equipping a room with advanced electronics is not enough. Making a room that is easy to use and versatile enough to keep pace with quickly evolving technology has become the ultimate goal.

The roomServer is specifically designed to meet this goal. It has an advanced and simple to use graphical user interface. Its openness and expandability allow technological advances to be incorporated quickly, and to be mastered easily by everyone. It is an embedded system and as such offers unprecedented chances to administer distributed physical work environments of people from virtual ones and vice versa.

Since each individual room has a roomServer of its own assigned to it, this makes it possible to control autonomously a room and all its devices in an efficient way. Several of these roomServers can be networked via the Intranet/Internet, which makes it possible to administer a set of rooms, or buildings, and to cluster them into a virtual building under a unified application view. All the steps required to set up the appropriate work environments and to establish the necessary network links between distributed users are automated in a way that communication and collaboration in virtual teams is just a mouse-click away.

The roomServer is an autonomous installation unit (typically located in the doorframe of the room), which is connected to the Intranet/Internet. It is based on a modern hand-held PC with a touch-sensitive LCD display as local interface and has integrated loudspeakers and a microphone. A smartcard reader and so-called single-chip PCs are being attached to it.

The roomServer provides a flexible distributed framework architecture (Figure 5.2-3) implemented by means of JAVA and Web technology, which can easily be tailored to the needs of its users. Through a standard Web browser, any roomServ-

5 Maintenance and Facility Management

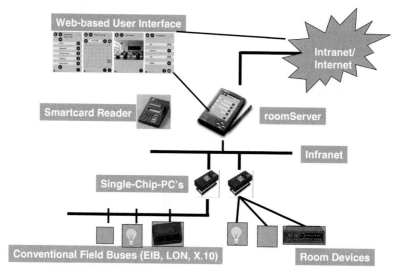

Fig. 5.2-3 Overall architecture

er may be accessed remotely in the same way as locally. The Web-based user interface shows a simple and easy to understand, graphical user interface (GUI). Each of the icons on it leads to one of the software modules that make up the system. When an icon is selected, either an action is immediately initiated, or a control panel opens to allow the selection of options.

Some sample screenshots of the user interface are shown in Figure 5.2-4. For example, users can control room devices such as blinds (open, close, move up and down). A calendar is provided so that rooms (eg, meeting rooms) can be reserved. The panoramic view is a very intuitive user interface element. Users can easily select the room devices with which they want to interact by simply pointing to them on a full 360° panoramic photograph (see Figure 5.2-4).

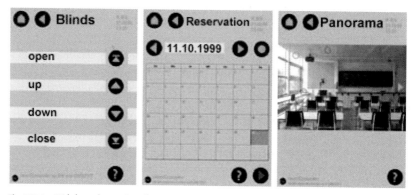

Fig. 5.2-4 Web-based user interface

5.2.5
Single-chip PCs

The different functions in a single room (eg, light, climate, heating) are controlled via so-called single-chip PCs, which are connected to the roomServer via a local Ethernet. These single-chip PCs are small innovative and low-cost devices, which combine all components of a PC on just a single chip. They provide digital and analog I/O interfaces, serial interfaces, and timers.

These days the (traditional) field-bus world is being turned on its head. Recently the 'new' Ethernet field bus has been the subject of discussion in building automation systems. Ethernet is not new, of course. It is a comparatively old technology that has been tried, tested, and used in millions of applications and is also extremely practical. However, Ethernet recently seems to have gained in popularity in building automation technology.

As a communication protocol, TCP/IP has become widely accepted. It is available and may also be used in building automation technology, so that the same infrastructure could be used for operation and control of room devices as well as for exchanging data with facility management systems or for accessing the functions and services of room devices from any computer on the Intranet or the worldwide Internet, wherever this may be required.

Connecting the single-chip PCs to the Ethernet allows access to all room devices connected to it by using the TCP/IP protocol. All that a roomServer connected to the same network needs to know is the IP address of a specific single-chip PC. A roomServer can then download programs to it and activate them, and monitor variables and thus the status of room devices. Data can be sent from the roomServer to a single-chip PC, which can be used by all types of operators in the single-chip PC, eg, inputs, outputs, flags, registers, timers, and counters.

The single-chip PCs use a specific TCP/IP stack together with the necessary software to realize a Web server. Of course, this is not an Internet server with megabytes of memory and simultaneous access by thousands of users as is required with Internet providers. On the contrary, a small Web server is provided with a maximum of about 5–10 simultaneous accesses and, as the 'disk space' access to the I/Os as well as, if appropriate, flags or registers of a subordinate software program.

The services provided by the single-chip PCs (eg, switching a light on or off) can then be accessed from the roomServer via a standard Web browser by using HTML, CGI scripts or JAVA programs. Service providers for underlying devices and networks translate the high-level operations on the single-chip PCs into corresponding commands sent to the connected room devices.

The advantage to the developer is that the way an application accesses a room device is always the same, regardless of the underlying protocol that is used to talk to the device. By shielding applications from the underlying heterogeneity of building automation components and protocols and providing a Web-based interface for controlling room devices, the single-chip PCs will help to get applications that use room devices developed quickly and easily.

5.2.6
Advanced Architectures

The deployment of roomServer networks and smart controllable room devices will allow the development of compelling new facility management applications that move us closer to the vision of the 'cooperative building' that enhances users' comfort, convenience, and ease of use. Every day more and more electronic devices are being introduced into buildings and home environments. At the moment most of these devices are isolated and can only provide the specific services for which they were designed.

However, one problem in developing such applications is that there are numerous competing building automation protocols (eg, X.10, LON, EIB) and different application programming interfaces (APIs) for accessing room devices. Since building automation systems are likely to remain heterogeneous, with multiple incompatible protocols used to discover and control different types of devices, application programmers are faced with the prospect of using multiple protocol-specific APIs with no common underlying programming model.

In addition, electrical installations in rooms and buildings are becoming more and more complex these days. In the past it was often sufficient to switch devices on and off. Today developers of facility management applications have to deal too much with the protocol-specific details of the room devices and networks that application uses. Furthermore, when new room devices are being installed, developers have to modify their application and/or install the appropriate drivers to use them, ie, there is no standard software infrastructure to shield applications from these low-level details.

To control the different devices in a room (such as lights, blinds, and heating), there are at present many conventional island solutions, which require great efforts with wiring (with a large number of cables) and configuration. Moreover, these systems often cannot be installed, configured, and managed independently of each other. Changes in the usage of rooms and buildings often require rewiring and reconfiguration of existing technical components, which results in great effort and expense. In addition, in modern facility management systems it is nowadays unavoidable also to provide functions for operating, controlling, and managing as well as for reporting, billing, and accounting.

As part of the future consumer computing and communication landscape, ubiquitous computing, spontaneous networking, and invisible computing refer to the vision of billions of small devices connected with each other in order to cooperate intelligently. Ubiquitous computing generally refers to the idea of many network-connected devices and software services working together simply and reliably. By connecting devices and software services to a network, and allowing them to be combined in some way, much more powerful services can be provided. Today, one of the most prominent technologies to realize this version is Sun's JINI.

The usage of JINI (JAVA Intelligent Network Infrastructure) is intended to provide a set of services that will make it easy to develop applications that discover and control room devices.

With JINI [3], the development of a distributed system and therefore the usage of interconnected devices and services becomes more simplified. JINI brings to the network the facilities of distributed computing, network-based services, seamless expansion, reliable smart devices, and ease of administration. Connecting a JINI-capable device (such as the single-chip PCs or the roomServer) to the network, switching it on, and starting a JINI service allow direct access to this service or device from virtually everywhere in the Intranet/Internet.

The usage of JINI results in the roomServer appearing as a set of distributed services. Its interfaces are always present, simply and uniformly, regardless of how they are implemented or where they are physically located. Thus, adding new devices and services to the roomServer simply means plugging them in. JINI simplifies not only connecting the single-chip PCs to the roomServer, but also the interconnection of the roomServers themselves (eg, to manage a set of rooms or buildings). JINI makes it possible to create a powerful infrastructure of interconnected distributed devices and services, which is very robust, flexible, and scalable.

5.2.7
Security Aspects

The need for confidentiality and privacy is gaining more and more in importance, particularly in open networks such as the Internet. Up to now security aspects have been largely ignored in facility management (FM) systems. Little by little they are coming into the consciousness of system operators, designers, and manufacturers. Connecting the various facilities within an office building to open nets (eg, the Internet) is playing a rapidly increasing role in today's FM systems. Security problems are especially evident when field buses are being linked to local area networks (eg, Ethernet). In the most field bus standards, security aspects are hardly foreseen, which in principle involves risks. New solutions are required here. In this section an overview of the security requirements of FM systems and of applicable security functions will be given.

In FM systems, which are connected to open networks, the security technologies must cover different aspects of security, such as confidentiality, integrity, and authentication. Therefore, different modules of security mechanisms must be available. In general, FM systems (like any other distributed system) have the following security requirements:

- access control to prevent unauthorized access to a device or service;
- authentication to confirm the identities of the communicating partners;
- data confidentiality to protect data against bugging and to provide traffic flow confidentiality;
- data integrity to protect data against loss and manipulation;
- non-repudiation to provide proof of origin and delivery of data.

The basic building blocks meeting these requirements are encryption, authentication, certification, and integrity preservation. Relevant methods to fulfill these security requirements are as follows:

- secret key encryption (the most commonly used methods are DES (data encryption standard) and IDEA (international data encryption algorithm), both operating on blocks of 64 bits length);
- public key encryption (eg, RSA (Rivest, Shamir, and Adleman) and Diffie-Hellman);
- hybrid encryption (combination of the above two methods);
- consistency checking.

The integration of security functionality can be built up on two different layers:

- security in the transmission or networking layer, ie, security is already provided by the networking protocol used (eg, Secure Sockets Layer (SSL)); an additional data manipulation by security applications is not necessary;
- security in the data layer, ie, before data are transmitted from a sender to a receiver they will be manipulated by the appropriate security functions in the application. The security functionality can either be applied to the application, or the application itself is designed to gain security for other programs, eg, the Secure Shell (SSH) [4].

One of the drawbacks of network layer security mechanisms is the need for secure underlying transport protocols, which are not available at present. IPv6 will provide this functionality in the near future. The advantage of data layer security is that the transmitted data can be subdivided into parts with sensitive and insensitive data with respect to human perception. In comparison with providing security on the data layer, all transmitted data are protected in the network layer. This tends to give problems when transmitting huge amounts of data (which is rather untypical for FM systems). The network layer is not capable of subdividing the data stream into parts with a higher need for protection and parts with a lower or no need for protection. Implementing security functions in the data layer has the advantage that only some parts of the data need to be protected and so the amount of time spent on protecting them can be extensively reduced. A general survey of security and cryptographic methods is given in [5].

The security policies for FM applications are not focused on optimal protection of highly confidential data, but rather than on protecting data and services against illegal access. Therefore, the security methods needed here have to be fast, with respect to real-time requirements, and to be cheap to implement in order to supply an emerging market of embedded devices and systems. The expense to break into such a system needs not to be very high, but it should be more expensive than the legal access to the provided services. In all these FM applications, the cryptographic functionalities must cover different aspects of security, such as confidentiality, integrity, and authenticity. Therefore, different modules of encryption mechanisms must be available. An elegant way to combine these modules could be a scaleable security gateway, providing different security functionalities, adaptive to the requirements of specific applications and the properties of special forms of multimedia data.

As the services provided (ie, controlling the devices within a room or a building) could be accessed from any computer on the same Intranet or Internet, one must prevent users or computers from accessing such services without having the proper rights. Access to a single room or a building should only be granted under specific circumstances and with prior authorization (ie, services in a room must only be accessible via a defined gateway or proxy, which encapsulates them). Where possible, direct access to the facilities has to be prevented by applying standard procedures such as local 'Firewalls', mutual authentication and securing the communication channels (eg, by using SSL).

In addition to unauthorized access to the services, unauthorized access to a room itself must be prevented under certain circumstances. Therefore, measures for the secure identification and authorization of people have to be implemented. For this purpose, digital identity cards (DICs) can be applied as a secure means for identification, authentication, and authorization of people together with privacy-enhanced auditing, and location management.

A smartcard reader connected to the roomServer in combination with security policies implements means for controlling access to a single room (locking/unlocking doors) and access to the services provided by the roomServer (and also their appearance to the outside).

We have developed DICs which are based on the specification of identity cards for the public sector and government departments [6]. As a technological basis we chose dual-interface smartcards (offering two interfaces, one with contacts and the other contactless) with an integrated crypto-controller. They provide means for the secure identification and authentication of people, allow the secure storage of important data (eg, private signature keys), and the secure performance of crypto-graphical functions (eg, data encryption).

Basically, the DIC has the same functions as a paper-based identity card. For example, it contains information about its cardholder, his/her citizenship, his/her name, date of birth, department information, a unique card number, and period of validity. Furthermore, it carries digitized pictures of the cardholder and of his/her signature. In addition, authentication data are stored on the card. Such authentication data consist of the roles of the cardholder together with the implicitly assigned authorizations to such roles. These authorizations can be granted permanently or temporarily.

In addition to the pure identity card functions and the authorization information, the DIC provides signature functionality according to the DIN specification for digital signatures [7]. For example, this would allow a cardholder to access electronic documents from a roomServer and sign them by using his digital signature.

Location management will play an increasing role. In the FM context it splits up into two fields, location management for fixed and mobile devices (eg, locating a specific device within a building) and location management of persons (eg, locate team-mates in order to communicate with them). A specific security problem of location management is that of allowing the subjects of such a system to retain control over the distribution of the information about their location, eg, in order to prevent others from location tracking. Therefore, concepts for controlling personal reachability are being provided while maintaining a high degree of privacy and data protection.

The information about the location of a person is sent via the Intranet from the roomServer to the location information server (LIS). The LIS hands over the data to the personal security manager (PSM) of the respective person. The PSM of a person then stores the location information cryptographically so that only that person can read it. A request about the current location of a person within a building can be handled only via the LIS and the respective PSM.

5.2.8
Conclusion

It has been shown that new flexible forms of individual and joint work require future workspaces to be developed by bringing together information and communication technology, architecture, design, and management of buildings. Physical workspaces (rooms) and physical objects have to be integrated into virtual work environments and vice versa. This is motivated by the realization that modern forms of work and organization require desk sharing by different users, frequent and dynamic reconfiguration of workplaces, and easy adaptation of rooms to varying requirements. Offices need to be equipped with information and communication technology, devices to be located, rooms to be reserved, furnished, set up for a specific meeting, or re-set to a previous state.

Thus, a new interdisciplinary research and development program has been initiated by the GMD Institute for Secure Telecooperation (SIT), which is carried out together with partners from universities, research institutes, industry, and user communities. The so-called Cooperative Buildings (COR) research and development program is located at the intersection of information technology, work organization, and architecture. It takes an integrated and systematic approach to support various aspects of collaboration among distributed people in order to develop future work environments. It brings together researchers and practitioners from a wide range of disciplines who develop and use innovative work environments by making use of recent advances in information and communication technology, new concepts in work organization, and a comprehensive perspective on office building design.

Attention is focused on ways and means to enable and support flexible forms of communication and collaboration for a variety of groups ranging from small local teams to large distributed organizations, utilizing a seamless integration of physical and digital objects, of real work spaces, and virtual information spaces embedded in real architectural environments. The emergent need to integrate the physical workspaces with the virtual ones is particularly addressed by the WWFM project, which is carried out by GMD in close cooperation with development and application partners from industry and academia. Interested parties from user groups, industry, and research are invited to contribute to the COR R&D program and to use its results.

Although JINI has been around for more than a year now, the technology has not been deployed widely in FM. Nevertheless, the discussion and development of JINI technology continues at an even increasing pace. One of the reasons for the reluctance of industry to introduce JINI widely may be its relatively high computing requirements (full-blown Java virtual machine) and its lack of security features.

The implementation of PSMs and local and central location managers in conjunction with appropriate public key cryptography methods can serve to handle privacy protection, eg, to prevent unauthorized knowledge about where people are located and/or have been located in the past, the resources they are using and/or have been using in the past, etc.

5.2.9
References

1   BAHR, K., BURKHARDT, H.-J., HOVESTADT, L., REINEMA, R., Integrating Virtual and Real Work Environments. In: *Proceedings of IEEE SoftCOM'99, Split, Croatia,* October 1999.
2   REINEMA, R., BAHR, K., BANKLOH, M., BURKHARD, H.-J., SCHULZE, G., *Cooperative Buildings – Workspaces of the Future,* in: *Proceedings of World Multiconference on Systemics, Cybernetics, and Informatics (SCI 98), Orlando, Florida,* July 1998, Vol. 1, pp. 121–128.
3   *Why JINI Technology Now?;* Palo Alto, CA: Sun Microsystems, 1999.
4   YLONEN, T., *The SSH (Secure Shell) Remote Login Protocol;* http://www.cs.hut.fi/ssh/RFC, 1995.
5   SCHNEIER, B., *Applied Cryptography,* 2nd edn.; New York: Wiley, 1996.
6   *Teletrust TTT-AG2 DIA, Digital Identity Card Specification Version 3;* 1998 (in German).
7   *DIN NI 17.4, Specification of Interfaces for Smartcards with Digital Signature Functions,* 1998 (in German).

6 System Technologies

6.1
Sensor Systems in Intelligent Buildings
HANS-ROLF TRÄNKLER, OLFA KANOUN, *University of the Bundeswehr Munich, Institute for Measurement and Control, Neubiberg, Germany*

6.1.1
Introduction

Sensor systems constitute the essential element for automatic control in intelligent buildings and entail many advantages for the environment.

For the users of intelligent buildings, sensor systems contribute to the improvement of the quality of life through enhancement of comfort, safety, and security. Buildings become more adaptable for the individual needs of the users. The users feel more secure through system assistance by the control and alarming of security-relevant situations. Annoying tasks can be easily accomplished through the simple input of commands to the sensor actuator systems.

For the environment, the use of sensor systems in buildings implies an efficient management of resources. This especially refers to the energy consumption by heating of buildings, which is generally reduced by using insulating materials in the building structure [1]. This usually results in restrictions of thermal comfort and indoor air quality, and can lead to mold growth in the corners and behind furniture [2]. The use of sensor systems permits, in contrast, a reduction in the waste of energy through demand-controlled heating and ventilation processes. Thereby, a low energy consumption is achieved without any restrictions on thermal comfort or indoor air quality for the users.

For the wide application of sensor systems in intelligent buildings, developers are challenged to present novel solutions allowing the needed functions to be realized within the special requirements for this mass market. Different technical requirements depending on the specific application should be fulfilled regarding the necessary precision, resolution, reliability, and dynamic behavior.

Concerning the purchase costs, the ideal price of many sensors and components for sensor systems in intelligent buildings has still not been reached. Sensor systems are at present generally implemented in offices and industrial buildings, in which high security measures are essential owing to the prerequisites of assurance or legislation. Therefore, the market for sensor systems is still restricted at present and the costs of the components are generally high. For wider use, including the private home sector, the components should be competitive in price and performance in order to gain the acceptance of the users. This will contribute significantly to an acceleration of the trend towards decreasing costs.

The production and operating costs of the systems should also be critically considered. Especially a long lifetime and high reliability are indispensable for the reduction of maintenance costs and to disburden the user from any annoying effects during operation.

In the next section an overview of the applications and advantages of sensor systems in intelligent buildings is given. The requirements for sensor systems in intelligent buildings are formulated in Section 6.1.3.

In Sections 6.1.4 and 6.1.5 sensor systems for special applications including safety, security, convenience, and comfort are presented in detail. Most of them have been developed in recent research projects [3–7] and the novel achievements in this field are described. An overview of the main future trends in this field is presented in Section 6.1.6.

6.1.2
Sensor Applications in Intelligent Buildings

The applications for sensor systems in intelligent buildings can be classified in four categories, according to the benefits for the user (Figure 6.1-1). Sensor systems are generally used to fulfill the following requirements:

- control of the basic building mechanisms
- security and safety for persons and buildings

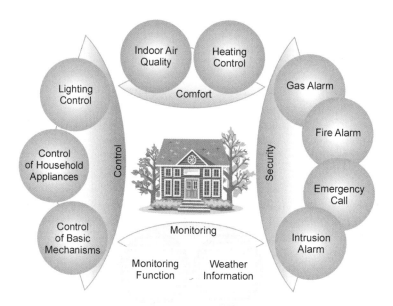

Fig. 6.1-1 Applications for sensor systems in intelligent buildings

- comfort and convenience for human beings
- monitoring of the important information for the user.

For the security of human beings and of buildings, sensor systems represent key components for the detection of dangerous situations. The detection of definite concentrations of combustible gases or gas mixtures is indispensable for the prevention of explosion risks. The simultaneous detection of smoke and temperature increase could signal an alarm for a fire in its early stages, so that major damage can be avoided.

The health and well-being of the residents of intelligent buildings depend significantly on the quality of the air and climate in the inside of the building. Sensor systems contribute significantly to the achievement of comfort for users by the realization of room occupancy-driven heating and ventilation.

Through the use of medical sensors, elderly and handicapped people can be more independent of the help of other people, so that the costs of special care or medical surveillance could be significantly reduced. Special sensor systems could be employed outside and inside the building in order to alarm intrusion or to manage the access of persons to the building.

In general, the use of sensor systems for safety can help significantly to avoid damage to persons and goods and allow users to have a feeling of security, which can enhance their quality of life.

The user would like to know about the situation in the building or outdoors. Therefore, sensor systems for monitoring weather information and running functions in the building are very important. User-friendly interfaces are therefore very important for the acceptability of sensor systems in intelligent buildings.

6.1.3
Requirements for Sensor Systems in Intelligent Buildings

The development of sensor systems for special applications requires the right choice of sensor devices, technology, signal processing, and network system. New developments of sensor systems consequently profit from the continuing technical progress in these fields.

For the development of sensor systems for intelligent buildings, essential prerequisites for acceptance by the end-user should be considered. Owing to the large number of sensors used for the automatic control of a whole building, consideration of costs and expenditure is important. The user does not like to be obliged to pay high costs for the procurement of the system or to be disturbed by annoying maintenance processes during operation. For this purpose, many aspects should be taken into consideration in the choice and development of the necessary components. Sensor systems should generally fulfill many requirements, such as:

- low costs
- small dimensions

- ease of installation (plug and play)
- easy and low-cost operation
- ease of retrofit in existing houses
- independence of particular manufacturers.

For the majority of applications, several sensors are simultaneously needed in order to realize one main function. Therefore, it is preferable to gather integrable sensors in multi-sensor modules (MSMs), which could be realized in microsystem technology, so that both manufacturing and installation costs could be significantly reduced [3].

Low-cost operation involves not only reduced power consumption, but also high reliability and long-term stability, which contribute significantly to a reduction in maintenance costs. The power consumption of sensor systems during operation can be significantly reduced through the use of microsystem technology, an intermittent operation mode and management of operating modes of individual system units [6].

6.1.4
Sensor Systems for Safety and Health

The feeling of security for users in intelligent buildings can enhance their quality of life. Sensor systems represent the key element for the detection of dangerous situations and helps significantly to avoid damage to persons and goods. With early detection of these hazards, corresponding alarms could be signaled and some special measures could be taken, eg, extinguishing a fire or closing windows. Figure 6.1-2 shows an example of a scenario of technical alarms in a smart home [3]. Depending on the expected danger situations, sensor systems should be conceived and placed in the building.

In implementation, some sensors could be used at the same time for different applications (Table 6.1-1). Therefore the sensors for different applications could be integrated together in MSMs, which help to reduce significantly the total costs of components and installation and maintenance of the system [3].

6.1.4.1
Fire Detection

Fire detection in good time can contribute significantly to the success of rescue operations and to limiting the degree of damage. In Germany, fires in private home lead each year to approximately 600 dead, 6000 injured, and damage costing about 800 million Euro [10].

The immediate reaction and the reliability of fire detectors are very important. Sensor systems in a building could provide information concerning the presence of persons and their health situation, so that rescue measures could be more effective.

Fig. 6.1-2 Scenario of technical alarms [8]

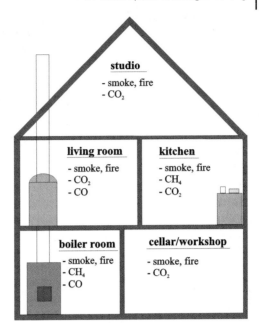

Tab. 6.1-1 Sensors for different applications in smart buildings [9]

Sensors for/applications	Air temperature	Air humidity	IR radiation	Gas detection
Heating/air condition	✓	✓		✓
Presence/intrusion			✓	
Gas monitoring	✓	✓		✓
Early fire detection	✓	✓	✓	✓

False alarms should be reduced in order to protect the user and the alarm reception offices from annoying effects. State-of-the-art fire detectors, such as optical (light scattering) and ionization detectors, are designed to perceive smoke particles and are therefore unable to differentiate between fire and disturbing event particles, such as cigarette smoke and disco-fog particles [9]. Statistics in Switzerland on data collected during 5 years (between 1992 and 1997) shows that only 11% of fire alarms were real [11]. This is due to the increasing sensitivity of up-to-date commercial fire detectors, in order to be able to detect smoldering fires earlier [12]. The detection of smoldering fires is one of the most difficult problems. They can lead to major damage, because no flame exists and the development of the burning process takes place very slowly without producing a lot of smoke particles in the early phase of the fire.

Significant advances were made recently in the field of early fire detection [9, 12, 13]. The main idea is to improve the performance of fire detectors by adding

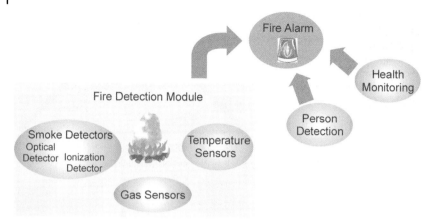

Fig. 6.1-3 Sensor systems for fire detection

gas sensors to them (Figure 6.1-3). Generally, when a fire begins, a transformation of energy and matter takes place [9]. The material transformation is characterized by the generated products, ash on the one hand and escaped gases and smoke particles on the other [14]. Since during the initial fire phase the gases diffuse more rapidly than smoke particles, the deployment of an array of gas sensors contributes to an improvement in detection speed and reliability [15].

Gas detection is nowadays performed using different types of gas sensors based on different physical principles. After investigations and simulations accompanied by tests and experiments in a fire laboratory [16], a gas sensor array composed of semiconductor metal oxide sensors for CO, H_2, and NH_3 was chosen [12]. These sensors are low priced and have different cross-sensitivities towards other gases. Therefore, they are able to detect some characteristic gas mixtures during fire development.

The additional use of sensor arrays in early fire detection systems requires a signal-processing method that is more sophisticated and difficult than the usually used simple threshold-based algorithms [17]. The signal processing should be able to discriminate between fire, non-fire, and disturbing event situations by identifying fire signatures from measured sensor responses (Figure 6.1-4).

A feature extraction unit is required in order to reduce the dimensionality of the measurement space and to extract suitable information characterizing fire situations. Then the extracted features have to be classified in order to estimate the class to which the measured data belong and to know if an alarm should be sent to the fire service. Investigations showed that neural networks are suitable for pattern recognition in the case of fire detection [13]. With the neural network in Figure 6.1-4, the signals were mapped to three possibilities characterizing the situations: fire, no fire, and disturbing event.

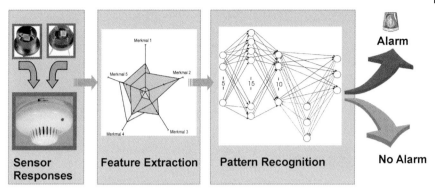

Fig. 6.1-4 Structure of a sophisticated fire detection algorithm [15]

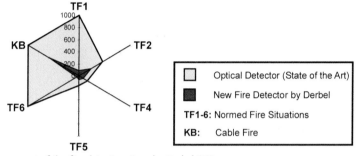

Fig. 6.1-5 Improvement of the fire detection time by Derbel [15]

Figure 6.1-5 shows the improvement achieved in fire detection times for several test fires (DIN EN54 and cable fire) with a commercial optical detector (O-Detector) and the new fire detection method by Derbel.

6.1.4.2
Gas Detection

Gas detection is needed for many security-relevant functions in intelligent buildings. The concentration of toxic and flammable gases should be continuously observed in order to alarm the user in danger situations and to reduce the risk of health damage or explosions through appropriate measures such as switching off the power supply in order to avoid sparks or opening windows in order to reduce the concentration of some gases.

In definite places in a building, eg, a kitchen or a parking garage, toxic gases such as CO (carbon monoxide) resulting from an open fire and higher concentrations of CO_2 (carbon dioxide) are expected. These gases need to be measured and to be critically considered in ventilation processes.

For the detection of flammable gases, pellistors are generally used. These are based on the measurement of the heat released during the burning process of a gas sample as a measure of the concentration of the existing flammable gases. The released heat is detected through measurement of the increase in electrical resistance within the pellistor.

For the detection of toxic gases, gas sensors based on different principles, eg, metal oxide, electrochemical cells, and infrared sensors, are commercially available and could be used in intelligent buildings [4]. Other sensors have been developed such as the quartz microbalance (QMB) and surface acoustic wave (SAW) sensors and are coming onto the market.

Table 6.1-2 shows a comparison of the properties of gas sensors based on different principles. As can be seen, the main problems with gas sensors are the cross-sensitivity and the long-term stability.

Although metal oxide sensors have good sensitivity, the problem of cross-sensitivity makes the determination of the concentration of only one gas difficult. Electrochemical cells, in contrast, are selective, but they have problems concerning long-term stability. They have a relative short lifetime and must be frequently calibrated.

Figure 6.1-6 shows the different effects influencing the properties of gas sensors. Because of the fluctuation of several factors during the production process, semiconductor elements show a certain specimen scattering. Temperature, humidity, and aging processes influence the behavior of most gas sensors. Aging processes affect the sensor properties such as the sensitivity in the case of electro-

Tab. 6.1-2 Properties of gas sensors

Gas sensor	Pellistor	Metal oxide	Electro-chemical cells	IR	QMB	SAW
Sensitivity	−	++	++	++	−	++
Dynamic behavior	+	+	0	+	+	+
Cross-sensitivity	−	−	+	++	−	−
Long-time stability	+	0	0	++	0	0
Costs	+	+	0	−	0	0

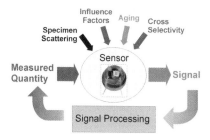

Fig. 6.1-6 Importance of signal processing in the case of gas sensors

chemical cells. The cross-sensitivity towards other gases and the influence of temperature and humidity are significant, eg, in the case of metal oxide sensors.

Especially for intelligent buildings, the reliability of the measurement signals should be critically considered. Therefore, special measures should be taken in order to improve the properties of gas sensors.

The precision of gas concentration measurement could be improved through accurate sensor signal modeling and corresponding signal processing. For instance, using the model by Horn (Equation 6.1-1), the temperature dependence of metal oxide sensors could be taken into account, so that an accurate calculation of the gas concentration could be made [18].

$$G = g \frac{B + Ax_g}{1 + Ax_g} \qquad (6.1\text{-}1)$$

where x_g=gas concentration, G=conductance, A and B=temperature-dependent parameters, and g=conductance referring to the charge carrier number.

Many advances in the development of electrochemical cells have been achieved in recent years. Their lifetime could be extended to 5 years. Recent investigations [19] have shown that the aging process of electrochemical cells results from the reduction of the active area (Figure 6.1-7). Therefore, by simultaneously measuring the impedance of the electrochemical cell, this effect could be compensated through signal processing, so that the real gas concentration could be measured, despite changes in the effective area.

Infrared (IR) sensors generally use an IR source together with an interference filter which is used to select the wavelength corresponding to the measured gas. The selectivity of this kind of sensor was recently improved by using a laser optical system. A

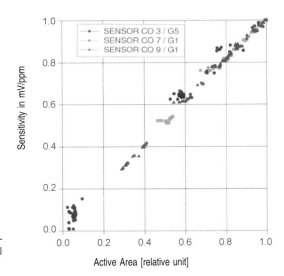

Fig. 6.1-7 Dependence of the sensitivity decrease of electrochemical cells on the active area [19]

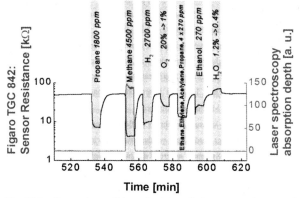

Fig. 6.1-8 Comparison of the selectivity of a laser optical system for the detection of methane with a SnO₂ metal oxide gas sensor [20]

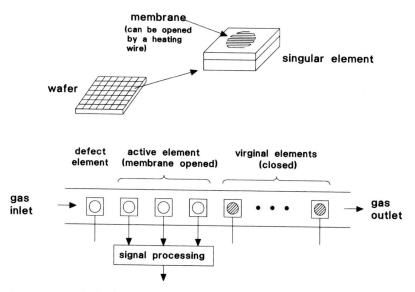

Fig. 6.1-9 Use of redundant, successively activated sensors for the improvement of reliability [21]

laser diode having the desired wavelength λ is used instead of an IR source [20]. Figure 6.1-8 shows the response of the laser optical system for methane ($\lambda = 1.651$ μm) compared with a metal oxide sensor by successive exposure to several gases.

The use of several redundant gas sensors has many advantages for applications with high reliability requirements. By comparison of the sensor signals, occurring failures could be recognized. Signal processing measures can be used in order to eliminate defect sensors. A sensor array of redundant, successively activated sensors is shown in Figure 6.1-9. The lifetime of this sensor system is a multiple of the lifetime of individual sensors.

6.1.4.3
Intrusion and Person Detection

Presence detection of persons plays a key role in a comprehensive control and protection system for home automation. Generally intrusion detection and access management are needed. For intrusion detection, sensor systems have to inform the user about the state of windows, doors, entrances, and exits of the building at any time. In public buildings, such as in companies, access management is generally needed that allows only authorized persons to enter particular sectors. In private homes the recognition of individuals can be employed to personalize the settings of heaters and lights when entering a room. The information about the presence of persons is also useful for other subsystems such as heaters and light switches to optimize the energy consumption in rooms.

Presence detectors should fulfill strict requirements since false alarms are costly and greatly reduce the confidence of and acceptability by the users. Special requirements are, for instance:

- high sensitivity and reliability of presence detection
- evaluation of the direction of motion (person enters or leaves the rooms)
- short response time (especially for door openers, light switches).

In addition to the sensor systems for the position detection of windows and doors, special presence detection sensors should be installed in a building. Several sensor classifications for presence detection can be distinguished:

- portal sensors (doorway, passageway, elevator entrance, etc.)
- zone (room) sensors and wide area supervision covering multiple zones [22]
- person identification systems such as camera systems, ID tags, fingerprint sensors or speaker-dependent speech processing.

To detect the presence of persons in a defined area, motion detectors are widely used. Besides passive IR detectors, ultrasound and microwave devices can be employed, which have a very good resolution. The Doppler effect on which these methods are based can be explained as follows: a reflector or an object moving with velocity v causes a frequency shift f_D between the transmitted and the received signal, which is inversely proportional to the wavelength λ of the signal (Equation 6.1-2). Therefore, the motion direction and the velocity of a reflector could be calculated from the evaluation of the frequency shift f_D.

$$f_D = \frac{1}{2\pi} \frac{d\varphi}{dt} = \frac{2}{\lambda} v \cos \alpha \qquad (6.1\text{-}2)$$

where α=angle between the direction of motion and the normal to the receiver.

The volumetric surveillance of large rooms is possible using only one ultrasound or microwave sensor, since the sent waves could fill a whole room independently of its shape due to numerous reflections from the walls. From the echo, targets can be classified with simple signal processing.

False alarm sources by ultrasonic detectors are generally the high sensitivity to noise and thermally induced air turbulence and movements of hanging curtains and plants.

Microwave devices may detect motion outside the room of observation, because they pass through walls without much attenuation. They could be misled by other electromagnetic fields (mobile telephones, etc.) or running fluorescent lamps. Large metallic objects can cause unexpected blind spots in the supervised field. Multi-path propagation causes confusion in the recognition of the right signal, because every reflected signal would be interpreted as a new object.

In Table 6.1-3, application-relevant features of ultrasound and microwaves are listed. For ultrasound sensors a high frequency is generally desired in order to achieve a better resolution. However, owing to the absorption of ultrasound in air, which increases approximately with the square of frequency, the frequency is limited to a few dozen kHz ($\lambda \approx 1$ cm). For microwave sensors, applications are restricted to the ISM bands (2.4, 5.6, 9.3, 24 GHz).

The reliability of Doppler-based sensors depends largely on the capability to distinguish target signals from an event produced by noise or other disturbances. In recent research [23], the confidence level of Doppler-based detectors was improved by a combined microwave-ultrasonic multi-sensor system (Figure 6.1-10). In this case noise and other disturbances affect the two waves differently. The Doppler shifts (f_{d_u} and f_{d_m}) caused by motion of the same target object are, however, inversely proportional to the wavelengths λ_u and λ_m of the transmitted ultrasonic and electromagnetic waves (Equation 6.1-3), and therefore the new sensor system can achieve better properties than an ultrasound or a microwave detector alone.

Tab. 6.1-3 Application-relevant features of ultrasound and microwaves detectors [24]

	Ultrasound	Microwaves
Sensor effect	Large (small wavelength) (low propagation velocity, $c = 3.4 \cdot 10^2$ m/s)	Smaller (usually larger wavelength) (velocity of light, $c = 3 \cdot 10^8$ m/s)
Doppler frequency	$f_D/v = 200$ Hz/(m/s) for $f_{US} = 34$ kHz ($\lambda = 10$ mm)	$f_D/v = 200$ Hz(m/s) for $f_{UW} = 30$ GHz ($\lambda = 10$ mm)
Maximum range	10 m	>10 m
Reflection	High for all solid materials	High for metals, water, living tissue, low for materials with small ε
Axial resolution	High (bandwidth >10%)	Low (bandwidth <1%)
False alarm sources	Air motion, sound sources, motion of curtains, etc.	Penetration through walls and windows, luminescent lamps

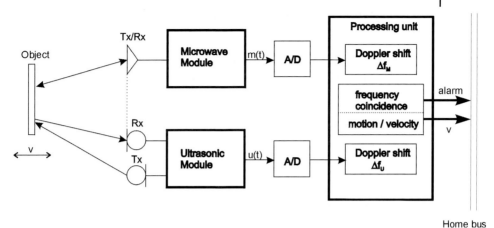

Fig. 6.1-10 Block diagram of the ultrasonic-microwave multi-sensor system for motion detection [23]

$$\frac{f_{d_u}}{f_{d_m}} = \frac{\lambda_m}{\lambda_u} \qquad (6.1\text{-}3)$$

A compact, low-cost sensor for the robust measurement of the presence and velocity of objects, using 24 GHz microwaves and 40 kHz ultrasound, has been designed (Figure 6.1-11).

The coherence of ultrasonic and microwave Doppler signals as a function of time, named 'Doppler frequency coincidence', is the primary criterion to discriminate the motion of target objects from false targets caused by interference phenomena and noise. It performs like an adaptive frequency-sensitive filter: the signal returning from a target object is enhanced, whereas uncorrelated noise and clutter components are suppressed.

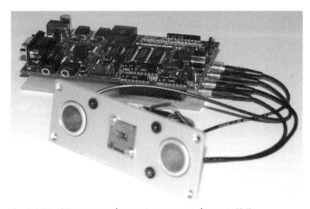

Fig. 6.1-11 Microwave-ultrasonic presence detector [24]

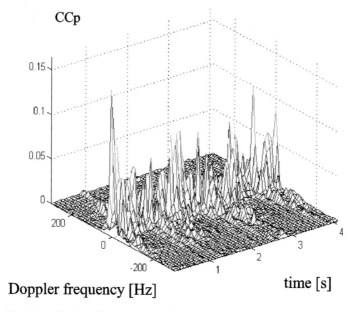

Fig. 6.1-12 Results of Doppler coincidence threshold detection in the presence of distortions [23]

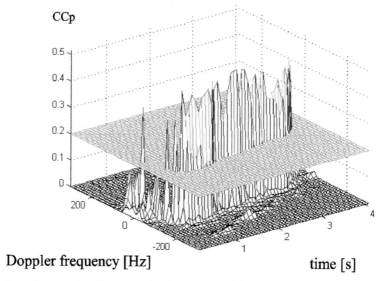

Fig. 6.1-13 Results of Doppler coincidence threshold detection for motion of a person [23]

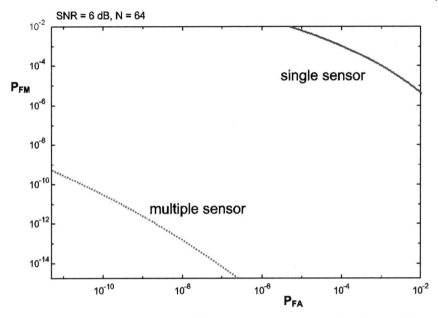

Fig. 6.1-14 ROC curves. P_{FA} = probability of false alarm, P_{FM} = probability of false dismissal [25]

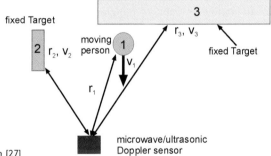

Fig. 6.1-15 Measurement situation [27]

In the presence of distortions, the cross-spectral density (CC_P) has a low level and shows no dominant frequency (Figure 6.1-12). If the same frequency is dominant in both the (frequency-normalized) ultrasonic and microwave Doppler spectra, CC_P has a significant maximum which will be detected by threshold detection (Figure 6.1-13).

With the data fusion performed [25], the performance is significantly improved compared with multi-sensors based on complementary sensor outputs, such as conventional ultrasonic-passive infrared (US-PIR) dual sensors or combinations of US and charge-coupled device (CCD) cameras [27]. The detector capabilities are characterized by receiver-operating curves (ROCs) which assess the true detection

Tab. 6.1-4 Motion detector performance analysis [24]

Effect	Ultrasonic	Microwave	Ultrasonic and microwave
Air motion	++	–	–
Curtains, etc.	+	+	–
Luminescent lamps	–	+	–
Penetration through walls	–	++	–
Moving person	++	++	++

rate P_D versus the corresponding false alarm rate P_{FA} for a variety of decision thresholds. In Figure 6.1-14, P_{FA} is shown against the probability of false dismissal, $P_{FM}=1-P_D$, at a certain detection threshold, depending on the signal-to-noise ratio (SNR) (Neyman-Pearson criterion). The results give evidence of the superior performance of a multiple sensor in comparison with a single sensor (microwave or ultrasonic).

The sensor system was investigated in several relevant test cases, such as the situation is sketched in Figure 6.1-15.

Table 6.1-4 gives an overview of the performance achieved by a combined ultrasound-microwave Doppler sensor compared with an ultrasound or a microwave detector. The new sensor system can detect moving persons and is not influenced by air motion, motion of hanging curtains, luminescent lamps, and penetration through walls [26, 27].

6.1.4.4
Sensor Systems for Health Safety

The opportunity to improve the quality of life represents a challenge for the home automation market. Especially for help-needing and elderly persons who live alone, an emergency call system can be integrated in the home automation system. This service allows such persons to live in their familiar surroundings, improving their quality of life and reducing thereby the otherwise accruing costs related to accommodation in a care home.

The main components of an emergency call system are:

- presence detectors for the estimation of the patient's location in the building
- distributed talking positions to an emergency call head office
- an emergency call central unit for communication with an emergency call head office
- manually controlled emergency call releases
- equipment for the measurement of medical data which can be used by the patient him- or herself
- devices allowing an automatic monitoring of vital parameters.

Equipment for medical diagnosis is generally needed for a tele-diagnosis which helps significantly to reduce the visits of medical personnel and doctors. This can be used by the patient him- or herself and send the data using the home bus to the central communication station in the building, which send them automatically via telephone or the Internet to an emergency call head office or to the treating doctor.

For a person who lives alone, it is generally not sufficient to have manually controlled devices or to rely on the presence of detectors of the building system in the case of a serious emergency call. These should be necessarily complemented with health care devices, which can rapidly send an emergency call automatically in case of unconsciousness. Therefore, the use of a sensor system which is always carried by the patient is indispensable. This system should be able to measure vital parameters in order to detect safety-relevant conditions. Some of the important vital parameters are:

- respiration
- concentration of oxygen in arterial blood
- blood pressure
- temperature of the body
- movements
- pulse rate

For acceptance by the users, this device should be wireless and energy autonomous in order to provide unhindered mobility of the supervised person. It should be light and easily carried. The user should be able to ignore it and to live normally without being obliged to take it down in any situation during the whole day.

Several possibilities have recently been realized in terms of easy wearable devices. One example is a hand transceiver (Figure 6.1-16) containing miniaturized sensors which measure the pulse rate, respiration, and the concentration of oxygen in arterial blood [28].

A further example is the MIT ring (Figure 6.1-17). Powered by a tiny battery, the ring works by manipulating light [30]. A light-emitting diode in the ring con-

Fig. 6.1-16 Miniaturized emergency call transceiver [29]

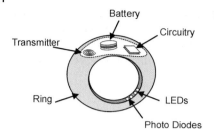

Fig. 6.1-17 MIT-ring for healthcare [30]

tinuously emits light into the finger. Some of that light is reflected off the blood in the finger, and is in turn captured by a photodiode. Such a ring could measure pulse rate and the potential cardiac condition from the size of blood vessels. The detection of the rate of blood flow can be used for the determination of blood pressure. The final signals are transmitted via an embedded antenna to receivers distributed in the building.

6.1.5
Sensor Systems for Heating, Ventilation, and Air Conditioning (HVAC) and Comfort

Sensor systems contribute significantly to discharging the user from some tasks which could be automated. An easily usable building system therefore provides several profitable functions for the users, which can considerably improve their quality of life.

A good quality of air and climate in the inside of the building contribute significantly to the well-being of residents in intelligent buildings, where sensor systems play a key role. They are indispensable in collecting necessary information HVAC.

6.1.5.1
Convenience and Easy Usability

The realization of convenience refers primarily to the basic equipment of the building system. It can also be diversely emphasized and may be carried on until self-learning building systems are developed, which consider the personal preferences of all identified persons in the building. The control of components in the building system could be simplified using speech control or generally remote control and thereby becomes a natural aspect.

For instance, the lighting control in the building could be coupled to the brightness or to the presence of people on floors and staircases. Lighting quality and quantity related to reflected and indirect glare and also illuminance and contrast values have a direct impact on the well-being, motivation, and productivity of persons in the building [31]. The optimal settings are generally individual and sophisticated systems should therefore optimally be combined with person detection, so that the personal optimal settings can be considered.

A further example is access to the building or to particular sectors in the building. This could be regulated without the necessity for keys, using fingertip sensors systems or image and voice recognition. These systems provide the user with undisturbed mobility in the building and contribute significantly to a feeling of well-being.

Some inconvenient tasks, which otherwise necessarily require the presence of persons in the building, such as garden irrigation, could be automatically achieved, demand controlled, if sensor systems measuring the soil water content are used [32].

Generally, the use of sensor systems and components for convenience is very individual and depends on the readiness of the user to invest in the building system. Nevertheless, a certain degree of convenience is offered in every building equipped with sensors and actuators.

6.1.5.2
Thermal Comfort

Thermal comfort is one of the most important factors for the well-being of persons in a building. A large number of parameters are responsible for the perception of comfort inside buildings [33]. These physical and physiological parameters, which determine the state of comfort, are illustrated in Figure 6.1-18.

Primary and dominant parameters were defined in [8] as:

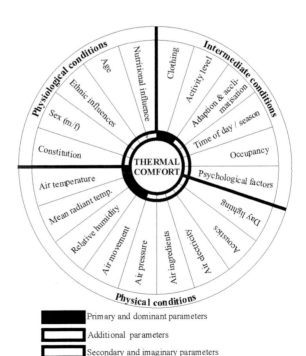

Fig. 6.1-18 Various physical and physiological parameters determining the degree of thermal comfort [34]

■ Primary and dominant parameters
□ Additional parameters
□ Secondary and imaginary parameters

- air temperature
- mean radiant temperature
- air humidity
- air motion
- clothing
- activity level.

The realization of thermal comfort is primarily related to a suitable heating concept taking several factors into account. Because heating is closely related to energy consumption and therefore also to environmental aspects, several heat insulation prescriptions are regulated by a law.

In Germany, 30% of the total energy consumption results from residential buildings, and 77.5% of that is the annual heating energy consumption [35]. This energy consumption can be reduced by about 50% by means of requirement-led peripheral heating and ventilation control. Heat consumption for the realization of comfort in a room is subject to temporal fluctuations and depends on the weather data and its position in the building. Automatic control is generally necessary and must be designed in such a way that an optimum between comfort needs, heating costs, and thermal losses is realized [36].

6.1.5.3
Indoor Air Quality

The indoor air quality (IAQ) is relevant for well-being. Inadequate ventilation in buildings can lead to serious problems, such as the sick building syndrome (SBS), building-related illnesses (BRI) and mildew [37].

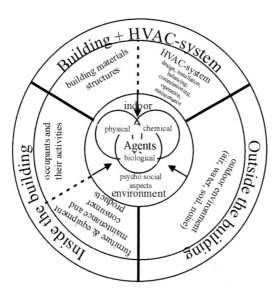

Fig. 6.1-19 Interactions of factors determining the indoor environment [37]

Fig. 6.1-20 Sensor systems for indoor air quality

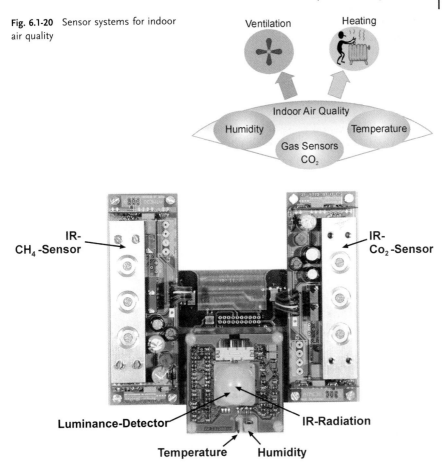

Fig. 6.1-21 Multi-chip module for indoor air quality [8]

Figure 6.1-19 shows the complexity of factors determining the indoor environment. The IAQ results from constant emission sources such as building materials, furniture, and technical equipment. Therefore, the hygienic aspect of the building physics assumes a constant air exchange of 0.5 ach (= air change per hour) to minimize the concentration of this inevitable indoor pollution. Von Pettenkofer [38] and Fanger [39] described a correlation between the CO_2 concentration and the pollution of indoor air by occupants and their activities. The CO_2 concentration is therefore one of the main quantities that should be considered in a demand-controlled ventilation system.

The control of the indoor air quality needs the use of sensor systems allowing the measurement of CO_2, temperature, and humidity. In order to guarantee thermal comfort for the users, both the ventilation and heating system should be considered within the control of indoor air quality (Figure 6.1-20).

An example of an MSM for the control of indoor air quality is shown in Figure 6.1-21. It is composed of two optical gas sensors for the detection of CO_2 and

CH₄ (methane), a temperature sensor, a humidity sensor, and sensors for presence detection. All sensors are coupled to the micro-controller interface, so that the signal processing is done within the sensor module. The necessary signals are sent on the building bus to the corresponding ventilation and heating equipment.

6.1.6
Future Trends for Sensor Systems in Intelligent Buildings

In order to realize safety, convenience, and comfort for the users of intelligent buildings, many sensor systems distributed over all the building are necessary (Figure 6.1-22). The main feature of the total system is its individuality. Every building will contain a certain selection of systems and components which were chosen by the users according to their individual requirements. The total system should therefore be maintained flexibly changeable, because the users' requirements change with time. In the future users will look after this new market sector in order to be able to adapt the building to their new requirements.

In the future many targets could be aimed for in the development of sensor systems for intelligent buildings, such as:

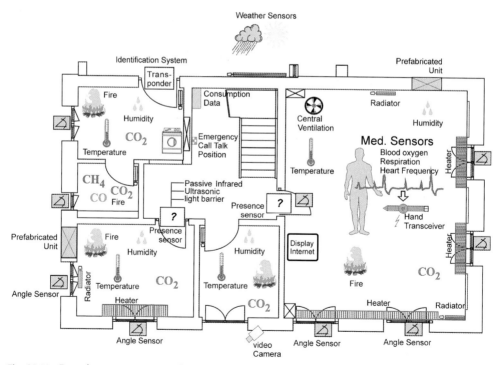

Fig. 6.1-22 Example scenario in an intelligent building

Fig. 6.1-23 Cross section of a prefabricated unit [8]

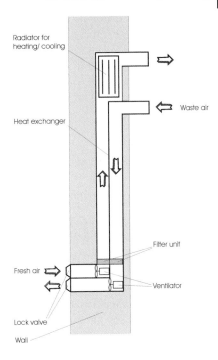

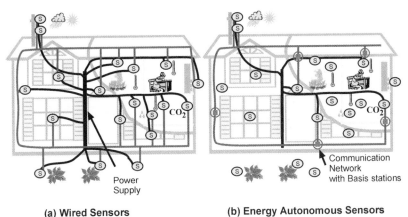

Fig. 6.1-24 Reduction of installation costs through the use of energy-autonomous sensors

- reduction of energy consumption
- reduction of installation and maintenance costs
- widespread use of wireless sensors
- use of sensor systems allowing self-monitoring and self-calibration
- increasing integration of sensors and actuators with the building matter (Figure 6.1-23).

One of the main important improvements for the reduction of installation costs is the use of wireless and energy-autonomous sensors. With the growing complexity of the sensor systems in intelligent buildings, the electric wiring of the system becomes a complex problem, which can cause difficulties with the system handling and affects its long-term reliability. If every sensor needs a network and a power supply wire connection, a large cable harness will be necessary. Energy-autonomous sensors and micro-systems (Figure 6.1-24), operating without wire energy supply and data transmission, in contrast yield considerable advantages. These imply greater convenience and higher mobility. The installation expense can be substituted by an easy parameter configuration. Components could be easily moved or added to the system. The modernization and retrofitting of existing buildings according to changing requirements become easier.

6.1.7
Acknowledgments

Many sensor systems described in this section were developed in the research projects VIMP [3], tele-Haus [5] and IWO-BAY [7]. The authors thank the BMBF (Bundesministerium für Bildung und Forschung (Federation Ministerial Department for Education and Research)) and the BFS (Bayerische Forschungsstiftung (Bavarian Research Foundation)), who supported these projects.

The authors would like also to acknowledge all colleagues who have provided illustrations of their work for this section.

6.1.8
References

1 Holz, R., Hourigan, A., Sloop, R., Monkman, P., Krarti, M., *Build. Environ.* 32 (1997) 31–43.
2 Clarke, J.A., Johnstone, C.M., Kelly, N.J., McLean, R.C., Anderson, J.A., Rowan, N.J., Smith, J.E., *Build. Environ.* 34 (1999) 515–521.
3 Schneider, F., Tränkler, H.-R., *MST News* 3 (1998) 33–36.
4 Doll, T., Eisele, I., Tränkler, H.-R., *Intelligentes Gas-Multisensorsystem*; 1–6: Geronimo Verlag, 1998.
5 Tränkler, H.-R., Schneider, F., *tele-Haus, Research Report*; Munich: 1–10, 2001.
6 Grassnick, R., Tränkler, H.-R., Geissler, C., Heinze, M., A Contribution to Energy-Saving Food Storage in Private Households, in: *Proceedings of Smart Systems and Devices*; 27–30 March 2001, Hammamet, Tunisia.
7 Tränkler, H.-R., Schneider, F., *Intelligente Hausinstrumentierung: IWO-BAY, Research Report*; University of the Bundeswehr Munich, 2000.
8 Schratt, L., Horn, M., Tränkler, H.-R., Decentralized, demand controlled heating and ventilation in the research project IWO-BAY, in: *Proceedings of Microtec 2000*; 25–27 September 2000, Hannover: Expo 2000, B. 6.2.3.
9 Schratt, L., Derbel, F., Tränkler, H.-R., in: *Proceedings of Sensor '99*; 18–20 May 1999, Nürnberg, 1999, Vol. 1, pp. 303–309.
10 Groh, H., Heimraucher retten Leben – eine ZVEI-Initiative, booklet *Heimraucher retten Leben*; Fachabteilung Sicherheitssystem, 1998.

11 Hess, E., Alarmstatistik – ungewollte und erfolgreiche Alarmmeldungen von Brandmeldeanlagen – eine 10-jährige Studie, *Brandmeldeanlagen*; Cologne: 1998.

12 Derbel, F., Performance Improvement of Fire Detection Systems by Means of Gas Sensors and LVQ Neural Networks, in: *Proceedings of ACIDCA 2000*; Monastir, Tunisia, 2000.

13 Derbel, F., Tränkler, H.-R., Fire detection with gas sensors and neural networks, in: *Proceedings of CESA 1998*; Hammamet, Tunisia, 1998.

14 Kohl, D., Gefahrenmelder auf der Basis von Gas-Multisensorsystemen, presented at the VDI/VDE-Workshop Messung von Luftschadstoffen, Anforderungen und neue Wege, Bad Aibling, Germany, 1995.

15 Derbel, F., Smart Sensor System zur Brandfrüherkennung und neuartiges Verfahren zur Brandsimulation, *Dissertation*; Munich: Institut für Meß- und Automatisierungstechnik, Universität der Bundeswehr München, 2001.

16 Derbel, F., Horn, M., Tränkler, H.-R., System Identification Techniques for Simulating Fire Detector Signals, in: *Proceedings of the 16th IEEE Instrumentation and Measurement Technology Conference (IMTC)*; Venice, 1999, pp. 898–903.

17 Derbel, F., Tränkler, H.-R., Erkennung von Gefahrenzuständen mit Gassensoren, in: *XI Meßtechnisches Symposium*; Rostock-Warnemünde, 1997, pp. 145–156.

18 Henze, G., Köhler, M., Lay, J.P. (eds.), *Umweltdiagnostik mit Mikrosystemen*; Weinheim: Wiley-VCH, 1999.

19 Makadmini, L., Erhöhung der Genauigkeit und Zuverlässigkeit amperometrischer Gassensoren, *Dissertation*; Munich: Institut für Meß- und Automatisierungstechnik, Universität der Bundeswehr München, 1997.

20 Mágori, E., Entwicklung eines Sensorsystems zur Detektion von Gasen mit NahInfraRot-Laserdioden, *Dissertation*; Munich: Institut für Meß- und Automatisierungstechnik, Universität der Bundeswehr München, 2000.

21 Roth, M., Endres, H.E., Tränkler, H.-R., US Pat. 09/142634.

22 Hashimoto, K., Tsuruta, T., Morinaka, K., Yoshiike, N., *Sens. Actuators* **79** (2000) 46–52.

23 Ruser, H., Mágori, V., *Sens. Actuators A* **67** (1998) 125–132.

24 Ruser, H., Mágori, V., Phase Coincidence between Ultrasound and Microwaves – a Powerful and Flexible Multisensor Principle, in: *Proceedings of Eurosensors XIV*; 27–30 August 2000, Copenhagen, Denmark, 2000, T2P36.

25 Luo, R.C., Kay, M.G., *IEEE Trans. Syst. Man Cybernet.* **19** (1989) 901–931.

26 Ruser, H., Mágori, V., Sweep linearization of a micro-wave FMCW Doppler sensor by an ultrasonic reference, presented at the IEEE International Frequency Control Symposium, Orlando, FL, 1997.

27 Ruser, H., Jena, A.v., Mágori, V., Tränkler, H.-R., A low-cost ultrasonic-microwave multisensor for robust sensing of velocity and range, in: *Proceedings of Sensor 1999*; Nürnberg, 1999, Paper C3.3.

28 Tränkler, H.-R., Engelen, R., Flaschke, T., Schratt, L., Emergency Call and Hazard Warning System in the Project VIMP, *Intell. Living*, 10–11 November 1997.

29 Schneider, F., Tränkler, H.-R., *Das Intelligente Haus*; Munich: Pflaum Verlag, 2001.

30 Thomson, E.A., Asada, H.H., Yang, B.-H., MIT ring monitors patients' vital signs, *MIT News*, April 1997.

31 Brügge, B., Pfleghar, R., Reicher, T., Internet Framework for Cooperative Buildings, in: *Proceedings of the EIB Conference*; 4–5 October 2000, Munich.

32 Flaschke, T., Derbel, F., Tränkler, H.-R., Determination of the Soil Water Content by Impedance Measurements, in: *Proceedings of Sensor 1999*; Nürnberg, 1999, Paper C2.5.

33 Fanger, P.O., *Thermal Comfort – Analysis and Applications in Environmental Engineering*; New York: McGraw-Hill, 1972.

34 Frank, W., *Raumklima und thermische Behaglichkeit. Berichte aus der Bauforschung*; H. 104, Berlin: Ernst & Sohn Verlag, 1975.

35 Geiger, B., Hess, H., *Energiewirtschaftliche Daten im Jahrbuch 99*; Düsseldorf: VDI Gesellschaft Energietechnik, VDI-Verlag, 1999, pp. 267–283.

36 Grassnick, R., Tränkler, H.-R., Occupancy-led Individual Room Control, in: *Proceedings of Smart Systems and Devices*; Hammamet, Tunisia, 2001.

37 Bischof, B., Dompke, M., Schmid, W. (eds.), *Sick Building Syndrome: Forschung und Erkenntnisumsetzung*; Karlsruhe: Verlag C. F. Müller, 1993.

38 Pettenkofer, M. v., *Über den Luftwechsel in Wohngebäuden;* Munich: Cotta'sche Buchhandlung, 1958.

39 Fanger, P. O., *Energy Build.* **12** (1988) 21–39.

6.2
System Technologies for Private Homes

Friedrich Schneider, Lars Binternagel, Yuriy Kyselytsya, Wolfgang Müller,
Thomas Schlütsmeier, Bernhard Schreyer, Rostislav Stolyar
Kay Werthschulte, Günter Westermeir, and Dirk Wölfle,
*Technische Universität München, Lehrstuhl für Messsystem- und Sensortechnik,
München, Germany*
Thomas Weinzierl, *Weinzierl Engineering, Tyrlaching, Germany*

6.2.1
Introduction

During the last few years there have been very rapid developments in information and communication technologies. The living and working aspects of people are increasingly influenced by all kinds of electronic components and systems, which open up new, far-reaching possibilities with a great variety of functionality. There have been also great efforts by industry and by research institutes to bring information and automation technologies into the private home ('Intelligentes Haus' or 'Smart Home'). Nevertheless the breakthrough in the market could not be achieved in Germany. But also in Europe the situation is similar through the EU-sponsored different research projects within ESPRIT. The same is true in this field in the USA and in Japan. There are several reasons for this situation – on the one hand there are technology reasons, and on the other hand it is due to the market situation:

1. Insufficient system-oriented integration;
2. No suitable sensors and actuators for the private home;
3. Hurdles with the installation and start-up operation of these systems both for end users and for craftsmen;
4. No MMI (man-machine interface) which is easily adaptable to the needs of the end user;
5. The long lasting discussion about the 'Convergence' towards a unified home bus system;
6. Too expensive for a general installation of bus technologies in the private home and in apartment houses;
7. No existing case studies for the long-term benefits of such systems.

On the other hand, there is great know-how in this field within Germany (and Europe) in many companies and research institutes. The following sections are intended to give an overview of completed and on-going projects of the Lehrstuhl für Messsystem- und Sensortechnik of the Technische Universität München in the area of system integration and software management systems for private homes. These projects have been and are mostly system oriented and illustrate some of our acquired experience during the last 7 years.

In Section 6.2.2 the requirements for home automation systems are discussed. An important part in home automation is the microcontroller level (Section 6.2.3). Nowadays there is no chance of developing proprietary software systems in this field. Therefore, one has to adapt hard- and software to the various hardware platforms and operating systems which are widely used in computer technologies (Section 6.2.4). Sections 6.2.5–6.2.7 contain the necessary application programs for different lifetime phases of a home automation system, starting with the configuration tools in Section 6.2.6. Here the ETS-Tool (EIB Tool Software; EIB = European Installation Bus), sold by EIBA, is explained briefly, together with upcoming new installation modes. The bus monitor and service programs in Section 6.2.5 serve as tools for installation of all devices in a system and for fault analysis. The man-machine interface (Section 6.2.7) is the most important interface to the user of a home automation system. The different hardware devices and their functional possibilities are discussed, reaching from PDAs (personal digital assistants) to the screens of TV sets. The software for design and operation of these devices has to be highly adaptable to the user and must be ergonomically designed. However, there should be not only locally oriented access to the home automation system but also tele-access via the Internet or telephone. For the latter services it is extremely important to use standard technologies such as Internet browsers or Java components. The article is concluded by an outlook on necessary developments (Section 6.2.8).

6.2.2
Requirements in Home Automation Systems

These requirements have to be derived from the benefits that one will gain by installing such a system. These benefits are not only personally oriented but also affect the economy in a country. Only if the personal benefits of at least one application field are evident to the occupant of a house will he be ready to invest a certain amount of money. Elder or convalescent people can stay longer in their own home if they have an emergency call service and tele-services installed. Security and safety are very important features, too. For this purpose sensors for the intelligent detection of presence are needed to activate warnings and alarms against intrusion. Of course, one can use the same sensors for presence-controlled heating, air conditioning or lighting.

Besides the exchange of sensor values and control information with its low data rate, communication with medium and in the future even with high data rates will be desirable, eg, for speech transmission or for monitoring purposes by video. On the other hand, people can feel safe if household appliances are supervised. Tele-control, Internet connections and ISDN services can bring benefits to the home which will be widely accepted by the users.

The increase in comfort achievable by tele-control and automation is an important factor for aged people, but also for those who are absent all day. Saving of money by reduced energy and water consumption and by reduced insurance rates are further incentives for private investment in intelligent home automation.

There will also be benefits to the economy of a country. If one really can open the market for home automation one can expect positive effects on the labor market, since there will be new jobs in both production and services (planning, consulting, installation and maintenance). In the field of health services there can be money savings if by using information and automation technologies aged people can stay longer in their own home, or if the status of sick people can be tele-monitored. This will result in a demand-oriented optimization of medical care (care-on-demand). Another important factor is the saving of energy for heating and air conditioning in connection with the recycling of heat energy and demand-oriented water consumption. The protection of our natural resources will affect our economy positively and will finally lead to an essential reduction of CO_2 emissions. The German Federal Government agreed in 1992 to reduce all CO_2 emissions in the year 2015 by 25% with respect to the levels in 1980.

If it is possible to define attractive application packages in home automation where the benefits are so evident to the users that they will invest in them, then companies are able to sell them for a modest price. To do so the home automation system must be an open system in hard- and software, where all components can be bought from various companies and ideally the installed home automation system can be supplemented later by additional features without problems.

In our opinion, the following application areas are attractive for many few occupants:

- Health Care Systems
  This application consists of a combination of noninvasive medical sensors, which monitor vital parameters, and of tele-communication units for tele-control and maybe tele-diagnosis. Supplemented by an emergency call system with the possibility of directly talking to an emergency service station from each room there can be emergency reaction 24 h per day. Or there can be direct dialling to the personal doctor's practice or to some relatives.
- Security and Safety
  This can be subdivided into the security of persons (intrusion alarms, emergency calls), the surveillance of all kinds of household appliances and the safety of the house through gas warning systems and early fire detection.
- Saving of Energy and Resources
  Intelligent heating and air conditioning management saves energy. Only automatically working single-room temperature control reduces energy consumption without loss of comfort. It should depend not only on temperature but also on air quality and on the number of persons present. Further, showing the current power consumption in connection with meaningful statistical functions will lead to conscious use of resources such as heat and water.
- In-house Communication
  In addition to the exchange of low data rate status and control information there should be higher data rate communication such as speech transmission and video transmission for monitoring purposes (supervising of entrance doors, supervision of babies and children). In the near future there will even

be the need for high data rate communication as a backbone for all kinds of services including the connection to the Internet and other information services from outside the home.
- Comfort
  With attractive monitoring and operating devices together with the necessary sensors and actuators the comfort in a house will be increased. If these features are combined with some sense of delight, younger technically oriented people maybe encouraged to buy such systems.

If one nowadays looks at the home (or building) automation market, one will find many components for a great variety of functions. Therefore, it cannot be the aim to invent a complete new home automation system, but rather one has to look for easy and attractive combinations of all the components. For this one needs an universal approach on the system side. There have to be microcontroller units which can manage intelligent sensor processing and transmitting via a bus system, but there also have to be PCs for complex data processing and management, for instance for the configuration and parameterization of the home system. On the operating system level it can be convenient to have pre-programmed functions in a microcontroller as in ContROS, but one has also to use all available standard operating systems such as Windows, Linux, and the embedded versions of them. Since home automation systems are decentralized networks, the connection and use of bus systems is an integral part of each system. Therefore, the access to the EIB (European Installation Bus) but also via TCP/IP and Ethernet to the World Wide Web are essential, and gateways in between are necessary to exchange data. The presentation of data depends very much on the user and on the purpose for which the data are collected. Thus a PDA (personal digital assistant) or in the future even a mobile phone can be used for alarms and for operating. If the user prefers a better visible presentation he will use a touch screen, maybe in the entrance of the house, or the television set in the living room, or the screen of his PC at home or at the workplace, where he will receive the information from his home either via ISDN or via the Internet.

The overall software structure will be a key element in such a system. On the one hand software technologies are developing very fast (see the very fast changing versions of Windows!), but on the other hand one may install the software in a house with life cycles of at least 20 years. Further, the configuration tool can be very different from the visualization tool during normal operation. Or the tool for investigating faults is used only by some service operatives and in most cases not by the end user. Nevertheless, it is necessary that at least the database is the same for all applications since otherwise the handling of a home automation system over its life cycle beginning from planning up to maintenance and extensions will be extremely difficult and expensive. Before reaching the goal that the software in home automation systems is as interchangeable as EIB hardware components there will still be much work. The following sections report on the efforts of our group to develop such a system.

6.2.3
Microcontroller Level

6.2.3.1
Realization

All components in a home automation system should be inexpensive, with low power consumption and small in size but nevertheless powerful in functions. Bus coupling units (BCUs) for the EIB system have very limited resources for user application programs and computing performance. These resources are sufficient for simple applications like push-button sensors. If more sophisticated implementations are required, an external microcontroller is necessary.

6.2.3.2
Choice of the Microcontroller

For cost reasons, small 8-bit microcontrollers should be sufficient for the required tasks. Since one serial connection is used for the connection to the bus coupling unit, communication applications commonly need a second UART (universal asynchronous receiver-transmitter). Programming for EIB applications is not necessarily fixed to one processor family if a higher programming language such as C is used.

A short selection of reasonable controllers is as follows:

- 8051 family (Infineon, Dallas, Analog Devices, etc.)
- Atmel AVR
- Microchip PIC.

6.2.3.3
Bus Connection with BCU

External microcontrollers are connected to the BCU over the so-called physical external interface (PEI). A serial protocol with hardware handshake is used. BCUs (Figure 6.2-1) transmit data over the bus system according to the ISO/OSI reference model for protocol stacks. The external microcontroller is able to do the bus access in different protocol layers, dependent on the communication demands. Most common is the 'Link Layer' with access to all protocol features and the 'Application Layer' with communication handling by the BCU itself.

6.2.3.4
Bus Connection via TP-UART (Figure 6.2-2)

Another way to get a connection to the EIB is with an especially designed IC from Infineon (previously a department of Siemens) offering access in a smaller packet size. The TP-UART-IC (TP=Twisted Pair) (Figure 6.2-3) is a transceiver, which supports the connection of microcontrollers of sensors, actuators, or other

6 System Technologies

Fig. 6.2-1 BCU in flush-mounting box

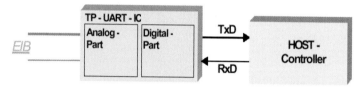

Fig. 6.2-2 Structure of the TP-UART

Fig. 6.2-3 TP-UART-IC

applications to the EIB. The integrated circuit (IC) generates further a stabilized 3.3 or 5 V supply to be used by a host controller and it does not need any external coils or transformers (except for some resistors and capacitors), thus realizing an EIB connection in a small package.

The TP-UART chip consists of two main parts: the digital part (UART interface) and the analog part (analog circuit part).

The analog part is directly connected to the EIB; the digital part is connected like a standard UART chip (two signal lines: TxD and RxD) to the host controller.

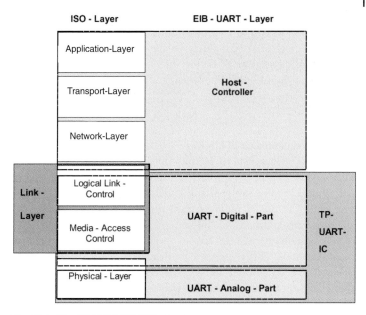

Fig. 6.2-4 The TP-UART ISO/OSI structure

It is possible to isolate the TP-UART electrically from the host controller with optocouplers in the TxD/RxD lines.

The TP-UART can handle the EIB telegrams up to Link Layer. The relation to the ISO Reference Model is illustrated in Figure 6.2-4.

The TP-UART contains the full Physical Layer and the incomplete Link Layer. The Link Layer has two sub-layers: Media-Access Control and Logical Link Control. The Logical Link Control is not complete; one part of it has to be handled by the host controller. That means that the TP-UART is not able to establish a connection alone, because eg confirmations have to be sent on the Link Layer. To realize the complete EIB function the not-included layers have to be implemented into the host controller.

6.2.3.5
Bus Coupling with RF-UART

In 1996 the EIBA (European Installation Bus Association) decided to realize a wireless solution for the EIB (RF) in addition to the two available communication media twisted pair (TP) and powerline (PL). In the next 2 years a group of experts developed the EIB RF system. Bosch Telecom was the leader of the development group. However, now, at the beginning of 2001, the RF components are still not available because of different problems in the past including the selling of Bosch Telecom and focusing of the EIBA on different tasks, especially the integration of BatiBus, EHS and EIB to the new 'Konnex' standard supporting all different media and protocols of the three bus organizations. Now it is planned to release a new RF specification in December 2001.

The basics of work will surely be the same in one year and that is the reason why a short description of the experience of the Technische Universität München (TUM) with the EIB RF system using a prototype implementation of RF-UART is given.

6.2.3.5.1 Requirements of EIB-RF

The EIB-RF system has to be fully integrated in the system structure of the EIB. That means that the system has to support the logical EIB topology, the configuration with ETS, the same protocol and data structure and should have a similar data rate. The use of radio as transmission medium makes it very important to guarantee the reliability of the communication. Especially the collision treatment is very important: EIB-RF has to use a medium access protocol to detect and avoid collisions. This requires a confirmation of the correct reception from the receiver to the transmitter (ACK). The realization of an access procedure and the confirmation of the reception, as well as the transmission of status messages, require a bi-directional communication.

6.2.3.5.2 Radiofrequency Specifications

The RF-UART uses the frequency range 868–870 MHz. This frequency range is reserved all over Europe for short-range devices and is forbidden for radio amateurs and for industrial, medical and scientific purposes. The main rules for using this frequency range are special limitations concerning the radiated power and the maximum transmission duration (duty cycle limitation). The whole frequency range is divided into four subranges according to the duty cycle. EIB-RF-uses the VLDC (very low duty cycle) and the LDC (low duty cycle) frequency ranges. The maximum permitted radiated power in these ranges may be 25 mW and the duty cycle is limited to 0.1 or 1%, respectively.

The data rate on the medium is 38.4 kbit/s. The modulation is binary frequency shift keying and the channel coding is Manchester. The communication range is >100 m in the free field.

6.2.3.5.3 Aspects of the System Reliability

The reduction of collision probability is done by a CSMA/CD method. If an RF device is going to send, it first checks on its center frequency and data rate if anything else is sending (also including foreign systems). If the device finds anything, it has to wait. If the channel is free the device can send the telegram. The transmission has been free of collision if the telegram is confirmed with a special response telegram.

This is the usual way in a point-to-point connection. However, in a point-to-multipoint connection acknowledgement of all addressed devices does not exist. The sending device knows about confirmations just from the so-called retransmitter. Every device can be configured as a retransmitter, but this makes sense only in a one- or two-floor house. The retransmitters repeat the telegram, thus transmitting it to the receiver and acknowledging reception to the sender at the same time. The scheduling is done by using retransmitter IDs during the configuration. In a configuration with two retransmitters, the sending device receives twice the telegram it sent and only knows that the two retransmitters have received the telegram. The probability of getting a telegram from a sender increases by use of the retransmitters, but it is not guaranteed to receive the telegram.

6.2.3.5.4 EIB-RF Frame

The EIB-RF frame (Figure 6.2-5) is similar to the EIB-PL frame. The usual EIB telegram frame is wrapped with a synchronization field, an RF control field with domain address (for different channels) and a check field. The frame core is identical with the EIB data frame in EIB twisted pair and power line systems.

6.2.3.5.5 Interfacing the RF-UART

Like the TP-UART, the RF-UART just includes the Physical Layer and the Link Layer (Figure 6.2-6). It uses the FT1.2 protocol to communicate with a host controller. This is a special serial protocol with high reliability. By using different services of the RF-UART it is possible to:

- configure the domain address
- configure the retransmitter functionality
- reset the RF-UART
- communicate with other devices via radiofrequency in point-to-point and broadcast connections

These services are services of the Link Layer. The standard EIB group communication is only possible if one implements the EIB protocol stack completely including the different tables to use group addresses and communication objects. Network Layer, Transport Layer and Application Layer and the tables have to be rebuilt in a host controller. Based on the EIB specification it is possible to create the same application programming interface (API) for the sensor or the actuator application as included in a 'traditional' BCU. The application on the bus node com-

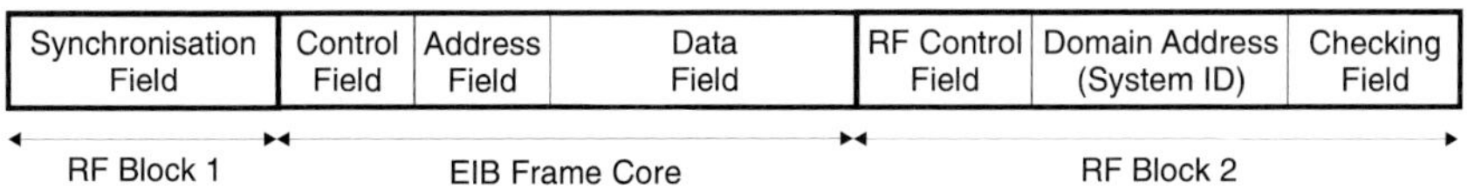

Synchronisation Field	Control Field	Address Field	Data Field	RF Control Field	Domain Address (System ID)	Checking Field

RF Block 1	EIB Frame Core	RF Block 2

Fig. 6.2-5 The frame structure of the EIB-RF system

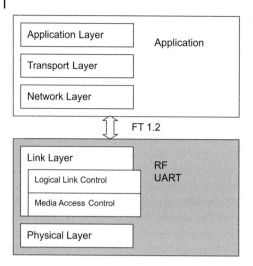

Fig. 6.2-6 The layer structure

municates with the bus in the same manner as the applications in the BCUs of EIB on twisted pair media and power line systems by writing to and reading from items in the communication object table. Supporting the PEI one can use common EIB application modules such as switches, etc.

6.2.3.6
Operating System ContROS

ContROS (microcontroller real-time operating system) is an operating system for 8-bit microcontrollers of the 8051 family and is written completely in the programming language C. It uses cooperative time-slice multitasking with surveillance of the real-time condition.

It is necessary for a flexible use of the microcontroller to be able to write one's own application programs. The realized program normally consists of several thousand bytes of assembler code. These data then should be transmitted from the PC to the microcontroller over the EIB bus. Since the EIB is too slow for the transmission of such an amount of data, another method was realized, in which applications are created by connecting precompiled software modules. These modules are all included in the operating system and cannot be changed by the user. When configuration is finished, the resulting data of the connections are sent over the EIB to the microcontroller. With this method the amount of data can be reduced to some hundred bytes.

6.2.3.7
Feature Controller

To establish a feature controller for EIB, ContROS is the ideal platform. It includes logical and timer functions and is 'programmed' in a graphical way: every

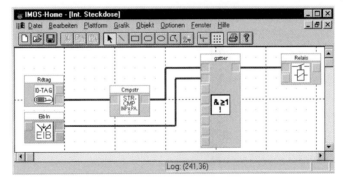

Fig. 6.2-7 The IMOS configuration tool

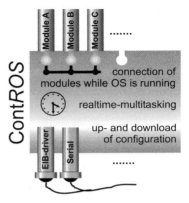

Fig. 6.2-8 ContROS functionality

software module has a corresponding symbol that can be placed in the configuration editor. After placing and configuring all modules, the logical connections between the individual modules must be set up. This is done by drawing connection lines between the modules (see Figure 6.2-7). A data type check is done automatically to ensure the compatibility of the exchanged data. Additionally the sequence of modules can be defined. The modules will be called by the multitasking management of ContROS in the right order (see Figure 6.2-8).

6.2.3.8
Intelligent Outlet

The 'intelligent outlet', a development of the research project VIMP (Verteilte Intelligente Mikrosysteme für den privaten Lebensbereich; distributed intelligent microsystems for private homes, a project sponsored by BMBF and VDI/VDE-IT Teltow, realized by partners from industry and universities), was designed to allow the surveillance of household appliances. It has the size of a double outlet. One part is the outlet; the other contains the electronics, including a miniaturized power sensor (see Figure 6.2-9). If the power consumption exceeds a certain limit,

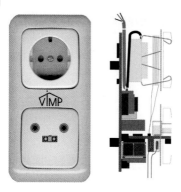

Fig. 6.2-9 The intelligent outlet

a malfunction of the appliance is detected. The decision depends on the type and operating state of the appliance. It is important to know which appliance is connected to the outlet. If a small ID tag (small transponder) is mounted on every plug used, the 'intelligent outlet' can identify the appliance by reading the identification number of every plug. Further there is a 16 A relay in the outlet to switch on or off power if the controlling unit notices a dangerous condition. Finally the 'intelligent outlet' fits into the EIB system and fulfills all necessary requirements for bus devices. Further, the application in the outlet is programmable in a flexible way by transmitting the configuration to the intelligent outlet via the EIB bus using ContROS. By choosing powerline transmission for EIB, the 'intelligent outlet' can be changed to an 'intelligent adapter plug'. Thus stationary mounting is no longer needed and the number of possible applications increases.

6.2.4
Operating System's Level

6.2.4.1
Introduction

Finding appropriate ways of connecting the EIB to computers has always been a major problem in the past. If an RS-232 interface is used, a special byte-by-byte handshake is needed between the computer and the connected BCU. Additionally, this handshake is time critical. Therefore, it is not possible to use a standard driver for the serial port to access the EIB from an application. Further, the trend in PC development according to the specifications of Intel ('PC 98') will not provide legacy devices such as the serial or parallel port any longer. Since software drivers for serial communication are always restricted to specific operating systems and not always available, different hardware solutions for a more general connection of devices to the EIB are needed.

Many interfaces have been released on the personal computer market during the last few years, but not each hardware interface is suitable. Additionally, the costs of hardware and the support under the different versions of Microsoft Windows play an important role in choosing the appropriate interface.

Which kind of hardware device driver should be developed and which software style should be used for device drivers since the software is improved with every new version of Windows? Unfortunately, not all driver styles are downward compatible to older versions, and even they are not the same in recent NT and 95/98 versions.

6.2.4.2
Interfaces

6.2.4.2.1 **The Serial Port**

Today, most applications use the serial port to access the EIB hardware. Because of the importance of this interface today, we have developed appropriate drivers for Windows 95/98 and for NT 4.0. However, for the future a more intelligent interface and a higher transmission speed are needed. We have decided to use the universal serial bus (USB). For a wireless alternative we decided to use the infrared medium IrDA (infrared data association). However, to connect the EIB via USB or IrDA it means more effort in hard- and software than simply using the serial port.

A very cheap and easy way to build an interface between the EIB and a PC is to use the serial port RS-232. A standard BCU provides a serial interface on its physical external interface (PEI) which can be connected directly to a serial port of a PC by adding line drivers. BCU1 provides a four-wire serial protocol with combined software/hardware handshake and BCU2 additionally supports a two-wire RxD/TxD port with FT1.2 protocol.

All operating systems have a standard driver for the serial port, but these drivers are not able to handle time-critical handshake protocols and to build a reliable connection. Appropriate 32-bit drivers are needed. So the challenge was to build one driver that fits in the architecture of Windows 95/98, Windows 2000 and Windows NT 4.0. In all operating systems the standard serial protocol for other applications should still be available. In Windows 95/98 the serial port is handled via a so-called virtual device driver (VxD). The problem with these operating systems is that there can be only one driver that accesses one serial port.

The solution was to write a new driver for the serial port that replaces the original driver. To combine all three protocols in one driver we decided to design it object-oriented in C++. In a base class CPort all functions are implemented which are common for both protocols. One derived class CSerialPort processes the standard serial communication and the classes CEIB1Port and CEIB2Port handle the two EIB protocols. To connect the operating system to the driver the class member functions of the port classes have been translated to standard C calls. The handshake protocol to access the EIB bus has been realized in a state machine. The usage of the driver by an application or a dynamic link library is exactly the same as for the standard serial port. It is only necessary to set one more parameter to declare which protocol to use.

6.2.4.2.2 The Universal Serial Bus

The USB, which was first established in 1994, is nowadays part of most PCs. Because of its appropriate price and high bandwidth it is suitable for keyboard, mouse, audio and even video applications. Transmission speeds up to 12 Mbps and the possibility of connecting up to 127 devices while the system is running (hot plug and play) are the most important features. In addition, the USB is able to supply small devices with a voltage of 5 V.

We developed a hardware interface between the EIB bus and the USB bus, which is supplied with power by the USB and will be explained in Section 6.2.4.4.

6.2.4.2.3 The IrDA Interface

As the amount of hardware interfaces and the need for peripheral devices grow, the number of cables also increases. Wireless transmission is sometimes more convenient. Recently some interesting solutions based on infrared or radiofrequency have been introduced.

In PC technology the simplest wireless interface is the IrDA port. It is supported by Windows 95/98 and 2000. The connection via IrDA is based on a complex software stack. It prevents loss of data during interruption of transmission and interference by other infrared devices. IrDA allows transmission speeds from 9.6 kbps up to 4 Mbps in specification 1.1. This software stack is already included in Windows, so one can easily use this interface in a PC. In recent years the IrDA interface was included in nearly every portable computer and handheld device.

An IrDA-EIB device will be explained below.

6.2.4.3
The IrDA-EIB Interface

The EIB modem (see Section 6.2.4.6) can also be realized with infrared connection to the PC instead of the serial cable. We built a cable-less serial interface for BCUs installed in flush-mounting boxes. It works like an EIB modem, but uses infrared light for data transmission.

For the hardware interface between the EIB bus and infrared media an 8-bit microcontroller circuit is used. The physical layer of the IrDA port is realized by using a standard transceiver that can be accessed through a serial port of the controller. In the firmware of this interface the IrDA protocol stack is integrated to establish a connection. It was not necessary to implement the full stack, but a superset of the so-called IrDA lite stack. The IR connection supports the IrCOMM 3 wire raw protocol; automatic baudrate detection supports baudrates from 9600 to 38 400 baud and the packet size can vary from 64 to 512 bytes. The prototype is shown in Figure 6.2-10.

Fig. 6.2-10 IrDA-EIB modem

6.2.4.4
USB-EIB Interface

The hardware interface between the EIB bus and the USB bus is supplied with power by the USB. It is realized with an 8-bit microcontroller circuit connected to an USB controller and an EIB interface on-board. The data transfer is handled via a memory-mapped access. The USB controller indicates data coming from the PC by transmitting an interrupt.

The USB driver is always a part of the operating system and processes the USB protocol for all devices connected to the USB bus. For each device there has to be one individual driver, a so-called client device driver. We have developed such a client device driver for our interface. It has to be in the format of a WDM (Windows Driver Model) driver and can be used with Windows 98 and Windows 2000. If the hardware and the driver stack are running, it is easy for an application to access the USB device.

As illustrated in Figure 6.2-11, the EIB-USB bridge can be connected to the EIB in two ways: via BCU and via TP-UART.

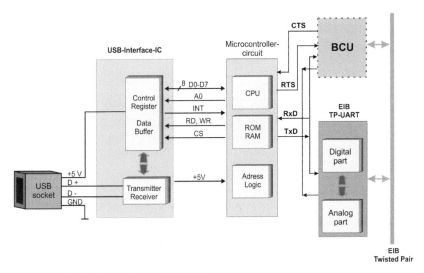

Fig. 6.2-11 Block scheme of the EIB-USB bridge

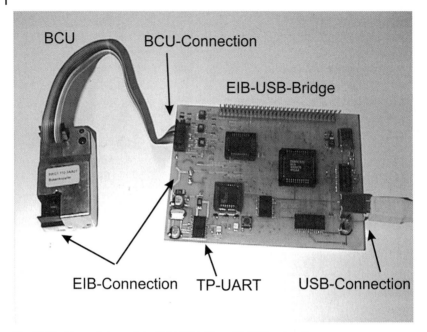

Fig. 6.2-12 The prototype of the EIB-USB bridge with BCU connection

BCU The microcontroller is connected to BCU1 with four signal lines (RxD, TxD and two handshake lines) or to BCU2 with two lines (RxD and TxD). The microcontroller receive telegrams from the PC via the USB controller and sends them to the BCU or the other way from BCU to the PC. To avoid a telegram's loss, the EIB-USB bridge has two buffers for sending and receiving, each one with a capacity of ten telegrams.

TP-UART This IC has no complete EIB protocol stack; therefore, it is necessary to implement it (Network Layer, Transport Layer and Application Layer) in the firmware of the EIB-USB Bridge.

Both versions are designed for an isolation voltage of 5000 V from the EIB. The firmware is written in C language and makes it possible to transfer the source code to other platforms. The prototype (see Figure 6.2-12) is realized with the possibility of switching between BCU and TP-UART interfaces.

6.2.4.5
Bluetooth

In addition to IrDA, another standard has been formed to achieve wireless high data rates for short-distance data transfer applications. Bluetooth was released in 1998 by Nokia, Ericsson, IBM, Toshiba and Intel. In December 1999, other companies followed.

Bluetooth was created to provide a wireless system for ad hoc connectivity. It uses the license-free ISM frequency 2.45 GHz for data transmission. The range of a bluetooth device with 0 dBm transmitter output power is about 10 m. Therefore, bluetooth is competing with applications which are currently running with IrDA. Bluetooth provides isochronous and asynchronous data connections with transfer rates of up to 723 kbps. At the time of writing there are only a few devices on the market. Some solutions such as bluetooth PCMCIA extension cards will be very useful also for controlling the EIB using any kind of PDA or Web-Pad. In future this will be a good choice for portable devices even though some problems have to be solved. As the ISM frequency at 2.4 GHz is license free, other wireless applications are also running on the same frequency. Wireless local area network (LAN) is operating at this frequency and it is severely interfered with by Bluetooth.

6.2.4.6
The EIB Modem

The EIB modem, which has nothing in common with a modem as it is used for data transmission on telephone lines, consists of an 8-bit microcontroller. The specific controller we use is an 8051 derivative with two built-in serial UARTs and has a very low power consumption. Figure 6.2-13 shows the structure of the EIB modem hardware.

Because of possible different voltage levels of the EIB installation and the connected computer, an optical isolation of the RS232 lines is essential. In our case, the optocouplers are supplied by the signal lines, whereas the microcontroller is supplied by the EIB BCU. It is only allowed to drain about 10 mA from the BCU, so the whole circuit's power consumption has to be very low.

The EIB modem allows three different operation modes, which can be selected by simple serial commands. The so-called ETS mode just builds a simple bridge between the connected computer and the BCU for using standard EIB software. The other two modes allow ASCII telegram communication either with or without CTS/RTS hardware handshake. In this case the transfer rate is 19 200 bps instead of 9600 bps. This higher data rate is necessary for converting the bytes of EIB telegrams to a space-separated hexadecimal representation. Every transmitted byte will

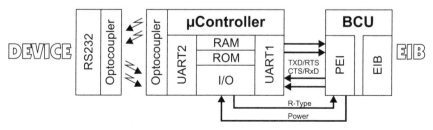

Fig. 6.2-13 EIB modem hardware structure

be expanded into three characters. Every whole telegram ends with a ⟨cr⟩ and ⟨lf⟩ character, so the transmission can be controlled by a simple terminal software.

No data will be lost because of the doubled speed. The normal BCU uses 9600 bps, but is not able to handle 100% of the bus traffic; in fact it is much slower and has only a buffer for one telegram. For this reason, the type resistor of the BCU is switched. If no computer is connected to the EIB modem, the controller switches the resistor off and the BCU leaves the listen mode, which clears its receiving buffer. Now no buffer overflow occurs in the bus coupling unit because of unhandled telegrams.

The EIB modem will look like a commercial standard serial adapter. In fact, it should be compatible with such an adapter but offer an additional, simpler access to the EIB for other applications.

6.2.4.7
Software Interfaces

6.2.4.7.1 Connecting Applications to the Different Interfaces
In addition to the development of the bus access, it is necessary to develop a comprehensive set of software modules to allow different kinds of applications comfortable use of the above-described EIB interfaces. Of course, it is important to allow multiple access to the bus by different applications at the same time. In using the DCOM (Distributed Component Object Model) a component software architecture was developed where different applications may even be distributed on different computers, which are connected via a LAN. To meet these requirements we have designed a stack of dynamic link libraries and a server application described in the following. The complete architecture is shown in Figure 6.2-14.

6.2.4.7.2 The Access DLL
The Access DLL is a dynamic link library that encapsulates the hardware ports and provides asynchronous bus access for its clients. There is a port class in object-oriented design for each supported hardware port, one for the serial port, one for the USB interface and one to access the bus system via IrDA. The Access DLL exports functions for opening and closing a port and also for sending and receiving telegrams to/from the EIB bus.

6.2.4.7.3 The Service DLL
Most applications for the EIB do not only need a single telegram interface as it is provided through the Access DLL. Many programs have to deal with complex standard functionality provided by the EIB bus such as reading the error flags in a BCU or downloading an application program. For these recurrent actions the Service DLL consists of a set of so-called service machines encapsulated in service ports. One service port, for example, allows downloading of an application program into a BCU. The client of this DLL only has to call a start function supplying the physical address

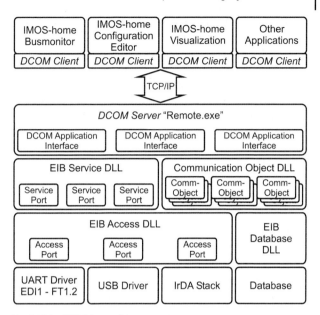

Fig. 6.2-14 IMOS-*home* driver structure

of the target device and an array of bytes containing the user program. The service port then creates all necessary EIB telegrams, sends them to the bus, and waits for the desired response. In communication with the calling client the service port gives progress information, possible error codes, and the end message if the service succeeds. Service ports are always connected with an access port out of the Access DLL. That means opening a service port always leads to opening an appropriate access port to allow communication with the EIB bus.

6.2.4.7.4 The Communication Object DLL

In dealing with a running EIB installation, communication objects play an important role. These are, in the sense of the EIB, the data interfaces for components communicating with each other via the bus. Each communication object is also the place of storage to hold the actual value of the object and therefore it owns some bytes of memory.

In standard components, such as a switch actuator or a light sensor, these communication objects are implemented in the BCU. However, as the memory of a BCU is limited, the number of objects is also limited. For example, to connect a visualization tool there may be the need for more than 1000 communication objects. There is no BCU that has sufficient memory to hold so many objects.

To solve this problem, our Communication Object DLL has been implemented. It contains the framework for the organization of the communication objects as well as the object itself to be used by application programs. Each communication object represents also an object in the sense of object-oriented programming. After loading the DLL the communication objects will be created dynamically by a function call. The structure of the objects can be read out of the project database. For this purpose the DLL opens the database, reads the table that describes the communication objects, and closes the database afterwards.

To realize the communication to the bus system, the Communication Object DLL loads the Access DLL and opens the wanted port. It registers itself as a client for all group-addressed telegrams coming from the bus. When a telegram arrives from the bus it will be routed to the affected objects and interpreted. After the new value is copied into the communication objects the clients will be notified using windows messages. A complex set of functions for managing the states of the objects and for error processing guarantees a high speed of reaction while minimizing the bus load.

6.2.4.7.5 The DCOM Server

Multiple applications should be able to communicate with the bus system at the same time. The Access and the Service DLLs, of course, may be opened simultaneously by different programs. However, according to the new architecture of 32-bit windows, every application 'sees' its own instance of the DLLs. That implies that sharing one port between different applications is not possible in standard programming. To avoid the conflicts of cracking the separated address spaces of different applications, it was decided to use the DCOM concept. All the functionality that one needs to share one object is supplied. The server, called Remote.exe, acts as a stand-alone server program, which is able to load all three described DLLs. It offers the above-mentioned services in three DCOM interfaces, one for the functions of the Access- DLL, one for the Service DLL and one for the Communication Object DLL. These interfaces can be used on the same computer by different client applications. A fourth interface has been implemented to realize the callback to the applications.

However, DCOM also allows application programs to be distributed on different computers. The communication between one computer connected to the EIB and the second with the application program is done via a TCP/IP network connection. The server computer is identified via its computer name or the IP address.

6.2.4.7.6 The Client Programs

For applications which want to communicate with the EIB, there are now different possibilities to connect. The first possibility, of course, is to open a hardware driver directly. However, this is not a smart solution, because here all the code one needs to manage communication has to be implemented in the application. Therefore, it is much better for programs which need a telegram interface to EIB to use the Access- DLL. The API of this DLL is much more comfortable and the functionality to open different kinds of hardware ports is already included.

If an application needs to use standard services it is suggested to open the Service DLL. Complex state machines can be started with the call of a single function.

For programs such as visualizations the Communication Object DLL supplies a comfortable API. Values of the process image can be read and manipulated without ever interpreting or creating a single EIB telegram.

However, all these solutions have one important drawback. Opening a port out of one of these DLLs, loaded directly into the address space of the application, means opening it exclusively. Certainly another application can also load the same DLLs but already opened hardware ports cannot be accessed twice.

The solution is to use DCOM server Remote.exe. This program offers all functions out of the described DLLs via its DCOM interfaces. Now multiple applications can use the same hardware ports not only on the same computer but also on any computer connected via a LAN or the Internet. Of course, to use the functions of the DLLs via a network, a DCOM connection has to be established.

6.2.4.8
Accessing the EIB with Windows CE and Other Operating Systems

Microsoft's Windows CE is an operating system for the area of embedded systems. There are many hardware platforms on which Windows CE runs, eg, MIPS, ARM, SH3, and SH4. All these platforms can vary widely in their interface circuitry regarding the serial port. The implementation of an EIB serial device driver for Windows CE is in fact not a problem but because of the above-mentioned variety one implementation is only running on one specific hardware.

A possible solution is to use a standardized serial PCMCIA card and create an EIB driver that uses the PCMCIA driver of the system. Such a driver also has the advantage that it can be implemented as a loadable module that is independent of the kernel. The drawbacks are the need for additional hardware and software. A PCMCIA interface and the PCMCIA driver stack are necessary. Another drawback is again the dependence on the serial PCMCIA card. One specific EIB driver works only with one specific ASIC of the serial card. If there is a new revision of the ASIC a new EIB driver must be implemented.

In other operating systems the drawbacks of the implementation of an EIB serial driver are similar. To avoid these software problems, the EIB modem was developed that provides platform-independent and operating system-independent access to the EIB (see Section 6.2.4.6).

There are a lot of hardware and/or platforms on the market that may need access to the EIB for control, visualization, gateway, or other purposes. In the field of workstations, for example, there are SUN workstations with their operating system Solaris, PCs with Linux, Windows, BeOS, OS/2, MACs with MAC OS, RS6000 with AIX, etc. The variety of embedded devices running a RTOS (real-time operating system) is even higher. Just to mention some, there are RTOS like QNX, VxWorks, Nucleus, OSEK. If one looks at the huge variety of embedded devices running a non-RTOS or just proprietary software, the idea of writing a device driver for each one of them fades very fast.

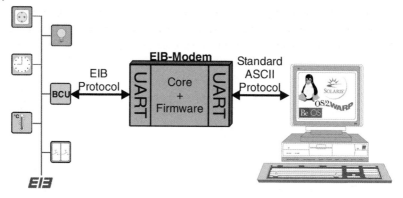

Fig. 6.2-15 EIB modem connects host system to the EIB

The EIB modem contains two serial ports and a core segment where the firmware is stored (see Figure 6.2-15). The first serial port is connected to the BCU on the EIB and the second port is plugged to a serial interface of the host system. Therefore, one can use standard ASCII protocol on the embedded system to access the EIB.

6.2.5
Bus Monitoring and Service Programs

6.2.5.1
Bus Monitoring with Different Interfaces

By using the EIBService.dll and the EIBAccess.dll, a bus monitor program can communicate with the different hardware interfaces such as serial, USB, and IrDA.

In addition to a solution with one PC, the special DCOM architecture of the EIBRemote.exe and EIBMonitor.exe makes it possible to monitor the EIB locally (direct connection of PC to EIB) or remote via Intranet/Internet.

6.2.5.2
Bus Monitor Structure and Functionality

6.2.5.2.1 Services from EIBAccess.dll
The EIBAccess.dll includes methods to use the different interfaces to the local BCU. It is possible to read different parameters from the local BCU such as mask version, the serial protocol type. It is also possible to select the OSI layer of the BCU.

6.2.5.2.2 Services from EIBService.dll
The methods included in EIBServices consist of special tasks. It is possible to find all EIB devices in a line/area or one can read and write the physical address of

any EIB device connected to the bus. It is also possible to read the memory of an EIB device and write to it (upload/download). With methods of the EIBService.dll there is the possibility of resetting an EIB device or just clearing the error flags and switching the programming LED.

Another part of the services is macros. A macro consists of several telegrams to perform complex tasks or a telegram pattern and is just a comfortable way to provide an easy use of one macro instead of many separate telegrams or completely self-composed telegrams especially in bus monitoring.

6.2.5.2.3 EIBRemote.exe

The EIBRemote.exe works as a DCOM server and combines the services of EIB-Access.dll and EIBService.dll to an enhanced interface. Any application and especially the bus monitor acts as a DCOM client and can communicate with the DCOM server local or remote via Intranet/Internet. With a connection to the project database one can get the interpretation of the sent and received telegrams and release this information to the clients, especially to the bus monitor.

6.2.5.2.4 EIBMonitor.exe as Bus Monitor Program

The EIBMonitor.exe is the appropriate DCOM client to get full access to the EIB (Figure 6.2-16). With the interpretation of the telegrams in the DCOM server one

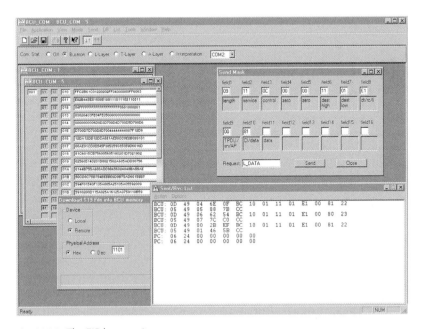

Fig. 6.2-16 The EIB-bus monitor program

can obtain interpreted information about the telegram traffic, thus making error detection easier. Inside the monitor each received telegram gets a time stamp and is stored in a list to be checked for errors. To send telegrams one can fill in the appropriate fields or use telegram patterns for special purposes and add the rest.

6.2.5.3
Future Work: Interpretation and Test Management

The bus monitor can find out how many devices are in a line or in an area. With the decoded list the interpretation of incoming telegrams is done. One new service will extract the 'group communication channels' and should test the decoded channel (checking the sender and the receiver). It seems of interest to build test sequences for every channel to get a functional evaluation of the whole installation. Combined with report generation, a reliable distributed network is created by scheduling these tests at different times in a year. The next step is to find out the EIB device that produces an error. In combination with the remote access a technician is able to check the installation from his office, find the EIB device that does not work correctly (if any) and just has to change this component or the configuration of that component.

6.2.6
Configuration of Home Automation Systems

6.2.6.1
Introduction

In modern systems of building automation it is not enough to connect the new device electrically with the bus, there is also the need for allocating a unique address, depending on the possibilities of the installed system, and for programming its logical connections, that means configuring the device.

'Configuring the device' or configuration means the description of the components which are installed in a system and the functions which can be used by the system. The configuration of a system itself is the description of data flows and the producing of the logical connections between the single modules, eg, through parameterization of communication channels. It is also the creation of the parameters for signal processing. In addition, signal processing algorithms can be inserted during the configuration process if it is not already included in the notion 'programming', eg, choice and connection of the different software modules for the wanted algorithms. A comprehensible description of the algorithms for an adequate platform can then be loaded as a download in the decentralized component.

For configuration of EIB systems the unique tool ETS (EIB Tools Software), shown in Figure 6.2-17, is applied. It is able to configure the whole system as well as every EIB device certificated by the EIBA owing to the database of the de-

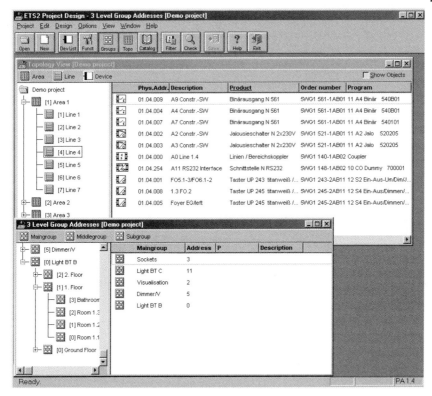

Fig. 6.2-17 The ETS software

vices' properties (product database) and the fact' that all devices are based on the standard BCU, which is described in Section 6.2.3.3.

The whole of the information that was obtained in the process of system configuration is written in the project database, with the possibility of changing or extending it.

It was mentioned above that the configuration of every EIB device starts with an assignment of a unique physical address (see Figure 6.2-18), which corresponds to the location of the device in the installed EIB system and consists of the number of the area, of the line and of the device in the line.

To have the possibility of distinguishing the necessary device from another in the address-programming process, every standard BCU has a button, which switches it into the 'Program Mode'. Only in this mode does the BCU accept requests, which assign it a new address. After appropriating the device with the physical address its further configuration is possible on the basis of the connection-oriented communication.

The next important steps of the configuration are:

Physical address																
Byte 1								Byte 2							
7	6	5	4	3	2	1	0	7	6	5	4	3	2	1	0
area number				line number				device number							

Fig. 6.2-18 Structure of the physical address field

- parameterization and loading of the application
- generation and loading of the address table
- modification of the association table.

A more detailed description of each of the steps and parameters of the configuration can be found in [2].

6.2.6.2
Easy Configuration

Experience shows that in EIB installations the costs for the configuration of the system are high. Additionally to the installation costs there might be expenses for the training and the additional equipment such as the ETS (EIB tool software) and a portable computer.

For home automation systems making their way into installations with low functionality there must be a way to make installation easier and cheaper. Specifications made for the Konnex Technology address this problem by introducing new ways of installation – the easy configuration modes. Konnex Technology is the merger of the three big European bus systems for installation, the European Installation Bus (EIB), the BatiBUS and the European Home System (EHS). This merger was made to ensure one common system of home automation in Europe. The three associations representing the three different bus systems – the EIB Association (EIBA), the BatiBUS Club International (BCI) and the European Home Systems Association (EHSA) – work together to a common standard represented by the Konnex Association.

In the specification of the Konnex Association there are three different configuration modes for configuring a bus system:

- S-mode (system mode)
- E-modes (easy modes)
- A-mode (automatic mode).

The S-mode (system mode) represents installation in the system mode with the use of a software tool such as the ETS for the EIB. This allows installers trained with the ETS to install home automation systems as they are used. This makes the system mode useful to these installers, who are trained with the ETS.

The E-modes (easy modes) are different configuration modes with the goal of easing the configuration of bus systems. Here usually no software tool is needed and the installation is faster and easier than with the ETS. This installation mode only allows limited functionality compared with the S-mode. There are mainly two different E-modes: one based on logical tags and one based on a controller. In the E-mode based on logical tags the communication objects are set using the links indicated by the logical tags. The logical tag could, eg, be implemented by a coding wheel. This configuration mode is easier than the S-mode, but limits the possibilities of a bus installation.

The controller-easy mode is based on one central bus device, the controller, which is able to configure new devices. Here, the installer chooses the devices that should communicate, eg, by pressing a button on each device. The controller reads the information stored in the device (the device descriptors), checks whether they fit together (have compatible communication objects), and sets the association tables and communication addresses accordingly. This is easier than configuring every device by ETS, but introduces an extra bus device, the controller, which is not needed in the other modes.

The A-mode is based on runtime-interworking standards of European Home System Specification 1.3a (EHS 1.3a), which is a plug and play standard implemented in existing EHS devices. The standard of EHS 1.3a was adapted to be compatible with the other bus systems and for runtime-interworking to function between different bus devices.

All these modes set the future in the installation of European bus systems and will be implemented in bus devices which fulfil the Konnex specifications.

In the project IWO-BAY at the Lehrstuhl für Messsystem- und Sensortechnik there are algorithms developed for the easy configuration. These are essential for configuring the RF-based bus system developed at the Lehrstuhl für Messsystem- und Sensortechnik (see Section 6.2.3). The algorithms will be implemented on a central base station, which configures the bus units and gives the installer a graphical interface.

In the beginning a new easy configuration component has the following information pre-configured:

- application ID
- application program
- filled communication object table
- empty address table
- empty association table
- physical address 0xFFFF
- all error flags are set (the application program is stopped)
- AddressTableLength = 1 (no group communication is allowed).

The process of configuring a new easy configuration component via the base station is as follows:

1. The base station is triggered to look for new components in the system.
2. The component is activated by pressing its Learn button. It connects to the base station which assigns a physical address from the database.
3. The base station adds the new component to its database.
4. Other new components can be added (steps 1–3).

5. The base station asks the installer what information to fill into address tables and association tables, ie, which components have to be joined together.
6. The installer connects the different components together and the base station configures the different components according to it.
7. The base station makes a test of the configuration and adds the information to its database.

This algorithm is an example of the integration of installation routines in a home automation system. It uses the graphical user interface provided by the base station normally only used for controlling the home automation system to give the installer easy ways of configuring a bus system. Another way of configuring a bus system is via the Internet. This is described in the next section.

6.2.6.3
Configuration via the Internet

Another future plan for bringing the EIB to every home is the ability of configuring the EIB via the Internet. This functionality is implemented in the iETS (Internet EIB Tool Software). It consists of a server and a client part. The iETS server is installed on a computer directly linked to the EIB system and to the Internet. Thus the iETS client can access the server via the Internet to configure the EIB system. The client consists of a normal ETS, for which there is an option pack adding Internet functionality. The iETS server has to be bought as an extra tool from EIBA.

This tool makes it possible to do all the configuration work via the Internet without being present at the EIB installation. This means a great benefit and facilitation when existing EIB installations are altered, because the installer can do the reconfiguration from a distant point.

6.2.6.4
The IMOS Tool

The software IMOS-*home* (integrated management of open systems for the private home) represents the concept of an unified tool for the configuration of different hardware platforms, including devices that work on EIB.

The graphical interface of IMOS-*home* shown in Figure 6.2-19 offers the user the possibility of visualizing his distributed automation system. The control of the program is based on an interface that is easily understood by the user and hides the large and usually more complex part of the hardware configuration for different devices of the distributed system. The user has the possibility of building his own system by those hardware devices, which are stored in the library, or of constructing his own hardware device on the basis of signal-processing algorithms. With the aid of simple manipulations of the graphical objects on the screen the user has the possibility of fast and flexible configuration of signal flows in the distributed system, which helps in saving time in designing and programming the system.

6.2 System Technologies for Private Homes | 539

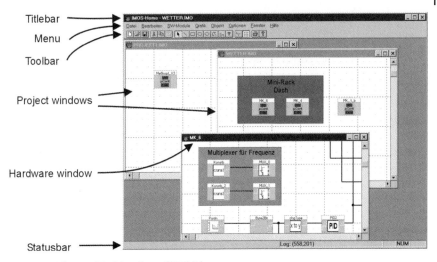

Fig. 6.2-19 The graphical interface of IMOS-*home*

After the step of the graphical configuration of the distributed system is completed, the software package IMOS-*home* transforms the visual information into logical connections of different devices between themselves and prepares the necessary data and algorithms for downloading them into devices.

6.2.7
Visualization and Tele Services

6.2.7.1
Possibilities of Visualization

For controlling purposes and visualization a variety of devices can be taken. For the home automation area the availability of devices which are capable of displaying status information about a building is very important. Also the device has to be more or less mobile. The choice has to be made according to the needs of the application. For displaying the status of a whole building a big display is needed. To access only a few functions a small PDA is sufficient.

A television set can be found in almost every household and is very useful for displaying information. The resolution of the display is acceptable but for controlling and sending commands it needs some enhancements such as an additional remote control. There are products which modulate the displayed information into the video signal of a television set. All the controlling functions have to be done by additional hardware which is connected to the EIB and the television.

More useful is a PC, which is more flexible and can be easily adapted to all needs. A PC is equipped with an external display. A laptop already has an integrated display but still is too heavy to be carried around all the time. For mobile applications with the need for a big display Web-Pads are suitable. They also provide touch screens as a means for user inputs and several slots such as PCMCIA slots for extension cards.

If a big display is not needed, PDAs are the best choice. They are developed to be small, lightweight and with low power consumption for standalone battery supply.

6.2.7.2
Video and EIB

EIB is particularly suitable for controlling actuators and visualization of all the sensors of a building. For visualization inside a building and especially security aspects an additional component is needed, ie, the faculty to display still images or motion pictures.

There are only a few possibilities for displaying images on computer-based hardware. To convert the analogue video signal into a digital byte stream some kind of hardware is needed. For the use of PCs there are several grabber cards on the market which fulfill this task. The visualization software using grabber cards has to recognize and handle access to the manufacturer-dependent hardware. This is normally done for Windows operating systems by using software libraries shipped with the product. That means that an own access method has to be developed for every manufacturer or even product type of the grabber card.

However, besides doing the A/D conversion inside a computer, this task can be accomplished by the camera itself or an external module which provides a standardized protocol to transfer the digital video images. There are interfaces to USB, IEEE1394 and TCP/IP. Some cameras provide a web server which generates HTML documents, picture streams or only single pictures. The camera server is TCP/IP based and communicates using the HTTP protocol. HTTP GET commands are used to request single pictures or video streams. The pictures of the used video server are encoded using the JPEG format. The server supports also multipart MIME types (x-mixed replace) as defined by Netscape. Multipart MIME types are well defined but this special subtype (x-mixed replace) is only implemented by Netscape. Every new (JPEG) picture is separated by a boundary described in the HTTP header.

The visualization software shown in Figure 6.2-20 establishes a connection to the server, starts the request and parses the received data. The JPEG picture is decompressed and displayed after reception. The GET request for the video server can be parameterized as shown in Figure 6.2-20. The dialog is used to enter the IP address of the video server, to supply the GET string and to choose the camera.

The disadvantage of this server is that for each video stream a single TCP/IP connection is established and kept open. To reduce the needed bandwidth for transmission of the video streams it will only be displayed and opened if the current document of the visualization software containing the video is on the top and shown. If another application gets into the foreground the video stream is released. The video picture can be placed on any page of the EIB control document.

The software has been installed on computers with touch screens in a building near Munich which serves for research purposes. In this case the camera is con-

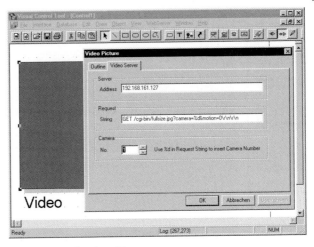

Fig. 6.2-20 Configuring video stream

trolled by the EIB. The camera can be moved horizontally and vertically and switched off if not needed.

6.2.7.3
Visualization Software

The visualization is a software-based tool to display and control a whole home system within only one device. The solution described here is written as PC software for Windows, but it does not look like this. The program does not show any typical Windows elements such as a title bar or scroll bars. The user finds a design that may look like that in Figure 6.2-21. He can choose between different views by pressing a button on the bar at the left of the screen.

With this tool one can not only access the information on the home bus, it also offers the function of a telephone which can be used to communicate with the telephone at the door. In addition it is possible to show the pictures of several cameras. These functions will be shown in the following sections.

6.2.7.3.1 Controlling the Home

To control the home system the software shows different graphical elements on the surface. Some of them allow user input, some display the actual values and states of the sensors and devices in the house. For easy handling it is very important that all user input can be done with a mouse or using a touch screen. Figure 6.2-21 shows a page for lighting control.

While the structure of the home system is stored in a relational database, the surface of the user interface is stored in several files, one for each page. This software is developed especially to be used with the European Installation Bus EIB.

Fig. 6.2-21 The lighting control

To access the information on the bus, the visualization device is connected to the bus system via the serial port or the USB port using an appropriate adapter.

6.2.7.3.2 Telephone Application

To use the visualization tool as a telephone, the device has to be connected to the S0 bus, the ISDN line. The device behaves like a normal telephone in the system and can be used like this. For the telephone application a special view is reserved. It appears automatically when a phone call arrives or somebody takes the handset.

6.2.7.3.3 Displaying Outdoor Cameras

For security aspects it is very useful to survey, for example, the entrance of the house with the help of video cameras. To get the pictures into the software the program uses the so-called streaming JPEG format. The video stream comes from a video server, which is able to grab up to four analogue video streams and convert it to the streaming JPEG format. The connection between the video server and the visualization device is done through a LAN via the TCP/IP protocol. It is also possible to use pivoted cameras. The move commands are transmitted via the EIB bus.

6.2.7.3.4 Design of the User Interface

As most buildings are planned individually, the user interface also has to be designed specially for each house. That means that one not only need the visualization tool but also an editor to design the surfaces.

To design the user interface the configuration database of the bus system is used. In this database it is easy to find which devices are available and which functions in the net are to be displayed and controlled. The user can choose which items he wants to see later on his screen and also select an appropriate outline. For example, he wants to see the temperature of the living room. Then he has to look for this sensor in the database and click on its output. Then he can decide whether he wants to see the temperature as a number, as a bar or both.

In the same way, the elements of the telephone application and of the camera display can be added and adapted. The design criteria are totally free. For example, one can do it in an architectural style where one shows the outline of the building and places the controls on it, or one can plan a special surface for each person living in a house. One can design an individual look.

Of course, this high flexibility in design is difficult for many people because they do not know how to start. In this case one can use a design assistant which does the searching in the database and suggests a graphical representation to the user. In standard systems such as a four-room flat with standard equipment it is also possible to take a template supplied with the program and adapt it to the project. Figure 6.2-22 shows the design editor running.

After designing a user interface surface, it can be stored in a file and downloaded into the different devices as described below.

6.2.7.3.5 Implementation and Hardware

The visualization is programmed for Windows 32-bit, so it runs on every standard PC with this operating system. However, normally one does not have to think about a PC tower with monitor, mouse and keyboard. They are all built as embedded PCs, which do not look like a computer at all. For example, Figure 6.2-23 shows an embedded PC device for wall mounting. It is directly connected to the bus and uses, in addition to some keys, a touch screen for user input. On the left side is the handset used for the telephone.

Figure 6.2-24 shows the visualization on a TV set. In this realization the software runs on a set-top box that does the connection between the home system and the TV set. In this solution one have all features except the telephone application.

6.2.7.4
Special Applications and Clients for Visualization

As the time between the consumption of energy in private homes and the billing for the consumption normally averages 1 year, the inhabitant gets no feedback about where and when he consumed the energy. There is no feedback to let him know where he could save energy and in consequence costs for energy.

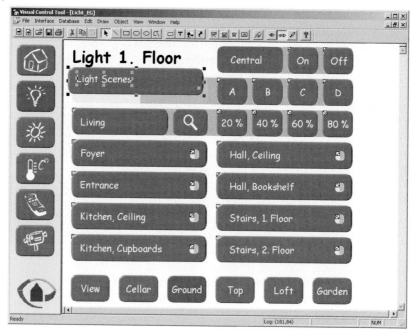

Fig. 6.2-22 The design editor

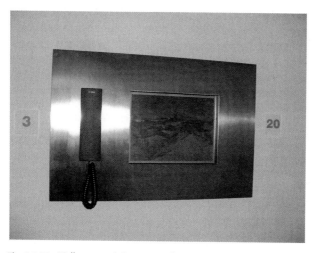

Fig. 6.2-23 Wall-mounted device (Siedle)

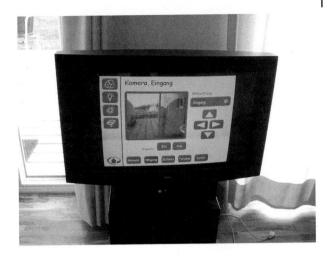

Fig. 6.2-24 Set-top box for television (Loewe-Opta)

This leads to the idea of 'energy monitoring', the visualization of energy consumption in private homes.

The largest amount of energy consumed by a household is used for heating purposes. As the consumption is not located in one place, a decentralized measuring system and bus system is needed.

The institute of Messsystem- und Sensortechnik has developed hard- and software concepts that allow the visualization of the energy consumption in a temporal resolution of about 1 h. The necessary visualization units receive the consumed energy in form of a counter value from heat cost allocators with a wireless interface (see Figure 6.2-25). The visualization units calculate the amount of energy in every room of a flat according to configuration parameters such as the normal power of the radiator and the thermal coupling between the sensor and the radiator (k_C) and several other constants that depend, eg, on the structural shape of the radiator. The visualization of those values is done by a graphical LCD that is controlled by a microcontroller. Every flat has one of these visualization units and every unit is coupled to the EIB via an included BCU. The current counter values are transmitted over EIB to a central house unit that saves and compresses the values. This central unit also calculates building-specific mean values that are transmitted back to each visualization unit. The schematic system structure is shown in Figure 6.2-26.

The visualization itself is done by bar charts with a corresponding text legend as shown in Figures 6.2-27 and 6.2-28. Figure 6.2-27 shows the current consumed energy values in kW h and Figure 6.2-29 shows an evaluation over the year (black bars) and the previous year (white bars).

Having such information, it is possible for the inhabitant to recognize energy waste and localize the reason for the waste (eg, slightly opened window with turned-on radiators or a damaged radiator valve).

546 | 6 System Technologies

Fig. 6.2-25 Heat cost allocator

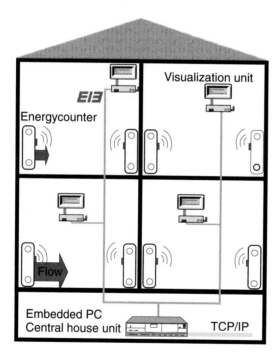

Fig. 6.2-26 System structure for energy monitoring

Fig. 6.2-27 Consumed energy values

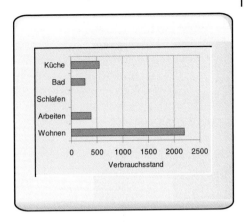

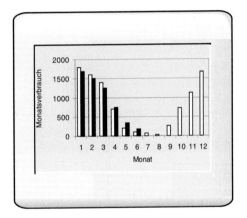

Fig. 6.2-28 Comparison with previous year

The central house unit has a built-in database structure with automatic compression and reduction of the received data to safe memory. The unit also supports TCP/IP for remote access. The data can be safely transmitted via different available hardware interfaces (Ethernet, ISDN, modem) to the energy service provider that bills the consumed energy to the inhabitant.

6.2.7.5
Access Technologies

Nowadays for most people it seems to be very important to get access to their EIB installation by the Internet, in different ways, eg, ISDN, LAN, etc. It is possible to use special network applications or standard Internet browsers which are available all over the world. In this case there are many things to be thought of, because of the different features of today's Internet browsers. There are some technologies such as DHTML, ActiveX, Applets, Java Script or CGI which could be used for controlling an automated building. Unfortunately, not every browser interpret this

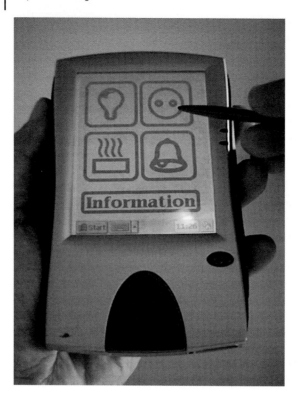

Fig. 6.2-29 Windows CE-powered PDA with pocket browser

code in the same way or even rules the technology. Whereas sending commands to the EIB can be easily achieved by HTML, using formulas with POST tags, which will be interpreted by a simple server application, receiving actions from the installation is rather complicated. If the state, eg, of a lamp changes, the server has to inform the client browser, so it can switch an image or the color of a button. Because standard HTML provides a one-way communication (client pull), this is not possible. Some browsers, such as Netscape Communicator, allow a server push mechanism, which holds an open connection, so the server is allowed to send a new document if any state changes, but not every HTML client. Efforts are being made by the WWW Consortium to make this mechanism standard. A simple solution is the automated page reload in defined time periods, where the whole page is reloaded by the browser after expiration of a predefined time. Another method with GET tags is described in the following section.

The most powerful technology is definitely provided by Java applets. To use applets, a so-called virtual machine has to be installed in the browser. The applet which is loaded from the web server is executed in its own part of the HTML page. This application can set up its own connection to an EIB server for controlling actuators and restating values of sensors. Each applet could visualize one or more devices of the building installation. Applets seem to be the best solution for our needs if a powerful computer is available. The current state for small PDAs with Windows CE or Palm OS browsers does not provide applets, which leads us to use a CGI method.

6.2.7.6
Use of PDAs with HTML and CGI

Our aim was to develop PDA software for wireless controlling devices in the home (see Figure 6.2-29). To provide one software solution for different hardware platforms we decided to use simple HTML pages with GET tags.

The PDA is connected to the server with a TCP/IP connection by the related operating systems. This could be done by infrared light over the IrDA stack, wireless LAN cards or soon by Bluetooth. The HTML server is located on a Windows PC, which is connected to the EIB installation by a serial interface.

Whenever a button on the HTML page is pressed, a GET request from a form tag is sent with all necessary parameters (new value and number of the communication object) to the web server. Now a small CGI program, in our case a simple exe-file, translates the request and sends the information to a OLE server (see Figure 6.2-30). Here the required telegrams for the EIB installation are built and sent over the serial port.

Because of the GET request, the client browser is waiting for a new HTML page from the server. The server now sends a new page with the current states of the devices, which it fetches from the OLE server. These states are provided as pictures which also are used as buttons. The user now gets a response of the state of the devices every time he switches an object. This technique also allows combination with an automated reload timer.

6.2.7.7
Standard Browser and EIB. The EIB Web Server

Personal computers and small electronic devices such as laptops, mobile phones and PDAs can be used to get control over the EIB. This can be done by develop-

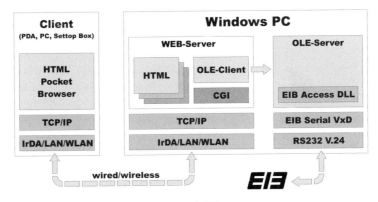

Fig. 6.2-30 Controlling EIB via HTML and CGI

ing special software together with the use of embedded systems. As an example of such a solution an implementation for an existing installation of a building for experimental research which is part of the project tele-Haus sponsored by the BMBF (Bundesministerium für Bildung und Forschung, the German Ministry for Research) will be shown with a practical presentation. The software enables a Windows-based computer to be run as an HTML-compatible web server. The EIB installation can be controlled and visualized by a standard Internet browser.

Another approach is to connect the EIB to a residential gateway which unifies the access methods (Figure 6.2-31). This gateway not only provides a standardized medium of communication but also gives the chance to access the functions of a private home using the capabilities of a WAN. Therefore, all services starting from EIB for home automation, integration of PDAs, and other devices into an enclosing system architecture has to be developed.

The kind of connection to the external user can roughly be distinguished by three main aspects: security, transmission speed, and the protocol which is taken. The most secure way to transfer data is a point-to-point connection. This can be achieved by using two modems to connect two PCs. All other techniques especially using the Internet result in a transmission medium which carries more than one connection in parallel. The transmission speed has to be adapted to the amount of data to be transferred.

The last aspect is very important regarding the level of compatibility and the requirements for development efforts. The Internet protocol is (TCP/UDP-) IP. These protocols are very widespread and adopted in almost every hardware platform. The alternative is to take a proprietary protocol. This reduces the alternatives for using standard applications to display the data remotely too much and is not taken into account. Even PDAs can perform a TCP/IP protocol to access the Internet. Therefore, a TCP/IP connection is used to get access to the gateway which is placed inside the house.

An IP-based protocol can be run either by using a direct connection to the Internet or a dial-up connection through ISDN or analogue telephone line as men-

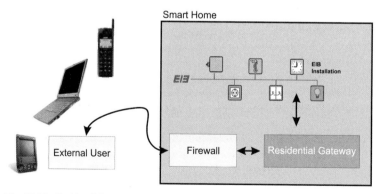

Fig. 6.2-31 Residential gateway

tioned above. If more bandwidth is needed, one can think about xDSL connections.

The gateway shown in Figure 6.2-31 performs several tasks. It has to provide access to the EIB installation and allow multiple access paths to the building. In order to be independent of special software for display purposes at the client side – the gateway will be the server – a standard Internet browser will be used. On each PC an Internet browser can be found which uses the Hyper Text Transfer Protocol (HTTP). Therefore, the gateway will have to run parts of the HTTP (1.0/1.1) at the server side.

The web server has access to the EIB using the EIB driver software. The presentation of the web interface can be edited locally with a set of given elements: simple graphical elements such as circles, rectangles, pictures in JPEG and BMP graphics format, and active components represent the state of an EIB object (eg, output of a binary actuator).

The EIB driver software provides the EIB access through a serial port and even does the communication object management. The web server is notified by changes and updates its views. If in the other direction the state of an object is changed by the user, a request is sent to the driver. The internal image of the communication object is updated after receiving the confirmation of the sent group telegram.

The communication between the Internet browser on the client side and the web server rests upon TCP/IP-socket connections. As the server-side software is written in Visual C++, the socket functions are encapsulated in Microsoft Foundation Classes (MFCs). To get a notification about client requests a listening socket is created at the start-up time of the web server. The web server can be activated and deactivated in runtime.

If a remote connection is established the server gets a Windows message on the listening socket. This socket spawns a new instance of the request socket object. This instance is now responsible for all following actions. After accepting the incoming connection and creating the new instance, the listening socket is available for new requests. After fulfilling the request the created request socket is killed. The general procedure for a response to a made request is shown in Figure 6.2-32.

The client's first GET request will initiate an authorization procedure. After the client's authentification the server provides an HTML document.

The server produces three different areas by using frames at the browser (Figure 6.2-33). The leftmost frame shows a document called view manager. This view manager stores links along with pictures. The user can change the current document which is held in the frame at the right by clicking a link at the view manager. Another way to change documents is to place links to another document within a document.

At the bottom on the left side a status frame is placed. In this status frame the result of pending operations is shown.

All frames are built dynamically, which means during runtime in the memory. In general there are some solutions to transform the content of a document to an Internet browser. Only the most important elements of a document should be

552 | 6 System Technologies

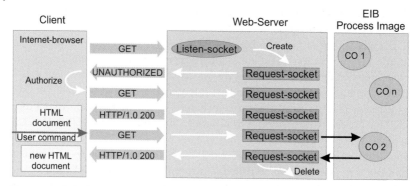

Fig. 6.2-32 Web server responding to incoming requests

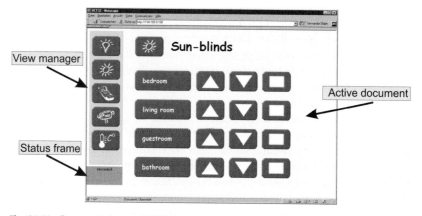

Fig. 6.2-33 Representation as HTML document in Netscape. Output produced by web server

listed: text, pictures, links and EIB objects (eg, switches, bars, text). Each element in the document can be translated in primitive HTML code extended by JavaScript, but this would result in a loss of the displayed information. Another way is to transform each element to a graphical representation and store it in a common spread picture format. As a result, a collection of pictures has to be assembled according the constraints given by the document. Still there would be a problem in placing all these elements accordingly. At the visualization software all elements can be placed in any order, which means that they can even be stacked, eg, a text element can be placed upon a vertical bar.

Therefore, the whole document is realized in one picture. The picture is converted into a common used graphics format, in this case portable network graphics (PNG). In comparison with PNG, JPEG offers a higher compression ratio but the images also become blurred depending on the compression ratio. Each element places additional information in the form of HTML or JavaScript code. The active areas in a document such as EIB objects or links are defined as sensitive areas using the ⟨map⟩ tag in the HTML document. The JavaScript code is mainly needed for active EIB objects. By clicking on one of these sensitive areas the corresponding JavaScript is executed.

6.2.7.8
Security Aspects of Tele-services Using HTTP

Using a direct Internet access also brings some drawbacks. All data passed over the network can be manipulated and analyzed. Therefore, some security mechanisms have to be provided.

Using standard Internet browsers allows some basic authentication procedures. The simplest one is basic authentication. If a client starts his first request the server answers with status 401 (authentication required) to challenge the authentication of the client. The drawback of basic authentication is the transmission of the user name and password in unencrypted form. This only provides a very poor protection. The digest access authentication is extended to encryption of user name and password. Nevertheless, both mechanisms are quite weak. The basic authentication scheme is vulnerable to man-in-the-middle attacks and replay attacks. Once the password and user name is sent as clear text an eavesdropper can access all requested pages. The digest authentication scheme prevents the password and user name from being divulged but still all requested data are transferred as clear text and can be eavesdropped.

This lack of security can only be prevented by implementing algorithms such as Secure Socket Layer (SSL). Secure Socket Layer is quite secure when using 128 bit RSA keys. It is invulnerable to man-in-the-middle attacks and eavesdropping. None of the encryption keys used (RSA, IDEA, or MDA) can be gained. At the client side an additional mechanism (certificates) ensures that the server responses are not replayed. As some of the encryption algorithms used are licensed (RSA in the USA, IDEA in Europe), commercial products have to be licensed.

Another way to protect the contents of a web page is to use SHTTP. This protocol is a superset of HTTP and can use several encryption algorithms in a similar way to SSL.

Another way is to use a dial-up connection to connect to the server. This ensures that (almost) nobody can listen to the network traffic between the two computers.

Similar to this approach, a server side IP list could be managed which lists trusted clients. This is only a small extension due to the fact that IP addresses can be faked.

6.2.7.9
Using Applets, Java in Tele-services

In its beginnings, Sun Microsystem's Java was intended to be a programming language for small devices. Soon it was discovered as a platform-independent programming language not only for small applications. First of all Java is an interpreter language, which means Java is not executed directly by the CPU. In between the compiled code (byte code) generated by a Java compiler and the CPU lies the Java virtual machine (JVM). The JVM interprets Java byte code. The JVM is platform dependent but offers a common interface to Java applications. In general, the byte code can be run on any JVM which refers to the same specification.

There are different JVMs which differ in the given class collection and interfaces to resources of the computer. There are JVMs for embedded applications such as Personal Java by Sun or Java Microedition (based on the Connected Limited Device Configuration) for PDAs and mobile devices and, of course, Java Standard Edition. All of these products include the Java Compiler and additional tools for development of Java programs. For executing translated byte code, the JVM for the specific operating system and a set of class files are sufficient to run an application.

Java Internet applications are programmed as applets. The applets are downloaded by the Internet browsers and executed by the built-in JVM. Like every application that is downloaded and executed, applets have to follow security constraints. There are several restrictions for Java applets. Applets are not able to access files of the local filesystem which are stored at the client side. They can only connect to the origin server where they were downloaded. Java is a powerful programming language and it could easily be misused. These are only a few mechanisms to prevent misuse. The JVM of the browser provides more less obvious procedures to ensure the privacy of data stored on the executing computer.

The advantage of Java applets used together with Internet browsers is their availability. Every Internet browser can execute applets (when configured to do so). However, there are also some restrictions. The browsers are equipped with proprietary JVMs which should follow the specifications, but often some method calls do not work on all browsers. Therefore, the applets should be tested on several browsers to assure that they are working correctly.

Using applets together with an Internet browser allows a connection to the server in both directions. A connection can be held by the applet all the time to the server using sockets (TCP/IP endpoint) to transfer data to and from the server. This connection is useful in home automation applications. It can be used to inform the browser about changes in the process image that have occurred at the server side.

Java has developed a programming language not only for Internet applications. Single applications can also be run not using the applet 'model'. As they are not intended to be downloaded, they offer more flexibility and less security constraints.

6.2.8
Outlook

In November 2000 the congress and fair 'e/home' took place in Berlin. The organizers were content about the resonance of this event. The papers presented demonstrate that the aspect of an 'electronic home' with possibilities to communicate from outside the home was very strongly represented. Home automation, on the other hand, was only a small part in this congress and the emphasis was laid on the applied bus systems. Nevertheless, all the papers showed clearly that this field is growing dynamically. On the other hand, there was no proposal for an overall system and probably there is no chance of having such a system in the near future. And, of course, the occupant of the future home was not questioned as to which parts he will accept and which not. However, there is no chance of having two or three different communication systems in the private home if the complexity of technology is not hidden. So the electronic home with home automation included will have success on the market only if there is a combination of techniques which fulfill the needs of the occupants and which can be handled easily without long training periods. On the other side there is the widespread enthusiasm about all kinds of communication (mobile phones, Internet services, E-mail, etc.). If one can use it for introduction into the 'intelligent home' there will be a good chance that people will accept it and also will invest in it.

What has to be done technically to form the prerequisites? First it is necessary to bring together the different technologies – the information technologies with their rapidly growing possibilities and the building technologies with their numerous sensors and actuators. There has to be a communication infrastructure within the house which makes possible the different services from inside and outside the house. The basic idea behind it is to connect different subsystems (with high and low data rates) via a fast backbone and by intelligent micro-gateways. With such a communication structure one can connect conventional control applications with high-end applications in the field of audio/video (eg, via HAVI). By such a strategy all EIB components which are already on the market can be used and do not have to be redesigned. On the other hand, if there was to term a new bus system for washing machines, dishwashers, etc., it can be integrated into such a home system. The same is true for the connection between TV sets, video recorders, Hifi equipment, etc., where the definition of HAVI as a bus system makes it necessary that, eg, the state of a switch can be shown on the screen of the TV set. Finally, one can introduce via the same backbone all the Internet technologies with its wide spread browsers as user interfaces and with embedded HTTP servers. Since the bus interfaces as an electronic component can be produced at very low cost if the number is high, the future home automation system does not depend so much on the installed physical layer. With this structure one gains a lot of flexibility particularly with respect to installations into existing homes.

In addition, one should not underestimate the necessity for standardized software interfaces in the future instead of having only standardized bus systems as in the past. To define standardized software interfaces on the basis of widespread standard operating systems such as Windows 2000, Windows CE, Linux, etc., and to use modern software technologies such as Object-oriented Programming with Java, etc., will be a great challenge.

If one looks at the sensors and actuators, which are nowadays on the market, many of them were not developed for the home market. Therefore, there will be the need for further components either to widen the scope or to replace not well suited ones. In certain areas such as medical sensors miniaturized microsystems are essential. However, also in many other fields one needs microsystem sensors to design compact sensors with very low energy consumption and low price if the numbers produced are high. Therefore, microsystem technologies are crucial for the home automation market. On the other hand, the number of units is not nearly as high as with RAM chips. The solution to this problem cannot be the monolithic structure of an IC, but a multi-chip module (MCM), which is stackable. Then one can concentrate one function (eg, the microcontroller with the storage ICs) in one layer. By piling up several layers one can design diverse sensors out of a basic variety of modules. The development of such design kits for sensors (and actuators) will hopefully produce sensors which are well adapted to the private home. A typical example where diversified sensor principles should be used is the detection of presence (IR, US, RF). This can be combined with new technologies for identification (transponder, fingerprints and image processing). To extract the necessary information from the sensor signal one needs complex and µC-oriented signal processing algorithms. One important aim will be the transfer of these complex systems to low-cost versions for the consumer market. Also, new sensors for measuring the current energy consumption and non-invasive medical sensors are not only a technical challenge but will decide on the future market of home automation.

Adaptable software packages for planning, configuration, installation, and maintenance are a necessity to increase the acceptance both for end users and for craftsmen. With IMOS-*home* a comprehensive concept was developed within the research project VIMP (see Section 6.2.3). It comprises all lifetime phases of a software project, starting with the configuration but also including daily operation and maintenance. It has to be supplemented and broadened to the new software technologies such as electronic device description, embedded Internet applications with Java and Jini, plug-and-play technologies, etc.

A very sensitive area for the acceptance will be the MMI. Here one needs attractive, highly adaptable, ergonomically designed interfaces (displays), which can be easily altered for different groups of end users (elder people, computer freaks). They have to be self-explanatory and easy to memorize. In the development of these MMIs the emphasis will be laid on interfaces which are designed by unified principles and which will be shown on technically and optically attractive displays and service terminals.

In conclusion, it can be stated that there are still many questions to be answered in the field of home automation, but if industry and research institutes work together trustfully there is a great chance of opening up a very big market like the electronic components and information technologies in the car market.

6.2.9
Internet Addresses:

Homepage of the BatiBUS Club http://www.batibusc.com
Homepage of the EHS Association http://www.ehsa.com
Homepage of the EIB Association http://www.eiba.com
Homepage of the Konnex Association http://www.konnex-knx.com.

6.2.10
References

1 BERNERS-LEE, R., FIELDING, R., FRYSTYK, H., *RFC1945: Hypertext Transfer Protocol – HTTP/1.0*, Aay, 1996.
2 BINTERNAGEL, L., MÜGGE, G., *Energie-Monitoring – Visualisierung des Energieverbrauchs*; HLH, Vol. 51 No. 9 (2000) pp. 82–83.
3 BINTERNAGEL, L., MÜLLER, W., STOLYAR, R., SCHNEIDER, F., *New Technologies for Connecting EIB to Computers and Energy Monitoring*, 3 EIB Scientific Conference, Munich, 5 October 2000.
4 BINTERNAGEL, L., SCHNEIDER, F., in: *Das intelligente Haus – Arbeiten und Wohnen mit zukunftsweisender Technik*; SCHNEIDER, F., TRÄNKLER, H.-R. (eds.); Munich: Pflaum, 2001.
5 BINTERNAGEL, L., SCHNEIDER, F., Tagung Sensoren und Meßsysteme 2000, Ludwigsburg 13 March 2000, in: *VDI-Bericht*, No. 1530.
6 DIETRICH, D., KASTNER, W., SAUTER, T. (eds.); *EIB. Gebäudebussystem*; Heidelberg: Hüthig, 2000.
7 EDDON, G., EDDON, H., *Inside Distributed COM*; Redmond, WA: Microsoft Press, 1998.
8 EIBA, Brüssel, *EIB Handbook*, Issue EIB 2.21; Brussels, 1995.
9 EIBA, Brüssel, *EIB Handbook 3.0*; Brussels, 1998.
10 FRANKS, J., HALLAM-BAKER, P., HOSTETLER, J., LEACH, P., LUOTONEN, A., SINK, E., STEWART, L., *RFC 2069: An Extension to HTTP: Digest Access Authentication*; 1997.
11 FREED, N., BORENSTEIN, N., *RFC2046: Multipurpose Internet Mail Extensions (MIME) Part Two: Media Types*; 1996.
12 GLATZER, W., *Revolution in der Haushaltstechnologie*; Frankfurt/Main: Campus, 1998.
13 HEAP, N.W., *An Introduction to OSI*; Oxford: Blackwell Scientific Publications, 1993.
14 Internet Society, *Hypertext Transfer Protocol – HTTP/1.1*; 1999.
15 Internet Society, *RFC 2617: HTTP Authentication: Basic and Digest Access Authentication*; 1999.
16 JEANROND, P., HORST, H., ROHRBACHER, H., *EIB-Gebäudesystemtechnik: die zukunftssichere Elektroinstallation*; Düsseldorf: Pflaum, 1996.
17 KELLNER, C., Grafische Konfiguration verteilter Systeme. Dissertation am Lehrstuhl für Messsystem- und Sensortechnik der TU-München, in: *Fortschritt-Berichte VDI*, Reihe 10, No. 593, Düsseldorf: VDI, 1999.
18 KOON, J., *USB Peripheral Design*; San Diego: Annabooks, 1998.
19 KREUZBERG, J., *Handbuch der Heizkostenabrechnung, Zentralheizung und Fernwärme*; Düsseldorf: VDI, 1998.
20 KUPPLER, F., *Heizkosten richtig Erfassen und Verteilen, Geräteübersicht – Praxiserfahrungen – Kostenverteilung, Kontakt & Studium*; Band 132, 2. Auflage, Esslingen: expert, 1999
21 MÜGGE, G., *Vergleich verschiedener Heizkostenverteilsysteme*; HLH 44 (1993), 1998, No. 2, pp. 77–81, No. 3, pp. 153–157.
22 MÜLLER, W., KELLNER, C., *ContROS – Das Betriebssystem für MCMs*, Tagungsband des VIMP-Abschlußseminars, Neubiberg bei München, 4. Dezember 1998; Munich: Lehrstuhl für Meß und Automatisierungstechnik, Universität der Bundeswehr München, 1998, pp. 48–54.
23 MÜLLER, W., KÄMPF, W., Die intelligente Steckdose – Ein Schritt zum intelligenten Haus. *Tagungsband der 3. ITG/GMA-Fachtagung Sensoren und Meßsysteme*, Bad Nauheim, 9–11 March 1998.
24 MÜLLER, W., Analyse der Leistungsprofile elektrischer Hausgeräte zur Zustands- und Fehlererkennung, *Tagungsband des XIV. Messtechnischen Symposiums*, Wien, 28–30 September 2000.
25 MÜLLER, W., in: *Das intelligente Haus – Arbeiten und Wohnen mit zukunftsweisender Technik*; SCHNEIDER, F., TRÄNKLER, H.-F. (eds.); Munich: Pflaum, 2001.
26 ONEY, W., *Systemprogrammierung für Windows 95*; Redmond, WA: Microsoft Press, 1996.
27 SCHNEIDER, F., TRÄNKLER, H.-R. (eds.), *Das intelligente Haus – Arbeiten und Woh-*

28 SCHNEIDER, F., *nen mit zukunftsweisender Technik*; Munich: Pflaum, 2001.
28 SCHNEIDER, F., Decentralized intelligent microsystems for the private home (VIMP), a project of the BMBF; *2. EIB-Scientific Conference*; TU München, Munich, 1999.
29 SCHNEIDER, F., *Feldbusse – Grundlagen und Anwendungsgebiete*; OTTI-Technologiekolleg, Regensburg, 20–21 September 2000.
30 SCHNEIDER, F., WESTERMEIR, G., WÖLFLE, D., *Heimautomatisierung – Elektronikmarkt der Zukunft?*; Vortrag am 10.11.2000 auf der e/home, Kongressmesse Intelligentes Heim, Berlin, 9 – 10 November 2000.
31 Siemens, *Technical Data Sheet for the EIB-TP-UART-IC*; Siemens, 2000.
32 SUVAK, D.D., WANG, L., *Minimal IrDA Protocol Implementation (IrDA Lite)*; Walnut Creek, CA: Infrared Data Association, 1996.
33 Telefonaktiebolaget LM Ericsson, International Business Machines Corporation, Intel Corporation, Nokia Corporation, Toshiba Corporation, *Specification of the Bluetooth System Version 1.0 B*; 1999.
34 *USB Universal Serial Bus Specification Ver. 1.1*; 1998, http://www.usb.org.
35 WEINZIERL, T., *Gebäudesystemtechnik*; Vortrag bei der Tagung Datenübertragung auf Niederspannungsnetzen (Powerline) am 16.05.2000 im ‚Haus der Technik', Essen.
36 WEINZIERL, T., Grafisches Bediensystem für einen Hausbus, itg Fachtagung 1998; in: *itg Fachbericht 148*; Bad Nauheim, 1998, pp. 515–520
37 WEINZIERL, T., in: *Das intelligente Haus – Arbeiten und Wohnen mit zukunftsweisender Technik*, SCHNEIDER, F., TRÄNKLER, H.-R. (eds.); Munich: Pflaum, 2001.
38 WEINZIERL, T., in: *EIB Proceedings, Part 2*; Brussels: EIBA, 1999, pp. 25–36.
39 WEINZIERL, T., *Integriertes Managementkonzept für die Gebäudesystemtechnik, Dissertation*; Munich: Lehrstuhl für Messsystem- und Sensortechnik, TU-München, 2000.
40 WEINZIERL, T., SCHNEIDER, F., IMOS-home: The Software Concept for Home Automation Within VIMP; 2 EIB Scientific Conference, TU München, Munich, 1999.
41 WERTHSCHULTE, K., *Das Haus der Zukunft*; Vortrag auf der Messe Light & Building, Frankfurt, 22 March 2000.
42 WERTHSCHULTE, K., SCHNEIDER, F., Providing Access to an EIB Installation Using Internet Browser Technology, 3 EIB Scientific Conference, TU-München, Munich, 2000.
43 WESTERMEIR, G., WEINZIERL, T., SCHNEIDER, F., in: *EIB Proceedings Part 3*; Brussels: EIBA, 2000, pp. 133–145.
44 WESTERMEIR, G., WEINZIERL, T., SCHNEIDER, F., Tagung Sensoren und Messysteme 2000, Ludwigsburg, 13 March 2000, in: *VDI-Bericht*, No. 1530; 2000, pp. 33–42.
45 World Wide Web Consortium, *PNG (Portable Network Graphics) Specification, Version 1.0*, Recommendation, 1 October 1996.
46 Zentralverband Elektrotechnik- und Elektronikindustrie/Zentralverband der Deutschen Elektrohandwerke, *EIB: Handbuch Gebäudesystemtechnik, Grundlagen*, 4. überarbeitete Auflage, Frankfurt, 1997.

List of Symbols and Abbreviations

Symbol	Designation	Section
a	constant	4.1
A	area of building component	2.6
A	area of heat exchange	2.6
A	sensor area	4.1
A_S	surface area of outer wall	2.4
AER	air exchange rate	2.4
B	blind position	2.4
c	propagation velocity	6.1
c	specific heat capacity of air	2.4, 2.6
c	specific heat capacity of water	2.6
C_{comf}	weighting coefficient for thermal discomfort term	2.2
C_D	photodiode capacitance	4.3
C_p	specific heat	4.1
C_{pheat}	weighting coefficient for heating energy term	2.2
C_S	storage capacitance	4.3
CC_P	cross-spectral density	6.1
CO_2	CO_2 concentration	2.4
d	particle radius	4.1
d	thickness	4.1
d_0	normal diameter of a human being	2.1
D^*	normalized detectivity	5.1
$e(k)$	error	2.4
E	electric field	4.1
E	error	2.1
E_i	incident light energy	4.1
E_s	scattered light energy	4.1
E_{sol}	horizontal global solar radiation	2.2
E_v	Raman shift	4.1
f_D	Doppler shift	6.1
f_D	frequency shift	6.1
Δf	bandwidth	4.1

List of Symbols and Abbreviations

Symbol	Designation	Section
$F(\omega,t)$	perturbation	4.1
g_W	window transmission coefficient	2.4
G	conductance	2.3
HQ	elevator travel height	4.3
I	current	2.3
I	light intensity	4.1
I	number of ions created per unit time in unit volume	4.1
ΔI	increase in air temperature	2.1
I_M	supply current	2.5
I_{photo}	photocurrent	4.3
I_S	signal current	2.5
IZ	rope safety factor	3.3
j	number of building part	2.6
J	cost	2.2
k	heat transfer coefficient	2.4
k	time index	2.4
k	time step	2.2
k	transmission coefficient	2.6
K	constant	4.1
K_P	proportional gain	2.4
m	fixed mass	4.1
$\dot{m}$	mass flow	2.6
m_e	exhaust air flow rate	2.1
m_f	fresh air flow rate	2.1
MG	mixed gas signal	2.4
n	emission exponent	2.6
n	equilibrium ion concentration	4.1
n	number of meters in bus system	2.5
n	refractive index	4.1
N	number of smoke particles per unit volume	4.1
NEP	normalized equivalent power	4.1
p	pyroelectric constant	4.1
P	polarization	4.1
ΔP	total incident power	4.1
P_2	exhaust air pressure	2.1
P_5	fresh air pressure	2.1
P_8	supply air pressure	2.1
P_D	true detection rate	6.1
P_{FA}	false alarm rate	6.1
P_{FM}	probability of false dismissal	6.1
P_{heat}	heating command	2.2

Symbol	Designation	Section
$P(x, y, t)$	image brightness function	2.1
q_d	heat flow of air draft	2.4
Q	absorption cross section	4.1
$\dot{Q}$	heat emission	2.6
Q	heat loss/gain	2.4
Q	heat of adsorption or reaction	2.3
$\dot{Q}_0$	design heat emission	2.6
r	particle radius	4.1
R	counting rate	2.6
R	resistance	2.3, 2.5
RH_m	room relative humidity	2.1
s_w	surface area of window	2.4
S	sensitivity	4.1
$S(t)$	illumination function	4.1
Δt	sampling time	2.4
T	temperature	2.4
TCO	temperature coefficient	2.7
T_{flow}	flow temperature of heating fluid	2.2
T_m	indoor temperature	2.2
T_m	room temperature	2.1
T_{out}	outdoor temperature	2.2
T_{ret}	return temperature of heating fluid	2.2
T_s	sensor temperature	2.7
T_s	supply air temperature	2.1
$T_{setpoint}$	indoor setpoint temperature	2.2
T_{umrt}	unirradiated mean radiant temperature	2.1
u	manipulated variables	2.4
U	voltage	2.5
ΔU	change in signal voltage	4.1
$U(t)$	voltage across flakes forming a capacitor	4.1
v	speed change	2.1
v	velocity	6.1
v_{rm}	localized air speed	2.1
V	heated room volume	2.6
V	voltage	2.3
$\dot{V}$	volume flow rate	2.6
V	volume of room	2.4
V_{ref}	reference voltage	4.3
w	nominal values	2.4
x	element	2.4
x^*	optimal decision	2.4
x_a, y_a, z_a	camera-oriented coordinate	2.1

List of Symbols and Abbreviations

Symbol	Designation	Section
x_b, y_b, z_b	camera-oriented coordinate	2.1
x_w, y_w, z_w	world coordinates	2.1
X	set of possible solutions	2.4
X	average flame radiation	4.1
X	complex impedance	2.3
X_a, Y_a	frame memory coordinate	2.1
X_b, Y_b	frame memory coordinate	2.1
Y	complex admittance	2.3
α	angle between direction of motion and normal to receiver	6.1
α	heat transition coefficient of air	2.4
α	ion recombination rate	4.1
$a(k)$	proportionality constant	2.4
β	mean air renewal rate	2.6
ε	insulation parameter	2.4
ε	optical constant	2.3
ε_p	dielectric constant	4.1
θ	scattering angle	4.1
θ	temperature	2.6
$\theta(t)$	temperature of pyroelectric flakes	4.1
λ	cost-comfort weighting factor	2.4
λ	wavelength	4.1, 4.3, 6.1
μ	fuzzy function	2.4
μ_{Tm}	degree of membership	4.1
ν	frequency	2.3
ρ	density of air	2.6
ρ	density of water	2.6
σ	conductivity	2.3
τ	time constant	2.4
φ	relative humidity	2.4
φ	work function	2.3
Φ_i	mean brightness in a room	2.4
ω_0	natural pulsation	4.1
τ_{el}	electrical time constant	4.1
τ_{th}	thermal time constant	4.1

Abbreviation	Explanation	Section
AC	air conditioning	2.1
A/D	analog/digital	6.2
AHU	air-handling unit	2.1
ALUS	Automatisches Luft-Umschalt-System	2.3

Abbreviation	Explanation	Section
ANN	artificial neural network	2.2
ANSI	American National Standards Institute	3.1
API	application program(ming) interface	3.4, 5.2, 6.2
ARC	alarm receiving center	4.1
ARQ	automatic repeat request	3.2
ASC	application specific controller	3.1
ASHRAE	American Society of Heating, Refrigeration and Air Conditioning Engineers	3.1
ASIC	application-specific integrated circuit	1.1, 2.6, 3.3, 4.4, 6.2
ATM	automatic teller machine	4.1
BAS	building automated system	2.1
BCH	Bose-Chaudhuri-Hocqenghem	3.2
BCI	BatiBus Club International	6.2
BCU	bus coupling unit	3.1, 6.2
BMS	building automation system	2.7
BPMS	Body Pressure Measurement System	3.4
BRI	building-related illnesses	6.1
CAL	common application language	3.2
CCD	charge-coupled device	2.1, 3.3, 4.1, 4.3, 6.1
CCITT	Comité Consultatif International Télégraphique et Téléphonique	3.1
CCTV	closed-circuit television	2.1, 3.2, 4.1, 4.3
CDMA	code division multiple access	3.2
CDS	correlated double sampling	4.3
CEBus	consumer electronics bus	3.2
CEN	Comité Européen de Normalisation	3.1
CGI	Computer graphics interface, computer generated image	6.2
CIE	control and integrating equipment	4.1
CIM	computer integrated manufacturing	3.1
CM	control monitoring	5.1
CMMS	computerized maintenance management system	5.1
CMOS	complementary metal oxide semiconductor	2.7, 4.1, 4.3
COR	Cooperative Buildings	5.2
CPU	central processing unit	4.1
CSEM	Centre Suisse d'Electronique et de Microtechnique	2.2
CSMA/CD	carrier sense multiple access with collision detection	3.2
CV	code violation	3.1
DAI	distributed artificial intelligence	3.1

Abbreviation	Explanation	Section
DCOM	Distributed Component Object Model	6.2
DHTML	dynamic HTML	6.2
DIC	digital identity card	5.2
DR	dynamic range	4.3
DSP	digital signal processor	4.1
DXF	Drawing Interchange Format	3.2
EAM	enterprise asset management	5.1
ECE	electrochemical etching	2.7
EDPC	electronic data processing center	2.5
EHS	European Home System	3.2, 6.2
EHSA	European Home System Association	6.2
EIB	European Installation Bus	1.1, 3.1, 6.2
EIBA	European Installation Bus Association	3.1, 6.2
EIBnet	EIB on Automation Net	3.1
EIB-PL	EIP – Power Line	6.2
EMC	electromagnetic compatibility	3.1
EMI	electromagnetic interference	3.3
ERP	enterprise resource management	5.1
ETS	EIB Tool Software	3.1
FAN	field area network	3.1
FAR	false accept rate	4.1, 4.2
FCC	Federal Communications Commission	3.2
FDM	frequency division multiplexing	3.2
FDMA	frequency division multiple access	3.2
FEMA	failure modes and effects analysis	5.1
FET	field effect transistor	2.5
FM	facility management	5.2
FPN	fixed-pattern noise	4.3
FRR	false reject rate	4.1, 4.2
FSR	force-sensing resistor	3.4
FTP	File Transfer Protocol	3.1
GAN	global area network	3.1
GPS	Global Positioning System	3.2
GUI	graphical user interface	5.2
HAN	home area network	1.1
HAVI	home audio video interoperability	3.2
HCA	heat cost allocator	2.6
HES	home electronic system	3.2
HFC	halofluorocarbon	4.1
HISS	Human Interface Supervision System	5.1
HTML	hypertext markup language	2.1
HTTP	Hypertext Transfer Protocol	3.1, 6.2

List of Symbols and Abbreviations

Abbreviation	Explanation	Section
HVAC	heating, ventilation, and air conditioning	1.1, 2.1, 2.2, 2.3, 2.4, 2.6, 2.7, 6.1
IAQ	indoor air quality	2.1, 6.1
IC	integrated circuit	3.1, 4.1, 6.2
IDLH	immediately dangerous to life or health	4.1
IEC	International Electrical Commission	3.1, 4.3
IMOS	integrated management of open systems	6.2
IMOS-home	IMOS for the private home	6.2
IP	Internet Protocol	3.2
IPC	industrial personal computer	3.1
IR	infrared	2.1, 2.3, 3.2, 4.1, 4.3, 6.1
IrDA	Infrared Data Association	3.2, 6.2
IRFPA	infrared focal plane array	4.1
ISDN	Integrated Services Digital Network	6.2
ISM	industrial, scientific, and medical	3.2
ISO	International Organization for Standardization	3.1
ITO	indium tin oxide	4.1
JINI	Java Intelligent Network Infrastructure	5.2
JIPM	Japanese Institute for Plant Maintenance	5.1
JVM	Java virtual machine	6.2
LAN	local area network	3.1, 3.2, 4.1, 6.2
LC	inductive-capacitive	2.1
LCD	liquid crystal display	6.2
LDC	low duty cycle	6.2
LDI	L-chip driver interface	3.1
LED	light-emitting diode	3.3, 4.1, 4.3, 4.4
LEL	lower explosion limit	4.1
LFA	local feature analysis	4.1
LIS	location information server	5.2
LON	local operating network	3.1, 3.3
MAC	maximum allowable concentration	4.1
MAC	medium access layer	3.2
MAP	manifold absolute pressure	2.7
MAP	manufacturing automation protocol	3.1
MCM	multi-chip module	6.2
MCSA	motor current signature analysis	5.1
MCV	microcontroller unit	2.7
MEMS	microelectromechanical systems	2.7
MFC	Microsoft Foundation Class	6.2
MIMOSA	Machinery Open Systems Alliance	5.1
MMI	manmachine interface	6.2

Abbreviation	Explanation	Section
MOS	metal oxide semiconductor	4.3, 5.1
MS/TP	masterslave/token passing	3.1
MSM	multi-sensor module	6.1
NETD	noise equivalent temperature difference	4.1
NIR	near-infrared	4.3
NTC	negative temperature coefficient	2.3
ODP	ozone depletion potential	4.1
OEE	overall equipment efficiency	5.1
OEM	original equipment manufacturer	3.3
OFDR	optical frequency domain reflectometry	4.1
OLE	object linking and embedding	6.2
OSI	open system interconnection	3.1
OWS	operator workstation	2.1
PAM	plant asset management	5.1
PAS	plant automation system	5.1
PBC	performance-based code	4.1
PCA	principal component analysis	2.4
PCB	printed circuit board	2.7
PD	photodiode	4.3
PDA	personal digital assistant	6.2
PdM	predictive maintenance	5.1
PEI	peripheral external interface	6.2
PIN	personal identification number	4.1, 4.2
PIR	passive infrared	4.1, 4.3, 6.1
PL	powerline	6.2
PLC	programmable logic controller	3.1
PMV	predicted mean vote	1.1, 2.1, 2.2
PNG	portable network graphics	6.2
PPD	predicted percentage dissatisfied	2.1
PPTCB	pulsed pressure, temperature cycling, with bias	2.7
PSM	personal security manager	5.2
PSTN	public switched telephone network	4.1
PZT	lead zirconate/titanate	4.1
QMB	quartz microbalance	6.1
RAM	random access memory	4.1
RAS	remote access system	4.1
RC	resistance-capacitive	2.1
RCFA	root cause failure analysis	5.1
RCM	reliability centered maintenance	5.1
rf, RF	radiofrequency	3.2, 4.3, 5.1, 6.2
RH	relative humidity	2.3
ROC	receiver operating curve	6.1

Abbreviation	Explanation	Section
RTD	resistance temperature detector	2.1
RTOS	real-time operating system	6.2
SAP	services access point	3.1
SAW	surface acoustic wave	6.1
SB	start bit	3.1
SBS	sick building syndrome	6.1
SC	switched capacitor	4.3
SFOE	Swiss Federal Office of Energy	2.2
SI	system integrator	3.1
SMPT	Simple Mail Transfer Protocol	3.1
SNR	signal-to-noise ratio	3.2, 4.3, 6.1
SP	service provider	3.1
SRD	short-range device	2.5
SSH	Secure Shell	5.2
SSL	Secure Socket Layer	5.2, 6.2
SU	service user	3.1
SWAP-CA	Shared Wireless Access Protocol – Cordless Access	3.2
TCP	Transmission Control Protocol	3.1
TCP/IP	TCP/Internet Protocol	6.2
TDMA	time division multiple access	3.2
TIF	tacho interface	3.3
TLV	threshold limit value	2.3, 4.1
TOC	total organic carbon	2.3
TP	twisted pair	6.2
TPM	total productive maintenance	5.1
TPMS	tire pressure monitoring system	2.7
TWA	time-weighted average	4.1
UART	universal asynchronous receivertransmitter	6.2
UDP	User Diagram Protocol	3.1
UEL	upper explosion limit	4.1
UPnP	universal plug and play	3.2
US	ultrasonic	2.1, 6.1
USB	universal serial bus	6.2
UV	ultraviolet	4.1
VAV	variable air volume	2.1, 2.7
VDMA	Verband Deutscher Maschinen- und Anlagenbau e.V.	2.3
VIMP	Verteilte Intelligente Mikrosysteme für den privaten Lebensbereich	6.2
VLDC	very low duty cycle	6.2
VLL	vertical link layer	3.1
VOC	volatile organic compound	2.3

Abbreviation	Explanation	Section
VQ	vector quantization	4.1
WAN	wide area network	3.1, 4.1
WDM	Windows Driver Model	6.2
WWFM	worldwide facility management	5.2

Index

absorption
- fire sensing 311 ff
- gas sensing 342
- intrusion sensing 350

accelerometers 286
access control 308, 399 ff
Access DLL 528
access methods
- air conditioning 30, 91
- elevators 264, 276
- home automation 547
- M-bus 136
- pressure sensors 173, 187

acoustic monitoring
- elevators 281
- industrial installations 452, 454

active damping system 283, 286
active illumination 421
active intrusion sensing systems 356
active mode, acoustic monitoring 455
active ride control 285
active smart food systems 9
actuators
- fieldbus 203
- pressure sensors 196

addressable systems 333
adressing meters 133
aerosols 308 f, 329
air conditioning
- energy allocation 161
- intelligent 29 ff
- private homes 512

air exchange rates 113
air-flow sensors 47
air handling unit (AHU)
  40 ff, 49, 52, 56
air pressure sensors 173 ff
air quality 85
air temperature 67
AIRBUS 320 216
airtecture 14 f
alarm receiving centers (ARCs) 379

alarm systems 307 ff
- flame sensing 333
- intelligent buildings 47, 486 ff
- load sensing 429

algorithms
- energy management 108 f
- signal processing 384

allocators 546
ammonia 338, 343, 345
anemometers 34, 162
anisotropic etching 180
ANSI/EIA 709 Standard 221
application layer
- fieldbus 208 ff, 229
- M-bus 141

application programming interface
  (API)
- private homes 519
- sensing chair 297
- worldwide facility management 476

application specific controller
  (ASC) 228
application specific integrated circuit
  (ASIC) 265, 319
applications
- elevators 262
- fire extinguishing 379
- flame sensing 334
- intelligent buildings 486 ff
- pressure sensors 175, 185
- private homes 512, 528, 543
- thermal flow sensors 166

arrays
- air quality 91 ff
- intrusion sensing 351

artificial neural networks (ANNs) 65
artificial skin 293
assessment, M-bus 146
attacks see: security
authentication 399 ff
automated video surveillance cameras
  410

automatic control   485
automatic meter reading   127 ff, 169
automatic modes, private homes   536
automatic processing, sensing chair   293
automatic repeat request (ARQ)   248
automatic teller machine (ATM)   355, 363
automation systems, private homes   512
Automatisches Luft Umschalt-System (ALUS)   92
automotive applications, pressure sensors   185
automotive imagers   417
autozero facility   188
availability
– fieldbus   207
– incremental encoders   276
– industrial installations   451

back scattering   313
bands, magnetic   282
bandwidth
– accelerometers   289
– wireless networks   246
baseline drift, air sensors   91
batteries
– gas sensing   346
– smoke sensing   319
– wireless networks   247
Bernoulli equation   33
bidirectional transmission, M-bus   147
billing service provider   129
bimorph accelerometers   356
biometric authentication   399 ff
biometric reading   368
biometric sensors   425
birefringence   17
blackbody radiation   327
blind systems, energy management   116 ff, 123
blood oxygen concentration   501
blood pressure   501

Bluetooth   257, 526
body pressure measurement system (BPMS)   297
body temperature   501
bolometers   348
Bose–Chaudhuri–Hocquenghem (BCH) coding   248
bridge layers   209
bridge transducer model   182
brightness
– energy management   118
– light emitting diodes   429
– optical flow   43
– smart cameras   410
Brillouin effect   324
broadcast connections   229
browsers   549
building automated systems (BASs)   46, 174
building automation and control (BACnet)   225
building automation protocols   476
building modules   66
building related illness (BRI)   504
buildings, intelligent   3–25
bulk micromachined pressure sensors   180
bus applications   131
bus coupling unit (BCU)   515, 519 f, 529 f
bus interfaces   187
bus monitoring   532

cable extension position transducer   277
cable network, heat cost allocators   169
calibration
– camera systems   41
– gas sensing   346
– load sensing   429
– pressure sensors   182, 190
calorimeters   89, 340
CAMAC device   214

camera systems
– air conditioning 41
– home automation 542
– intelligent buildings 495
– intrusion sensing 354
CAN, fieldbus 215
capacitive sensors 33 f, 89
– pressure 33, 183
car optical switches 265
carbon-based combustion 327
carbon dioxide 85 ff, 91, 96 ff, 513
carbon monoxide 338, 342, 345, 490
carcinogenic constituents 86
catalytic devices 340
central meter reading 131
central process unit (CPU)
– emergency handling 375
– identification sensing 365
certification, EIB 220
chair, sensing 293 ff
charge coupled devices (CCD) 41, 499
– elevators 280
– intrusion sensing 354
– smart cameras 410
chemical etching 180
chemical transduction 17
chip photomicrographs 417
chlorine 343
chlorobenzenes 87
chloroform 87
CIF imagers 414 ff
client programs, private homes 530
climate dynamics 104
climate module 65 f
closed circuit television (CCTV)
  41, 354, 409
clustering, scene spots 45
code division multiple access (CDMA)
  249
codings, wireless networks 248
coexistence, radio signals 248
cooling effect, air conditioning 52
combustion gas 311, 338 ff
comfort
– air conditioning 36

– energy management 107 ff
– HVAC control 51
– intelligent buildings 6, 485, 503,
  514
commissioning, pressure sensors 174
communication basics, fieldbus 206 f
communication master, M-bus 135
communication object DLL 529
communication systems 10, 513
compatibility, air conditioning 30
complementary metal oxide
  semiconductor (CMOS)
– pressure sensors 181
– smart cameras 410 ff
– smoke sensing 319
computer-assisted methods, signal
  processing 385
computer simulations
– air conditioning 53
– wireless networks 244
computer vision
– HVAC control 40
– sensing chair 293
computerized maintenance
  management system (CMMS) 459
condition monitoring 451, 458
configurations
– fieldbus 211
– home automation 534
– M-bus 145
construction site safety 427 ff, 437 f
consumer electronics bus (CEBus)
  253
contact cards 365
contact sensors 293
contactless linear incremental encoders
  278
contacts, intrusion sensing 356
contaminants, air quality 87
control
– air conditioning 53
– elevators 264, 283
– fieldbus 203
– home automation 520, 541
– intelligent buildings 15, 48, 485 ff

- NEUROBAT 63 f, 66 ff
- thermal comfort 104
control and indicating equipment (CIE) 375
ContROS 514, 522
convenience, intelligent buildings 6, 502
converters, smoke sensing 319
cooling cycles, pressure sensors 193
coronal discharge monitoring 457
correlated double sampling (CDS) 414
correlation algorithms, elevators 281
corrosion 452 f
costs
- air conditioning 30
- intelligent systems 487
- NEUROBAT 65 f
Cotton–Mouton effect 17
counter electrodes 343
counterweights, elevators 261, 273

dampers, pressure sensors 195 ff
damping systems, elevators 283
data bus, meter reading 130 f
data link layers 208 ff, 251
data networks, wireless 258
data processing 302, 428, 440
data stream, fieldbus 209
data transfer via data bus 130
data transmission links, elevators 276
data transmission via radio 147
dead imprints recognition 369
decentralization 206
degradation, industrial installations 454, 456
demand side management, fieldbus 235
detection characteristics
- gas sensing 345
- smoke sensors 411
deterministic code, magnetic sensors 282
dew sensors 16, 36

digital identity card (DIC) 479
diodes 412
dioxanes 87
discharge, industrial installations 455
displaying outdoor cameras 542
distometers 279
distributed component object model (DCOM) 528
district heating 160
disturbance signal, intrusion sensing 358
domestic buildings 104
domotics 6
door zone detection 263, 269
doping, metal oxides 345
Doppler effect 17
- intrusion sensing 347, 357
- motion sensors 495
drifting 30, 175
dry fire extinguishing systems 376 f
dual interface, identification sensing 373
dynamic range, smart cameras 413

easy configuration, home automation 536
eccentricity 453
economics
- elevators 275
- energy management 107 f
edge pixels 44
EEPROM see: memory cards
eigenface 370
eigenspace methods, sensing chair 296
elastic waves, intrusion sensing 355
electric testing 452, 456
electric traction 261
electrical potential differences 133
electrical sensors, intelligent buildings 16
electricity meters 47
electrochemical cells
- air quality 89

– gas sensing   343
– intelligent buildings   492
electrochemical etching (ECE)   180
electrodes, gas sensing   343
electrolyte sensors   89
electromagnetic compatibility (EMC)
– fieldbus   216
– flame sensing   336
electromagnetic flow meters   34
electron mobility   345
electronic heat cost allocators   168
electronic noses   95
electronics, smoke sensing   318
elevators   261 ff
embedded systems, fieldbus   204
emergency calls   374 ff, 486, 500, 512
emergency terminal speed limiting
   device   269
emission exponent, energy
   allocation   161
encoders, incremental   264, 276
energy consumption
– air conditioning   58
– intelligent buildings   485, 507
– metering   159
– NEUROBAT   65, 81 f, 85
energy control   16, 22, 27–196
– home automation   513, 545
energy cost allocation   159 ff
energy management, optimal   103 ff
energy saving   186
energy suppliers, residential
   buildings   129
enterprise resource planning (ERP)
   460
environment
– air conditioning   30, 56
– flame sensing   336
– pressure sensors   186
equipment, load sensing   428 f
equipment failure profile, industrial
   installations   452
erosion   452 f
errors, wireless networks   247
etching   180

Ethernet
– building automated systems   48
– wireless networks   254
– worldwide facility management   475
ethylbenzenes   87
ethylene oxide   338, 343
Ettingshausen effect   17
European home system (EHS)   253
European Installation Bus (EIB)
   13, 204, 218 ff
– private homes   514, 517, 540
– EIBAccess.dll   532
– EIBMonitor.exe   533
– EIBnet, fieldbus   227
– EIBRemote.exe   533
evacuation systems   374
evaporative heat cost allocators   167
expert systems, industrial
   installations   458
explosive gas sensing   338 ff
extinction, smoke sensing   313 f, 321
eye retina, identification sensing   369

face recognition
– access control   399, 402
– identification sensing   370
facility management   449–482
– fieldbus   234
– intelligent buildings   6, 23
factory calibration procedures, pressure
   sensors   190
failure
– elevators   274 ff
– industrial installations   452
– M-bus   140
– uplift shoring systems   432
failure modes and effects analysis
   (FEMA)   462
false accept rate (FAR)   368, 403, 407
false reject rate (FRR)   368, 407
Fanger equation   50, 68, 87
Faraday effect   17
ferrography   455
Festo airtecture hall   14

fiber lasers   324
field area networks (FAN)
– fieldbus   205 ff, 211 ff, 230 ff
– pressure sensors   175
fieldbus systems   22, 203 ff
field effect sensors   89
field measurements   436
field-oriented vector control, induction motors   265
field testing   386
Figaro sensors   89, 345
file transfer protocol (FTP)   230
filters
– optical   348 ff
– pressure sensors   175
finger ring, health monitoring   501
fingerprint recognition
– access control   399, 402
– identification sensing   369
– intelligent buildings   495
– smart cameras   425
fire alarm   486, 488 f
fire extinguishing   376
fire pulsation   389
fire sensing   39, 308 ff
firefly   257
firewall   550
fixed pattern noise (FPN)   413
flame detectors   388
flame sensing   311, 328 ff
floor receivers   155
flow sensors
– air conditioning   33
– building automated systems   47
– energy allocation   164 ff
fluidic muscles   15
fluorescence sensors   89
food systems   7
force sensing resistors (FSRs)   293, 301
formwork failure   427 ff
forward scattering   313
Fourier transformation   392
frames, M-bus   142
frequencies, wireless networks   255

frequency band   333
frequency division multiple access (FDMA)   249
frequency division multiplexing (FDM)   255
frequency shift, motion sensors   495
Fresnel optics   349, 353 f
friction effects   17
fuel flow meters   47
functional description, elevators   261
functionality, bus monitoring   532
fusion algorithms   358
FuturElife smart building   12 ff
fuzzy logic supervisory control   103 ff
fuzzy logics   385 f, 466

galvanometric effect   17
gas detection
– air quality   89
– intelligent buildings   486, 489 ff
– safety   308, 338 ff
gas heating, energy allocation   160
gateway solution   231
glassbreak detectors   358
global area networks (GAN)   205
global positioning systems (GPS)   250
governor roller, elevators   270
GPIB device   214
gradient constraint equation, air conditioning   42
grain boundaries   344
gray level images   293, 301

hackers see: security
Hall effect   17
halofluorocarbons (HFCs)   378
Halon system   377 f
hand geometry   405
hand transceiver   501
Handle   297
hands-free cards   368
hardware, home automation   543
health care systems   513

health monitoring 488 f, 500 ff
heat cost allocators (HCAs) 167 ff, 546
heat meters 162 ff
heat sensing 323
heating control
– energy allocation 160
– intelligent buildings 486 ff
– private homes 512
– self-commissioned 63–84
heating cycles, pressure sensors 193
heating systems 120, 123
heating temperature, energy management 110
heating-ventilation-air conditioning (HVAC) 4, 21, 27–196, 502
heptan fire 330
hidden Markov modeling (HMM) 303
hierarchies, fieldbus 206
high-resolution scanner 406
high-rise elevators 261 ff
high-security domains 399
high-sensitivity systems, smoke sensing 321
high-speed image sequences 417
home area network (HAN) 5, 10
home audio video interoperability (HAVI) 253
home electronic system (HES) 253
Horn model 493
Hostaflon ET 14
hot-air turbulences 358
hot-wire anemometers 34
human interface supervision system (HISS) 465
human sensory systems 465 ff
humidity
– air conditioning 29, 35, 51, 57, 84 f
– intelligent buildings 489, 504
– pressure sensors 175
– thermal comfort 104
hybrid systems
– flame sensing 334
– intrusion sensing 359
– M-bus 144

hydraulic elevators 261
hydrocarbon fires 327, 389
hydrocarbons 84 f
hydrogen 343, 345, 490
hydrogen chloride 338, 343, 345
hydrogen cyanide 338, 343
hydrogen fires 327
hydrogen sulfide 332, 337, 339
hygrometers 18, 35
hypertext markup language (HTML) 48
hypertext transfer protocol (HTTP) 230, 540

identification systems 363 ff, 401
illumination
– CMOS imagers 421
– smart cameras 410
image processing, HVAC control 40
image recording 466
image transfer 47
imaging 10, 410 f, 415 ff
imbalance, industrial installations 453
immediately dangerous to life or health (IDLH) 338
implementation, home automation 543
in-building networks 241 ff
in-house systems 144
incremental encoders 264, 276
indium tin oxide (ITO) mirrors 350
indoor air quality (IAQ) 53, 85
– conditioning 37
– intelligent buildings 485 ff, 504
– NEUROBAT 67
indoor scenes, smart cameras 410
inductive pressure sensors 33
industrial installations 451 ff
industrial networks 213
industrial personal computers 204
industrial prototype 75
industrial scientific medical (ISM) uses 256
information systems 22, 201–304

infrared barriers 356
infrared connections, wireless networks 242 ff
infrared data association (IrDA)
– private homes 524 f
– wireless networks 257
infrared motion sensors 38
infrared optical switches 270
infrared region diodes 319
infrared sensors
– air quality 89
– elevators 279
– flames 327
– gas 341
– intelligent buildings 489, 492
– safety 309
infrared standards 256
infrared thermography 452, 455 f
installation
– elevators 275 f, 291
– gas sensing 346
– heat cost allocators 168
– industrial 451 ff
– intelligent buildings 488, 507
– load cells 428
– M-bus 137
– pressure sensors 173
integrated circuit 515
integrated management of open systems (IMOS) 538
integration 18
– expert systems 460
– system-oriented 511
intelligence, perceptual 293
intelligent buildings 3–25, 485 ff, 511
– smart cameras 409
intelligent development, signal processing 384
intelligent outlet, private homes 521
interactive systems, flame sensing 333
Interbus 215
interfaces
– fieldbus 214, 236
– identification sensing 373
– man-machine 511

– pressure sensors 186
– private homes 515 f, 523 ff, 532 f
– sensing chair 297
– standardization 18 f
– worldwide facility management 474
interferences
– air conditioning 30
– M-bus transmission 152
interferometers 17
international organization for standartization (ISO) 204 f
Internet
– HVAC systems 46
– intelligent buildings 5 ff
– private homes 514, 528, 550
– worldwide facility management 474
Internet connection 230
Internet protocol 243
interoperability 223, 232
intrusion detection
– intelligent buildings 486 f, 495
– safety 308, 347 ff
ionization detectors 489
ionization smoke sensors 39, 309, 316 f
iris scanner 406
ISDN, home automation 548
ISO/OSI layers model
– fieldbus 207 ff
– wireless networks 251

JINI 253
Johnson noise 17
junctions, pressure sensors 180

Karhunen–Loeve expansion 296
Kerr effect 17
knowledge-based authentication 400
Kohonen network 388 ff

laboratory room characteristics   87
laboratory work   430
Langmuir probe   17
laser optical sensors   279
layers
– fieldbus   207 ff
– M-bus   138
– private homes   520
lead zirconate/titanate (PZT)   348, 355
leak detection   455
level converter, M-bus   136
life safety/security   305–448
life time, gas sensors   495
light attenuation   388
light barriers   263
light emitting diodes (LEDs)
– elevators   281
– load sensing   429
– smart cameras   423
– smoke sensing   319
light extinction, smoke sensing   321
light scattering sensors   329, 489
light scattering/absorption, fire sensing   311 ff, 320 ff
lighting control   486, 512
linear beam detectors
– flame sensing   327
– safety   309
– signal processing   387
link layers
– M-bus   141
– private homes   520
lithium titanate   348
live recognition   369
load management   6
load sensing
– elevators   263, 271
– improved construction site safety   427 ff
local area network (LAN)
– fieldbus   205
– identification sensing   372
– private homes   528, 548
– wireless networks   250

local feature analysis (LFA)   370
local operating network (LON)
– elevators   263
– fieldbus   221
location information server (LIS)   480
logistics, load sensing   438
long-frame, M-bus   142
long-term air quality evaluation   95
long-term repeatability   175
long-term studies, private homes   511
LonTalk protocol, fieldbus   221
LonWorks
– fieldbus   212 ff, 221 f
– wireless networks   243, 253
loudspeaker systems   374
low cost heating controllers   63
lower explosion limit (LEL)   339
lubricant analysis   452, 454 f

M-bus (metering bus)   134 f
machinery information open system alliance (MIMOSA)   460 ff
magnetic contacts, intrusion sensing   356
magnetic inductive flow sensors   165
magnetic sensors
– elevators   270, 282
– intelligent buildings   16
magnetic strips, identification sensing   364
maintenance   449–482
– air conditioning   30
– industrial installations   451 ff, 461 f
– intelligent buildings   23
malfunctioning, building automated systems   47
man-machine interface   511
manifold absolute pressure map systems (MAP)   185
markets
– intelligent buildings   6
– pressure sensors   185
marking tape   278
mass-sensitive sensors   89

mathematical description, thermal systems   65
maximum allowable concentration (MAC), gas sensing   338
measurement ranges, air conditioning   30
mechanical flow sensors   164
mechatronic systems   261
medium access layer protocol   257
megahertz ranges, M-bus transmission   153 ff
membranes, translucent   15
memory cards, identification sensing   365
metal oxide semiconductor (MOS) transistors   412
metal oxide semiconductor sensors   466
metal oxide sensors
– air quality   89
– gas   344
– intelligent buildings   492
meter reading, residential buildings   127 ff
metering, energy costs   159
methane   345, 490, 494
microbolometers   351
microcontroller level, private homes   515
microcontroller unit (MCU)   189, 198
microelectromechanical systems (MEMS)   179
microprocessor cards   365
microsystems, intelligent buildings   8
microwave detection
– intrusion sensing   358
– safety   308
microwave devices   496
Mie scattering   311 f, 315 ff
miniaturization   18 f
mirrors, intrusion sensing   349, 353
misalignment, industrial installations   453
MIT ring   501

mobile devices   11
modems, European Installation Bus   527
modes, home automation   536
monitoring, acoustic/ultrasonic   452 ff
motion control
– air conditioning   37 f
– elevators   273
– intelligent buildings   495
– smart cameras   419
motor current signature analysis (MCSA)   457
mounting orientation, pressure sensors   173, 177
multi-level networks   206
multi-objective fuzzy optimization   106
multicast connections, fieldbus   229
multicriteria detectors, flame sensing   329
multiple sensors, intelligent buildings   499
multiresolution, fire detection   387
multisensor systems   20, 358
multitrack code, magnetic sensors   282

natural gases, fire extinguishing   378
near-infrared range (NIR)   412, 422
negative temperature coefficent (NTC)   163
Nernst effect   17
network layer, fieldbus   206, 208 ff, 229
networks
– intelligent buildings   5
– wireless   241 ff, 250 f
neural fuzzy systems   466
neural networks
– identification sensing   370
– NEUROBAT   63–84
NEUROBAT, heating control   22, 62–82

Neuron C 222
Neymann–Pearson criterion 500
nitric oxide 338
nitrogen 84 f
nitrogen dioxide 338, 343
nodes, fieldbus 204 f, 222 f, 233
noise, sensing chair 299
noise equivalent power, intrusion sensing 349
noise equivalent temperature difference (NETD) 351
non flaming fires 327
noncontact cards, identification sensing 367

occupant sensors
– active illumination 421
– air conditioning 37
octanes 87
odometers 276
odor sensation 95
office room characteristics
– air quality 86
– NEUROBAT 72
– thermal comfort 104
ohmic resistor 317
open-loop control 63
open system interconnection (OSI) 207 ff
operating systems, private homes 520 ff, 531 f
operator workstation 48
optical dynamic range 410
optical extinction, smoke sensing 321
optical filtering 348 ff
optical flow, velocity fields 42
optical frequency domain reflectometry (OFDR) 326
optical light scattering, smoke sensing 318 ff
optical properties, smoke particles 314
optical sensors 16, 489
optical switches 264, 270

optochemical sensors 89
optomechanical sensors 276 ff
orifice plates 33, 162
original equipment manufacturer (OEM) 282
oscillations
– Mie domain 315
– signal processing 391
outdoor climate
– NEUROBAT 65
– thermal comfort 104
outdoor scenes, smart cameras 410
outdoor temperature 57
output signals, pressure sensors 175
outside systems 144
overall cell reactions, gas sensing 343
overall equipment efficiency (OEE) 463
overvoltage protection 134 f, 141
oxygen 84 f
ozone depletion zone (ODP) 378

packaging, pressure sensors 183
palladium catalyst 340
parasitic bipolar transistors 412
particle detection 308
passive infrared radiation (PIR) 347, 419
passive mode, acoustic monitoring 455
passive smart food systems 9
passwords, access control 402
pattern recognition
– access control 399
– human sensoring 466
– identification sensing 369 ff
– intelligent buildings 491
– sensing chair 293
pellistors 340, 492
Peltier effect 17
people counters, infrared 38
perceptual intelligence 293
performance 19 f
– air conditioning 30

- energy management   107
- M-bus transmission   151
- NEUROBAT   69 ff
- sensing chair   300
PERI shoring   442 ff
peripheral external interface (PEI)   515
personal digital assistant (PDA)   514, 549 f
personal identification   40, 69, 495
personal identification number (PIN)   363, 400
personal security manager (PSM)   480
phosphine   343
photoacoustic cells   341
photoelectric smoke sensors   39, 309
photogate devices   412
phototubes, flame sensing   328
physical effects, intrusion sensing   351
physical layers
- fieldbus   208 ff, 229
- wireless networks   251
physical specifications, M-bus   138
physical transduction   17
physical workspace, worldwide facility management   469
physiological response, sensing chair   295
piezoelectric semiconductor elements   164
piezoelectricity   17
piezoresistive accelerometers   287
piezoresistive pressure transducer   181
Pitot tubes   33
pits   453
pixels   42, 413
plant asset management (PAM)   460
plant automated system (PAS)   460
plant availability   451
platinum catalyst   340
platinum sensors   163
plug and play   233 ff, 488
pn-diodes   412
pn-junctions   180
pneumatic building structures   14 f

Pockels effect   17
point extinction detector, smoke sensing   323
polarization   314
PolyRex multisensor detector   330
portable devices   10
portable network graphics (PNG)   552
position sensors   277, 290
positioning, wireless networks   249
potassium hydroxide   180
potential barriers, gas sensing   344
powerline, private homes   517
PPTCB (pulsed pressure, temperature cycling with bias)   184
Prandtl tube   176
predicted mean vote (PMV)   22
- air conditioning   50, 57
- HVAC control   50, 57
- NEUROBAT   67
predicted percentage dissatisfied (PPD)   53
predictive maintenance (PdM)   451
presentation layer, fieldbus   210
pressure
- air conditioning   29
- sensing chair   293 f
pressure sensors
- air conditioning   31
- building automated systems   47
- HVAC   173 ff
principal component analysis (PCA)   296
process area, fieldbus   203
process parameter control   452, 456
processing   18
processing time, fieldbus   207
processing unit, access control   401
processor card, identification sensing   366
product visions   7
Profibus   215
profiles, fieldbus   223, 232
programmable logic controllers (PLCs)   204

protocols
- access control   400
- EIB   219
- private homes   523, 540, 550
- simple mail transfer protocol (SMTP)   230
- token ring   254
- transmission control (TCP)   230
prototype realization, NEUROBAT   75 f
proximity cards, identification sensing   373
psychrometers   35
pulse rate monitoring   501
pyroelectric sensors   327, 349, 351 ff
pyrolysis   311

quartz microbalance
- air sensors   95
- intelligent buildings   492
- safety   309

radiant temperature, air conditioning   51
radiation sensing   308, 327
radiator heat emission   160
radio propagation prediction tool screenshot   245
radio transmission, heat cost allocators   170
radiofrequency connections, wireless networks   242 ff
radiofrequency specifications   518
Raman scattering   324
random access memory (RAM)   366
ranges
- explosion/combustion   339
- pressure sensors   173, 177
- wireless networks   244
Rayleigh scattering   311 ff
read out, CIF imager   416
read-out device with wake-up signal   150

reading
- heat cost allocators   169
- identification sensing   364
- magnetic sensors   282
real-time conditions, fieldbus   211
real-time expert systems   459
receiver operating curves (ROCs)   499
receiving systems, meter reading   155
reciprocating equipment   453
recliner, sensing chair   294
recognition *see:* pattern recognition
redundant systems, elevators   275
Reed–Solomon coding   248
Reed switch   356
reflective mode, acoustic monitoring   455
reliability
- elevators   275
- industrial installations   452
- intelligent buildings   485
- M-bus transmission   152
- pressure sensors   184 ff
- private homes   518
- wireless networks   247
reliability-centered maintenance (RCM)   461 f
remote control, wireless networks   257
remote meter reading   128, 130
repeatability
- air conditioning   30
- elevators   264
- pressure sensors   175
repeater, M-bus   137
residential gateway   550
resistance temperature detectors (RTDs)   30
resistance temperature sensors   163
resistive sensors, air conditioning   35
resistors, sensing chair   293, 298
resolution
- incremental encoders   276
- intelligent buildings   485
- intrusion sensing   351

resonance pattern, elevators   271
respiration   501
response range, gas sensing   346
response time   30, 174 f
responses
– intelligent buildings   491
– smoke sensors   310, 316
retina, identification sensing   369
retro reflectors   322
reverse voltage protection, M-bus   140
Reynolds number   164
ride control, elevators   285
Righi–Leduc effect   17
rigidity, elevator ropes   271
roller guide shoes   286
room characteristics
– air conditioning   55
– NEUROBAT   73
– thermal comfort   104
room segmentation   349
room surveillance   496
roomServer   472
root cause failure analysis (RFCA)   462
rope sets   261, 271
rotating equipment   453
roughness   453
routers, fieldbus   211 ff
runtime measurements   281

safety   305–448
– elevators   261, 266 ff, 273, 291
– improved constructions   427 ff
– intelligent buildings   6, 23, 485, 488
– pressure sensors   186
– private homes   513
Sagnac effect   17
scattering, fire sensing   311 ff
scene brightness, air conditioning   43
scene spots fuzzy clustering   45
Schlieren see: temperature fluctuations
season study, NEUROBAT   71
security   305–448
– biometric systems   405

– fieldbus   234
– intelligent buildings   6, 16, 23, 485
– private homes   513
– tele services   553
– wireless networks   258
– worldwide facility management   477
Seebeck effect   17
Seebeck voltage   351
seismic microphones   355
selectivity, gas sensing   345
self-commissioned heating control   63–84
self-diagnosis/adaption   4
semiconductors   344
sensing chair   293 ff
sensing floor   293, 301 ff
sensitivities
– air sensors   30, 93
– pressure sensors   174
sensor head configurations   429
sensor sheets   298
sensory inspection   452, 457
serial port   523
Service DLL   528
service user/provider, fieldbus   208
services access points (SAPs)   208
session layer, fieldbus   208 ff
setups, pressure sensors   188
seven-layer model
– fieldbus   208
– wireless networks   251
shaft information systems, elevators   261, 264 f, 271 f, 284 f
shape recognition   424
shocks, elevators   264
shoring systems   427 ff, 439 f
short-circuit protection
– M-bus   141
– meter reading   134 f
short-frame, M-bus   142
sick building syndrome (SBS)   3, 86, 504
signal conditioning, pressure sensors   182
signal controlling, NEUROBAT   70

signal evaluation, air quality 91
signal processing 383 ff
– intrusion sensing 358
– safety 308
– transduction 17 f
signal to noise ratio (SNR)
– intrusion sensing 349
– smart cameras 413
– wireless networks 244
silane 343
silicon diaphragm pressure sensors 175, 179 f
simple mail transfer protocol (SMTP) 230
simulation study
– air conditioning 57
– NEUROBAT 69 ff
– supervisory control 119 ff
single-chip PCs 475
single-station detection, flame sensing 332
sitting posture, sensing chair 299
slowdown monitoring, elevators 267
smart cameras 409
smart cards 364, 399
smart food systems 9
smart home 511
smart sensors 464
smart systems 20 f
smoke detectors 308 f
smoke particles
– flame sensing 329
– intelligent buildings 489
smoke sensors 38
smoldering 311 f, 329 ff, 489
solar radiation
– air conditioning 57
– energy management 118
– NEUROBAT 65, 69
solid-state imaging 411
sound propagation 309
sound waves 281
specifications
– active damping system 287
– air conditioning 29 f

– air sensors 91
– M-bus 138
spectral analysis 388
spectrographic analysis 455
spectroscopy 18
speed, wireless networks 246
speed measurements, elevators 264
sprinkler systems 377
spots fuzzy clustering 45
spread spectrum, wireless networks 246
staff member monitoring 399
standard nickel elements 69
standardization 18
standards
– emergency handling 376
– fire extinguishing 379, 382
– flame sensing 336
– gas sensing 347
– identification sensing 373 f
– intrusion sensing 361 f
– meter readings 154 f
– wireless networks 250
static flow sensors 164
static sitting posture 299
stereoscopic camera systems 41
stochastic modeling, sensing chair 293
stochastical transmission, M-bus 148
Stokes scattering 325
stopping devices, elevators 266
styrenes 87
sulfur dioxide 338, 343
sun radiation 308, 327
supervisory control, thermal comfort 103 ff
surface acoustic wave (SAW) sensors 492
surface micromachining 181
surface mounting, sensing chair 293
surge compression testing 457
switched capacitor technique 414
switches 264, 270, 356
system complexity, fieldbus 231

system concepts
– flame sensing   332
– intrusion sensing   359
system technologies   482–558
– private homes   511–558

tacho interface (TIF)   265
tampering, intrusion sensing   359
technical specifications
– air sensors   91
– M-bus   138
tele services   539, 553
telecommunication   469
telegram formats, M-bus   142
telemetry, wireless networks   258
telephone applications   542
temperature effect, pressure
    sensors   192 f
temperature fluctuations   308, 326 f
temperature sensors
– air conditioning   30
– building automated systems   47
– energy allocation   163
– heat cost allocators   169
– intelligent buildings   16, 489, 504
temperature
– air conditioning   29, 51, 57
– energy allocation   160
– energy management   117 ff
– fire sensing   323
– health monitoring   501
– NEUROBAT   63 ff
– pressure sensors   175
tension elements   15
test results
– NEUROBAT   76
– supervisory control   119
text-dependent voice recognition
    371
thermal comfort
– intelligent buildings   485, 503
– optimal   103 ff
thermal flow sensors   165
thermal noise   17

thermal systems, mathematical
    description   65
thermistors   31
thermocouples
– air conditioning   30
– energy allocation   163
– intrusion sensing   347
thermoelectric detectors   324
thermoelectric gas sensors   351
thermography, infrared   452, 455
thermophysical room characteristics
    73
thermopiles   352
thin-film technology   163
three-electrode electrochemical sensor
    344
threshold limit value (TLV)
– air quality   86
– gas sensing   338
time check systems   399
time constants, pressure sensors
    187
time division multiple access
    (TDMA)   249
tire pressure monitoring system
    (TPMS)   185
token ring protocol   254
tool software, fieldbus   220, 224, 232
topology
– fieldbus   212, 222
– M-bus   137
– meter reading   132
total organic carbon (TOC), air
    quality   86
total productive maintenance (TPM)
    461 f
touch sensors   293
toxic constituents, air quality   86
toxic gases   338 ff, 492
traction machines   261, 271
traction sheave   264, 270
transduction principles   17 f
transfer protocols, fieldbus   230
transistors   412
translucent membranes   15

transmission characteristics
- energy management   118
- meter reading   133 ff
- private homes   518
transmission control protocol (TCP)   230
transport layer, fieldbus   208 ff, 229
transport-oriented protocols   211
transportation   22, 201–304
tribology   455
tubine flow meters   34
tunneling FAN protocols   231
twisted-pair UART (TP-UART)   507

ultrasonic flow meters   35, 164
ultrasonic monitoring   452, 454
ultrasonic motion detectors   37, 357
ultrasonic passive infrared (US-PIR) dual sensor   499
ultrasonic sensors   16, 496
ultrasound detectors   308
ultraviolet detection   309, 327
unidirectional transmission, M-bus   148
unified home bus systems   511
uniform code, magnetic sensors   282
universal plug and play (UPnP)   243, 253
universal series bus (USB)   524 f
uplift, shoring systems   431, 435
upper explosion limit (UEL)   339
user datagram protocol (UDP)   230
user interfaces, home automation   543
user module, NEUROBAT   66

van der Pol oscillator   391
vanadium oxide (Vox)   351
vanes   266
variable air volume (VAV)   31, 52
- pressure sensors   176, 195
vector quantization, voice recognition   371

velocity field computation   42
velocity profile, air flow   162
ventilation rate   88, 105
ventilation systems   120
Venturi meters   34, 162
vibration analysis   451 f
vibration microphones   355
vibrations, elevators   264, 271, 285
video cameras   409, 540
virtual link layer (VLL)   226
virtual links, wireless networks   252
virtual workspace   469
visual reading, heat cost allocators   169
visualization, home automation   539, 543
vital parameters, health monitoring   501
voice evacuation system   374
voice recognition   9, 371
volatile organic compounds (VOCs)   86 ff, 91, 98 ff
volume flow   160, 176
vortex shedding meters   34

warm-up effect   194
water heating systems   63
water temperature   160
wearable devices, health monitoring   501
Web servers   549
Weibull curve   452
wet-fire extinguishing systems   376 f
Wheatstone bridge   340
wide-area network (WAN)
- alarm receiving   380
- fieldbus   205
wire sound waves, elevators   281
wired links   252
wireless communication components   430, 441
wireless devices, flame sensing   335
wireless in-building networks   241 ff
wireless products   258

wireless sensors 464
wireless supports 10
wireless transmission 524
wiring, fieldbus 204 f, 236
wiring mistakes, pressure sensors 174
Woltman counter 164
woodsmoldering 330

workspaces 469
world wide web 467
worldwide facility management (WWFM) 469 ff

xylenes 87